With compliments

ELECTRICAL MACHINES AND CONVERTERS
Modelling and Simulation

PROGRAMME COMMITTEE

Chairman:
H. BUYSE, Professor, UCL, Belgium

Members:
Z. BINDER, representative of IFAC, INP Grenoble, France
M. GARRIDO, INIC, Portugal
W. GEYSEN, KUL, Belgium
G. MAGGETTO, VUB, Belgium
R. LE DOEUFF, ENSEM, France
D. W. NOVOTNY, University of Wisconsin, USA
R. PONCELET, ULB, Belgium
V. RAJAGOPALAN, University of Québec, Canada
J. ROBERT, ULg, Belgium
J. C. SABONNADIERE, INP, Grenoble, France
B. TRANNOY, LEEI-ENSEEIHT, France

STEERING COMMITTEE

Chairman:
J. ROBERT, Professor, ULg, Belgium

Members:
M. CRAPPE, Faculté Polytechnique of Mons, Belgium
E. GUDEFIN, ENSEM, Nancy, France
W. LEGROS, ULg, Belgium
P. PIROTTE, ULg, Belgium
J. C. PONCELET, General Secretary of AIM, UNERG, Belgium
P. VAN REMOORTERE, Ecole Royale Militaire, Belgium
M. VAN STEENKIST, Gent University, Belgium
Ph. VIAROUGE, Laval University, Canada

ELECTRICAL MACHINES AND CONVERTERS

Modelling and Simulation

Proceedings of the IMACS International Symposium
Liège, Belgium, 17-18 May, 1984

organized by

A.I.M.
Association des Ingénieurs Electriciens
Sortis de l'Institut Montefiore
Liège, Belgium

edited by

H. BUYSE
Départment d'Electricité
Université Catholique de Louvain
Louvain-la-Neuve, Belgium

and

J. ROBERT
Institut Electrotechnique Montefiore
Université de Liège
Liège, Belgium

1984

AMSTERDAM · NEW YORK · OXFORD

ISBN: 0 444 87596 4

Published by:
ELSEVIER SCIENCE PUBLISHERS B.V.
P.O. Box 1991
1000 BZ Amsterdam
The Netherlands

Sole distributors for the U.S.A. and Canada:
ELSEVIER SCIENCE PUBLISHING COMPANY, INC.
52 Vanderbilt Avenue
New York, N.Y. 10017
U.S.A.

Library of Congress Cataloging in Publication Data

IMACS International Symposium (1984 : Liège, Belgium)
Electrical machines and converters.

"Organized by A.I.M., Association des ingénieurs électriciens sortis de l'Institut Montefiore, Liège, Belgium."
Includes index.
1. Electric machinery--Mathematical models--Congresses.
2. Electric current converters--Mathematical models--Congresses. 3. Electric machinery--Data processing--Congresses. 4. Electric current converters--Data processing--Congresses. I. Buyse, H., 1941- . II. Robert, J., 1925- . III. Association des ingénieurs électriciens sortis de l'Institut Montefiore. IV. Title.
TK2189.I53 1984 621.31'042'0724 84-13671
ISBN 0-444-87596-4

PRINTED IN THE NETHERLANDS

FOREWORD

The modelling and simulation of electrical machines and converters are topics which seem of great interest today, especially in view of the industrial needs and also due to the ever increasing potential of the digital computers.

The basic idea, however, is not new since the first modelling methods are the equations written at the beginning of the century. The main purpose of these equations was to describe various types of machines in order to predict essentially their steady state behaviour. Many refinements were added and, at the beginning of the twenties reasonable methods were available to calculate the performance of the machines and to help designers.

At that time a new trend began to emerge, due to the ever increasing constraints existing in the power systems. Transient, as well as subtransient reactances were defined, unbalanced operation was more deeply investigated. In some particular domains equivalent circuits were developed, matrix and tensor analyses were used.

A.C. and D.C. network analysers were expanded. In some cases mechanical differential analysers were used.

New electronic analog computers, which appeared essentially in the forties and fifties prompted many engineers to use such calculating machines to study the transient performance of various types of electric machines used either in power systems or in control applications.

With the advent of digital computers continuous simulation programs were evolved and a new trend appeared : the use of very sophisticated programs, such as E.M.T.P. to predict the transient performance of power systems, or, to analyse for example, the transient operation of large induction machines.

Some discrepancies between calculated and measured performances now lead many engineers to search for improvements of the methods which are available today.

For example, new components were introduced in Park's models. Similarly new measuring techniques are needed to obtain the values of all the necessary parameters.

These efforts result in methods which could be named "third generation modelling".

This includes also advanced work in two particular domains.

The first is a direct result of the ever increasing use of semi-conductor devices associated with all types of electrical machines. This trend leads to the development of more and more simulation programs which include the machines and their electronic drives. In general these problems are very complicated for two main reasons : First, the number of possible configurations is high since it depends on the state of all semi-conductors. Secondly, the currents and voltages contain rather large harmonic components and the models of the machines should be designed to represent their effects.

The modelling of these systems is often believed to be the most practical way of predicting their performance. This, however, is not necessarily true, since global methods have been developed using the theory of electrical machines with sliding contacts. These methods are, in fact, the generalization of the old theories of D-C machines "seen from the outside", in which all transient effects in the rotor coils due to commutation are ignored. The results obtained by such approximations are very representative of the actual performance of the machines.

The other domain which has become more and more important is the study of electromagnetic fields in the machines.

Not too long ago the only practical way for calculating these fields was by using graphical methods and each calculation involved many hours of tedious work with a pencil and, principally, an eraser.

Rheographic devices were then used and consisted mainly of electrolytic tanks with automatically driven probes. They gave very interesting results but necessitated a good deal of experimental work.

With the advent of digital computers this domain of activity started to expand. Finite difference equations were used and the theory of finite elements was developed and we now have at our disposal very sophisticated programs which in seconds calculate all we need.

This is in fact very slow and for transient problems we would like to obtain solutions in milliseconds rather than in seconds.

And now, what is the future ? No doubt the computer improvements will help. Parallel processors will permit higher computing speeds, but the problems of stability of the solutions are to be mastered.

On the other hand one has always to search for more efficient methods of writing the equations and of solving them.

The non-linear properties of iron should also be more easily modelled.

The papers which are assembled in this book represent good samples of the efforts made in all the domains we have mentioned above.

If we work on new modelling and simulation methods we can with confidence say : "The future : we are working on that".

H. BUYSE J. ROBERT

TABLE OF CONTENTS

SESSION 5B - Simulation and modelling of converters (II)

Chairman: R. LE DOEUFF

SESSION 6 - Modelling of synchronous machines and converters (I)

Chairman: B. TRANNOY

SESSION 7 - Modelling of synchronous machines and converters (II), simulation of power generation systems

Chairman: M. CRAPPE

Session 1

IDENTIFICATION OF THE PARAMETERS OF ELECTRICAL MACHINES

Chairman: R. PONCELET

ELECTRICAL MACHINES AND CONVERTERS - Modelling and Simulation
H. Buyse and J. Robert (editors)
Elsevier Science Publishers B.V. (North-Holland)/© IMACS, 1984

EXPERIMENTAL DETERMINATION OF LARGE TURBOGENERATOR PARAMETERS

Michel CRAPPE, Marc DELHAYE, Michel RENGLET (Faculté Polytechnique de Mons)
Moha NACIRI (Faculté Polytechnique de Mons and Ecole Mohammedia de Rabat)
Luc SOENEN (Laboratoire Belge de l'Industrie Electrique)

Belgium

This paper is dealing with the experimental determination of large turbogenerator two axis model parameters (PARK's models).
After a general review of our works in that field, we particularly deal with DC decay test at standstill (step response) for determining d and q axis parameters.
We especially present theoritical aspects, results and comments about the DC decay test performed at the field side, for d-axis parameters determination.
Obtained values are used for reconstruction of experimental signals from tests performed in the network and at the manufacturer's testing bench.

TERMINOLOGY - SYMBOLS - UNITS

We use IEC terminology and symbols for the impedances and for the time constants [1][2]. For the elements of d and q axis equivalent circuits, we use inductances and resistances suffixed as shown in figure 1.

Suffixes : a : armature winding
d,q : direct, quadrature axis
m : magnetising
f : field winding
k : damper winding

All the impedances are expressed in per unit and time constants in seconds.

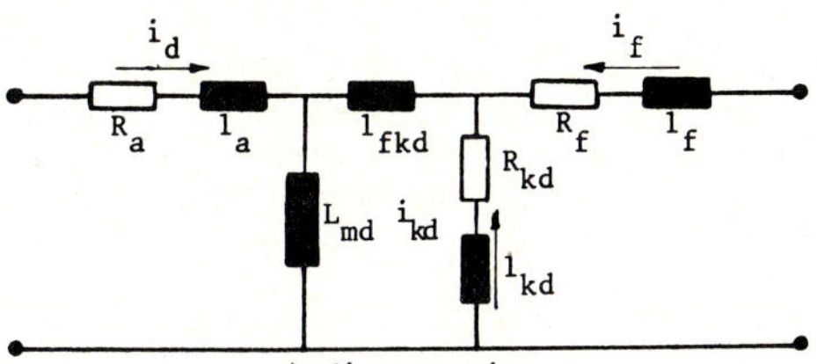

a) direct axis

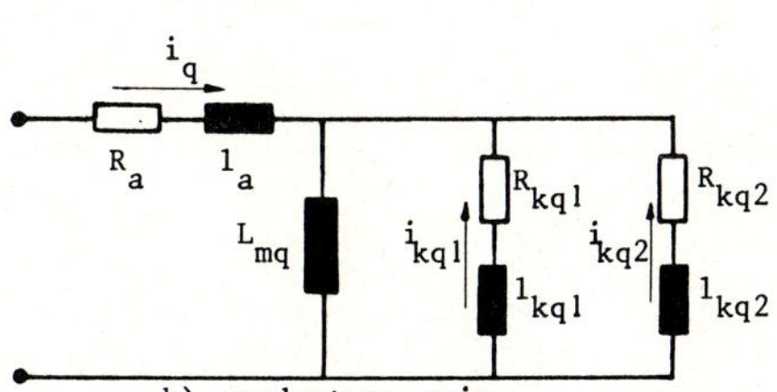

b) quadrature axis

Figure 1 : Equivalent circuit notations
(a) : direct axis
(b) : quadrature axis

1. INTRODUCTION

This paper is dealing with the experimental determination of large turbogenerator models, in order to compute the dynamic behaviour of large power systems.

In practice, lumped element linear PARK's models are universally used with generally one damper circuit on d-axis and two in q-axis. These models are usually enhanced with CANAY's mutual leakage inductance [3][4].
However, there is no consensus about the methodology for parameters determination [3].
Numerous methods are propounded for the experimental determination of these parameters [1][2], but, in practice, the analysis of the corresponding signals and the interpretation of their results are still setting problems.
The application of IEC recommended methods can give large discrepancies between the values obtained by different methods [5].
For these reasons, we have entered upon a critic and systematic study of these experimental methods [6] - [10].
The purpose of this study is to optimize the experimental conditions and the analysis process, and to quantify the confidence which can be attributed to the results.

Presently, we are working on the following methods :

- sudden three-phase short circuit test;
- DC decay test in armature winding at standstill;
- DC decay test in field winding at standstill;
- field extinguishing test with armature on short or open circuit, machine running at rated speed;
- frequency response at standstill.

For each method, we systematically analyse the signals from both armature and field terminals. The experimentation is carried out in two steps : firstly, the tests are studied on three-phase micromachines (220 V, 2 kVA, 1500 r.p.m.), and after that, performed on large turbogenerators. We use systematically a computer aided analysis of the experimental data [7].

The experimental signals are identified to analytical expressions, where machine parameters appear; these expressions are deduced from the model.

We have performed and analysed DC decay test in armature winding at standstill on several large turbogenerators of the Belgian network (39.2 MVA, 337.5 MVA and 1330 MVA), and on the 1330 MVA turbogenerator sudden three-phase short circuit tests for five different initial voltages (0.1 - 0.2 - 0.3 - 0.4 - 0.5 U_N) [8].
Besides, all the mentionned methods have been performed on a 361.4 MVA turbogenerator at the manufacturer's testing bench (February-March 1983).
A comparative study of these last test results will be presented in a 1984 CIGRE report [10].

In the present paper, we are essentially dealing with DC decay test at standstill (step response method) for determining d and q axis parameters. Application limits of these methods are put forward. Discrepancies between obtained values are explained by considering signal shape and physical phenomena.
We show advantages to perform the DC decay test at the field side for d-axis parameters determination, in comparison with the test performed at the armature side.
Finally, we present a method for obtaining "compromise values" which allowed us to compute very satisfactory restitution of experimental signals.

2. DATA PROCESSING METHOD

For understanding purpose, we give hereafter the principle of our general data processing method (for more informations : cfr. [7]).

The experimental data processing is performed in two steps : an automatic analysis is realised following the usual methods (IEC recommandations). The results from this analysis are then used as initial conditions for the identification by least square algorithm. We identify experimental data with analytical expressions, which contain machine parameters, by means of regression computer programs. These programs, implementing the LEVENBERG-MARQUARDT's least square algorithm, use iterations on the JACOBIEN matrix of the function : sum of square deviations between experimental data and values given by the analytical expressions.
All the computer programs are implemented on a DEC VAX-11/780 (2.5 Mbytes central memory). These programs include several specific subroutines in order to analyse experimental signals automatically (starting time detection, offset detection, local quadratic curve fitting, parasite elimination, ...). From statistical informations, we also quantify the quality of the signal analysis, and determine the statistical confidence of the results [7].

3. DC DECAY TEST IN THE ARMATURE WINDING AT STANDSTILL (IEC 34-4-A, clause 59)

This test is particularly easy to perform, even in the power station [7][8]. It gives the step response in both axis and the recorded signals are therefore a sum of exponentials.
The method used for analysing the signals is the identification of the exponential terms, from which we determine the d and q axis parameters. The number of exponential components is equal to the number of required circuits for representing the machine in the considered axis. This test then allows to investigate the complexity of the mathematical model to be retained.
We have performed numerous such tests on turbogenerators and micromachines. They were analysed and we conclude the following [9][10] :

- This test is very sensitive to the influence of the magnetic phenomena. We have systematically obtained very low values of magnetising inductances L_{md} and L_{mq}, and also the open circuit time constants T'_{do} and T''_{do}. This is explained by the low permeability values during the test. These low values are characteristic of magnetic recoil curves of very low magnitude due to practical current limitations.

- The identification of the exponential terms is very sensitive to the signal shape, and besides the actual signals are unfortunately distorted by non linearity effects. Therefore, the recorded signals are not exactly a sum of exponentials. Difficulties are especially met for tests performed in the quadrature-axis where the iron effect of the rotor steel body is not masked by a field winding as for the direct-axis.

- The analysis of the sole field current leads systematically to inconsistent results. Simultaneous processing of both armature and field currents recorded in a direct-axis test is so sensitive to the weighting of these two signals that the test is very difficult to be applied.

- The magnetising inductances L_{md} and L_{mq} must be adapted to the magnetic level, anyway.

- The obtained values for the reactances X'_d, X''_d, and X''_q, and for the time constants T''_d, T''_{do}, T''_q and T''_{qo} are systematically excessive.

- In the q-axis, physically unexpected results have been obtained, such as a leakage reactance greater than magnetising reactance [7].

4. DC DECAY TEST IN FIELD WINDING AT STANDSTILL [9]

4.1. Introduction

To avoid problems met in the previous method (DC decay test in armature winding), we propose to carry out the DC decay test at the field side for direct-axis parameters determination.
This test is performed in two sequences : armature on open circuit and on short circuit. It brings more information than the previous test, for which it is not realistic to open the field winding.
The test applied at the field side brings several other advantages in comparison with the test performed at the armature side [9] :

- on a magnetic level standpoint, quasi normal values are reached because a large variation of the magnetic field can be imposed;
- more simple equations in order to calculate the machine parameters as showed hereafter;
- reduction of the requirement for the short-circuit contactor resistance (cfr. figure 2).

4.2. Principle and wiring diagram

The principle of this test is analogous to the previous one. It is also easy to perform in practice. DC voltage is applied to the field winding, machine at standstill. The rotor is placed so that the magnetic field axis be in quadrature with one armature phase-axis (phase A on the figure 2). Only one quantity must therefore be recorded at the armature side (a voltage for the test with armature on open circuit and a current for the test with armature on short circuit).
Hereafter, we present the theoritical aspects for the usual model, enhanced with the field-damper leakage reactance l_{fkd}. All the expressions can be generalised for models with more damper circuits [9].

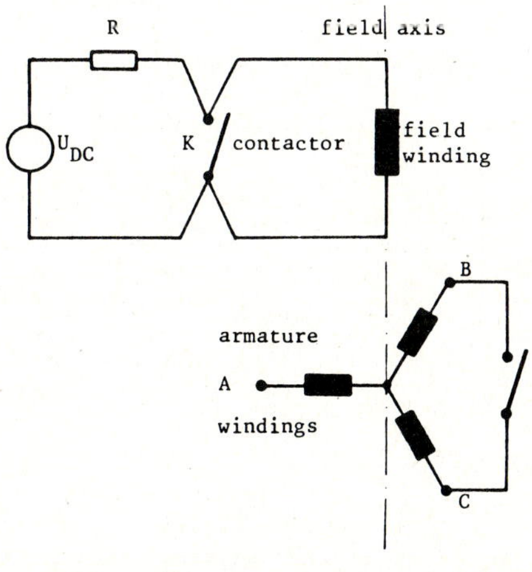

Figure 2 : DC decay test in field winding, machine at standstill - wiring diagram, armature on open circuit and on short circuit

4.2.1. Test with armature on open circuit

We record the whole process of field current decay and the armature voltage, after field winding has been short-circuited (K contactor closed).
Theoritically, from the analysis of these two signals, it must be possible to determine the following parameters :

$$T'_{do}\ ,\ T''_{do}\ ,\ X_{md}\ ,\ T_{kdo}\ \text{and}\ T_x = l_{kd}/r_{kd}$$

From those parameters, we can compute directly all the elements of the direct-axis equivalent circuit for the model without the mutual keakage inductance l_{fkd} (R_a and l_a must be known).

4.2.2. Test with armature on short circuit

In this case, we record the whole process of the field current decay and the armature current. From the recorded signal analysis, it is theoritically possible to determine the time constants T'_d and T''_d.

Finally, from both previous test results and from the armature leakage inductance l_a, we are able to compute all the d-axis parameters of the PARK's model, enhanced or not with the field-damper mutual leakage inductance l_{fkd}.

4.3. Theory of the DC decay test in the field winding, machine at standstill

4.3.1. Test with armature on open circuit

From the general PARK's equations in Laplace transforms, we obtain the following expression for the field current Laplace transform after field winding has been short-circuited :

$$i_f(p) = \frac{i_{fo}}{p} + i'_f(p)$$

$$= \frac{i_{fo}}{p} - i_{fo} \cdot \frac{(1 + pT_{kdo})}{p(1 + pT'_{do})(1 + pT''_{do})}$$

$$= i_{fo} \cdot \frac{T'_{do} + T''_{do} - T_{kdo} + pT'_{do} \cdot T''_{do}}{(1 + pT'_{do})(1 + pT''_{do})} \qquad (1)$$

where i_{fo} is the initial field current;
i'_f is the surimposed field current change.

The recorded field current has thus the following form :

$$i_f(t) = i_{fo} + i'_f(t) = i_{fo} \cdot \sum_{i=1}^{2} A_i \cdot e^{-t/T_i}$$

or, in Laplace transform :

$$i_f(p) = i_{fo} \cdot \sum_{i=1}^{2} \frac{A_i \cdot T_i}{1 + pT_i}$$

$$i_f(p) = i_{fo} \cdot \frac{A_1T_1 + A_2T_2 + pT_1T_2(A_1 + A_2)}{(1 + pT_1)(1 + pT_2)} \qquad (2)$$

Consequently, the time constants T'_{do} ($\equiv T_1$) and T''_{do} ($\equiv T_2$) are directly obtained from the field current analysis which gives the A_i and T_i parameters. We also obtain the value of T_{kdo} from the following expression :

$$T_{kdo} = T'_{do}(1 - A_1) + T''_{do}(1 - A_2)$$

The following condition holds :

$$A_1 + A_2 = 1$$

The armature voltage change is expressed by :

$$u_d(p) = i_{fo} \cdot L_{md} \cdot \frac{1 + pT_x}{(1 + pT'_{do})(1 + pT''_{do})} \qquad (3)$$

where $T_x = l_{kd}/r_{kd}$.

As the recorded voltage has the form :

$$u_d(t) = \sum_{i=1}^{2} A'_i \cdot e^{-t/T_i}$$

(notice the same time constants for field current and armature voltage), and

$$u_d(p) = \frac{A'_1.T'_{do} + A'_2.T''_{do} + pT'_{do}.T''_{do}.(A'_1 + A'_2)}{(1 + pT'_{do})(1 + pT''_{do})} \qquad (4)$$

from identification side to side of (3) and (4), we obtain :

$$L_{md} = \frac{A'_1 \cdot T'_{do} + A'_2 \cdot T''_{do}}{i_{fo}}$$

$$T_x = \frac{T'_{do} \cdot T''_{do} \cdot (A'_1 + A'_2)}{A'_1 \cdot T'_{do} + A'_2 \cdot T''_{do}}$$

From the armature voltage analysis which gives the A'_i and T_i parameters, we obtain thus :

$$T'_{do}\ ,\ T''_{do}\ ,\ L_{md} \text{ and } T_x = l_{kd}/r_{kd}$$

The d-axis parameters, l_f , l_{kd} , R_f and R_{kd} can be obtained from the four following equations for the model without mutual leakage inductance l_{fkd} ($l_{fkd} = 0$) :

$$T'_{do} + T''_{do} = \frac{L_{md} + l_f}{R_f} + \frac{L_{md} + l_{kd}}{R_{kd}}$$

$$T'_{do} \cdot T''_{do} = \frac{l_f \cdot l_{kd} + (l_f + l_{kd}) \cdot L_{md}}{R_f \cdot R_{kd}}$$

$$T_{kdo} = \frac{l_{kd} + L_{md}}{R_{kd}}$$

$$T_x = \frac{l_{kd}}{R_{kd}}$$

(cfr. analytical solution of those four equations with four unknowns on appendix 1).

4.3.2. Test with armature on short circuit

In this case, from the PARK's equation system, we obtain the following expression for the field current :

$$i_f(p) = \frac{i_{fo}}{p} - i_{fo} \cdot \frac{1 + p(T'_4 + T'_5) + pT'_5.T'_6}{p(1+pT'_1)(1+pT'_2)(1+pT'_3)} \qquad (5)$$

$$= i_{fo} \cdot \frac{T'_1+T'_2+T'_3-T'_4-T'_5+p(T'_1T'_2+T'_1T'_3+T'_3T'_2-T'_5T'_6)+p^2T'_1T'_2T'_3}{1+p(T'_1+T'_2+T'_3)+p^2(T'_1T'_2+T'_2T'_3+T'_3T'_1)+p^3T'_1T'_2T'_3}$$

where $T'_1,\dots,T'_6$ are depending of the equivalent circuit elements (l_f , R_f , ...).

The recorded field current has thus the form :

$$i_f(t) = i_{fo} \sum_{i=1}^{3} B_i \cdot e^{-t/T'_i}$$

Or, in Laplace transform :

$$i_f(p) = i_{fo} \sum_{i=1}^{3} \frac{B_i \cdot T'_i}{1 + pT'_i} \qquad (6)$$

By identification side to side of equations (5) and (6), we obtain :

$$B_1 + B_2 + B_3 = 1$$

$$T'_4 + T'_5 = \sum_{i=1}^{3} T'_i(1 - B_i) \qquad (7)$$

$$T'_5.T'_6 = T'_1.T'_2(1 - B_1 - B_2) + T'_2.T'_3(1 - B_2 - B_3) + T'_3.T'_1(1 - B_1 - B_3) \qquad (8)$$

From (7), (8) and the identified time constants T'_1 , T'_2 and T'_3, it is then possible to determine five elements of the d-axis equivalent circuit from the five equations given in appendix 2 [9].

The armature current d-axis component has the following expression :

$$i_d(t) = \sum_{i=1}^{3} B'_i \cdot e^{-t/T'_i}$$

The same identification method as for the field current, leads to :

$$\sum_{i=1}^{3} B'_i = 0 \quad ; \quad \frac{L_{md}}{R_a} = \sum_{i=1}^{3} B'_i \cdot T'_i \qquad (9)$$

$$T_x = \frac{l_{kd}}{R_{kd}} = \frac{T'_1T'_2(B'_1+B'_2)+T'_2T'_3(B'_2+B'_3)+T'_3T'_1(B'_3+B'_1)}{\sum_{i=1}^{3} B'_i \cdot T'_i} \qquad (10)$$

From (9), (10) and the identified time constants T'_1 , T'_2 and T'_3, we can obtain five equations similar to those established for the field current (appendix 2) from which we can determine five elements of the direct-axis equivalent circuit.

In conclusion, from the analysis of the field and armature current of the DC decay test in field winding with short-circuited armature, it is theoritically possible to determine seven elements of the direct-axis equivalent model (seven equations for eight unknowns values : R_a , R_f , R_{kd} , l_a , l_f , l_{kd} , l_{fkd} and L_{md}).

4.4. Practical use

4.4.1. Test with armature on open circuit

The field current analysis leads systematically to excessive values of the subtransient time constants. This is due to the shape of the signal

conditionned principally by the large time constants.
The armature voltage, on the other hand, contains much more information about the subtransient phenomena [9]. The voltage is an induced one, directly depending of the flux variation. The subtransient time constants obtained from this voltage analysis are much smaller.
The reconstruction of experimental signals from tests performed in the network and at the manufacturer's testing bench showed us that these smaller values are those which give the best results [10].
The simultaneous identification of the field current and the armature voltage in order to use concurrently all the information, gives results extremely depending on the weighting of the two signals between each other.

4.4.2. Test with armature on short circuit

From this test, the field current analysis leads to acceptable results. Generally, the analysis of the armature current gives unacceptable time constants, as for the field current analysis from test in the armature side (section 3). This is due to the combination effect of the signal shape and the identification process.

4.4.3. Combination of the results of tests with armature on open circuit and on short circuit

In practice, to determine the direct-axis parameters, we propose to take account of both tests with armature on open circuit and on short circuit. The combination of the results of these two tests can be made as follows :

$$T'_{do}\ ,\ T''_{do}\ ,\ L_{md}\ ,\ T_x$$

from the armature voltage analysis, test with armature on open circuit

$$T'_1\ ,\ T'_2\ ,\ T'_3$$

from the field current analysis, test with armature on short circuit

From these previous values, we determine T'_d , T''_d and L_d/R_a solving the following equations :

$$\frac{L_d}{R_a} = T'_1 + T'_2 + T'_3 - (T'_{do} + T''_{do})$$

$$T'_d + T''_d = \frac{T'_1.T'_2 + T'_2.T'_3 + T'_3.T'_1 - T'_{do}.T''_{do}}{L_d/R_a}$$

$$T'_d.T''_d = \frac{T'_1.T'_2.T'_3}{L_d/R_a}$$

We can now determine l_f , l_{fkd} , l_{kd} , R_f imposing the value of armature leakage inductance l_a, by solving the equation system :

$$T'_d+T''_d = \frac{1}{R_f}(l_f+l_{fkd}+\frac{L_{md}.l_a}{L_{md}+l_a})+\frac{1}{R_{kd}}(l_{kd}+l_{fkd}+\frac{L_{md}.l_a}{L_{md}.l_a})$$

$$T'_d.T''_d = \frac{1}{R_f.R_{kd}}\left[l_{kd}l_f+(l_{kd}+l_f)(l_{fkd}+\frac{L_{md}.l_a}{L_{md}+l_a})\right]$$

$$T'_{do}+T''_{do} = \frac{1}{R_f}(l_f+l_{fkd}+L_{md}) + \frac{1}{R_{kd}}(l_{kd}+l_{fkd}+L_{md})$$

$$T'_{do}.T''_{do} = \frac{1}{R_{kd}.R_f}\left[(l_{kd}l_f+(l_{kd}+l_f)(l_{fkd}+L_{md})\right]$$

$$T_x = \frac{l_{kd}}{R_{kd}}$$

These equations have an analytical solution without any approximation [9].
Nevertheless, there is a practical problem because of the sensivity of the results with respect to T''_{do} value, which is not determined with a good accuracy [9]. In order to circumvent this problem, we impose the value of the subtansient reactance X''_d. This one is determined with a good accuracy from the standstill test at rated frequency (IEC - 34-4, clause 44).

In the two equations for determining T'_d and T''_d values, we eliminate the product $T'_{do}.T''_{do}$ using the expression of the subtransient inductance :

$$L''_d = (L_{md} + l_a)\ \frac{T'_d\ .\ T''_d}{T'_{do}\ .\ T''_{do}}$$

It follows :

$$T'_d\ .\ T''_d = \frac{T'_1\ .\ T'_2\ .\ T'_3}{L_d/R_a}$$

$$T'_d + T''_d = \frac{T'_1.T'_2 + T'_2.T'_3 + T'_3.T'_1 - \dfrac{T'_1.T'_2.T'_3}{L''_d}\ .\ R_a}{L_d/R_a}$$

4.5. Results and comments about DC decay tests in the field winding of a 361.4 MVA turbogenerator (3000 RPM - 20 kV)

Hereafter, we present results from DC decay tests performed in the field winding of a 361.4 MVA turbogenerator at the manufacturer's testing bench. Two initial field currents have been imposed, 0.1 p.u. and 0.3 p.u. For each one, the tests with armature on open and short circuit have been realised two times.

Test (-)	i_{fo} (pu)	T'_{do} (s)	T''_{do} (ms)	T_x (ms)	T'_1 (s)	T'_2 (ms)	T'_3 (ms)	X_{md} (pu)
I	0.1	6.51	24	2.5	8.19	264	15.2	2.43
I'	0.1	6.40	46	-	8.12	262	15.5	-
II	0.1	6.44	12	2.6	8.18	258	14.0	2.43
II'	0.1	6.35	34	-	9.02	262	14.5	-
III	0.3	7.45	19	2.5	9.26	266	16.7	2.46
III'	0.3	7.41	48	-	9.20	260	16.3	-
IV	0.3	7.44	22	2.6	9.01	275	15.7	2.46
IV'	0.3	7.35	34	-	9.02	262	14.5	-

Table 1

Table 1 :
I', II', III' and IV':
results deduced from the analysis of the sole field current
I , II , III and IV :
results deduced from the simultaneous analysis of the signals from both armature and field terminals
"-" : means unacceptable identified value

Test (-)	T'_d (s)	T''_d (ms)	T'_{do} (s)	T''_{do} (ms)	T_x (ms)	X_d (pu)	X'_d (pu)	X''_d (pu)
I	1.09	15.4	6.51	24	2.5	2.69	0.46	0.29
I'	0.98	17.2	6.40	46	-	2.69*	0.42	0.15
II	1.06	13.9	6.44	12	2.6	2.69	0.45	0.49
II'	0.99	18.0	6.35	38	-	2.69*	0.42	0.20
III	1.18	16.3	7.45	19	2.5	2.72	0.43	0.37
III'	1.07	18.0	7.41	48	-	2.72*	0.39	0.15
IV	1.23	16.0	7.44	22	2.5	2.72	0.49	0.35
IV'	1.17	12.4	7.35	34	-	2.72*	0.43	0.16

* Imposed values

Table 2

This table 2 presents the turbogenerator parameters deduced from the table 1 values.

This last table shows the inconsistency of the results particularly for the subtransient reactance X''_d.
To circumvent this problem, we impose the value of the subtransient reactance X''_d as mentionned before, and we obtain the results of table 3.

Test (-)	T'_d (s)	T''_d (ms)	T'_{do} (s)	T''_{do} (ms)	T_x (ms)	X_d (pu)	X'_d (pu)	X''_d (pu)
(I)	1.06	15.9	6.40	23	2.5	2.69	0.45	0.31
(II)	1.02	14.3	6.35	20	2.6	2.69	0.43	0.31
(III)	1.15	16.6	7.41	23	2.5	2.72	0.42	0.31
(IV)	1.15	14.8	7.35	20	2.5	2.72	0.43	0.31

Table 3

(I) and (II) : results from tests at 0.1 p.u.
(III) and (IV) : results from tests at 0.3 p.u.

These last results show a very good consistency for all the parameters. They allow good reconstruction of experimental signals as showed below (section 4.6).
The results of the previous table 3 are uncorrected ones, directly deduced from the analysis process.
In order to compare the obtained values with the manufacturer's data, they have been adapted to the rating conditions of those last ones (temperature, resistance of armature and field circuits, magnetic level).

The corrected values are given on next table 4.

Test (-)	T'_d (s)	T''_d (ms)	T'_{do} (s)	T''_{do} (ms)	T_x (ms)	X_d (pu)	X'_d (pu)	X''_d (pu)
(I)	1.36	15.9	8.31	23	2.5	2.74	0.45	0.31
(II)	1.31	14.3	8.28	20	2.6	2.74	0.43	0.31
(III)	1.28	16.6	8.29	23	2.5	2.74	0.43	0.31
(IV)	1.29	14.8	8.25	21	2.5	2.74	0.43	0.31
(1)	1.34	57	7.27	84	-	2.74	0.50	0.34
(2)	1.21	50	6.83	57	-	2.74	0.48	0.42
(3)	1.04	35.0	7.98	43	-	2.74	0.40	0.32
(4)	1.43	11.0	9.52	15	2.7	2.74	0.41	0.29

Table 4

(1) : results from armature current analysis - DC decay test in armature winding at standstill
(2) : results from simultaneous analysis of field and armature current - DC decay test in armature winding at standstill
(3) : manufacturer's data - old method
(4) : manufacturer's data - new method

4.6. Reconstruction of experimental signals

The obtained values have been used for the reconstruction of experimental signals from sudden three-phase short-circuit tests performed at the manufacturer's testing bench and in the Belgium network.
All computations have been done by numerical integration of PARK's equations, with the use of EMTP program (Electromagnetic Transient Program).

A detailled comparative study of results from different methods will be presented in a 1984 CIGRE report [10].
In the present paper, short of space, we are only giving reconstruction obtained with the results from the DC decay test in field winding.

4.6.1. Sudden three-phase short-circuit test at the manufacturer's testing bench (0.1 initial voltage)

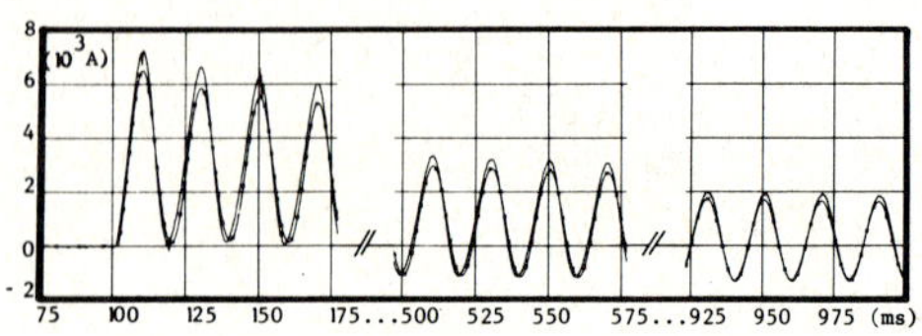

Figure 3 : Armature current (the computed current has been voluntary shifted from the measured one)

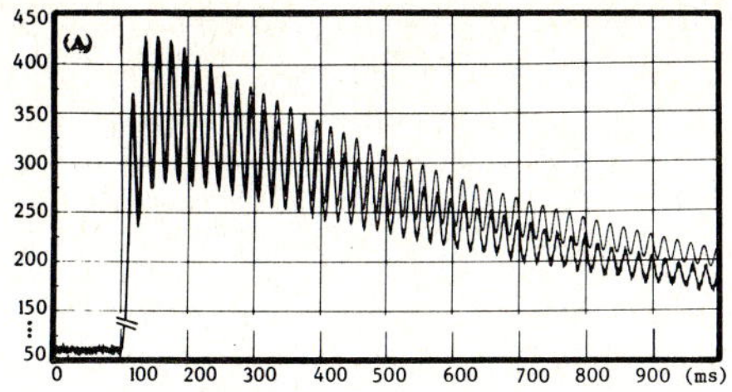

Figure 4 : Field current

These results are much beter than those obtained from the DC decay test in the armature winding, especially for the field current [10].

4.6.2. Three-phase short-circuit on the Belgian 150 kV network

We have computed the behaviour of an identical turbogenerator of the Rodenhuize power station during a voluntary effected short-circuit (metallic three-phase short-circuit on a tower 15 km from the turbogenerator; duration : 122 ms; generator at rated P and Q [11].

The comparison between computed and measured signals shown on the next figures 5 and 6, concerns the armature current and the mechanical torque between the generator and the low pressure turbine.
For all the computations, we have used q-axis parameters deduced from a DC decay test, except for X''_q deduced from a test at rated frequency at standstill.

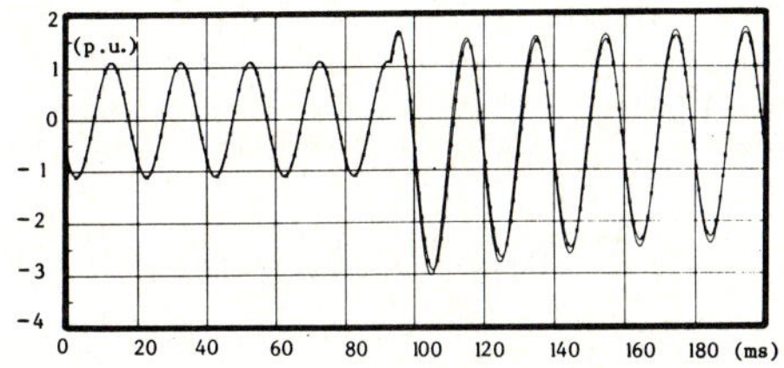

Figure 5 : Armature current

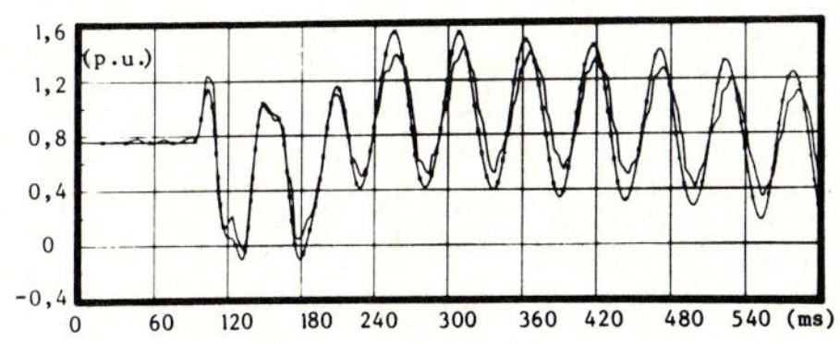

Figure 6 : Mechanical torque

These last results are very acceptable; they are similar to those obtained with parameters from other reliable methods of experimental determination [10].

5. CONCLUSIONS

The analysis of any experimental signal, sum of exponential terms, is a critical problem, and a great care must be taken in interpreting the results anyway.
For the synchronous machine test signals, non linear effects complicate this problem. A this standpoint, we have presented the results of a critic of the DC decay test in the armature winding which leads to excessive values for the subtransient parameters. Besides, this test is characterized by small incremental magnetic permeability, and requires consequently important magnetic corrections of the results.

DC decay test performed in the field winding permits to circumvent these problems :

- it works at quasi normal magnetic permeability and so requires only small magnetic corrections;

- it may be effected in two configurations, with armature on open and on short circuit; that increases the available information;

- the test with armature on open circuit brings an armature voltage signal very representative of the subtransient phenomena. The analysis of this voltage leads to better results of the subtransient parameters, quasi independent of the identification conditions.

Finally, very satisfactory parameter values have been obtained by imposing a reliable value of X''_d (obtained from a standstill test at rated frequency) and by an adequate combination. These parameters allowed good reconstructions of experimental signals from different tests performed at the manufacturer's testing bench and in the Belgian network.

REFERENCES

[1] Recommendations for rotating electrical mechinery.
Part 4 : Methods for determining synchronous machine quantities from tests (IEC, 34-4, 1967, 87 p.).

[2] First supplement to publication 34-4 (1967) : Unconfirmed test methods for determining synchronous machine quantities (IEC, 34-4-A, 1972, 96 p.).

Figure legend : ——— : measurement ; -x-x-x- : computation

[3] G. SHACKSHAFT,
Report on questionnaire seeking new informations on generator parameters for stability studies (July 1980, 15 p., CIGRE).

[4] M. CANAY,
Ersatzschemata der synchronmaschine sowie Vorausberechnung der kengrossen mit beispielen (Ph d thesis at Ecole Polytechnique de l'Université de Lausanne, September 1969).

[5] L. SOENEN, M. CRAPPE, M. DELHAYE,
Détermination des paramètres de la machine synchrone. Essais sur micromachine de la Faculté Polytechnique de Mons (Report 1/1100/801, March 1978, 45 p., LABORELEC, Belgium).

[6] M. CRAPPE, M. DELHAYE, L. SOENEN, F. VAN DE MEULEBROECK,
Modèles des machines synchrones et de leurs organes de commande. Où en sommes-nous ? (Cycle de conférences de la Société Royale Belge des Electriciens, December 1980, 35 p.).

[7] M. DELHAYE,
Etude, traitement et analyse de signaux pour la modélisation de PARK des alternateurs de petites, moyennes et grandes puissances (Ph D Thesis, Faculté Polytechnique de Mons, October 1982).

[8] M. CRAPPE, M. DELHAYE, M. NACIRI,
Experimental determination of large turbogenerator parameters (ICEM-82, Budapest, Proceedings - part 3, pp. 872-875).

[9] M. NACIRI,
Etude critique et comparative de différentes méthodes expérimentales de détermination des paramètres dynamiques des turboalternateurs de grande puissance (Ph D Thesis, Faculté Polytechnique de Mons, to be submited on February 1984).

[10] M. CRAPPE, M. DELHAYE, M. NACIRI, A. CRISPIN, Ph. LORENT, L. SOENEN,
Experimental determination of large turbogenerator dynamic parameters by computer aided analysis (CIGRE, 1984, report 38-10, 11 p., to be presented at the 1984 session).

[11] J.F. GOOSSENS, A.J. CALVAER, L.J. SOENEN,
Full scale short-circuit and other tests on the dynamic torsional response of Rodenhuize NR - 4 - 300 MW - 3000 RPM turbogenerator - Part I (pp. 4166 to 4179) - Part II (pp. 4174 to 4182) (IEEE Trans. PAS, Vol. PAS-100, n° 9, September 1981).

[12] J.J. MORE,
The Levenberg-Marquardt's algorithm. Implementation and theory, numerical analysis (Lecture notes 630, Springer, Dunod, 1977, pp. 105-116).

[13] M. WILLEMS,
Algorithmes de minimisation d'une somme de carrés (University of Mons, 1979, 101 pp.).

APPENDIX 1

$$L_{md} = (A'_1 . T'_{do} + A'_2 . T''_{do}) \cdot \frac{1}{i_{fo}}$$

$$R_{kd} = \frac{A'_1 . T'_{do} + A'_2 . T''_{do}}{T_{kdo} - T_x} \cdot \frac{1}{i_{fo}}$$

$$l_{kd} = T_x . R_{kd}$$

$$l_f = \frac{L_{md} R_{kd} [T'_{do} . T''_{do} - T_{kd}(T'_{do} + T''_{do} - T_{kdo})]}{L_{md}(T'_{do} + T''_{do} - T_{kdo}) - R_{kd}[T'_{do} T''_{do} - T_x(T'_{do} + T''_{do} - T_{kdo})]}$$

$$R_f = \frac{L_{md} + l_f}{T'_{do} + T''_{do} - T_{kdo}}$$

APPENDIX 2

$$T'_4 + T'_5 = \frac{L_{md} + l_f + l_{fkd}}{R_{kd}} + \frac{L_{md} + l_a}{R_a}$$

$$T'_5 . T'_6 = \frac{(L_{md} + l_a)(l_f + l_{fkd}) + l_a . L_{md}}{R_{kd} . R_a}$$

$$T'_1 + T'_2 + T'_3 = \frac{L_{md} + l_{kd} + l_{fkd}}{R_{kd}} + \frac{L_{md} + l_{kd} + l_f}{R_f} + \frac{L_{md} + l_a}{R_a}$$

$$T'_1 . T'_2 + T'_2 . T'_3 + T'_3 . T'_1 = \frac{l_{kd} l_f + (l_{kd} + l_f)(L_{md} + l_{fkd})}{R_{kd} . R_f}$$

$$+ \frac{L_{md} l_a + (L_{md} + l_a)(l_{fkd} + l_{kd})}{R_{kd} . R_a} + \frac{L_{md} l_a + (L_{md} + l_a)(l_{fkd} + l_f)}{R_a . R_f}$$

$$T'_1 . T'_2 . T'_3 = \frac{1}{R_a R_{kd} R_f} \Big[(L_{md} + l_a)(l_{kd} l_f + l_{kd} l_{fkd} + l_f l_{fkd})$$

$$+ L_{md} l_a (l_{kd} + l_f) \Big]$$

---oOo---

ELECTRICAL MACHINES AND CONVERTERS - Modelling and Simulation
H. Buyse and J. Robert (editors)
Elsevier Science Publishers B.V. (North-Holland)/ © IMACS, 1984

NEW PROCEDURE FOR THE IDENTIFICATION OF DYNAMIC PARAMETERS OF THE SYNCHRONOUS MACHINE BY THE THREE-PHASE SHORT-CIRCUIT TEST AND EXTENSION TO OTHER TESTS

Dr Ir J.C. MAUN

Electrical Engineering Dpt
Université Libre de Bruxelles, Belgique

The sudden three-phase short-circuit test on the synchronous machine at no load is studied here. We show that a complete analytical solution of the short-circuit currents can be derived.
After a discussion of the experimental problems, a digital procedure is described which considers the complete expressions of the currents. We point out that the signals identification algorithms give more precise results if we proceed to a simultaneous identification of several signals both at the stator and at the rotor.
This procedure is extended to other tests : the D.C. decay in the armature winding at standstill and the frequency response test at standstill.

1. INTRODUCTION

Most applications of the dynamic behaviour of the synchronous machine are currently studied starting from the two axis theory based on the PARK's transformation. As for any modelling, this theory builds upon an idealization of the synchronous machine and it is important, before developping test methods and parameters identification, to be conscious of the fundamental assumptions [4] .

For a treatment of the phase quantities as well as for a study in the two axis of the machine, a basic assumption consists in neglecting all non-linear effects and principally saturation. This allows the application of the superposition principle. In fact, saturation is an important factor and we take it into account by an adaptation of the parameters which however remain linked to a linear theory. Harmonics are also neglected but we know that one of the topics of machine design is to minimize their influence. Assumptions are generally put down to define the leakage inductances because stator windings are always distributed in this kind of machine. Moreover the rotor behaviour is supposed to be represented by a few equivalent circuits. This is only an approximation of the damper windings influence. And when the machine has a solid rotor, it must be modelled as an infinite number of electrical circuits where eddy currents flow. With saturation, the behaviour becomes non-linear and the modelling is not strictly true.

We see that the theoretical developments involve a set of assumptions which have been criticized by some authors.

Indeed it is clear that the current complexity of simulation programs, as well as the study of some particular problems, clearly demonstrates the limits of these basic assumptions and that a lot of work must be done for a more rigorus modelling of saturation and of the rotor iron behaviour in dynamic studies. Nevertheless, we think that local studies of flux and currents are not necessary for most of the present modelling applications.

Results given by the simulations will only be really accurate if we have a complete model as well as a set of adequate parameters at our disposal. Now we must point out that the test methods have but little evolved and that there is a gap between the performances of digital simulations and the accuracy of the classical graphical derivation of the dynamic parameters used by these programs.

2. A NEW APPROACH FOR THE PARAMETERS MEASUREMENT

The main criticisms for the standard test method are :

1 - The treatment is manual and graphic.
2 - The theoretical developments leading to the relations giving the parameters are based on analytical solutions.
This implies
- a lot of simplifying assumptions;
- the impossibility to consider models with enough electrical circuits.

3 - The treatment of these tests does not use all of the experimental signals and, sometimes it does not

consider all the available data.

4 - It is not uncommon to have a discrepancy between a simulation using the derived parameters and experimental results, particulary for the dynamic behaviour seen from the field circuit access [5].

Therefore, the three-phase short-circuit test and the other considered methods have been developed with the same approach

1 - The use of a digital data acquisition.
2 - A complete automatisation of the treatment based on the use of a mini-computer.
3 - The derivation of the response parameters by an optimal numerical identification.
4 - The use of all the data and of all the available signals for a simultaneous identification taking the common parameters into account.
5 - The developments of new methods to derive the parameters based on the complete analytical expressions and on the provided sensibility of the parameters.

3. THE SUDDEN THREE-PHASE SHORT-CIRCUIT TEST

The usual treatment of this test is well known [13]. It considers only the stator currents envelopes in a semi-log plot and gives, by a graphic method, most of the d-axis quantities of the machine. Unfortunately, the analytical expressions of the short-circuit currents displaying the transient and subtransient reactances and time constants, used in this method, are based on a lot of simplifying assumptions [10]. To point that out, it is necessary to achieve a more complete analysis of the short-circuit currents response.

3.1. Analysis of the currents response

Let us consider a synchronous machine in a steady state, at no load, with a constant rotation speed and a constant field voltage. The stator voltages are given by

$$\begin{aligned} v_{A0} &= v\sqrt{2}\cos(\omega_r t + \theta_v) \\ v_{B0} &= v\sqrt{2}\cos(\omega_r t + \theta_v - 2\pi/3) \\ v_{C0} &= v\sqrt{2}\cos(\omega_r t + \theta_v + 2\pi/3) \end{aligned} \qquad (\underline{1})$$

After a PARK's transformation and considering either that the speed ω_r is equal to the synchronous speed ω_c, or that the PARK's transformation coincides with the rotation speed, ω_r, we have the following equations

$$\begin{aligned} v_{h0} &= 0 \\ v_{d0} &= 0 \\ v_{q0} &= v\sqrt{3} \end{aligned} \qquad (\underline{2})$$

Let $t = 0$ at the instant of short-circuit. For this linear model the superposition principle can be applied so that the currents and the fluxes are the sum of

- the steady state values ;
- the response to an impulse $-v_{q0}$.

The equations for the superimposed quantities (') in the LAPLACE domain are given by

$$\begin{aligned} 0 &= R_a i'_d + p\Psi'_d - \omega_r \Psi'_q \\ -\frac{v\sqrt{3}}{p} &= R_a i'_q + p\Psi'_q + \omega_r \Psi'_d \\ 0 &= R_f i'_f + p\Psi'_f \end{aligned} \qquad (\underline{3})$$

with

$$\begin{aligned} \Psi'_d &= L_d(p)\, i'_d(p) \\ \Psi'_q &= L_q(p)\, i'_q(p) \end{aligned} \qquad (\underline{4})$$

Eliminating the fluxes between (3) and (4)

$$\begin{aligned} \mathrm{Den} &= R_a^2 + R_a p[L_d(p) + L_q(p)] \\ &\quad + (p^2 + \omega_r^2) L_d(p) L_q(p) \\ i'_d(p) &= \frac{\omega_r L_q(p)}{\mathrm{Den}} \cdot \frac{-v\sqrt{3}}{p} \\ i'_q(p) &= \frac{R_a + pL_d(p)}{\mathrm{Den}} \cdot \frac{-v\sqrt{3}}{p} \\ i'_f(p) &= -pG_f(p)\, i'_d(p) \end{aligned} \qquad (\underline{5})$$

these results are general, for any number of damper windings. To point out clearly the classical assumptions, let us develop these equations for a model with one damper winding on each axis, but the theory can be easely extended for a model with more damper windings.

In this case, the operational inductances and the operational factor are given by

$$\begin{aligned} L_d(p) &= L_d \frac{(1 + pT'_d)(1 + pT''_d)}{(1 + pT'_{d0})(1 + pT''_{d0})} \\ L_q(p) &= L_q \frac{(1 + pT''_q)}{(1 + pT''_{q0})} \\ G_f(p) &= \frac{L_{md}}{R_f} \, \frac{(1 + pT_{kd})}{(1 + pT'_{d0})(1 + pT''_{d0})} \end{aligned} \qquad (\underline{6})$$

The short-circuit currents can then be written, in the LAPLACE domain, in the form of a polynomial quotient in p. The coefficients are easy to derive from the machine classical parameters.

$$i'_d = \frac{N_d(p)}{D(p)} \quad \text{order } 3/6 \qquad (7)$$
$$i'_q = \frac{N_q(p)}{D(p)} \quad \text{order } 4/6$$
$$i'_f = \frac{N_f(p)}{D_f(p)} \quad \text{order } 2/5$$

The denominator $D_f(p)$ is identical to $D(p)$ in which the pole at zero is eliminated.
When computing the zeros of the denominator by a numerical method, for example a NEWTON-BAIRSTOW algorithm, we find that the transfer functions have one pole at zero, three real poles and two complex conjugate poles

$$0, -p_{c1}, -p_{c2}, -p_{c3}, -p_{c4} + j\omega_c, -p_{c4} - j\omega_c \qquad (8)$$

Let

$$T_{ci} = -1/p_{ci} \qquad i = 1,4 \qquad (9)$$

The expressions of the short-circuit currents in the time domain have the form

$$i'_d(t) = I_{d0} + I_{d1}e^{-t/T_{c1}} + I_{d2}e^{-t/T_{c2}} + I_{d3}e^{-t/T_{c3}} + [I_{d4}\cos(\omega_c t) + I_{d5}\sin(\omega_c t)]e^{-t/T_{c4}} \qquad (10)$$

with $I_{d0} = -v\sqrt{3}\,X_q/(X_d X_q + R_a^2)$

$$i'_q(t) = I_{q0} + I_{q1}e^{-t/T_{c1}} + I_{q2}e^{-t/T_{c2}} + I_{q3}e^{-t/T_{c3}} + [I_{q4}\cos(\omega_c t) + I_{q5}\sin(\omega_c t)]e^{-t/T_{c4}}$$

with $I_{q0} = -v\sqrt{3}\,R_a/(X_d X_q + R_a^2)$

$$i_f(t) = I_{f0} + I_{f1}e^{-t/T_{c1}} + I_{f2}e^{-t/T_{c2}} + I_{f3}e^{-t/T_{c3}} + [I_{f4}\cos(\omega_c t) + I_{f5}\sin(\omega_c t)]e^{-t/T_{c4}}$$

Let us now examine the different simplifying assumptions commonly used in the classical short-circuit analytic studies.

The first assumptions (H1) considers the low p.u. value of R_a compared to $L_d(p)$ and $L_q(p)$. We can always neglect R_a^2 in the above equations. Thus

$$I_{d0} = -v\sqrt{3}/X_d \qquad (11)$$

which is the basis of the well-known d-axis synchronous reactance measurement.

A second assumption (H2) neglects R_a in the numerator of $i'_q(p)$ in equation 5. This assumption put to zero the steady-state short-circuit current I_{q0}.

Hitherto, the short-circuit currents have always a form given by equations 10.

Let us make a third assumption (H3). Based on the fact that R_a is small, we can simplify the term $R_a^2 p[L_d(p)+L_q(p)]$ by neglecting the winding resistances. With this approximation

$$X_d(p) = X''_d \qquad (12)$$
$$X_q(p) = X''_q$$

The transfer functions order has decreased and is now

$$2/5 \quad \text{for } i'_d(p) \qquad (13)$$
$$1/3 \quad \text{for } i'_q(p)$$
$$1/4 \quad \text{for } i'_f(p)$$

Let

$$p_a = \frac{\omega_r R_a}{2}\left(\frac{1}{X''_d} + \frac{1}{X''_q}\right) = \frac{1}{T_a} \qquad (14)$$

It follows for the short-circuit currents in the time domain

$$i'_d(t) = I_{d0} + I_{d1}e^{-t/T'_d} + I_{d2}e^{-t/T''_d} + [I_{d3}\cos(\omega_c t) + I_{d4}\sin(\omega_c t)]e^{-t/T_a} \qquad (15)$$

$$i'_q(t) = I_{q1}e^{-t/T''_q} + [I_{q2}\cos(\omega_c t) + I_{q3}\sin(\omega_c t)]e^{-t/T_a}$$

$$i_f(t) = I_{f0} + I_{f1}e^{-t/T'_d} + I_{f2}e^{-t/T''_d} + [I_{f3}\cos(\omega_c t) + I_{f4}\sin(\omega_c t)]e^{-t/T_a}$$

with

$$\omega_c = \sqrt{\omega_r^2 + p_a^2} \qquad (16)$$

After these assumptions, we find the classical short-circuit time constants T'_d, T''_d, T''_q and the aperiodic time constant T_a again.

A fourth assumption (H4) simplifies these equations. It considers that

$$p_a \ll \omega_r \tag{17}$$

because the armature resistance is smaller than the subtransient reactances.

Then $\omega_c = \omega_r$ (18)

The following assumption is based on the common values of the synchronous machine time constants. If we consider (H5)

$$\frac{1}{\omega_r} \ll T'_{d0}, T''_{d0}, T'_d, T''_d, T''_{q0}, T''_q, T_{kd} \tag{19}$$

in the inverse LAPLACE transformation of equations 5, we find that

$$I_{d4} = I_{f4} = I_{q2} = 0 \tag{20}$$

in equations 15. Thus, this assumption eliminates any phase displacement in the alternative component of the short-circuit currents.

Finally, a sixth assumption (H6) is based on the difference between transient and subtransient time constants

$$T'_d,\ T'_{d0} \gg T''_d,\ T''_{d0},\ T_{kd} \tag{21}$$

Only in this case, we find the classical coefficients of the short-circuit currents again. And we have

$$i_d(t) = -v\sqrt{3}\left[\frac{1}{X_d} + \left(\frac{1}{X'_d} - \frac{1}{X_d}\right)e^{-t/T'_d} + \left(\frac{1}{X''_d} - \frac{1}{X'_d}\right)e^{-t/T''_d} - \frac{1}{X''_d}\cos(\omega_r t)\,e^{-t/T_a}\right] \tag{22}$$

$$i_q(t) = \frac{-v\sqrt{3}}{X''_q}\sin(\omega_r t)\,e^{-t/T_a}$$

$$i_f(t) = I_{f0} + I_{f0}\left(\frac{X_d - X'_d}{X'_d}\right)\left[e^{-t/T'_d} - \left(1 - \frac{T_{kd}}{T''_d}\right)e^{-t/T''_d} - \frac{T_{kd}}{T''_d}\cos(\omega_r t)\,e^{-t/T_a}\right]$$

To obtain this last form for the field current, it must be noted that there are assumptions in the transient reactance derivation.

The phase currents are given by the inverse PARK's transformation

$$i_A(t) = \sqrt{2/3}\,[\cos(\alpha)\,i_d(t) + \cos(\alpha + \pi/2)\,i_q(t)] \tag{23}$$

If we don't make any assumption, the phase current has the form

$$\begin{aligned} i_A(t) = {} & I_{A0}\cos(\omega_r t + \alpha_0 + \varepsilon_0) \\ & + I_{A1}\cos(\omega_r t + \alpha_0 + \varepsilon_1)\,e^{-t/T_{c1}} \\ & + I_{A2}\cos(\omega_r t + \alpha_0 + \varepsilon_2)\,e^{-t/T_{c2}} \\ & + I_{A3}\cos(\omega_r t + \alpha_0 + \varepsilon_3)\,e^{-t/T_{c3}} \\ & + I_{A4}\cos[(\omega_r - \omega_c)t + \alpha_0 + \varepsilon_4]\,e^{-t/T_{c4}} \\ & + I_{A5}\cos[(\omega_r + \omega_c)t + \alpha_0 + \varepsilon_5]\,e^{-t/T_{c4}} \end{aligned} \tag{24}$$

After all the simplifying assumptions, we have

$$\begin{aligned} I_A(t) = {} & -v\sqrt{2}\left[\frac{1}{X_d} + \left(\frac{1}{X'_d} - \frac{1}{X_d}\right)e^{-t/T'_d} \right. \\ & \left. + \left(\frac{1}{X''_d} - \frac{1}{X'_d}\right)e^{-t/T''_d}\right]\cos(\omega_r t + \alpha_0) \\ & + \frac{v\sqrt{2}}{2}\left(\frac{1}{X''_d} + \frac{1}{X''_q}\right)\cos(\alpha_0)\,e^{-t/T_a} \\ & + \frac{v\sqrt{2}}{2}\left(\frac{1}{X''_d} - \frac{1}{X''_q}\right)\cos(2\omega_r t + \alpha_0)\,e^{-t/T_a} \end{aligned} \tag{25}$$

For the classical treatment of this test, a last assumption is considered :

$$X''_d = X''_q \tag{26}$$

which eliminates the 2ω pulsation component.

3.1. Assumptions influence

To make clear the influence of these assumptions, a program has been written to compute the analytic expressions of the short-circuit currents i_d, i_q, i_f and i_A, for several assumptions from the exact solution to the classical response.

Different synchronous machines have been tested [1]. For example, Figures 1 and 2 show short-circuit currents of a 10 kVA salient pole laboratory machine. The analysis of the errors on the amplitudes, phases and time constants leads to some conclusions.

For small machines, it is clear that a parameter derivation from the classical short-circuit currents equations, used by the manual treatment, leads to important errors. Indeed, with the simplifying assumptions, errors are expected for all the parameters excepted the transient time constant.

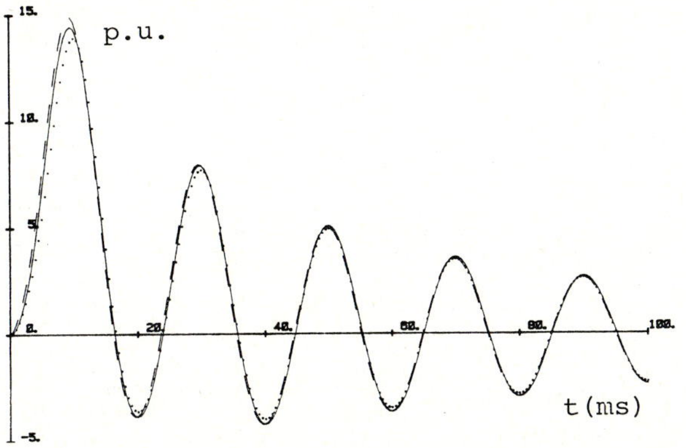

Figure 1 - Analytic simulation of $i_A(t)$

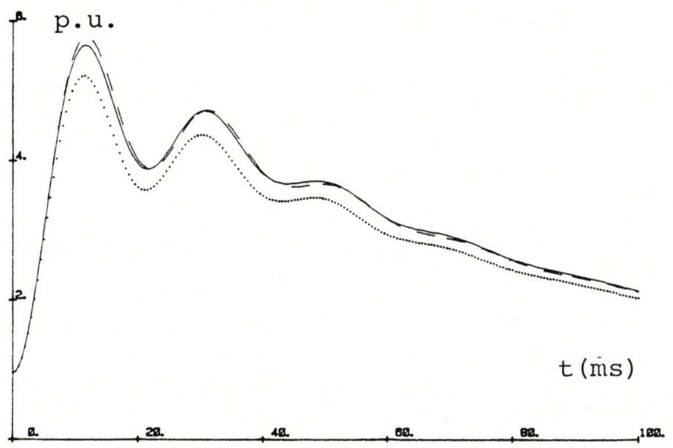

Figure 2 - Analytic simulation of $i_f(t)$
——— exact solution
----- after H1
..... after H4

For large synchronous machines, the influence of the assumptions is not so critical [10]. Nevertheless, inaccuracies of more than 10 % have been detected on the subtransient and aperiodic components amplitudes, principally for the field current. Moreover, considering the large value of the aperiodic time constant, a small difference between the forced pulsation ω_c and the steady-state pulsation ω_r leads to an important phase displacement in the alternating component of the field current. Thus, it will be difficult to achieve convergence of an identification algorithm based on a full treatment of the signal.

From these simulation results, a great interest appears to develop an identification method of the short-circuit currents using the exact analytical equations. This is the only means to improve basically the validity of the parameters deducted from this test.

3.3. Signals acquisition and conditionning

It must be pointed out that an identification to complex expressions like equations 10 and 24 will give accurate results only if the signals are recorded with great care. Therefore we have developed high performance digital data acquisition systems.

The rotation speed problem is more difficult to solve. Indeed, all the equations are based on a strictly constant rotation speed. Without this condition, it would be impossible to identify phase displacements between the different components of the signal. Therefore we have transformed the time scale in a fictitious one linked to the rotor position which is therefore necessary to measure.

All these steps are automatized as well as the determination of the short-circuit initial time and the suppression of the steady-state currents.

3.4. Identification of the currents parameters

It is possible to treat the phase signals directly or to transform these signals, since we dispose of the rotor position, and to identify the currents i_d and i_q. This transformation is not easy to implement, because any error on the relative amplitudes generates harmonics on the PARK's components. This procedure does not seem to be interesting especially as the phase currents have no more parameters than the currents i_d and i_q.

We identify the parameters of the signal analytic functions and not directly the classical machine parameters. Indeed this test excites poorely the q-axis of the machine and the sensitivity of the parameters of this axis is important compared to the d-axis.

The identification algorithm is a wheighted mean least squares algorithm using a GAUSS-NEWTON method.

The program performs an identification on

1 - one phase current only, for example $i_A(t)$;
2 - the field current only ;
3 - simultaneously the three phase currents ;

4 - simultaneously one phase current and the field current.

The simultaneous identification takes, of course, the common parameters of the different signals into account, principally the time constants and the forced pulsation. The program can treat 3000 samples of each signal and the results are display on a colour plotter or on a graphic screen.

These algorithms were first tested on signals generated by the exact analytic solutions in order to judge the method sensitivity to a lot of measurement errors and conditions : offset, initial time of short-circuit, sampling step, total time recorded. The conclusions of this preliminary study are :

1 - The identification of only one phase signal converges for a complete model only for initial values of the parameters very close to the exact quantities. This was expected because it is practically impossible to determine simultaneously the phase displacement ε_4 and the forced pulsation ω_c.

2 - The identification of the 2ω component is very important to obtain the convergence.

3 - The use of the field current only is not sufficient to identify all the parameters. In particular the subtransient components are not easely determined in the presence of an important alternating component.

4 - For the simultaneous treatments, the complete model described here is necessary to obtain the convergence.

5 - But for these treatments, the results obtained with a large domain of initial values are very accurate for all the parameters.

This method was then tested on laboratory machines (Figure 3 and 4). The conclusions of these treatments are the following :

1 - When we consider all the measurement samples and not only the envelopes, and when we identify simultaneously all the available signals, it is necessary to take into account the complete analytic expressions of the responses.

2 - On the other hand, the identification convergence of a complete model is only achieved for a simultaneous treatment of the signals, which leads to a more accurate identification of each signal.

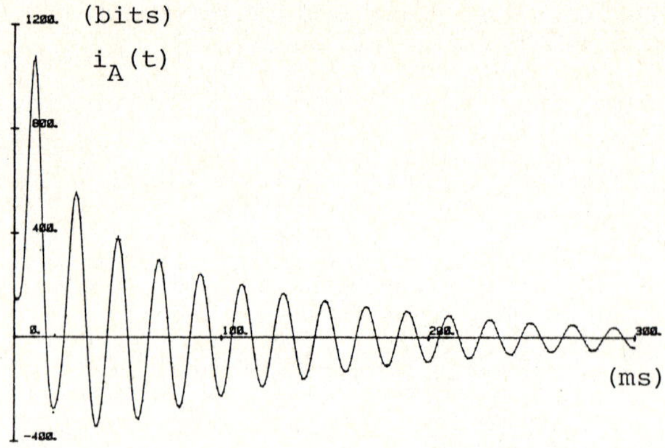

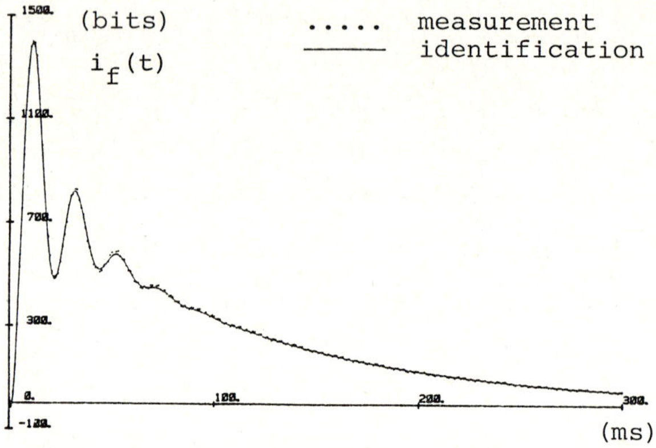

Figure 3-4 - Three-phase short-circuit test on a 14 kVA synchronous machine.

3.5. Derivation of the machine parameters

The number of parameters given by this identification process is greater than the number of independant machine parameters. Indeed, the constraints between the poles and the zeros of the transfer functions were not considered, and the amplitudes and phase displacements of the signals are not independant of each other.

Theoretically, it is possible to deduct a set of parameter for both axis. But the influence of the q-axis circuit is too small with this test to have a satisfactory accuracy by this way. An original method has then been developed [1] which consists of taking this influence into account under the form of corrections to derive the parameters of the common denominator of the transfer functions, namely the d- and q-axis short-circuit time constants. In spite of certain approximations in the treatment of the q-axis values, we obtain this way a greatly improved accuracy in the derivation of all the parameters.

4. EXTENSION OF THE PROCEDURE TO OTHER TESTS

4.1. The stator decrement test at standstill

The transient current is constituted by a sum of decaying exponentials. The rotor position allows to excite both the q- or the d-axis circuits. For the d-axis, it is fundamental to note that we dispose of a measure at the stator as well as at the rotor. The currents are given, for the d-axis, by

$$i_d(t) = \sum_{i=1}^{nd+2} I_{di}\, e^{-t/T_{di}} \qquad (\underline{27})$$

$$i_f(t) = \sum_{i=1}^{nd+2} I_{fi}\, e^{-t/T_{di}}$$

The transfer functions have the same poles.

The identification is achieved by an iterative mean least square method. This algorithm does not depend on a constant step sampling and can be easely extended to a simultaneous treatment of signals the transfer functions of which have the same poles.

Results obtained on 14 kVA laboratory machines show that the q-axis response contains three exponential components. Thus the solid rotor must be modelled for the q-axis by two damper windings. But it is principally the identification of the d-axis parameters which is improved by our procedure. Indeed, we find out, for the machine under study, that the transient armature current as well as the transient field current can be identified separately by three exponential components. Nevertheless, the identified time constants are far from being the same for the two signals. On the other hand, a simultaneous identification clearly shows the necessity of considering four exponential components. The identifications then prove more stable and accurate.

4.2. The standstill frequency response test

This test leads to a measurement of the harmonic impedances

$$v_d/i_d = Z_d(j\omega) = R_a + j\omega L_d(j\omega) \qquad (\underline{28})$$
$$v_q/i_q = Z_q(j\omega) = R_a + j\omega L_q(j\omega)$$

Unfortunately, there is a difficult problem of measurement accuracy for the d-axis, because the dynamic behaviour of the harmonic inductance is blinded by the armature resistance. The synchronous reactance and the transient time constants will be derived with a poor accuracy by this test. The use of the field current can resolve this problem. Indeed, the transmittance between the armature and the field current is given by

$$i_f/i_d = T(j\omega) = j\omega G_f(j\omega) \qquad (\underline{29})$$

The armature resistance does not step in this expression and an accurate identification will be possible at low frequency.

The frequency response identification is achieved by a mean least square iterative algorithm which allows a simultaneous treatment of the harmonic responses of two signals of which the transfer function poles are the same, like $Z_d(j\omega)$ and $T(j\omega)$.

The test results show that the identification can be done simultaneously on the two d-axis measures with a two damper windings model. We point out that the treatment of only the harmonic impedance, although it identifies the response accurately, leads to a model with one damper winding. The values of the synchronous reactance and of the time constants are then very different from the ones given by other tests or by the simultaneous treatment. This proves the necessity to take this transmittance into account for the treatment of the d-axis response. Moreover, it must be noted that, for the two axis, the algorithm puts a pole and a zero out of the studied frequency band although it was already very large - .04 Hz to 60 Hz. This phenomenon shows the limits of the two axis theory for the description of the dynamic behaviour of machines with solid rotor.

5. CONCLUSION

Although the developed treatments and algorithms lead to a more complete and reliable set of parameters, this improvement has not required a prohibitive increase of the test complexity. Moreover, the use of a standard computer system leads to an efficient test automatization. Thus, it is now possible to obtain fast the complete test results on site. The great care taken in this study of efficient identification algorithms, of the treatment of all the data and particularly of all the available signals leads to the derivation of a more reliable set of dynamic parameters. These tests show yet the limits of the two axis theory, which explain the variations between the parameters deduced from each developed test [9].

It also appears that none of these tests can be considered better than the others. The simplest tests, especially when the machine is at standstill, allow to deduce the parameters corresponding to the most complete models. But the synchronous machine is then placed in conditions far from a normal operating point. The sudden short-circuit test sets the machine in more normal conditions. This excludes some risks of errors but the tests are more difficult to treat and thus the model complexity which it is possible to use is more limited.

Finally, it must be pointed out that the methods developed in this study are adapted to treat the responses of the great turbo-generators. Our tests are being done on laboratory machines because it is easier to study on these the limits of the various methods. Moreover, the poles and the zeros of the transfer functions are closer for these small machines and they constitute an ideal test for the measurement and identification methods before applying them to great synchronous machines.

TABLE 1 - Comparison of results obtained on a 14 kVA salient poles machine

	Three-phase short-circuit	Stator decrement	Frequency response
X_d	1.103	1.071	1.080
X'_d	.226	.264	.256
X''_d	.169	.229	.219
X'''_d	___	.194	.201
T'_{d0}	522.7	444.8	470.4
T''_{d0}	15.86	57.1	56.5
T'''_{d0}	___	6.40	3.27
T'_d	107.3	109.5	111.6
T''_d	11.85	49.5	48.4
T'''_d	___	5.42	3.00
T_{kd1}	10.95	63.7	64.2
T_{kd2}	___	4.21	2.59
X_q	.650	.657	.630
X''_q	.491	.516	___
X'''_q	___	.403	.310
T''_{q0}	44.6	43.8	___
T'''_{q0}	___	4.82	6.26
T''_q	33.7	35.5	___
T'''_q	___	3.79	3.00

Reactances in p.u.
Time constants in ms

6. REFERENCES

[1] Maun, J.C., Contribution à la détermination des paramètres dynamiques de la machine synchrone, Thèse de Doctorat, U.L.B. (Janv. 1982).

[2] Maun, J.C., Measurement of the dynamic parameters of synchronous machines, I.E.E. Confer. Publication 213 (1982) 94-98.

[3] Maun, J.C., Bodson, M., La mesure des paramètres des machines synchrones par l'essai statique de réponse en fréquence, Revue E de la SRBE, X-101 (1982) 79-94.

[4] Adkins, B., Harley, R.G., The general theory of alternating current machines (Chapman & Hall, London, 1975).

[5] Canay, I.M., Causes of discrepancies on calculation of rotor quantities and exact equivalent diagrams of the synchronous machine, IEEE Trans, PAS-88 (1969) 1114-1120.

[6] Shackshaft, G., New approach to the determination of synchronous machine parameters from tests, Proc. IEE, 121-11 (1974) 1385-1392.

[7] Takeda, Y., Adkins, B., Determination of synchronous machine parameters allowing for unequal mutual inductances, Proc. IEE, 121-12 (1974) 1501-1505.

[8] Marxsen, A.L., Morsztyn, K., Analysis of synchronous machine transient tests using a minicomputer, Proc. IEE, 124-4 (1977) 377-380.

[9] Diggle, R., Dineley, J.L., Generator works testing - Sudden-short-circuit or standstill variable-frequency-response method, Proc.IEE, 128-4 (1981) 177-182.

[10] Razek, A.A., de Siqueira Campos,M., Poloujadoff, M., Etude du court-circuit symétrique brusque d'un alternateur à basse fréquence, RGE, 88-6 (1979) 457-462.

[11] Shackshaft, G., Implementation of a new approach to the determination of synchronous machine parameters from tests, Proc. IEE, 124-12 (1977) 1170-1178.

[12] Hallingstadt, O., Estimation of synchronous machine parameters, Modelling, Identification and Control, I-1 (1980) 1-15.

[13] C.E.I., Recommandations pour les machines électriques tournantes, Part 4 : méthodes pour la détermination à partir d'essais des grandeurs des machines synchrones, Publication 34-4 (Genève 1967).

ELECTRICAL MACHINES AND CONVERTERS - Modelling and Simulation
H. Buyse and J. Robert (editors)
Elsevier Science Publishers B.V. (North-Holland)/© IMACS, 1984

NEW ALGORITHM FOR THE DETERMINATION OF THE SYNCHRONOUS MACHINE TIME CONSTANTS USING THE SAMPLED OUTPUTS

RICHARD Jean-Pierre*, LESENNE Jacques**
*Laboratoire de Systématique,
**Laboratoire Applications des Redresseurs de Puissance
Université de Lille I
59655 VILLENEUVE D'ASCQ CEDEX - FRANCE

The time constants of the synchronous machine are proposed to be measured by means of a characteristic exponents identification method. It uses data collected by sampling any state output of a periodical coefficents linear process, and can be usefully applied to the Park's model of the synchronous machine in symetrical short circuit.

The proposed work permits to break free from several usual sources of error, such as the offset or drift of the measurement line. The exact knowledge of the short circuit application time is neither necessary.

1. INTRODUCTION

In recent years, considerable interest has been generated by questions relating to the modelling of synchronous machine and to the determination of the characteristic parameters from tests. This interest has arisen due to the increasing power of the new generators and to the fact that the power systems are operating closer to the stability margins : models have then to be more and more accurate.

Numerous papers have been devoted to that subject [1][2][3][4][5][6].

We then propose a new approach of the identification problem, based on mathematical properties of linear periodical coefficients differential equations.

The problem of such equation has not yet been solved, but FLOQUET [7] and LIAPUNOV [8] proved a basic result : there exists a change of variable -generally unknown- which leads to a constant coefficients systems. The eigenvalues, called characteristic exponents, are representative of the dynamic process behaviour.

After a detailled study of the mathematical problem, we apply our results [9] to the synchronous machine, in order to determine the characteristic exponents, which in this case are the time constants and pseudopulsations involved in all transients.

The usual synchronous machine model defined by Park involves five windings. After eliminating the variables associated to the damping circuits, the three currents I_d, I_q, I_F may be expressed in terms of three voltages V_d, V_q, V_F. All the admittances have the same denominator, the associated characteristic equation, and the expressions of $V_d(p)$, $V_q(p)$, and $V_F(p)$ depend on the studied transient [10].

The symetrical short circuit is generally used for the parameters determination. The recording of the field current, after substraction of the steady state, gives the signal caused by the disturbance. The numerical values, treated by the proposed resolution algorithm, lead to the three real roots and the two conjugate complex roots of the characteristic equation, and therefore to the time constants and pseudopulsations.

The presented method permits to eliminate errors inherent in the measurement line such as :
- the output offset of a recorder, by using the first order differences of signals.
- a drift error proportional to the time, by using the second order differences of signals.

Moreover, the exact knowledge of the disturbance application time is not necessary.

This method can also be applied to the study of the stator currents by use of differences of data considered with a displacement of one or several periods.

Lastly, the method permits to systematize the study of Park's model validity.

2. PRESENTATION OF THE BASIC PROPERTY

In this part, we propose to recall some results proposed in [9], concerning the linear periodical coefficients systems described by the following state space representation :

$$(1)\quad \overset{\circ}{X}(t) = \frac{d}{dt} X(t) = A(t) . X(t)$$

$$\begin{cases} X(t) \in \mathbb{R}^q \quad A(t) \in \mathbb{R}^{q\times q} \quad t \in [0,+\infty[\\ \exists T > 0, \quad t \geq 0,\ A(t+T) = A(t) \end{cases}$$

X = 0 is the equilibrium of (1) (stable or not)

The FLOQUET's theorem ensures the existence of a change of variable P(t), of period T, which leads from (1) to a constant coefficients system, characterized by a matrix M and a state vector U :

(2) $U(t) = P(t)\,X(t)$

(3) $\overset{o}{U}(t) = M\,U(t)$

$$\begin{cases} M \in \mathbb{R}^{q\times q},\ P(t) \in \mathbb{R}^{q\times q} \\ \det P(t) \neq 0 \\ P(t+T) = P(t) \end{cases}$$

The q eigenvalues λ_i of M are called characteristic exponents of (1), and their exponents $\mu_i = e^{\lambda_i T}$ are the characteristic multipliers. We are going to show that these quantities can be deduced from the sampled measurement of state outputs such as

(4) $s(t) = V^T X(t) \qquad V \in \mathbb{R}^q$

Throughout the paper, we will use the following notation for the variables sampled with a period T and starting from an instant τ.

(5) $s(\tau + nT) = s_n(\tau)$.

Basic Property (I)

Whatever the initial sampling time may be, all the outputs of (1) defined by (4) (5) are verifying a same recurrence equation of degree q, expressed by (6) :

(6) $s_{n+q}(\tau) + \sum_{i=0}^{q-1} a_i\, s_{n+i}(\tau) = 0$

Moreover, the zeros of the associated polynominal $p(\lambda)$

(7) $p(\lambda) = \lambda^q + \sum_{i=0}^{q-1} a_i\, \lambda^i$

are the characteristic multipliers of (1), i.e :

(8) $p(\lambda) = \det(\lambda I - e^{MT})$

Elements of proof

By integrating (3) over a period T, starting from an instant (nT+τ) and using (2), it comes out :

$X_{n+1}(\tau) = [P^{-1}(\tau)e^{MT}P(\tau)]\,X_n(\tau)$

and consequently, for any entire i ≥ 0,

(9) $X_{n+i}(\tau) = [P^{-1}(\tau)\,e^{iMT}P(\tau)]\,X_n(\tau)$

Let us define $p(\lambda)$ and the coefficients a_i by (8) and (7). Each equation (9), i = 0...q, is multiplied by the corresponding a_i, and by adding them together, it comes out

$X_{n+q}(\tau) + \sum_{i=0}^{q-1} a_i\, X_{n+i}(\tau) = 0$

since e^{MT} verifies its own characteristic polynominal [12]. Equation (6) is then obtained when left multiplying by V^T. The result does not depend on τ.

This property has now to be developped with regard to the characteristic exponents identification.

3. APPLICATION TO THE NUMERICAL IDENTIFICATION OF THE CHARACTERISTIC EXPONENTS

Based on property (I), we know that any state output of a linear periodical coefficients system verifies an unic recurrence equation. Measuring a sufficient number of outputs at a sufficient number of sampling times may lead to this equation, and therefore to the characteristic multipliers of (1). We consider in this part different cases, according as the measurement is exact, or presents an offset error, or a timeproportional error.

3.1. The mesures are supposed to be exact

3.1.a. One measured output. Determination of the system order

We suppose here that an unic output s(t) is exactly measured with a sampling period θ verifying T = kθ, where k is an entire number. The value of the initial sampling time is not important, and can be arbitrarily chosen equal to zero. Any time n can be considered as the begining of a line of measures at times nθ, nθ+T, nθ+2T, ...

It is then easy to construct a p order matrix S_p by taking p different values of n, corresponding to the different lines. If p is strictly superior to the order q of the system, then the determinant of S_p is zero. This is an available test of order for the system, and is going to be developped on a simple example.

Considering k = 2, the measured data are :
s(0) s(T/2) s(T) s(3T/2) s(2T) ...
and can be organized into the matrix S_4 :

(10) $$S_4 = \begin{bmatrix} s(0) & s(T) & s(2T) & s(3T) \\ s(T/2) & s(3T/2) & s(5T/2) & s(7T/2) \\ s(T) & s(2T) & s(3T) & s(4T) \\ s(3T/2) & s(5T/2) & s(7T/2) & s(9T/2) \end{bmatrix}$$

If the system is of order 3, then the determinant of S_4 is zero, since each line is composed of solution of a same recurrence equation (see property (I))

This permits to test the order of a given linear process with periodical coefficients :

Proposition II

Let q be the lowest entire such that, for any above constructed matrix S_p,
$\det S_{q+i} = 0 \quad \forall i > 0$
$\det S_q \neq 0$
then q is the order of the system.

The proof is in [9], which also approaches the observability problem, linked to the influence of the output and of the initial sampling times. Indeed it may happen that the whole system (1) is not observable by the outputs s(τ+nT).

The calculated order is then less than the real order of the whole system, but the corresponding characteristic exponents are still belonging to a subset of the M eigenvalues. In order to obtain the whole spectrum of M, changing of τ or s is then necessary.

In the general case, the matrix S_{q+1} has to be determined, and this involves at most $(q+1)^2$ measures of a same output and at least (2q+3) measures (for θ equal to T).

3.1.b. One measured ouput. Determination of the exponents

Once the order calculated by the above described means, the same matrix S_{q+1} allows the determination of the characteristic exponents of the process. If A is the vector composed by the (q+1) coefficients of the polynomial (8), it verifies

$$A = (a_o, a_1, \ldots, a_{q-1}, 1)^T$$

$$S_{q+1}\ A = 0 \quad (11)$$

The first q lines of equation (11) constitute a linear Kramer system, whose solutions are the a_i, given by inversion of S_q.

For example, the matrix S_4 described in (10) leads to

$$a_o s(0) + a_1 s(T) + a_2 s(2T) = - s(3T)$$
$$a_o s(T/2) + a_1 s(3T/2) + a_2 s(5T/2) = - s(7T/2)$$
$$a_o s(T) + a_1 (2T) + a_2 s(3T) = - s(4T)$$

The inversion of a S_3 type matrix gives a_o, a_1, and a_2.

The knowledge of A permits to calculate the zeros of $p(\lambda)$ (7), whose zeros are the characteristic multipliers μ_i. The λ_i are then calculable. ($\lambda_i = 1/T \text{ Log } \mu_i$)

The whole development of the method is summarized in the figure 1. It needs the measurement of a (q+1) order matrix, the inversion of a q order matrix, and a zeros calculation for a qth degree polynomial.

3.1.c. Several measured outputs

If several state outputs are measurable, it is possible to use them in the definition of the data matrix S_q. The lines of S_q are then corresponding to different (or not) outputs. This presents different advantages :

- it reduces the chances of non observability : even if each output, considered separately, does not lead the whole set of characteristic exponents, their association may do.
- the experimentation time is shorter, since several measurements can be taken in parallel. This can be important in the case of fastly converging (or diverging) outputs.

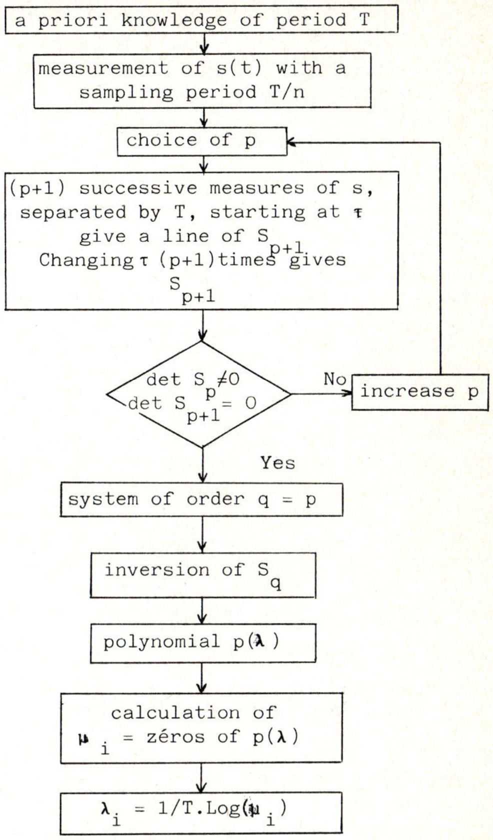

Figure 1

3.2. The measures are inacurate

In any case, the measured output s(t) must be a state component, and its equilibrium is then zero. This means that the permanent working conditions and a possible measurement error have to be eliminated.

3.2.a. Offset error

If the recorders are presenting unknown constant offset errors, the same method remains available when considering the first order differences of signals. It is then necessary to measure two data matrices of type S_{q+1}, which lines correspond to the same outputs with different initial sampling times. Their difference, denoted S^1_{q+1}, then verifies the equation (11). The algorithm of figure 1 is then preserved, and gives the actual values of λ_i when replacing S_q by S^1_q.

This needs to take two times more measures, compared to the case of an exact measurement.

3.2.b. Drift error

If the recorders errors are time-proportional, we have to construct two matrices of type S^1_{q+1} (which means four times more measures), and to calculate their difference S^2_{q+1}. The algorithm is then applied to this second order difference matrix S^2_{q+1}.

3.2.c. Steady state

If the permanent working conditions are non zero, unknow, but periodical, the state output can be obtained by making the difference between two matrices S_{q+1} of same construction, but separated by a period. This is, for example, the case of a current output presenting a sinusoïdal steady state.

4. PARK'S MODEL OF THE SYNCHRONOUS MACHINE

The Park's model of the synchronous machine is given by figure 2, the damper windings being modelised by two equivalent short circuited windings KD and KQ.
Park's transformation allows to characterize the idealized synchronous machine by the following set of constant coefficient differential equations (12). The Park's transformation is then the equivalent of FLOQUET's theory.

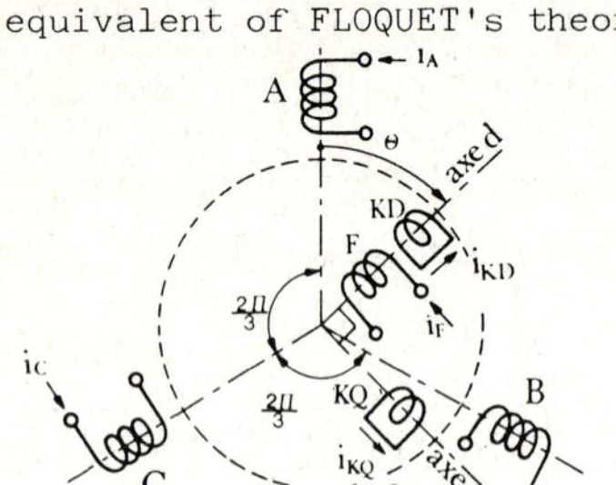

Figure 2

$$(12)\quad [V_d\ V_q\ V_F\ 0\ 0]^T = M\,[I_d\ I_q\ I_F\ I_{KD}\ I_{KQ}]^T$$

$$M=\begin{bmatrix} R_s+L_dp & -L_q\omega' & M_Fp & M_{KD}p & -M_{KQ}\omega' \\ L_d\omega' & R_s+L_qp & R_F\omega' & M_{KD}\omega' & M_{KQ}p \\ M_Fp & 0 & R_F+L_Fp & M_{FD}p & 0 \\ M_{KD}p & 0 & M_{FD}p & R_{KD}+L_{KD}p & 0 \\ 0 & M_{KQ}p & 0 & 0 & R_{KQ}+L_{KQ}p \end{bmatrix}$$

All the parameters used are well known.
From equation (12), we can derive [13] by directly inverting the impedance matrix of equation (13).

$$(13)\quad \begin{aligned} I_d &= Y_{dd}V_d + Y_{dq}V_q + Y_{dF}V_F \\ I_q &= Y_{qd}V_d + Y_{qq}V_q + Y_{qF}V_F \\ I_F &= Y_{Fd}V_d + Y_{Fq}V_q + Y_{FF}V_F \end{aligned}$$

Y_{ij} being the operational admittances of the idealised synchronous machine. The expression of those nine admittances are given in Table 1. The values of $V_d(p)$, $V_q(p)$, $V_F(p)$ depend on the transient being studied. The symmetrical three phases short circuit is generally used to derive the different time constants of the synchronous machine. For that transient we have :

$$V_d(p) = V_F(p) = 0$$

so that

$$I_F(p) = Y_{Fq}(p)V_q(p)$$
$$I_d(p) = Y_{dq}(p)\ V_q(p)$$
$$I_q(p) = Y_{qq}(p)\ V_q(p)$$

Thus recording of the field current $I_F(t)$, after substraction of the steady state, gives the signal $i_F(t)$ caused by the disturbance. The other currents measurable in windings A,B and C are composed of I_d and I_q combinations, with periodical coefficients : for example, the expression of $i_A(t)$ is :

$$i_A(t) = E_M.\left(-\frac{I_q(t)}{E\sqrt{3}}\sin\omega' t + \frac{I_d(t)}{E\sqrt{3}}\cos\omega' t\right)$$

The measurement of $i_A(t)$, $i_B(t)$, $i_C(t)$ or $i_F(t)$ permits the application of the above described method.

These numerical values treated by the proposed resolution algorithm lead to the three real roots and the two conjugate complex roots of the characteristic equation given by the expression of polynomial D(p) which appears in the denominator of all the admittances

$$D(p) = p^5 + \alpha_4 p^4 + \alpha_3 p^3 + \alpha_2 p^2 + \alpha_1 p + \alpha_0$$

the α_i being given in Table 1.

In different papers M. DELHAYE [14] has pointed out the numerous errors introduced by the measurement line and has proposed a statistical method to increase the accuracy. Among the different methods we find :

- the necessity of determining the exact instant when the short circuit appears.
- the influence of the offset of recorder.
  The errors due to those two causes of discrepancies disappear with the proposed method.

It is then possible to study the validity of PARK's model of the synchronous machine. We only have to determine whether or not the determinant is equal to zero or closed to that value. In that case we can use for that test the usual model.

If the value is far from zero we have to determine a new model, more complex but more accurate, by determining the order which leads to a determinant of null value. This will lead to a more complex representation of the damper windings as shown by figure 3 and will introduce sub-sub transient.

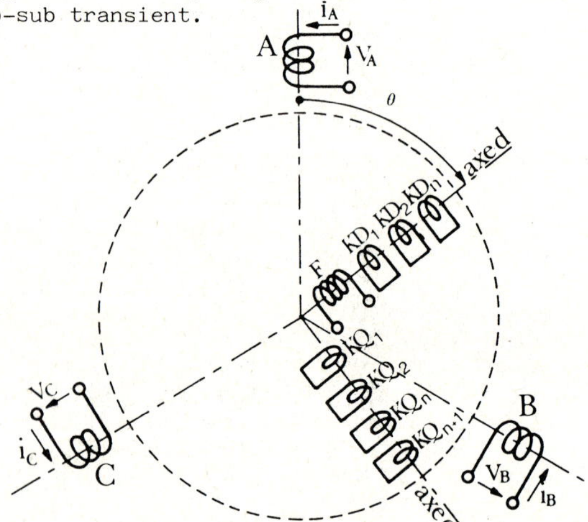

Figure 3

Table 1

$$a = T_{KD}/T''_{do}$$

$$B = \frac{X'_d\,[X''_d(X_d - X''_d) + (X_d - X'_d)(a^2 X'_d - 2a\,X''_d)]}{X_d\,X''_d\,(X'_d - X''_d)}$$

$$C = \frac{X'_d\,[X_d - X''_d + (X_d - X'_d)a(a-2)]}{X_d\,(X'_d - X''_d)}$$

$$P = R_S\left(\frac{1}{X''_d} + \frac{1}{X''_q}\right)$$

$$Y_{dd} = \frac{\omega'}{X''_d}\,\frac{1}{D(p)}\left[p^2 + \left(\frac{C}{T'_d} + \frac{X''_d}{X'_d T''_d}\right)p + \frac{X''_d}{T'_d T''_d X_d}\right] \times \left[p^2 + \left(\frac{R_S\omega'}{X''_q} + \frac{1}{T''_q}\right)p + \frac{R_S\omega'}{T''_q X_q}\right]$$

$$Y_{dq} = \frac{\omega'^2}{X''_d}\,\frac{1}{D(p)}\left[p^2 + \left(\frac{C}{T'_d} + \frac{X''_d}{X'_d T''_d}\right)p + \frac{X''_d}{T'_d T''_d X_d}\right]\left(p + \frac{1}{T''_q}\right)$$

$$Y_{dF} = -\frac{M_F\omega' T_{KD}}{R_F X_d T'_d T''_d}\,\frac{1}{D(p)}\left(p + \frac{1}{T_{KD}}\right) \times \left[p^3 + \left(\frac{R_S\omega'}{X''_q} + \frac{1}{T''_q}\right)p^2 + \left(\omega'^2 + \frac{R_S\omega'}{X_q T''_q}\right)p + \frac{\omega'^2}{T''_q}\right]$$

$$Y_{qd} = -\frac{\omega'^2}{X''_q}\,\frac{1}{D(p)}\left[p^2 + \left(\frac{1}{T''_d} + \frac{B}{T'_d}\right)p + \frac{1}{T'_d T''_d}\right]\left(p + \frac{X''_q}{X_q T''_q}\right)$$

$$Y_{qq} = \frac{\omega'}{X''_q}\,\frac{1}{D(p)}\left(p + \frac{X''_q}{X_q T''_q}\right)\left\{p^3 + p^2\left(\frac{R_S\omega'}{X''_d} + \frac{B}{T'_d} + \frac{1}{T''_d}\right) + p\left[\frac{1}{T'_d T''_d} + R_S\omega'\left(\frac{C}{T'_d X''_d} + \frac{1}{X'_d T''_d}\right)\right] + \frac{R_S\omega'}{X_d T'_d T''_d}\right\}$$

$$Y_{qF} = \frac{\omega'^3 R_S M_F T_{KD}}{X''_q R_F X_d T'_d T''_d}\,\frac{1}{D(p)}\left(p + \frac{X''_q}{X_q T''_q}\right)\left(p + \frac{1}{T_{KD}}\right)$$

$$Y_{Fd} = -\frac{M_F\omega' T_{KD}}{R_F X_d T'_d T''_d}\,\frac{1}{D(p)}\,p\left(p + \frac{1}{T_{KD}}\right) \times \left[p^2 + \left(\frac{1}{T''_q} + \frac{\omega' R_S}{X''_q}\right)p + \frac{\omega' R_S}{X_q T''_q}\right]$$

$$Y_{Fq} = -\frac{M_F\omega'^2 T_{KD}}{R_F X_d T'_d T''_d}\,\frac{1}{D(p)}\,p\left(p + \frac{1}{T''_d}\right)\left(p + \frac{1}{T_{KD}}\right)$$

$$Y_{FF} = \frac{B}{R_F T'_d}\,\frac{1}{D(p)}\left(p^4 + \delta_3 p^3 + \delta_2 p^2 + \delta_1 p + \delta_0\right)$$

with $\delta_3 = \frac{1}{B\,T'_d} + \frac{C\omega' R_S}{B\,X''_d} + \left(\frac{1}{T''_q} + \frac{\omega' R_S}{X''_q}\right)$

$$\delta_2 = \omega'^2 + \left(\frac{1}{BT'_d} + \frac{C\omega' R_S}{BX''_d}\right)\left(\frac{1}{T''_q} + \frac{\omega' R_S}{X''_q}\right) + \frac{\omega' R_S}{BT''_d X_d} + \frac{\omega' R_S}{X_q T''_q}$$

$$\delta_1 = \omega'^2\left(\frac{1}{T''_q} + \frac{1}{B\,T'_d}\right) + \frac{\omega' R_S}{B\,X_q}\left(\frac{1}{T''_d T''_q} + \frac{C\omega' R_S}{X''_d T''_q}\right) + \frac{\omega' R_S}{B\,X_D}\left(\frac{1}{T''_d T''_q} + \frac{\omega' R_S}{T''_d X''_q}\right)$$

$$\delta_0 = \frac{\omega'^2}{BT''_d T''_q}\left(1 + \frac{R_S}{X_d X_q}\right)$$

$$X_d = \omega' L_d \;;\; X'_d = \omega'(L_d - M_F^2/L_F) \;;$$

$$X''_d = \omega'\left(L_d - \frac{L_{KD}M_F^2 + L_F M_{KD}^2 - 2\,M_F M_{FD} M_{KD}}{L_F L_{KD} - M_{FD}^2}\right)$$

$$X_q = \omega' L_q \;;\; X''_q = \omega'(L_q - M_{KD}^2/L_{KD}) \;;$$

$$T'_{do} = L_F/R_F \;;\; T''_{do} = (L_{KD}/R_{KD})/(1 - M_{KD}^2/L_F L_{KD})$$

$$T'_d = (L_F/R_F)/(1 - M_F^2/L_d L_F) \;;$$

$$T''_d = \frac{L_{KD}}{R_{KD}}\left(1 - \frac{L_d M_{FD}^2 + L_F M_{KD}^2 - 2\,M_F M_{FD} M_{KD}}{L_{KD}\,(L_d\,L_F - M_F^2)}\right)$$

$$T''_{qo} = L_{KQ}/R_{KQ} \;;\; T''_q = (L_{KQ}/R_{KQ})/(1 - M_{QK}^2/L_q L_{KQ})$$

$$T_{KD} = (L_{KD}/R_{KD})/(1 - M_{FD}M_{KD}/M_F L_{KD})$$

The expression of polynomial D(p) which appears in the denominator of all the admittances is

$$D(p) = p^5 + \alpha_4 p^4 + \alpha_3 p^3 + \alpha_2 p^2 + \alpha_1 p + \alpha_o$$

with

$$\alpha_4 = \frac{1}{T''_q} + \frac{1}{T''_d} + \frac{B}{T'_d} + \omega' P$$

$$\alpha_3 = \frac{1}{T''_d T''_q} + \frac{1}{T'_d T''_d} + \frac{B}{T'_d T''_q} + \omega'^2\left(1 + \frac{R_S^2}{X''_d X''_q}\right) + \omega'\left[\frac{P}{T''_q} + \frac{P}{T''_d} + \frac{R_S}{T'_d}\left(\frac{B}{X''_q} + \frac{C}{X''_d}\right)\right]$$

$$\alpha_2 = \frac{1}{T'_d T''_d T''_q} + \omega'\left[\frac{P}{T''_d T''_q} + \frac{P}{T'_d T''_d} + \frac{R_S}{T'_d T''_d}\left(\frac{B}{X_q} + \frac{C}{X''_d}\right)\right] + \omega'^2\left(\frac{1}{T''_q} + \frac{1}{T''_d} + \frac{B}{T'_d}\right) + \omega'^2 R_S^2\left(\frac{1}{T''_q X_q X''_d} + \frac{1}{T''_d X'_d X''_q} + \frac{C}{T'_d X''_q X''_d}\right)$$

$$\alpha_1 = \frac{1}{T'_d T''_d T''_q}\left[\omega' P + T'_d + T''_q + B\,T''_d + \omega'^2\left(\frac{T'_d}{X'_d X_q} + \frac{T''_q}{X_d X''_q} + \frac{CT''_d}{X_q X''_d}\right)\right]$$

$$\alpha_o = \frac{\omega'^2}{T'_d T''_d T''_q}\left(1 + \frac{R_S^2}{X_d X_q}\right)$$

5. NUMERICAL SIMULATION

In order to illustrate the algorithm, we consider the simulation of (12) and (13) corresponding to the following numerical values :

X_d	2
X'_d	0,275
X''_d	0,2
X_q	2
X''_q	0,2
ω'	100 Π

T'_{do}	10
T''_{do}	0,044
T'_d	1,375
T''_d	0,032
T''_{qo}	0,08
T''_q	0,008
T_{KD}	0,02

This leads to the currents equations :

$$I_F(t)=i_{Fo}\left[1+6{,}247e^{-0{,}7181t}-2{,}3964817e^{-31{,}9105t}+0{,}074e^{-127{,}6629t}-(3{,}925\cos 295{,}5249t+1{,}443\sin 295{,}5249t)\,e^{-92{,}7082t}\right]$$

$$\begin{aligned}i_A(t)=&E_M\,[0{,}015\sin 100\Pi t-0{,}499\cos 100\Pi t]\\&+E_M\,[0{,}090\sin 100\Pi t-3{,}174\cos 100\Pi t]e^{-0{,}7181t}\\&+E_M\,[-0{,}088\sin 100\Pi t-1{,}431\cos 100\Pi t]e^{-31{,}9105t}\\&+E_M\,[-0{,}402\sin 100\Pi t-0{,}127\cos 100\Pi t]e^{-127{,}6629t}\\&+E_M\,[0{,}384\sin 100\Pi t+5{,}231\cos 100\Pi t]e^{-92{,}7082t}\cos 295{,}5249t\\&+E_M\,[5{,}253\sin 100\Pi t+1{,}424\cos 100\Pi t]e^{-92{,}7082t}\sin 295{,}5249t\end{aligned}$$

The model is a priori supposed of order 5 : therefore, the measurement has to include at least 6 periods of 20ms.

The numerical values of I_F and i_A are then proposed on table 2, which also includes the values of the difference of I_F and its steady state (1). We will denote :

$i_F(t) = I_F(t) - I_F(\infty) = I_F(t) - 1$

$\Delta I_F(t) = I_F(t+\Delta) - I_F(t)$ First order on I_F

$\Delta I_A(t) = i_A(t+T) - i_A(t)$ Period difference on I_A

$\Delta = 0{,}004s$. $T = 0{,}2s$

In a first time, the calculation is made on $i_F(t)$: the matrice S_6 is presented, and leads to the 3th, 4th, 5th and 6th order determinants. The sixth order one is near zero, so the fiveth degree polynomial associated to the five first rows of S_6 is calculated. This leads to the evaluations $\hat\lambda_i$ of the actual simulation time constants λ_i :

$\hat\lambda_1=-127{,}616$ $\lambda_1=-127{,}662$ (0,03%)
$\hat\lambda_2=-\ 31{,}904$ $\lambda_2=-\ 31{,}910$ (0,6%)
$\hat\lambda_3=-\ 0{,}715$ $\lambda_3=-\ 0{,}718$ (0,4%)
$\hat\lambda_4=-\ 92{,}73\pm 295{,}51j$ $\lambda_4=-92{,}71\pm 295{,}52j$ (0,02%)

The maximum error is of 0,6%, which is a very good approximation. Notice that the complex root imaginary part is obtained with an indetermination of kΠ, k entire.

In a second time, the same calculation is considered on ΔI_F data, and then on ΔI_A data. This allows the elimination of the steady states and of an eventual offset error. In these calculations, determinants are nearer zero, because data ΔI_F and ΔI_A are nearer zero, but the results λ_i are exactly the same than previously.

The evaluation algorithm is therefore very accurate.

TIME (S):	CURRENT IF:	CURRENT IA:	IF-STEADY STATE:
0	1	0	0
.002	1.67207	.734571	.672068
.004	3.21671	2.24211	2.21671
.006	4.91834	4.07028	3.91834
.008	6.23266	5.83418	5.23266
.01	6.90257	6.65593	5.90257
.012	6.94899	5.68222	5.94899
.02	5.40821	-3.71203	4.40821
.022	5.42737	-2.80458	4.42737
.024	5.62458	-.710861	4.62458
.026	5.90202	1.74348	4.90202
.028	6.16599	3.68404	5.16599
.03	6.35481	4.38572	5.35481
.032	6.44811	3.53039	5.44811
.04	6.35574	-3.91282	5.35574
.042	6.36051	-3.07685	5.36051
.044	6.39449	-1.06855	5.39449
.046	6.44677	1.32789	5.44677
.048	6.50344	3.20835	5.50344
.05	6.55288	3.87086	5.55288
.052	6.58871	3.07065	5.58871
.06	6.62935	-3.74645	5.62935
.062	6.6357	-2.96219	5.6357
.064	6.64549	-1.05697	5.64549
.066	6.65798	1.23328	5.65798
.068	6.67145	3.03591	5.67145
.07	6.68408	3.67091	5.68408
.072	6.69452	2.90544	5.69452
.08	6.71292	-3.60801	5.71292
.082	6.71466	-2.85463	5.71466
.084	6.71631	-1.01855	5.71631
.086	6.71795	1.19436	5.71795
.088	6.71945	2.93944	5.71945
.09	6.72057	3.55506	5.72057
.092	6.7211	2.81292	5.7211
.1	6.71639	-3.51238	5.71639
.102	6.71392	-2.77908	5.71392
.104	6.71115	-.989739	5.71115
.106	6.70812	1.16908	5.70812
.108	6.70487	2.87308	5.70487
.11	6.70136	3.47476	5.70136
.112	6.69759	2.74928	5.69759

CALCULUS ON (IF - STEADY STATE) DATA

MATRIX S6 (INCLUDING MATRIX S5) :

.672068	4.42737	5.36051	5.6357	5.71466	5.71392
2.21671	4.62458	5.39449	5.64549	5.71631	5.71115
3.91834	4.90202	5.44677	5.65798	5.71795	5.70812
5.23266	5.16599	5.50344	5.67145	5.71945	5.70487
5.90257	5.35481	5.55288	5.68408	5.72057	5.70136
5.94899	5.44811	5.58871	5.69452	5.7211	5.69759

```
3TH ORDER DETERMINANT= .397403
4TH ORDER DETERMINANT=-.110515E-2
5TH ORDER DETERMINANT=-.168108E-8
6TH ORDER DETERMINANT= .119824E-22
5TH DEGREE POLYNOMIAL:
-.993618E-3 +.274759E-1λ-.265795λ²+1.12733λ³-1.88347λ⁴+λ⁵
ROOT= .0779   POLE=-127.616   TIME CONSTANT= .783598E-2
ROOT= .5283   POLE=-31.9045   TIME CONSTANT= .313435E-1
ROOT= .9858   POLE=-.715089   TIME CONSTANT= 1.39843
COMPLEX ROOT=.147±.057 COMPLEX POLE= -92.73±295.51j
```

CALCULUS ON (△IF) DATA

```
3TH ORDER DETERMINANT= .219312E-2
4TH ORDER DETERMINANT=-.366271E-6
5TH ORDER DETERMINANT= .220249E-12
6TH ORDER DETERMINANT=-.122467E-27
5TH DEGREE POLYNOMIAL:
-.993618E-3 +.274759E-1λ-.265795λ²+1.12733λ³-1.88347λ⁴+λ⁵
ROOT= .0779   POLE=-127.616   TIME CONSTANT= .783598E-2
ROOT= .5283   POLE=-31.9045   TIME CONSTANT= .313435E-1
ROOT= .9858   POLE=-.715089   TIME CONSTANT= 1.39843
COMPLEX ROOT=.147±.057 COMPLEX POLE= -92.73±295.51j
```

CALCULUS ON (△IA) DATA

```
3TH ORDER DETERMINANT=-.008108
4TH ORDER DETERMINANT=-.618279E-6
5TH ORDER DETERMINANT= .2085E-10
6TH ORDER DETERMINANT=-.988292E-23
5TH DEGREE POLYNOMIAL:
-.993618E-3 +.274759E-1λ-.265795λ²+1.12733λ³-1.88347λ⁴+λ⁵
ROOT= .0779   POLE=-127.616   TIME CONSTANT= .783598E-2
ROOT= .5283   POLE=-31.9045   TIME CONSTANT= .313435E-1
ROOT= .9858   POLE=-.715089   TIME CONSTANT= 1.39843
COMPLEX ROOT=.147±.057 COMPLEX POLE= -92.73±295.51j
```

6. CONCLUSION

Based on this new theorical method, it is now possible to consider a systematic experimental study. Indeed, the numerical simulation has confirmed that the proposed algorithm enables the exact determination of the parameter as soon as the PARK's model is valid. On the opposite, if the sixth order determinant is not equal (or closed) to zero, this proves that PARK's model is unsuitable for the transient being studied. In this case, it is necessary to derive the order of the first non zero determinant, and then to evaluate the time constants corresponding to a new higher degree model, and to a more complex equivalent circuit.

If this method allows the elimination of some discrepancies due to the experimental measurement line (offset, drift, steady state, origin of time), the contraints linked to PARK's model are still remaining :

- the model has to be linear
- the angular speed must remain constant, in spite of the disturbance.

Moreover, our method introduces a new constraint : the sampling period must be equal to the rotating one (or a multiple). This can be easily solved by use of modern technologie circuits.

REFERENCES

[1] : CANAY I.M. "Causes of discrepancies on calculation of rotor quantities and exact equivalent diagram of the synchronous machine" IEEE Trans. PAS-88, 1969, pp 1114

[2] : TAKEDA Y., ADKINS B. "Determination of synchronous machine parameters allowing for unequal mutual inductances" Proc. IEE, Vol 121, N°12, December 1974, pp 1501-1504

[3] : WATSON W., MANCHUR G. "Synchronous machine operational impedance from low voltage measurements at the stator terminal" IEEE trans PAS-93, 1974, pp 777-784

[4] : SHACKSHAFT B. "New approach to the determination of synchronous machine parameters from tests" Proc IEE, n°11, November 1974, pp 1385-1392.

[5] : SHACKSHAFT B. Report on questionnaire on generator parameters for stability studies" CIGRE Study Commitee n°32, Working Group 03, April 1975

[6]: UMANS S.D., MALLICK S.A., WILSON G.L. "Modeling of solid rotor turbogenerators" IEEE trans. PAS-97, n°1, 1978, pp 269-291

[7] : FLOQUET "Sur les équations différentielles linéaires à coefficients périodiques" Annales Scient. de l'Ecole Normale Supérieure, T XII, 1883

[8] : LIAPUNOV A.M. "Problème général de la stabilité du mouvement" Original paper in Russian, Comm. Soc. Math. Kharkow 1892, English translation, Ann. Math Studies, Vol 17, Princeton 1949

[9] : RICHARD JP, LAURENT F. "Identification of dynamical properties of a linear process with periodical coefficients" Proc. of ACI'83 Symposium, Vol1, pp 8/7-8/11, 1983, Copenhagen, Denemark

[10] : LESENNE J., KOTNY JL., SEGUIER G. "Transmittance générale de la machine synchrone. Conséquence des simplifications usuelles." Proc of the 1rst Applied Modelling and Simulation Conference, Vol II, Lyon 1981 pp 238- 244.

[11] : LESENNE J., KOTNY JL., SEGUIER G. "Influence of the usual simplifying hypothesis on the study of the dynamical behaviour of the synchronous machine" MECO'81, Le Caire, Août-Sept. 1981, Actes pp 186-189.

[12] : GANTMACHER F.R. "Théorie des matrices" Tome 1 Dunod, Paris, 1966

[13] : LESENNE J. SEGUIER G. : "New simplified theory of Park's model of the synchronous machine" ICEM'82, Budapest, Sept 1982, Proc, Part 1, pp 47-50.

[14] : DELHAYE M. "Contribution à l'optimisation du dépouillement des oscillogrammes de court-circuit de la machine synchrone". Revue E, Vol IX, n°4-5, 1979, pp 78-95.

ELECTRICAL MACHINES AND CONVERTERS - Modelling and Simulation
H. Buyse and J. Robert (editors)
Elsevier Science Publishers B.V. (North-Holland)/ © IMACS, 1984

Modelling and model-reduction of the synchronous machine through singular perturbation

H.A. Nour Eldin, L.F. Lopez
Group of Automatic Control and Technical Cybernetics
University of Wuppertal, FRG

The singular perturbation methods are applied to a full 7th order model of the synchronous machine. A systematic and mathematically based order reduction is presented. A detailed three time scale analysis with the complete reconstruction of the "slow" and "high" speed dynamic components of all state variables of the full model is described. Depending on the kind and speed of the disturbance the introduction of optimized correcting factors, for frequency and damping of the simplified models, is needed.

1. Introduction

The dynamical behavior of the synchronous machine is described in the d-q reference frame by a nonlinear system of differential equations of the 7th order. Different state variables are used /1/,/2/,/3/.
Beside this "full" model, there exist reduced order models, which have been formulated using different approximations together with physical and empirical reasoning. Their scope of application was supposed to be, for example, in the dynamical simulation of the large electrical power networks or in the analysis and design of the machine control parameters and systems through modern control concepts.
Recently /4/, efforts to realize real time dynamic simulation of power networks using sets of micro-computers, show that the proper choise of the model used is crucial in achieving such a goal.
In this paper the singular perturbation methods /5/,/6/, are applied to the complete dynamical model of the synchronous machine. A systematic and mathematically based order reduction (5th, 3rd and 2nd order) with their range of application and validity is presented. Also the three time scale analysis with the reconstruction of the "slow" and "high" speed dynamic components of all state variables of the full model is described. It will be shown, that depending on the kind and speed of the disturbance, the introduction of optimized correcting factors for frequency and damping in the simplified models is needed.

2. State space representation of the synchronous machine (full model)

The complete model for a synchronous machine connected to an infinite bus with the constant voltage amplitude V_N through an external impedance $Z_e = r_e + jx_e$ is given in the d-q reference frame by the following set of nonlinear differential equations /7/.

Electrical system

$$\hat{L}_e \frac{d\hat{\underline{i}}}{dt} = R^*(\dot{\theta})\,\hat{\underline{i}}(t) - \underline{b}_f\, v_f^* - \underline{b}_N(\delta)\,|V_N^*| \qquad (2.1)$$

$$\underline{i}(t_0) = \underline{i}_0$$

Mechanical system

$$\delta(t) = \omega_0\, s(t) \qquad \delta(t_0) = \delta_0$$

$$s(t) = \frac{1}{T_A}\left(M_m(t) - \frac{1}{2}\,\underline{i}^t(t)\, M\, \underline{i}(t) \right) \qquad (2.2)$$

$$s(t_0) = 0$$

with

$$\hat{\underline{i}}^t(t) = (\hat{i}_f \;\hat{i}_d \;\hat{i}_D \;\hat{i}_q \;\hat{i}_Q) = (-i_f \;x_d i_d \;i_D \;x_q i_q \;i_Q) \qquad (2.3)$$

the state vector and

$M_m(t)$ = mechanical torque
T_A = mechanical time constant

the matrices of equations (2.1) and (2.2) are defined to:

$$\hat{L}_e = \begin{pmatrix} \hat{L}_e^d & 0 \\ 0 & \hat{L}_e^q \end{pmatrix} \quad \text{symmetrical positive definit}$$

$$R^*(\dot{\theta}) = (R_r + \dot{\theta} R_x)\,\omega_0 \;; \quad \dot{\theta} = 1 + s(t) \qquad (2.4)$$

$$M = \begin{pmatrix} 0 & M_1 \\ M_1^t & 0 \end{pmatrix}$$

with

$$\hat{L}_e^d = \begin{matrix} (1-\mu_d) & \alpha & (1-\sigma_{fD}) \\ \alpha & \alpha(1+\beta_d) & \alpha \\ (1-\sigma_{fD}) & \alpha & (1-\mu_f) \end{matrix}$$

$$\hat{L}_e^q = \begin{bmatrix} (1+\beta_q) & \gamma \\ \gamma & 1 \end{bmatrix}$$

$$R_r = \mathrm{Diag}\left(-\frac{(1-\mu_D)}{T_f^*} \;\; -\frac{\alpha}{T_d} \;\; -\frac{(1-\mu_f)}{T_D^*} \;\; -\frac{\gamma}{T_q} \;\; -\frac{1}{T_Q^*} \right)$$

$$R_x = \begin{bmatrix} 0 & 0 & 0 & 0 & 0 \\ 0 & 0 & 0 & 1+\beta_q & \alpha \\ -\gamma & -\gamma(1+\beta_d) & -\gamma & 0 & 0 \\ 0 & 0 & 0 & 0 & 0 \end{bmatrix} \qquad (2.5)$$

$$M_1 = \begin{bmatrix} -1/x_q & 0 \\ \frac{(x_q-x_d)}{x_q x_d} & 1/x_d \\ -1/x_q & 0 \end{bmatrix}$$

and

$$\alpha = (1-\mu_f)\,(1-\mu_d)\,(1-\mu_D)$$
$$\gamma = (1-\sigma_q)$$
$$\beta_d = x_e/x_d \qquad \beta_q = x_e/x_q$$
$$T_d = (r+r_e)/x_d \qquad T_q = (r+r_e)/x_q \qquad (2.6)$$
$$T_f^* = \omega_0\, T_f \qquad T_D^* = \omega_0\, T_D$$
$$T_Q^* = \omega_0\, T_Q$$

Finally, the vector in equation (2.1) are defined to:

$$\underline{b}_N^t(\delta) = (\;0\;\; \alpha\sin\delta(t)\;\; 0\;\; \gamma\cos\delta(t)\;\; 0) \qquad (2.7)$$

$$\underline{b}_f^t = \left(\frac{1-\mu_D}{T_f^*} \;\; 0\;\; 0\;\; 0\;\; 0 \right) \qquad (2.8)$$

and

$$v_f^* = \omega_0\, v_f \qquad \omega_0 = 2\pi 50 \text{ Hz}$$
$$|v_N^*| = \omega_0\, |v_N|$$

3. Eigen-axis model of the synchronous machine

One of the main difficulties in applying singular perturbation theory modelling, model reduction or model dynamic separation, is that the original system dynamic equation has to be described in a way, such that it is suitable for handling by singular perturbation.

For $|\theta(t)-1| < \varepsilon$, with $\varepsilon \simeq 10^{-2}$, the electrical part of the machine (equation 2.1) is linear. The solution of the generalized eigenvalue problem

$$R^* \underline{x} = \lambda \hat{L}_e \underline{x} \qquad (3.1)$$

with $\hat{L}_e$ = symmetrical, positive definit
$R^* = R_r + R_x$ = const.

leads to the eigenvalues of Table 1 (machine data given in the Appendix)

$\lambda_1 = -d_1^* + j\omega_1$	$\tau_1 = 0.247$ sec.
$\lambda_2 = -d_1^* - j\omega_1$	$f_1 \approx 50$ Hz
$\lambda_3 = -d_3^*$	$\tau_3 = 1.738$ sec
$\lambda_4 = -d_4^*$	$\tau_4 = 35$ msec
$\lambda_5 = -d_5^*$	$\tau_5 = 36.8$ msec

Table 1: Eigenvalues of the electrical part of the synchronous machine

The electrical part of the machine could therefore be splitted up into the following dynamic components:

1. A lightly damped 50 Hz, 2nd order subsystem represented by a complex pair of eigenvalues $\lambda_1, \lambda_2 = \bar{\lambda}_1$. This subsystem will be defined as the "equivalent d-q coupling".
2. A relatively "slow" subsystem of the first order defined by the real eigenvalue λ_3. It represents the "equivalent field".
3. Two relatively "fast" subsystems of the first order given by the eigenvalues λ_4, λ_5. They are defined as the "equivalent dampers".

Now, the electrical part of the machine (equation 2.1) could be transformed from the d-q coordinates to the eigen-axis coordinates, in order to obtain a suitable form for singular perturbation.
Hence:

$$\frac{d}{dt}\underline{i}(t) = \Lambda \underline{i}(t) - \underline{b}'_f\, v_f^* - \underline{b}'_N(\delta)\, |v_N^*| \qquad (3.2)$$
$$\underline{i}(t_0) = X^t\, \hat{\underline{i}}(t_0)$$

with

$\underline{i}(t) = X^t\, \hat{\underline{i}}(t)$ the new state vector (3.3)
X = orthonormalized real eigenvector (3.4) matrix from the real eigenvalue matrix relation $R^* X = L_e X\Lambda$

and

$$\Lambda = \begin{bmatrix} -d_1^* & \omega_1 & & & \\ -\omega_1 & -d_1^* & & & \\ & & -d_3^* & & \\ & & & -d_4^* & \\ & & & & -d_5^* \end{bmatrix} \quad ; \text{ real eigenvalue matrix} \qquad (3.5)$$

$$\underline{b}'_f = X^t L_e^{-1} \underline{b}_f \qquad (3.6)$$

$$\underline{b}_N'(\delta) = X^t \hat{L}_e^{-1} \underline{b}_N(\delta) \quad (3.7)$$

The mechanical system is now defined to:

$$\delta(t) = \omega_0 \, s(t) \quad ; \quad \delta(t_0)=\delta_0 \quad (3.8)$$

$$s(t) = \frac{1}{T_A}(M_m(t)-M_e'(t)) \quad ; \quad s(t_0)=0 \quad (3.9)$$

with

$$M_e'(t) = \frac{1}{2}\, \underline{i}(t)^t \, M' \underline{i}(t) \quad (3.10)$$

and

$$M' = X^t \, M \, X \quad (3.11)$$

A systematic order reduction could now be made through singular perturbation methods. The structure of the defined eigen-axis synchronous machine model is schematically reveled in Fig. 1. The decoupled representation of the electrical part of the machine simplifies the physical understanding of its characteristic motions.

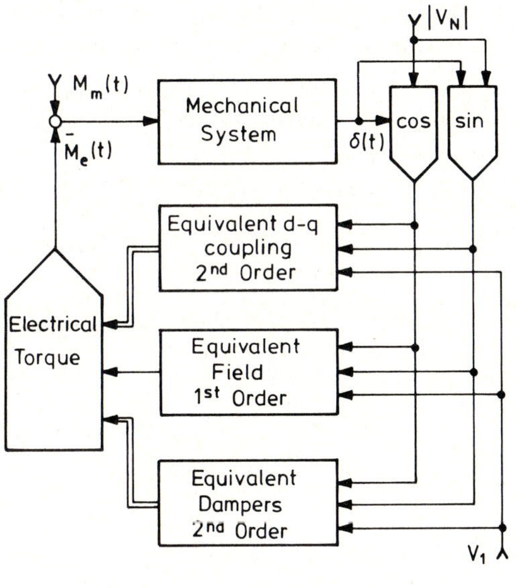

Fig. 1 Eigen-axis model of the synchronous machine

4. Model reduction of the synchronous machine

The complete machine dynamic could be interpreted as a nonlinear "three time scale"-problem. For synchronous machines with large field time constant, three distinct dynamical time behaviors are identified:

a. "Very" fast and lightly damped 50 Hz, 2nd order equivalent d-q coupling-system.

b. Two "fast" equivalent dampers.

c. One "slow" equivalent field and 2nd order "slow" mechanical system.

Taking into account the machine data, order reduction could be done. A simplified flow chart is given in Fig. 2. Starting from the eigen-axis 7th order model, different possibilities of order reduction could be found, depending on the specific machine parameters or the application range of the model.
In this paper path a-->b-->c will be illustrated. It corresponds to a commonly used order reduction sequence found in literature.
(For example /8/,/9/)

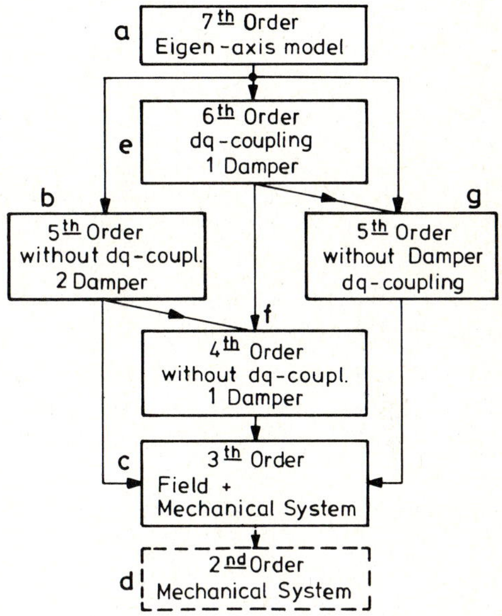

Fig. 2: Flow chart of the possible model-reduction steps

5. Nonlinear 5th order model(without d-q coupling)

In the first reduction step the equivalent d-q coupling subsystem is separated from the rest of the electrical system defined by equation (3.2).
Following the singular perturbation methods the fast transients of the machine in the d-q reference frame are "eliminated". It results in averaging the 50 Hz oscillatory mode, leaving in action the "slow" varying part of the states $i_1(t)$, $i_2(t)$. The obtained model is described by a nonlinear set of differential equations of the 5th order and two algebraic relations.
The original system is now approximated by

$$\begin{aligned} \underline{i}(t) &= \bar{\underline{i}}(t) + O(\mu_{dq}) \\ \delta(t) &= \bar{\delta}(t) + O(\mu_{dq}) \\ s(t) &= \bar{s}(t) + O(\mu_{dq}) \end{aligned} \quad (5.1)$$

where

$\bar{i}(t)$ = slow varying part of the electrical states

$\bar{s}(t), \bar{\delta}(t)$ = slow mechanical states

Note that according to the singular perturbation theory, the fast transients can be analyzed separately, resulting in a decoupled description of the original states in "slow" and "fast" modes as

$$\hat{i}(t) = X\ \bar{i}(t) + X_1\ \tilde{i}(\tau_{dq}) + O(\mu_{dq}) \qquad (5.2)$$

with

$X = (X_1 \mid X_2)$

$\dim X_1 = (5 \times 2)$

$\tilde{i}(\tau_{dq})$ = "fast" transients in the τ_{dq}-time scale

Numerical simulation results are illustrated in the figures found in the Appendix. Several cases are demonstrated:

Fig.5 : Mechanical torque step of -20% from the nominal value.
Fig.6 : Network-voltage step of -10% from the nominal value.
Fig.7 : Network-voltage step of -10% + "fast" system.

Note that due to the addition of the "fast" system to the reduced order model, the accuracy of the approximation of the mechanical state variables s(t) and δ(t) is increased (Fig.7). (The state space variables i_1 to i_5 correspond to the eigen-axis model (equation 3.3); the state space variables i_D to i_{FF} are defined in equation (2.3))

6. Nonlinear 3rd order model

Following path b-->c of Fig. 2 the obtained 5th order model can be further reduced to a nonlinear 3rd order model with four algebraic relations.
Depending on the machine data this reduction step may result in a very light damped system. In this case only the "slow" varying part of the input variables (or disturbances) $|V_N|$, v_f and $M_m(t)$ should be used. Numerical results are shown in Fig.8 for the case of a mechanical torque ramp (duration 2 sec).
If the speed of the disturbance is increased, the introduction of optimized correcting factors for frequency and damping of the "slow" dynamic are needed. This factors are situated in the mechanical part of the machine as shown in Fig.3.
The first factor SFAC corrects the damping, the second, DFAC the system frequency.
Numerically, gradient optimization methods could be used to find the optimal factors with known full reference model. Hybrid simulations could be done as well /10/.
Fig.9 illustrates simulation results of a sudden mechanical torque decrease of 10% from the nominal value with optimized correcting factors SFAC and DFAC. Notice that all state variables are reconstructed quite accurately. In Fig.4 the dependence of the correcting factors SFAC and DFAC from the relative mechanical torque step is shown. For "small" disturbances (<20% of the nominal value) only correction of damping is needed.
The "neglected" fast dynamic components could be reconstructed seperately.

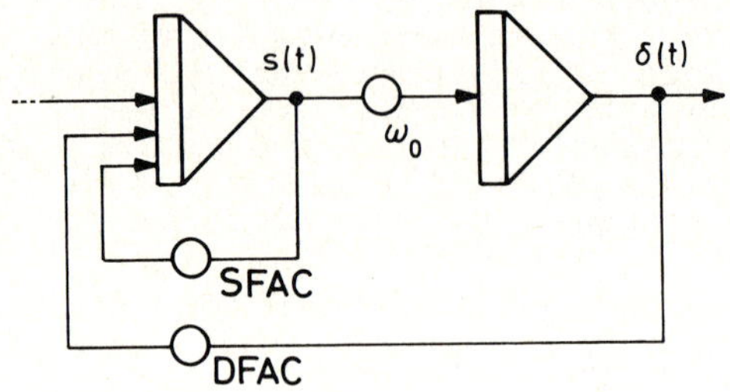

Fig.3: Correcting factors for frequency and damping

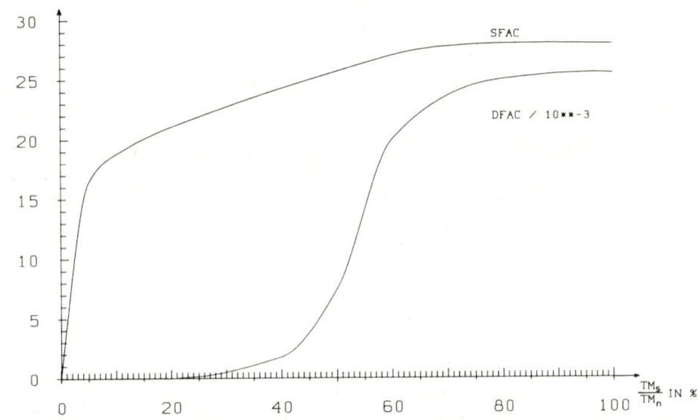

Fig.4: Dependence of the correcting factors from the mechanical torque input.

Literature

/1/ Th. Laible, "Die Theorie der Synchronmaschine im nichtstationärem Betrieb" Springer Verlag, Berlin 1952.
/2/ Anderson and Fouad, "Power System Control and Stability" IOWA State University Press, Ames, IOWA, USA.
/3/ Lerch E., Nour Eldin H.A., Wegmann P., Wehrli P., "Digitale Simulation der Synchronmaschine mit Zustandsraumdarstellung" ETZ-Archiv, Bd. 2 (1980), H.12, pp335-338.

/4/ K. Schmidt, W.M. Mansour, "Berechnung der dynamischen Stabilität in Energieversorgungsnetzen mit Parallelrechnern"
ETZ-Archiv, Bd. 4 (1982), H.4, pp133-139.

/5/ P.V. Kokotovic, R.E. O Malley Jr., P. Sannuti, "Singular perturbations and order reduction in control theory - An overview"
Automatica 12, Pergamon press, pp123-132 (1976).

/6/ J.H. Chow, J.J. Allemong, P.V. Kokotovic, "Singular perturbation analysis with sustained high-frequency oscillations"
Automatica 14, Pergamon press, pp271-279 (1978).

/7/ H.A. Nour Eldin, "Modelling and model-reduction of the synchronous machine through singular perturbation"
IFAC Workshop on singular perturbations and robustness of control, Ohrid/Jugoslavia, July 1982.

/8/ D. Nelles, "Die Beschreibung der Synchronmaschine für Ausgleichsvorgänge in Drehstromnetzen"
Wiss. Ber. AEG-Telefunken, 46 (1973).

/9/ S. Ahmed-Zaid, P.W. Sauer, M.A. Pai, M.K. Sarioglu, "Reduced order modelling of the synchronous machines using singular perturbation"
IEEE Transactions on Circuits and Systems, November 1982, Vol. CAS-29 (11), pp782-786.

/10/ L. Lopez, H. A. Nour Eldin, "Modelreduction of the synchronous machine through singular perturbation. An interactive hybrid simulation"
Interkama, Düsseldorf 9.-15.11.1983

Appendix

Generator data (p.u.):

$x_d = 2.7$	$x_q = 2.7$
$x'_d = 0.4$	
$x''_d = 0.2690$	$x''_q = 0.4230$
$r = 0.0015$	$x_{1\sigma} = 0.2120$
$\tau'_d = 1.0710$ sec	$\tau_j = 10.8$ sec
$\tau''_d = 0.0298$ sec	$\tau''_q = 0.0496$ sec

Network data:

$x_e = 0.3$

$r_e = 0.04$

Operating point:

$|V_N| = 1.$

$\cos\varphi = 0.8$

$|I| = 1.$

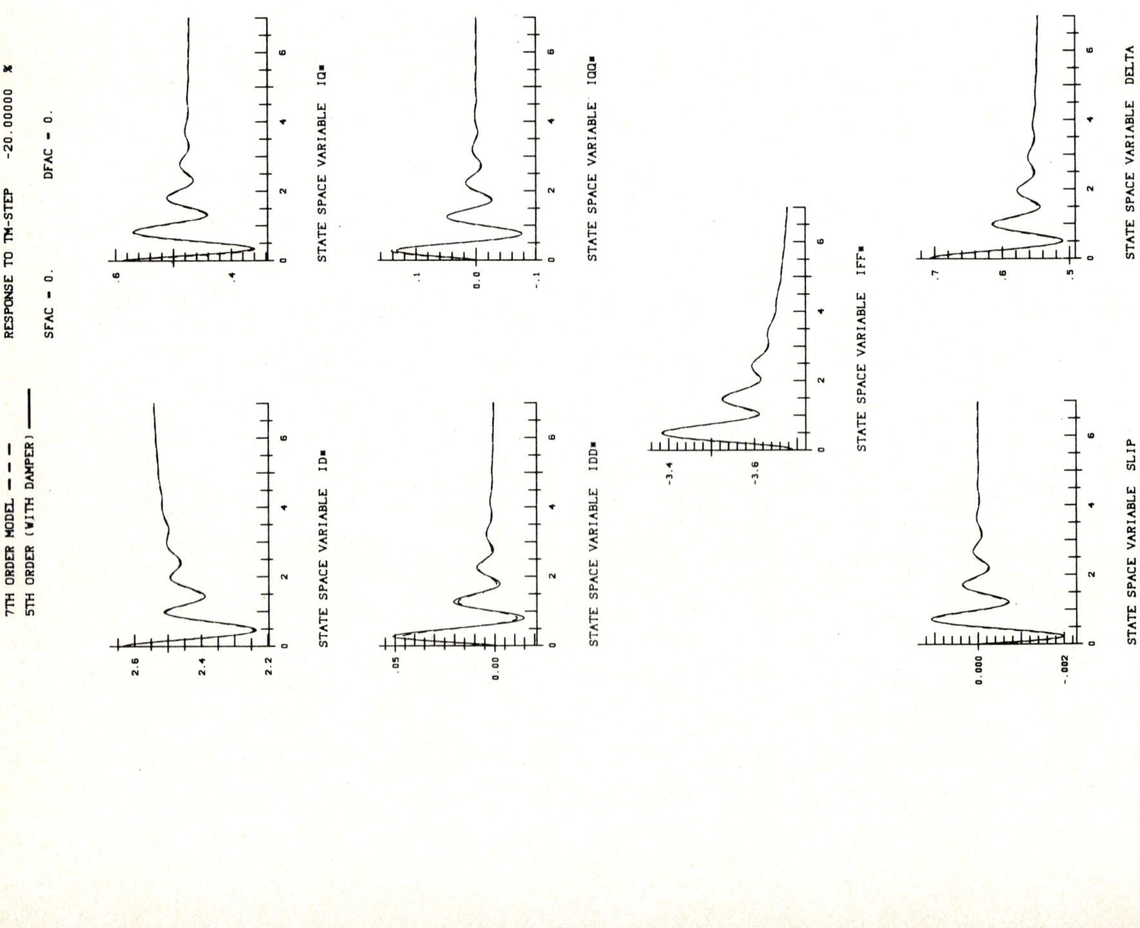

Fig.5: 5th order model; mechanical torque step response

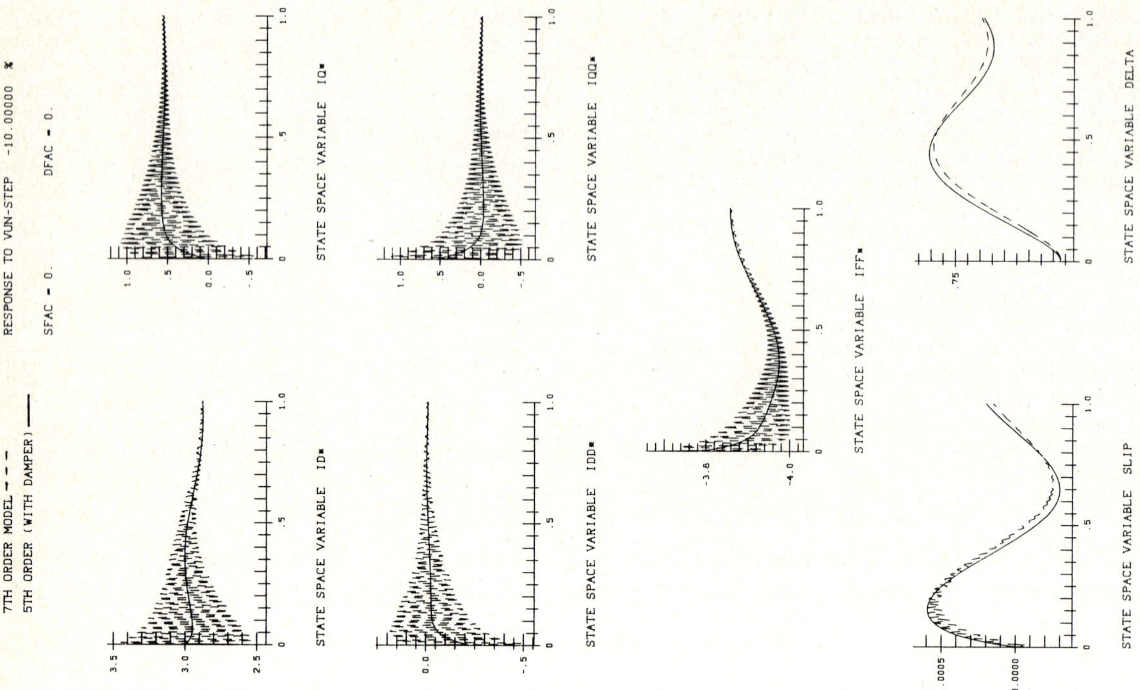

Fig.6: 5th order model;network voltage step(-10%) response

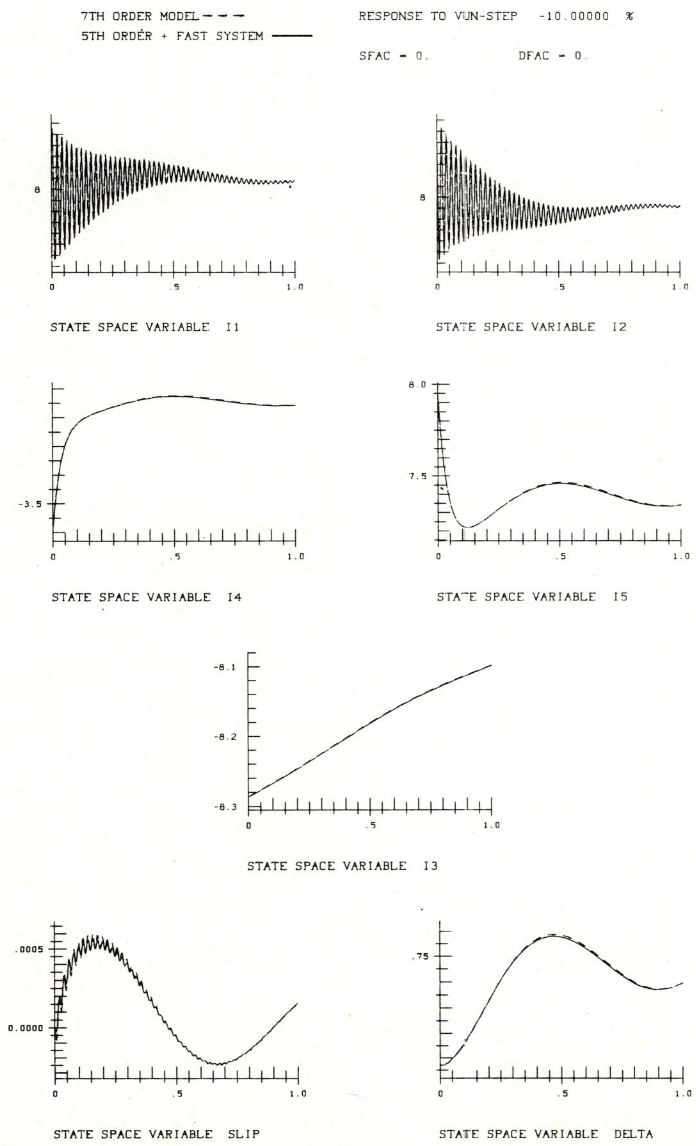

Fig.7: 5th order model; network voltage step(-10%) response + "fast" transients

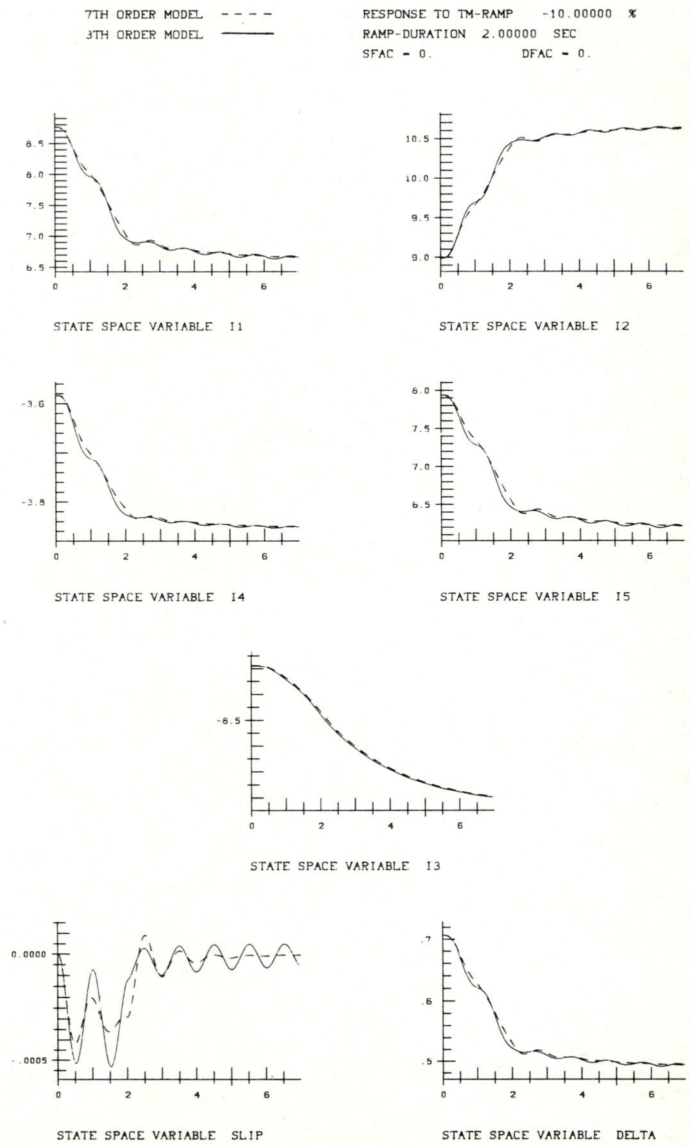

Fig.8: 3th order model; mechanical torque ramp response (duration 2 sec.)

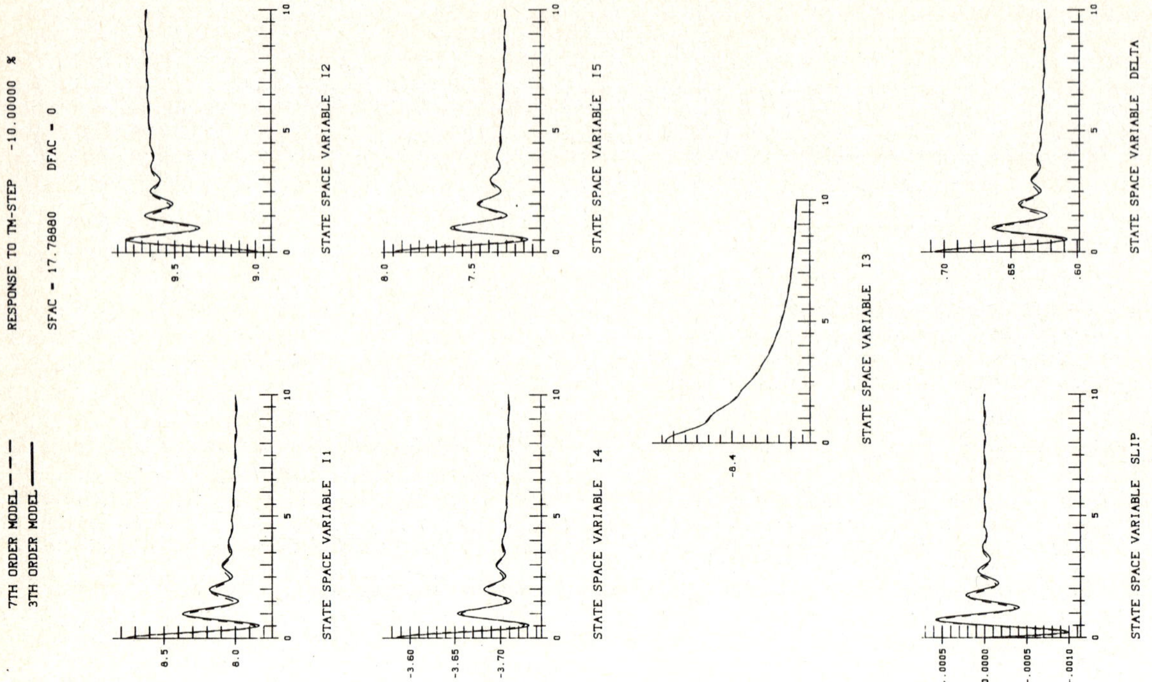

Fig.9: 3th order model; mechanical torque step response with optimized correcting factors

ELECTRICAL MACHINES AND CONVERTERS - Modelling and Simulation
H. Buyse and J. Robert (editors)
Elsevier Science Publishers B.V. (North-Holland)/ © IMACS, 1984

CHARACTERISTIC PARAMETERS OF THE COMPLEX MOTION MOTOR

Grzegorz Kamiński, Włodzimierz Przyborowski

Institut of Electrical Machines, Technical University
Warszawa, Poland

In the article will be presented the analysis of the characteristic physical and explaatation parameters in the dependence of some constructional knots - on the exaple of the complex motion spherical inductive motor. Ability of these parameters to the building different constrctions of such machines will be analysed.

1. INTRODUCTION

The present tendencies of constructional solutions of robots, gyroscopes, and manipulators of all kind require applying multi degree of freedom drivers. In this class of drivers multi degree of freedom electric motors can find wide application. Special attention should be given to the spherical ones. These motors can be constructed as DC, inductive or stepping ones. It should be underlined that gometrical structure of magnetic and electric circuits is an important factor that has influence on electromagnetic and first of all kinematic parameters of these machines. Therefore, the introducton to the formal analysis of these machines ounght to present detailed description of the structure and main constructional knots. The location inductor circuit, rotor, and kinematic knots seem to be very important with this type of machines of non-conventional kinematics. Also such parameters as: grove pitch, pole pitch, inductor field motion directions and rotor motion are of special importance and character.
This article presents the analysis of characteristic constructional, physical and exploitation parameters w which are relative to constructional solutions and their geometry.

2. MAIN CONSTRUCTIONAL KNOTS SOLUTION

In order to perform complex motion of the motor it is made of spherical link enfolded by casing which has spherical internal surface. Fig 1. In the casing we can distinguish external yoke and tooth zone, which constitute inductor generating magnetic field. The rotor of spherical geometry is situated concentrically in the casing so that seperating region measured radially constitutes air slit. The pole axis of the rotor can be situated in any direction to the pole axis of inductors.
If on the pole of axis of the rotor converter is placed /e.g. control rod/ the possibilities of angle range are limited to the angle zone determined by the free space, that is by cone angle γ round the pole axis.
It is possible to use different kinds of inductors depending on output parameters and the manner of supply.
Inductor´s elements are moduls made of packs of electrical siliceous sheets. Two ways of packing are possible:

Fig. 1

I Magnetic circuit of modul element of inductor is made of siliceaus sheets which dimensions change radially and are situated parallely to parallels. Ferromagnetic core has external yoke when parallel winding is used or internal one when coil winding is applied.

Moreover, in the proximity of the slit there is tooth zone meridional or parallely meridional when two windiugs are used.
II Magnetic circuit of a modul element of inductor is made of siliceous sheets which have nonvariable dimensions and is situated parallely to meridians. Ferromagnetic core and tooth zone are formed in analogical way.
In both cases, when such windings are used resulting in perpendicular one to another propagation of magnetic field tooth zone of winding placed at greater distance from rotor constitutes a yoke for winding placed closer to the rotor. Module elements are fixed to two rings situated in parallel planes. It is characteristic for the spherical motors that only one kind of inductor performs in a natural way one degree of freedom motion. Whereas inductors which have two perpendicular propagations of magnetic field give two degree kinematic.
It is also characteristic for these motors that a large number of them can be machines which have no axial symmetry in avrangement of inductors or even variably situated inductors. It gives great possibilities of electromagnetic power reception.
The analysis, in the first stape of research, of electromagnetic field kinematic of inductor makes it possible now to consider certain modul inductor that has unidirectional phase bands.
Formulating the problem.
Let´s consider a modul inductor that is in shape of a sector of the sphere, related to the set of coordinates xyz, when "OZ" axis is a pole axis of the motor. Module inductor makes one fourth part of the whole inductor.
In meridional direction the module inductor´s angle extent is

$$2\alpha = \alpha_0 = \pi - \gamma$$

The angle extent of a module inductor in the parallel direction is

$$2\beta = \beta_0 = \pi/2$$

γ - conical angle determinating control rod mothion zone /reception of power/

The considered inductor has winding which generates travelling electromagnetic field. Phase bands of this winding i.e. currents are distributed parallely in planes perpendicular to the pole axis. The described inductor has constant pole pitch τ_m in the direction of field wave propagation in sections generated by meridian planes in its total active zone. The spherical rotor of this motor, radius R, is smooth and displaces with angular velocity ω round ox axis.
For the defined geometry of the machine and the construction of inductor the direction of field propagation is tangential to meridians situated relative to the pole axis of the motor.
For defined module inductor it is possible to determine specific electric loading or main magnetic flow in the same way as one does it for classical engines. The principal difficulty is the choice of frame of coardinates which parametrises the machine and in which dimensions of machines parameters are constant.
Since the machine is geometrically strong nonlinear it follows that: while certain metrical parameters e.g. τ_m - pole pitch, t_m - tooth pich are constant, the same parameters in angular measure τ_φ, t_φ are functions of angles.
For instance in spherical coordinates $\hat{r}, \hat{\varphi}, \hat{\vartheta}$ we have $\tau_m = const$; $t_m = const.$

$$\tau_\varphi = \frac{\tau_m}{R\sin\vartheta} \quad ; \quad t_\varphi = \frac{t_m}{R\sin\vartheta}$$

Where of course $\tau_m = mqt_m$, $\tau_\varphi = mqt_\varphi$
and R - the radins of the sphere of inductor;
q - number of sides by a pole and phases;
m - number of phases.

Taking into consideration geometrical parameters of inductor and the number of poles the above can also expressed as follows:

$$\tau_m = \frac{R\alpha}{p} \quad ; \quad \tau_\varphi = \frac{R\alpha'}{p}$$

α - inductor´s span in meridional direction

From the above mentioned dependencies there follow natural premises to the constructional advices. The shapes of the groove and tooth zones should be such that at any crossing by plane ϑ = const. there is constant pole pitch and groove pitch.
It ought to be underlined here that arc pole pitch determines magnitude and direction of magnetic field velocity in inductor.
From the from of excitation flow function based on continous Fourier distribution it follows that discrete electric loading of phase bands currents is given by:

$$A = \sum_n \frac{4I}{bmn} \sin\frac{nb\hat{\varphi}}{2} \sin\frac{n\tau\hat{\varphi}}{2} \frac{\sin np\tau\hat{\varphi} \sin nqt\hat{\varphi}}{\sin n\tau\hat{\varphi} \sin\frac{nt\varphi}{2}}$$

$$\times \sin\left[n\hat{\varphi} \pm \frac{c}{\sin\vartheta} - \omega_n t\right] \qquad (1)$$

The above formula can be generalized on the case of a few module inductors distributed in axially symetric way. Limiting considerations to the first time-space harmonic expression /1/ one can rewrite in the from:

$$A = A_m \exp[p\hat{\varphi} - \frac{c}{\sin\hat{\vartheta}} - \omega t] \qquad (2)$$

It follows that specific current loading of parallely to the parallels' oriented plane phase bands are functions of variables so that in spherical coordinates excited flow spreads slantingly to parametric lines $\hat{\varphi}, \hat{\vartheta}$.
Simutaneously field velocity for very small but finite element of inductor is different on the perimeter of the active zone of inductor in the direction of $\hat{\vartheta}$ variable
So the kinematic of inductor is determined by the whde area of velocity field of magnetic field under the active zone of inductor.
Consequently in this motor there is possibility of geometrical regulation of velocity by changing the position of inductor.
It is worth analyzing the velocity field of travelling electromagnetic field in local coordinates in which the direction of the field is given by one parameter and the magnitude of velocity is constant. Simply, we shall consider the elementary inductor which location is given by meridian and parallel angles
Simultaneously from this follow the points of localization of velocities V_α V_β Fig. 2 which are oriented as tangential to parallels and meridians. Meridans are related to the presupposed axis of rotation OX.

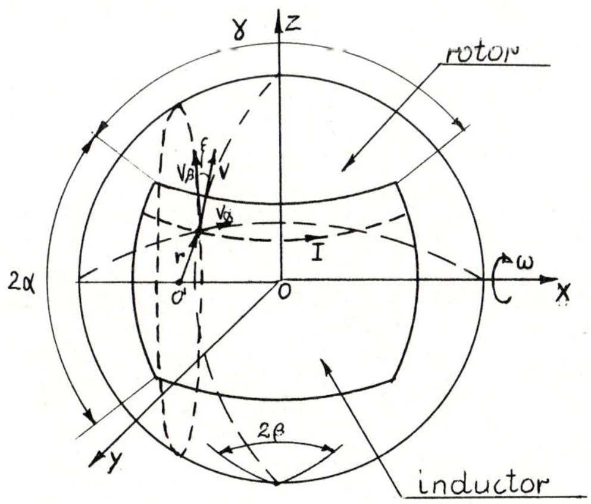

Fig 2

Basing on plane and spherical trygonometry we get

$$V_\alpha = V \sin\xi \quad ; \quad V_\beta = V \cos\xi \qquad (3)$$

V - linear velocity of magnetic field $2\tau m f$
ξ = arc sin[sin α sinβ]
The radius of parallel, on which V_β component of the field displaces, is:

$$r = R \sin\left\{\arccos\left[\frac{\sin\alpha}{\sin\gamma \cos\beta} + \mathrm{ctg}\gamma\, \mathrm{tg}\beta\right]\right\} \qquad (4)$$

γ - conical angle defining the zone of motion of control rod /reception of power/.
Now, it is worth introducing the notion of mean velocity and the mean radius of a parallel

$$V_{\alpha sr} = \frac{1}{\alpha_0 \beta_0} \int_{\alpha_1}^{\alpha_2} \int_{\beta_1}^{\beta_2} V \sin\xi(\hat{\alpha}, \hat{\beta})\, d\hat{\alpha}\, d\hat{\beta} \qquad (5)$$

$$V_{\beta sr} = \frac{1}{\alpha_0 \beta_0} \int_{\alpha_1}^{\alpha_2} \int_{\beta_1}^{\beta_2} V \cos\xi(\hat{\alpha}, \hat{\beta})\, d\hat{\alpha}\, d\hat{\beta} \qquad (6)$$

$$r_{\acute{s}r} = \frac{1}{\alpha_0 \beta_0} \int_{\alpha_1}^{\alpha_2} \int_{\beta_1}^{\beta_2} \sin\left\{\arccos\left[\frac{\sin\hat{\alpha}}{\sin\gamma \cos\hat{\beta}} + \mathrm{ctg}\gamma\, \mathrm{tg}\hat{\beta}\right]\right\} d\hat{\alpha}\, d\hat{\beta} \qquad (7)$$

here α_0, β_0 ares defining extent of inductor on the sphere meridian
$\hat{\alpha}, \hat{\beta}$ and parallel variables on the sphere.
Mean slip in the direction of motion of the rotor, when angular velocity ω is asumed, has the form

$$s_{\acute{s}r\beta} = 1 - \frac{\omega r_{\acute{s}r}}{V_{\beta \acute{s}r}} \qquad (8)$$

Meanwhile the local slip in the direction of motion of the rotor is expressed by the dependence

$$s = 1 - \frac{\omega R \sin\left\{\arccos\left[\frac{\sin\alpha}{\sin\gamma \cos\beta} + \mathrm{ctg}\gamma\, \mathrm{tg}\beta\right]\right\}}{V \cos[\arcsin(\sin\alpha \sin\beta)]} \qquad (9)$$

There can also be introduced the notion of local pseudo slip given by

$$s_p = \frac{V - V_\beta}{V} = 1 - \cos[\arcsin(\sin\alpha \sin\beta)] \qquad (10)$$

and mean pseudo slip

$$s_{p\acute{s}r} = \frac{V - V_{\beta \acute{s}r}}{V} \qquad (11)$$

The above dependences arc valid as mentioned when axis of rotation OX is fixed /one degree of freedom/.

In case of free suspension of rotor OX axis passes into OX′ axes of the movable system.
It is possible to search and determine this axis when motor is stable working /with one or a few module inductors/ by applying transformation to the non movable system xyz when Euler′s angles arc introduced.

Conclusions

The following conclusions follow from the presented above analysis:

- The application of inductor with parallel special electric load and with fixed pole pitch τ_m on the whole active zone, which is characteristic of its simplest construction, forms differential field of field velocities and consequently forces and moments.
- Local velocities and their mean values depend upon the inductor related to the axis of rotation of the spherical rotor.
- Installation of inductor symetrical to the equator gives compensation of velocity while any other location results in its formation.

It is characteristic for this motor that the slip of the field velocity named pseudoslip takes place. This phenomenon is manifested by inereasing the resultant slip when installation of inductor differs from symetrical one.
It effects exploatation parameters.

β_i	0°	15°	30°
V_{ro}	1.41	1.39	1.32
S	0.06	0.07	0.12

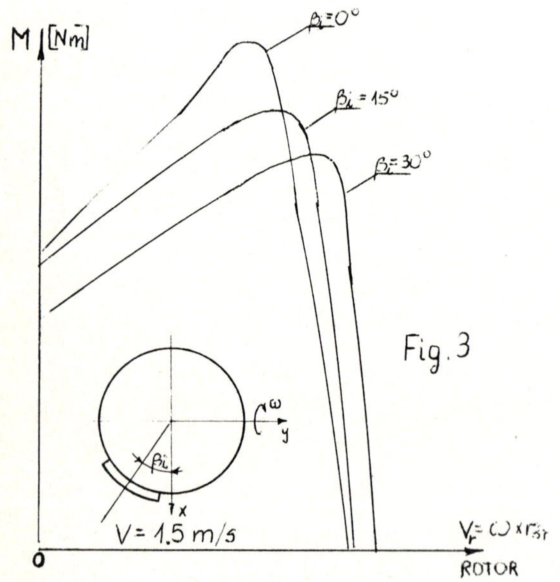

Fig. 3

When a few simetrically located inductors work the velocities and consequently moment of a couple of forces is possible to be compensated, but the slip of field of velocities still remains. Displacement of inductor from symetrical position toward the axis of rotation induces increase of angular velocity of a rotor.

REFERENCES

1 Kamiński,G., Kant,M., Virolleau,A... Silnik kulisty. PAT PRL/FRANCE 120585.
2 Miliach,A., Barabanow,A.,.. Trechstepiennyje elektriczeskie maszyny. Izd. Kijev 1979.
3 Kamiński,G., Sikora,J. Spherical motor. ICEM Budapeszt 1982.

Session 2

MODELLING OF INDUCTION MACHINES AND CONVERTERS (I)

Chairman: D.W. NOVOTNY

ELECTRICAL MACHINES AND CONVERTERS - Modelling and Simulation
H. Buyse and J. Robert (editors)
Elsevier Science Publishers B.V. (North-Holland)/ © IMACS, 1984

AUTOMATED SMALL SIGNAL ANALYSIS OF AC DRIVE SYSTEMS

D. S. Zinger
University of Wisconsin
Madison, Wisconsin

D. W. Novotny
University of Wisconsin
Madison,Wisconsin

ABSTRACT

A new formulation of the linearized small signal state variable matrix model for a converter driven electric machine is required when a system is modified or new systems are studied. To simplify this task an automated system to generate state variable models is developed. Rules that allow easy computer implementation are presented and typical applications are illustrated.

INTRODUCTION

Dynamic analysis of converter driven electric machines is often carried out using a linearized small signal state variable matrix model [1-5]. Whenever new systems are studied or existing systems are modified, a new state variable formulation is necessary. This requires considerable algebraic manipulation and the possibility of introducing errors in the model. An automated system for generating the state variable model has been developed which permits simply adding the desired control equations to preprogrammed machine-converter models.

The program APCAEM (Automated Perturbation Control Analysis of Electric Machines) is designed as a general purpose control analysis program for ac drive systems. Generalized rules are developed and implemented for creating state variable models using the basic machine-converter models as a starting point with the additional control laws entered in a simple format by the user. Presently the program contains VSI and CSI driven induction machine systems which include main flux saturation effects [6-8] as basic models. An interactive front end permits easy addition of most types of control laws, insertion of data, and selection of desired types of output. Present outputs include the state variable matrix entries, system eigenvalues, or any specified input-output transfer function or step response. The format of APCAEM allows for easy addition of new basic models or new output computations.

THEORETICAL BASIS FOR APCAEM

Generally three phase systems involving induction motors can be transformed into a set of orthogonal axes consisting of two phases and a zero sequence component. For three wire systems the zero sequence component is zero allowing the system to be modeled by two phases on a pair of orthogonal axes rotating at any arbitrary speed. With sinusoidal excitation, choosing the reference axes to rotate synchronously with the applied frequency transforms the steady state solution to DC.

The differential equations of a system derived in this manner are, however, nonlinear. For many purposes these equations can be linearized about a steady state operating point. The resultant small signal equations can generally be put in the form

$$\mathbf{P}\dot{\mathbf{x}} = \mathbf{Q}\mathbf{x} + \mathbf{R}\mathbf{u} \tag{1}$$

This form is very useful when adding control equations and can be easily transformed into the standard state equation form

$$\dot{\mathbf{x}} = \mathbf{A}\mathbf{x} + \mathbf{B}\mathbf{u} \tag{2}$$

The form of the model for an induction machine with main flux saturation included in the model is shown in the appendix. Additions to the basic small signal model can be made to incorporate voltage source or current source inverters or any of a large number of feedback or open loop regulators. However, when such a control is added, a modified system matrix has to be evaluated making it laborious to investigate many different specific types of controls. To automate this procedure it is necessary to find a general method of adding control equations to the given motor model. Since an inverter is often utilized in machine control, VSI and CSI inverter-motor models were incorporated as "basic models" in addition to the motor alone to avoid repeated need to add inverter constraints.

Allowable Alterations

The first thing to note when adding controls is that the motor is basically a fixed element; the motor itself is not changed by the addition of a controller. The motor outputs are utilized by the controller to modify the input signals. This means, as far as the equations are concerned, that constants (motor parameters) will remain constant. The only quantities that can be altered are states and inputs. For a system of the form of equation 1 this requires that the elements of the **P**, **Q**, and **R** matrices be rearranged and combined with other elements according to the changes in the states and inputs.

In spite of these restrictions there are still a large number of possibilities for changing the system. These mathematical control alterations include:

1) Increasing the number of states
2) Decreasing the number of states
3) Combining two states into one
4) Changing a state to an input
5) Increasing the number of inputs
6) Decreasing the number of inputs
7) Combining two inputs
8) Changing an input to a state
9) Combining a state and an input.

For purposes of simplification it is convenient to consider the direct elimination of a state or an input (items 2, 6, & 8) as basic changes in the physical system being modeled. Such changes will not be considered. Inputs, however, are allowed to be eliminated by being set to zero. Also controls can cause two states to be algebraically related thus removing a state by combining states into one. Controllers may also redefine inputs to be dependent on other inputs. This may cause an input to become a state or be eliminated. For a state to become an input (item 4) would mean being able to change a differential equation to an algebraic one. This is again interpreted as a change in the basic system and is not considered allowable.

After the cases noted above have been deleted from the mathematical control alterations the following allowable control alterations remain:

1) Increasing the number of states
2) Combining two states into one
3) Making a previous input dependent on a state or input
4) Addition of new inputs
5) Combining of a state with an input.

The equations that model a controller must then fit one or more of these allowable control alterations.

COMPUTER IMPLEMENTATION

Control Equations

The general form of the control equation is taken as

$$V_1[']=[E_2*]V_2['][+[E_3*]V_3]... \qquad (3)$$

where

V_n= A variable consisting of either an input, state, parameter, or temporary variable

E_n= An arithmetic expression consisting of addition, subtraction, multiplication or division of parameters and constants.

In this expression quantities enclosed in brackets ([]) are considered optional, the prime (') denotes differentiation and the asterisk (*) denotes multiplication.

This general expression requires a well defined set of rules to formalize its interpretation and reduce the number of ambiguous situations that may occur. Most of these rules follow normal algebraic interpretations. This allows the user the freedom of not being greatly concerned with the form of the equation. Those rules which differ from standard interpretation are defined in what appear to be the most logical and straightforward manner.

Before developing the general rules for interpreting a control equation it is useful to set up an example to use as an illustration of these rules. The example shown in figure 1 is a voltage source inverter (VSI) with volts/hertz control. With the small signal model of the VSI and motor available, the addition of the volts/hertz control requires only the equations:

$$\frac{dV_{ff}}{dt} = \frac{V_I - V_{ff}}{T} \qquad (4)$$

$$\omega = k_f \quad V_{ff} \qquad (5)$$

When rewritten in the form of the generalized control equation, the volts/hertz equations become:

```
VFF'=VI/T-VFF/T                (6)

W=KF*VFF                       (7)
```

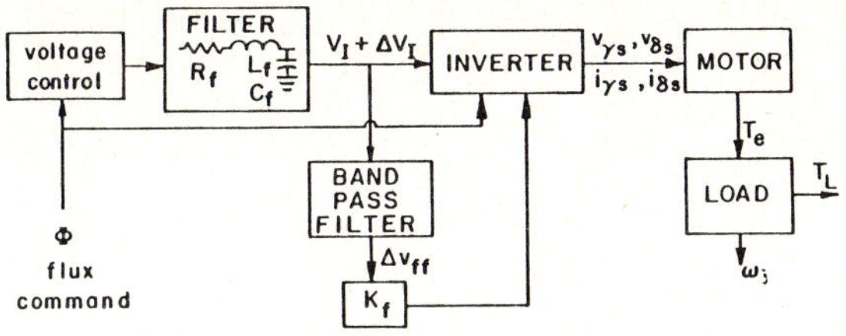

Figure 1: Block diagram of VSI-induction motor system with volts/hertz control.

Interpretation Rules

For automated interpretation of general control equations fairly extensive rules are required. These rules are formalized as follows:

1) A term is interpreted as a quantity separated by an equals sign (=) or by a plus (+) or a minus (-) sign not enclosed by parenthesis. Each term can consist of at most one variable multiplied (or divided) by an algebraic expression. Equation 6 thus has three separate terms, VFF', VI, and VFF/T. The terms of equation 7 are W and KF*VFF. Note that with this definition the expression (A+B)*V+C would have two terms, (A+B)*V and C.

2) All states and inputs must have been previously defined as being the appropriate type of variable. For the case being considered VI and VFF would have been previously defined as states while W would have been defined as an input.

3) Only one variable is allowed to be on the left side of the equation. Both equations 6 and 7 follow this rule. If equation 6 was rewritten as

VFF'+VFF/T=VI/T (8)

it would still be mathematically correct but would not be allowed because of this rule.

4) Any variable on the left of the equals sign not previously defined is considered a temporary variable. If for example, in equation 7, W was not previously defined as an input it would automatically be made a temporary variable.

5) All the components of terms on the right of the equals sign must have been previously defined. Thus for this example VI,T,VFF, and KF must have been defined as states, parameters, inputs, or temporary variables either directly or indirectly by a call for a specific machine model.

6) An input found on the left of the equals sign is transformed into a state. In the example system, even with W initially defined as an input, after reading in equation 7 W would be reinterpreted as a new state variable.

7) If the variable on the left of the equals sign is either undefined or a parameter and none of the terms on the right contain a state, input, or temporary variable, then the variable on the left is considered a parameter with a value equal to that of the expression on the right. Thus if the equation

TPI=2*3.14 (9)

was entered and TPI was not previously defined then TPI would be considered a parameter equal to 6.28.

8) The expression multiplying the variable in a given term consists of addition, subtraction, multiplication or division by parameters and constants. To interpret the expression different priorities are given to the various operations. The highest priority is given to the part of the expression inside of a set of parentheses. Next highest priority is given to multiplication and division with the lowest going to addition and subtraction. Thus an expression of the form

(A/B+C*D)*E (10)

would be evaluated by first dividing A by B and next multiplying C by D. The respective quotient and product are then added, and the result multiplied by E.

Assembly of Augmented System

Before the control equations are read, a basic system according to the users choice from those available is set up in the form

$$\mathbf{P\dot{x} = Qx + Ru + Tt} \quad (11)$$

For the given example the form set up (VSI-Induction motor) can be represented as shown in figure 2.

When a control equation is read following the interpretation rules developed, it is placed in a new row created at the bottom of each system matrix. Thus, when adding equations 4

$$\mathbf{I}p\begin{bmatrix} i_{\gamma s} \\ i_{\delta s} \\ i_{\gamma r} \\ i_{\delta r} \\ \omega_r \\ v_I \\ i_R \end{bmatrix} = \mathbf{A}\begin{bmatrix} i_{\gamma s} \\ i_{\delta s} \\ i_{\gamma r} \\ i_{\delta r} \\ \omega_r \\ v_I \\ i_R \end{bmatrix} + [\,\mathbf{b}_1 | \mathbf{b}_2 | \mathbf{b}_3\,]\begin{bmatrix} T_L \\ v \\ \omega \end{bmatrix}$$

Figure 2: Initial matrix arrangement for VSI-induction motor.

and 5 to the system of the given example, a new system, as shown in figure 3, will result. Notice that when equation 5 was placed into the system matrix, ω was changed from an input to a state following interpretation rule 6. The input vector associated with ω ($\mathbf{b}_3$ in figure 2) was moved from the input matrix to the state matrix.

If a new row contains only a nonzero element in the **T** matrix, the temporary variable associated with this element is considered to be a parameter equal to the value of the element. This allows the user to define new parameters dependent on other parameters. An example of a control equation that would create such a situation would be

A=VGSO+VDSO (12)

where A is defined as the sum of two steady state voltages.

After all equations are entered, the system matrices have been modified

$$\left[\begin{array}{c|cc} \mathbf{I} & \multicolumn{2}{c}{\mathbf{0}} \\ \hline \mathbf{0} & 1 & 0 \\ & 0 & 0 \end{array}\right] p\begin{bmatrix} i_{\gamma s} \\ i_{\delta s} \\ i_{\gamma r} \\ i_{\delta r} \\ \omega_r \\ v_I \\ i_R \\ v_{ff} \\ \omega \end{bmatrix} = \left[\begin{array}{ccccccc|c|c} & & & \mathbf{A} & & & & \mathbf{0} & \mathbf{b}_3 \\ \hline 0 & 0 & 0 & 0 & 0 & \frac{1}{T} & 0 & \frac{-1}{T} & 0 \\ 0 & 0 & 0 & 0 & 0 & 0 & 0 & -K_{ff} & -1 \end{array}\right]\begin{bmatrix} i_{\gamma s} \\ i_{\delta s} \\ i_{\gamma r} \\ i_{\delta r} \\ \omega_r \\ v_I \\ i_R \\ v_{ff} \\ \omega \end{bmatrix} + \left[\begin{array}{c|c} \mathbf{b}_1 & \mathbf{b}_2 \\ 0 & 0 \\ 0 & 0 \end{array}\right]\begin{bmatrix} T_L \\ v \end{bmatrix}$$

Figure 3: Matrix model of VSI-induction motor with volts/hertz control before reduction.

$$\mathbf{I}p\begin{bmatrix} i_{\gamma s} \\ i_{\delta s} \\ i_{\gamma r} \\ i_{\delta r} \\ \omega_r \\ v_I \\ i_R \\ v_{ff} \end{bmatrix} = \left[\begin{array}{ccccccc|c} & & & \mathbf{A} & & & & k_{ff}\mathbf{b}_3 \\ \hline 0 & 0 & 0 & 0 & 0 & \frac{1}{T} & 0 & \frac{-1}{T} \end{array}\right]\begin{bmatrix} i_{\gamma s} \\ i_{\delta s} \\ i_{\gamma r} \\ i_{\delta r} \\ \omega_r \\ v_I \\ i_R \\ v_{ff} \end{bmatrix} + \left[\begin{array}{c|c} \mathbf{b}_1 & \mathbf{b}_2 \\ 0 & 0 \end{array}\right]\begin{bmatrix} T_L \\ v \end{bmatrix}$$

Figure 4: Reduced matrix model of VSI-induction motor with volts/hertz control.

to include the information given in the control equations. This example system as it stands, however, is not of minimum order since the **P** matrix is singular. To remedy this situation rules have been developed to eliminate unnecessary matrix entries without losing the information in the control equations. These recombination rules can be listed as follows:

1) All temporary variables are eliminated by combining rows. Such a rule is needed since complicated equations are often broken into two or more simpler equations for ease of understanding. By combining the rows with common temporary variables, the unnecessary linking variable is eliminated and the more complicated equation is created from the simple ones.

2) All states with zero columns in the **P** matrix are eliminated by combining rows. In the example system this proves to be the case for ω. By multiplying the last row by each of the elements of the $\mathbf{b}_3$ vector and adding this to the appropriate row, the result would appear as in figure 4.

3) All equations with zero rows in the **P** matrix are eliminated by combining columns. The method of execution of this rule is similar to that of the previous rule with comparable results.

A full set of rules outlining how a control equation is to be written, interpreted, and combined into the system matrix has now been described.

Program Structure

The implementation of the rules set forth is most easily done in a structure designed for matrix manipulations and calculations. For the system to be more user oriented a different type of data structure would be useful. To try and make the program fit well with both types of data structures it was divided into two major sections, the calculation and interactive sections.

The calculation section reads in data from a highly formatted file, places the information in the proper matrix, and does the requested calculations. This section can be considered divided into two major subsections, one that performs the recombinations and one that does the proper output functions.

In the recombination subsection the formatted data is read and placed in the proper matrices. The matrices are then recombined using the rules already discussed. This generates a system that is easily used in calculations.

The subsection that does the output functions takes the matrices and performs the necessary operations to generate the user requested outputs. Presently outputs of eigenvalues, transfer functions, and step responses are available.

The calculation section requires data to be highly formatted. This is very inconvenient to the user who would prefer a more freely formatted input structure. To accommodate this an interactive program section was written to transform the data from a form more desirable to the user to one more easily manipulated by the calculation section. The interactive program section also aids the user with a series of prompts and menus.

ILLUSTRATIVE EXAMPLES

To verify the applicability of the control equation format of equation 3 to a wide range of problems, a number of examples were tested. These examples use different initial models to which control equations are added. Previous calculations done with standard programs were used to verify the operation of APCAEM.

The first example to be considered is the VSI system with volts/hertz control as shown in figure 1. Using the standard VSI model in APCAEM and neglecting the filter in the volts/hertz controller on the small signal model, the only additional control equation

SYSTEM PARAMETERS

$R_S = 0.1315\Omega$ $R_R = 0.0939\Omega$ $L_S = 0.0963H$ $L_R = 0.0963H$

$M = 0.09449H$ $P = 2$ $J = 10.64\ KgM^2$ $D = 0.0\ NMs$

$R_F = 0.338\Omega$ $C_F = 0.00665F$ $L_F = 0.00372H$ $K_F = 0.077$

V = 1328v
freq. = 60Hz
NO LOAD

	MOTOR ONLY	MOTOR AND FILTER	MOTOR, FILTER, AND VOLTS/HERTZ
POLES	-1.00	-0.442 ± j2.37	-0.874 ± j1.87
	-1.42 ± j15.6	-1.39 ± j16.6	-1.49 ± j15.8
	-0.477 ± j2.39	-1.86 ± j6.93	-1.32 ± j8.68
		-1.00	-1.00
ZEROS	-1.40 ± j15.0	-1.38 ± j16.6	-1.48 ± j15.8
	-0.996 ± j0.0895	-1.82 ± j7.00	-1.36 ± j8.64
		-0.0991 ± j0.127	-1.68
			-1.02

Table 1: Normalized Roots for VSI-Induction Motor System with Volts/Hertz control.

that is needed for APCAEM is:

$$W=KF*VI \quad (13)$$

The transfer function poles and zeros found using this method were compared with those found by Fukazawa [9]. The results show good agreement. A sample set of eigenvalues with the associated motor and control parameters is shown in table 1.

A more involved example is the addition of a voltage feedback loop to a CSI system. The block diagram of such a system is shown in figure 5. In this system the terminal voltage VSMX must be derived from the state variables IGS, IGR, and IDR. This voltage is then compared to a reference voltage VR (a new input variable) with the difference (error) used to control the original rectifier input voltage V. The control equations needed by APCAEM for this case are (A,B,C,DA,E,F, and G are temporary variables):

```
A=VGO*(RS*COSZGA+WO*LST*SINZGA)          (14)
  +VDO*(WO*LS*COSZGA-RS*SINZGA

B=VGO*(LS*COSZGA)-VDO*LST*SINZGA         (15)

C=VGO*M                                  (16)

DA=VDO*WO*MT                             (17)

E=VDO*MT                                 (18)

F=-VGO*WO*MT                             (19)

G=-VGO*(LS*IDSO+M*IDRO)                  (20)
   +VDO*(LS*IGSO+M*IGRO)

VSMX=A*IGS+B*IGS'+DA*IGR+C*IGR'          (21)
     +F*IDR+E*IDR'+G*W

V=GW*VR-GW*CDC*2*VSMX/(3*VSMO)           (22)
```

where

COSZGA = cosine of reference frame angle
IDSO = steady state stator delta axis current
IDRO = steady state rotor delta axis current
IGSO = steady state stator gamma axis current
IGRO = steady state rotor gamma axis current
LS = stator inductance
LST = saturated stator inductance
M = magnetizing inductance
MT = saturated magnetizing inductance
RS = stator resistance
SINZGA = sine of reference frame angle
VDO = steady state delta axis voltage
VGO = steady state gamma axis voltage
WO = applied stator frequency.

The temporary variables A, B, C, DA, E, F, and G have been employed to simplify the equations for the terminal voltage VSMX. Equation 22 is the feedback equation defining the original rectifier input voltage V of the basic model.

Eigenvalues found using these controls with the standard CSI model of APCAEM were compared with the results found by Kolb [8]. These results again proved to be favorable as shown in table 2. Additional tests up to a twelfth order system have been carried out with good results.

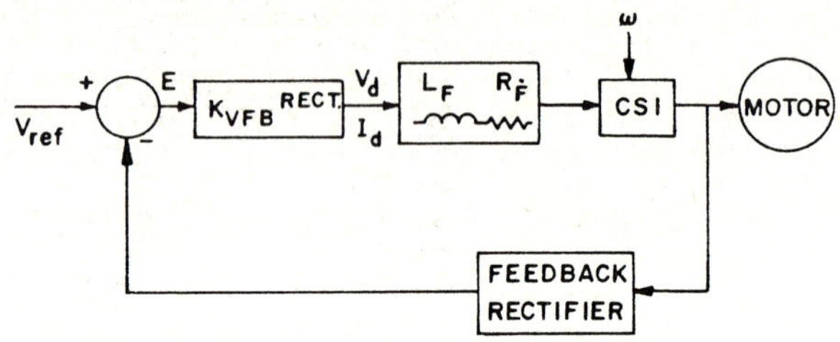

Figure 5: Block diagram of CSI-induction motor with stator voltage feedback.

SYSTEM PARAMETERS

$R_S = 0.05\Omega$ $R_R = 0.05\Omega$ $L_S = 0.01H$

$L_R = 0.01H$ $M = 0.0975H$ $P=2$

$J = 54\ K_gM^2$ $D = 0.0$ NMs $R_F = 0.05\Omega$

$L_F = 0.01$

V = 1910 v
freq. = 191 hz
slip = 0.017

NORMALIZED EIGENVALUES

-5.92

$-0.3659 \pm j0.5156$

-0.09761

Table 2: Normalized Eigenvalues for CSI-Induction Motor System with Voltage Feedback.

SUMMARY

To aid in the small signal analysis of converter driven electric machines generalized rules were developed to allow automated assembly of system state equations starting from a basic system model. These rules define a general form for control equations, provide specific interpretations for these equations, allow for the assembly into system matrices and subsequent reduction of the matrices to final form.

The rules were formalized in APCAEM, and a user oriented input section was developed to facilitate applying the program to real systems. Examples have shown how various basic systems can be modified with considerably less effort than the usual approach of formulating the state equations directly. The performance of APCAEM has been tested in a large number of example systems with excellent correspondence to independently generated results.

REFERENCES

[1] Rogers, G.J., Linearized Analysis of Induction Motors, Proc. IEE, Vol. 112, No 10 (Oct. 1965) 1917-1925
[2] Nelson, R., Lipo, T. A., and Krause P., Stability Analysis of a Symmetrical Induction Induction Machine, IEEE Transactions on Power Apparatus and Systems, Vol PAS -88, No 11 (Nov. 1969) 1710,1717
[3] Lipo, T.A. and Plunkett, A.B., A Novel Approach to Induction Motor Transfer Functions, IEEE Transactions on Power Apparatus and Systems, Vol PAS-93, No. 5 (Sept/Oct 1974) 1410-1418
[4] Lipo, T.A. and Krause, P., Stability Analysis of a Rectifier-Inverter Induction Motor Drive, IEEE Transactions on Power Apparatus and Systems, Vol PAS-88 (Jan 1969) 55-66
[5] Cornell, E.P. and T. A. Lipo, Modeling and Design of Controlled Current Induction Motor Drive Systems, IEEE Transactions on Ind. Appl., Vol IA-13 (July/Aug. 1977) 321-330
[6] Melkebeek, J., Magnetizing Field Saturation and Dynamic Behavior of Induction Machines; Part 1: Improved Calculation Methods for Induction Machine Dynamics, IEE Proceedings, Vol 130, Part B #1 (Jan 1983) 1-9
[7] Melkebeek, J., Magnetizing Field Saturation and Dynamic Behavior of Induction Machines; Part 2: Stability Limits of a Voltage-Fed Induction Motor and Self-Excited Induction Generator, IEE Proceedings, Vol 130, Part B #1 (Jan 1983) 10-17
[8] Kolb, T., Main Flux Saturation Modeling in Inverter Driven Machines, M. S. Report, University of Wisconsin (1981)
[9] Fukazawa, K., Influence of Volts/Hertz Control on Filter Impedance Effects in VSI-Fed Induction Machines, M. S. Report, University of Wisconsin (1979)

APPENDIX

The small signal model for an induction machine with main flux saturation can be placed in the form

$$\dot{\mathbf{x}} = \mathbf{Ax} + \mathbf{Bu} \tag{A1}$$

This model includes currents and voltages in a synchronous reference frame. The delta axis of this reference frame is aligned with the air gap voltage and the gamma axis leads the delta by 90 degrees.

The parameters in the model are as follows:

- R_S - stator resistance
- R_R - rotor resistance
- L_S - stator inductance
- L_R - rotor inductance
- M - magnetizing inductance
- P - number of pole pairs
- J - inertia
- D - damping
- γ - dynamic saturation parameter

The saturated inductances (LST, LRT, MT) are calculated using the dynamic saturation parameter as follows:

$$L_{ST}=L_S-M(1-1/\gamma) \tag{A2}$$

$$L_{RT}=L_R-M(1-1/\gamma) \tag{A3}$$

$$M_T=M/\gamma \tag{A4}$$

Steady state parameters used in the model include:

- ω_0 - applied stator frequency
- ω_{S0} - slip frequency
- I_{GS0} - stator gamma axis current
- I_{DS0} - stator delta axis current
- I_{GR0} - rotor gamma axis current
- I_{DS0} - rotor delta axis current
- ϕ - angle between terminal voltage and air gap voltage

The small signal state variables are as follows:

I_{GS} - stator gamma axis current
I_{DS} - stator delta axis current
I_{GR} - rotor gamma axis current
I_{DR} - rotor delta axis current
ω_R - rotor speed

The small signal inputs are as follows:

V - source voltage
T_L - load torque
ω - stator frequency.

The matrices and vectors of the state equation can be expressed as:

$$x = [\ I_{GS},\ I_{DS},\ I_{GR},\ I_{DR},\ R\ ]^T$$

$$u = [\ V,\ T_L,\ \omega]^T$$

$$A = \begin{bmatrix} a_{11} & a_{12} & a_{13} & a_{14} & a_{15} \\ a_{21} & a_{22} & a_{23} & a_{24} & a_{25} \\ a_{31} & a_{32} & a_{33} & a_{34} & a_{35} \\ a_{41} & a_{42} & a_{43} & a_{44} & a_{45} \\ a_{51} & a_{52} & a_{53} & a_{54} & a_{55} \end{bmatrix}$$

$$B = \begin{bmatrix} b_{11} & 0 & b_{13} \\ b_{21} & 0 & b_{23} \\ b_{31} & 0 & b_{33} \\ b_{41} & 0 & b_{43} \\ 0 & b_{52} & 0 \end{bmatrix}$$

The elements of the A and B matrices are described by :

$$a_{11} = -R_S C_{L1} \qquad a_{12} = \omega_0 L_{ST} C_{L1} + \omega_{SO} M_T C_{MA} \qquad a_{13} = -R_R C_{MA}$$

$$a_{14} = \omega_0 I_{DSO} - L_R C_{MA} I_{DRO} \qquad a_{15} = -M\ C_{MA} I_{DSO} - L_R C_{MA} I_{DRO}$$

$$a_{21} = -\omega_0 L_S C_{L2} - \omega_{SO} M\ C_{MB} \qquad a_{22} = -R_S C_{L2}$$

$$a_{23} = -\ \omega_0 M\ C_{L2} - \omega_{SO} M\ C_{MB} \qquad a_{24} = -R_R C_{MB} \qquad a_{25} = M\ C_{MB} I_{GSO} + L_R C_{MB} I_{GRO}$$

$$a_{31} = -R_S C_{MA} \qquad a_{32} = \omega_0 L_{ST} C_{MA} + \omega_{SO} M_T C_{L3} \qquad a_{33} = -R_R C_{MA}$$

$$a_{34} = \omega_0 M_T C_{MA} + \omega_{SO} L_{RT} C_{L3} \qquad a_{35} = -M\ C_{L3} I_{DSO} - L_R C_{L3} I_{DRO}$$

$$a_{41} = -\omega_0 L_S C_{MB} - \omega_{SO} L_R C_{L4} \qquad a_{42} = -R_S C_{MB}$$

$$a_{43} = -\ \omega_0 M\ C_{MB} - \omega_{SO} L_R C_{L4}$$

$$a_{44} = -R_R C_{L4} \qquad a_{45} = M\ C_{L4} I_{GSO} + L_R C_{L4} I_{GRO}$$

$$a_{51} = -P^2 M\ I_{DRO}/J \qquad a_{52} = P^2 M_T I_{GRO}/J \qquad a_{53} = P^2 M\ I_{DSO}/J$$

$$a_{54} = -P^2 M_T I_{DSO}/J \qquad a_{55} = -D/J$$

$$b_{11} = C_{L1}\ (I_{DSO} L_S - I_{DRO}\ M) + C_{MA}\ (I_{DSO} M + I_{DRO} L_R) \qquad b_{13} = C_{L1} C_1$$

$$b_{21} = -C_{L2}\ (I_{GSO} L_S + I_{GRO} M) - C_{MB}\ (I_{GSO} M + I_{GRO} L_R) \qquad b_{23} = C_{L2} C_2$$

$$b_{31} = C_{MA}\ (I_{DSO} L_S + I_{DRO} M) + C_{L3}\ (I_{DSO} M + I_{DRO} L_R) \qquad b_{33} = C_{MA} C_1$$

$$b_{41} = -C_{MB}\ (I_{GSO} L_S + I_{GRO} M) + C_{L4}\ (I_{GSO} M + I_{GRO} L_R) \qquad b_{43} = C_{MB} C_2$$

$$b_{52} = -P/J$$

where

$$C_{MA} = (M_T^2 M - L_{ST} L_{RT} M)/\Delta \qquad C_{MB} = (M^2 M_T - L_S L_R M_T)/\Delta \qquad C_{L1} = (L_{ST} L_R L_{RT} - M_T^2 L_R)/\Delta$$

$$C_{L2} = (L_S L_R L_{RT} - M_2 L_{RT})/\Delta \qquad C_{L3} = (L_S L_{ST} L_{RT} - L_S M\ M_T)/\Delta$$

$$\Delta = L_S(L_{ST} L_R L_{RT} - M_T^2 L_R) + M^2(M_T^2 - L_{ST} L_{RT}) \qquad C_1 = \frac{\sqrt{6}}{\pi}\cos\phi \qquad C_2 = \frac{\sqrt{6}}{\pi}\sin\phi$$

ELECTRICAL MACHINES AND CONVERTERS - Modelling and Simulation
H. Buyse and J. Robert (editors)
Elsevier Science Publishers B.V. (North-Holland)/© IMACS, 1984

DYNAMIC MODELS FOR VOLTAGE CONVERTER AND FREQUENCY INVERTER FED INDUCTION MACHINES

Dr. ir. J. A. Melkebeek

State University of Gent
Laboratory for Industrial Electricity
Gent, Belgium

In this contribution fundamental harmonic dynamic models for voltage converter fed and for frequency inverter fed induction machines are presented. The small signal induction machine model employed is a model which takes into account the main field saturation effect. The frequency inverter and voltage inverter models on the other hand are derived neglecting higher harmonics and assuming ideal characteristics of the power electronic devices. The combination of these converter models with the machine model turns out to permit a straightforward derivation of the dynamic drive characteristics. Experimental data are presented as well, corroborating the proposed models.

1. INTRODUCTION

The use of power electronic supplies for induction machines has been increasing very fastly in the last few years. Frequency inverters are employed in order to obtain accurate and at the same time high-efficiency speed control; voltage inverters were firstly used for speed variation in the low slip range but are now used more and more for energy conservation purposes.

Together with the advent of power electronic supplies the dynamic requirements for electric machine drives are becoming higher and higher as well. In addition, the manifold and quite distinct operating points involved tend to make the classical machine dynamic analysis methods too inaccurate. For example, in the classical induction machine dynamic analysis the magnetising field saturation effect is not properly represented.

In this contribution a recently proposed (and experimentally confirmed) dynamic induction machine model which allows more accurately for the magnetising field saturation is reviewed firstly. Next, the paper reconsiders some straightforward but fairly accurate dynamic models for frequency and voltage converters which have been presented in recent literature. These converter models are then combined with the saturated induction machine model. This combination turns out to result in simple and accurate converter-machine dynamic models.

2. INDUCTION MACHINE SATURATED MODEL

In recent years the dynamic behaviour of induction machine drives has been dealt with in numerous papers [1,2,3]. Despite the useful results, predicting most of the trends observed in practise, experiments have shown that the theoretical predictions concerning stability behaviour, which are based on the classical induction machine modelling are too pessimistic [2,3]. Recently, the cause of this discrepancy has been shown to be the inaccurate modelling of main flux saturation and a saturated induction machine model has been developed [4,5,6,7,8]. Experimental evaluation has indicated a considerable improvement on the classical modelling and in most cases a close correspondence between measurements and predictions.

Essentially, in the saturated model the inherent difference between the effect of incremental current changes in the main flux axis and in the axis in quadrature with main flux is taken into account. This is achieved in the following way:

1- the axes of the two axis model are rotated so as to align the flux axis with the negative y-axis

2- different magnetising inductances are used to correlate the incremental current changes in the two axes with their respective flux changes

A typical magnetisation characteristic for an induction machine is shown in figure 1. In steady state (air-gap flux Φ_{mo}, magnetising current I_{mo}) the chord slope inductance correctly represents the relationship between flux and current amplitudes: $\Phi_{mo}=L_{mo}I_{mo}$. This

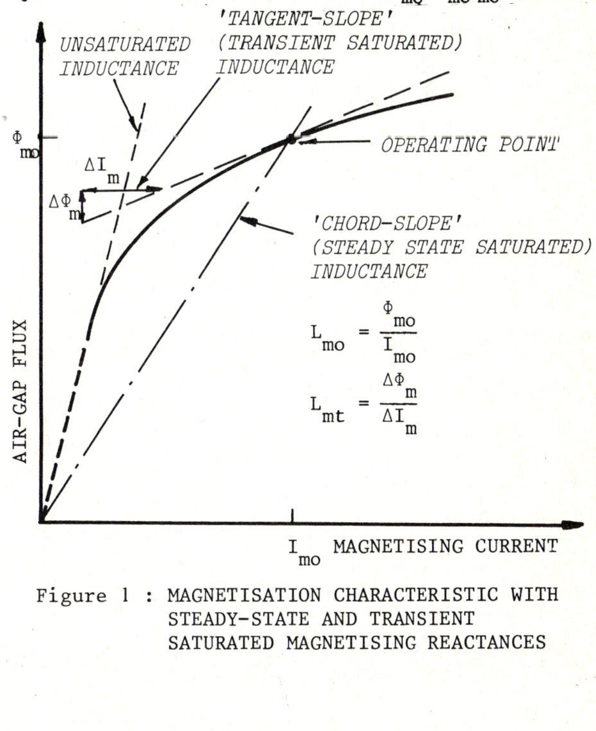

Figure 1 : MAGNETISATION CHARACTERISTIC WITH STEADY-STATE AND TRANSIENT SATURATED MAGNETISING REACTANCES

chord slope inductance is indeed commonly used for the steady-state modelling of saturated machines and gives fairly acceptable results.
In the classical dynamic modelling the chord slope inductance is also used to represent the incremental flux changes associated with incremental current changes. However, with the axes rotated as explained above, the incremental current changes in x and y axes correspond to changes in phase and amplitude of the flux respectively. For changes of the phase of the flux (x-axis) the chord slope inductance L_{mo} still represents the correct relationship between current and flux increments. For changes in the amplitude of the flux, however, it seems clear that the appropriate inductance is the 'tangent slope' inductance L_{mt} (which is proportional to the tangent to the magnetisation curve).
The resulting saturated small signal equations are given in Table I. In these equations all inductances with subscript 'o' represent the normal 'chord slope' inductances and correspond to either steady-state mmf terms or to mmf changes in quadrature with the main flux. The inductances with subscript 't' are the 'tangent' or 'transient' inductances, corresponding to incremental mmf variations aligned with the main flux. Of course, if all subscripts 't' are replaced by subscript 'o' the equations reduce to the classical equations. Besides the use of the tangent slope main field inductance another difference with the classical equations consists in the influence of the leakage division χ between stator and rotor. Contrary to the classical model the leakage division affects the equations since it determines the air-gap flux. However the effect is rather small and can mostly be neglected [6,7].
Combining the voltage-current equations of Table I with the equation of motion and writing them in per-unit with as reference values the rated (peak) values V_n, ω_n, T_n and the ideal short circuit currents $I_{ks}=V_n/\sigma\omega_n L_{so}$, $I_{kr}=I_{ks}L_{mo}/L_{ro}$ yields the machine equations expressed in terms of the minimal set of p.u. parameters given in

TABLE I : SMALL SIGNAL SATURATED INDUCTION MACHINE EQUATIONS

$$\begin{vmatrix}\Delta V_{sx}\\ \Delta V_{sy}\\ 0\\ 0\end{vmatrix} = \begin{vmatrix} R_s+L_{so}p & -\omega_o L_{st} & L_{mo}p & -\omega_o L_{mt}\\ \omega_o L_{so} & R_s+L_{st}p & \omega_o L_{mo} & L_{mt}p\\ L_{mo}p & -(\omega_o-\omega_{ro})L_{mt} & R_s+L_{ro}p & -(\omega_o-\omega_{ro})L_{rt}\\ (\omega_o-\omega_{ro})L_{mo} & L_{mt}p & (\omega_o-\omega_{ro})L_{ro} & R_r+L_{rt}p \end{vmatrix} \begin{vmatrix}\Delta I_{sx}\\ \Delta I_{sy}\\ \Delta I_{rx}\\ \Delta I_{ry}\end{vmatrix} + \begin{vmatrix} 0 & -L_{so}\Delta\omega & 0 & -L_{mo}\Delta\omega\\ L_{so}\Delta\omega & 0 & L_{mo}\Delta\omega & 0\\ 0 & -L_{mo}(\Delta\omega-\Delta\omega_r) & 0 & -L_{ro}(\Delta\omega-\Delta\omega_r)\\ L_{mo}(\Delta\omega-\Delta\omega_r) & 0 & L_{ro}(\Delta\omega-\Delta\omega_r) & 0 \end{vmatrix} \begin{vmatrix} I_{sxo}\\ I_{syo}\\ I_{rxo}\\ I_{ryo}\end{vmatrix}$$

$$(J/N_p)\dot{\omega}_r + T_\ell = T \quad ; \quad T = \tfrac{3}{2}N_p\,[L_{mo}(I_{syo}\Delta I_{rx}-I_{ryo}\Delta I_{sx}) + L_{mt}(I_{rxo}\Delta I_{sy}-I_{sxo}\Delta I_{ry})]$$

ω_o = steady-state stator frequency $\omega_{ro} = \dot{\theta}_o$ = steady-state rotor electrical speed
N_p = number of pole pairs
$I_{sxo}, I_{syo}, I_{rxo}, I_{ryo}$ = steady-state current in reference frame aligned with air-gap flux
R_s, R_r : stator and rotor resistances $L_{s\sigma}, L_{r\sigma}$: stator and rotor leakage inductances
L_{mo} = chord-slope magnetising inductance L_{mt} = tangent-slope magnetising inductance
$L_{so} = L_{s\sigma}+L_{mo}$ $L_{ro} = L_{r\sigma}+L_{mo}$ $L_{st} = L_{s\sigma}+L_{mt}$ $L_{rt} = L_{s\sigma}+L_{mt}$

TABLE II : MINIMAL SET OF P.U. MACHINE PARAMETERS

Parameter	Definition	Remarks
p.u. stator damping	$\rho_{so}=R_s/\sigma_o\omega_n L_{so}$	$L_{so}=L_{s\sigma}+L_{mo}$; $L_{s\sigma}$=stator leakage; $\tau_{so}=\rho_{so}^{-1}$ =p.u. time ct.
p.u. rotor damping	$\rho_{ro}=R_r/\sigma_o\omega_n L_{ro}$	$L_{ro}=L_{r\sigma}+L_{mo}$; $L_{r\sigma}$=rotor leakage; $\tau_{ro}=\rho_{ro}^{-1}$ =p.u. time ct.
total leakage coefficient	$\sigma_o=1-L_{mo}^2/L_{so}L_{ro}$	$L_{mo}=M_o$=magnetising field inductance (mutual inductance)
transient saturation parameter (‡)	$\gamma=L_{mo}/L_{mt}$	γ=1 for unsaturated cases and for classical model γ=2...5 in most cases
leakage division (‡)	$\chi=L_{s\sigma}/L_{r\sigma}$	$\chi\approx 1$ in most cases ; negligible effect on dynamics
p.u. supply frequency	$\nu_o=\omega_o/\omega_n$	ω_o=operating point supply frequency
p.u. slip frequency	$\nu_{so}=s_o\omega_o/\omega_n$	ω_o=operating point slip; $\omega_o s_o$=slip frequency $\omega_{ro}=(1-s_o)\omega_o$=rotor electr. speed; ν_{ro}=p.u. rotor speed
p.u. electromechanical time constant	$\tau_m=J\omega_n\omega_{so}/N_pT_o$	J=total inertia ; N_p=number of pole pairs
mechanical damping	$\delta_w=(dT_\ell/d\omega_r)(N_p/J\omega_n)$	T_o=operating point torque ; T_ℓ=load torque

Table II [6,7]. Compared to the classical equations two additional parameters are required, that is the saturation parameter $\gamma = L_{mo}/L_{mt}$ and the leakage division parameter $\chi = L_{s\sigma}/L_{r\sigma}$ (although the latter can mostly be assumed as unity). For an unsaturated machine $\gamma = 1$; normally γ ranges between 1 and 5, a value of 2...3 being most likely.

3. INVERTER INDUCTION MACHINE MODELLING

3.1 Steady-state inverter models

The two basic inverters treated in this contribution are illustrated in fig. 2. For the six step VSI case of fig. 2a each thyristor is conducting 180°. Therefore, with all thyristors 'on' at all times the inverter uniquely relates the DC supply voltage to the AC side. The short circuit current between two phases in each step is not directly related to the DC side but is determined by the load. The PWM voltage inverter is a modification of the VSI type with (e.g.) an additional short circuit between the three phases. For the CSI case of fig. 2b each thyristor conducts 120°. Because only 2 thyristors are on at all times the inverter uniquely relates the AC current to the DC current. The AC voltage in the open circuited phase in each step is not directly related to the DC side but is fixed by the load [9].

In [9,10] the steady-state fundamental harmonic equivalent circuits shown in fig. 3a and 3b (corresponding to fig. 2a and 2b respectively) have been derived. The PWM inverter is represented by the circuit of fig. 3a as well, with 'a' the PWM voltage ratio (a=1 for the 6-step inverter). On the right side of these circuits is the AC load, connected to the inverter terminals; on the left hand side is the DC source, referred to the AC side.
For the VSI/PWM case of fig. 3a the load voltage is completely constrained by the DC voltage, $V_{re} = V_e$, $V_{im} = 0$. Only the in-phase component of the load current is reflected to the DC side, $I_{re} = I\cos\phi = I_e$. The other component, which corresponds to the inverter short circuit current is represented by the current through the variable capacitor in parallel with the load. For the dual case of a CSI inverter the load current is completely determined by the DC source: $I_{re} = I_e$, $I_{im} = 0$. Only the in-phase component of the load voltage is related to the DC side, $V_{re} = V\cos\phi = V_e$. The other component, that is the voltage in the open circuited phase, is symbolized by the voltage over the variable series reactive element. In both circuits the variable reactive element is continuously cancelling the reactive part of the load and represents the inherent ability of inverters to supply (absorb) the reactive power of the load.
The main assumption these circuits are based on is the neglect of the invertor harmonics. The inverter is therefore treated as an ideal DC to sine wave converter, transforming the total DC input power to fundamental AC power. Because of the low-pass characteristics of electrical machines this assumption is quite acceptable for our purposes. More details about these inverter schemes can be found in Ref. 9.
The combination with steady-state machine models has been pointed out to yield accurate representations of the fundamental harmonic steady-state behaviour of inverter-machine drives [10,11]. Very useful results have been obtained, for example, for synchronous machines.

3.2 Dynamic inverter-induction machine model

The extention of the inverter representations to small signal dynamic models is very feasible. Indeed, the only important assumption concerns the neglect of the higher harmonics which is indeed common practise in dynamic analysis as well. The dynamic inverter models have already

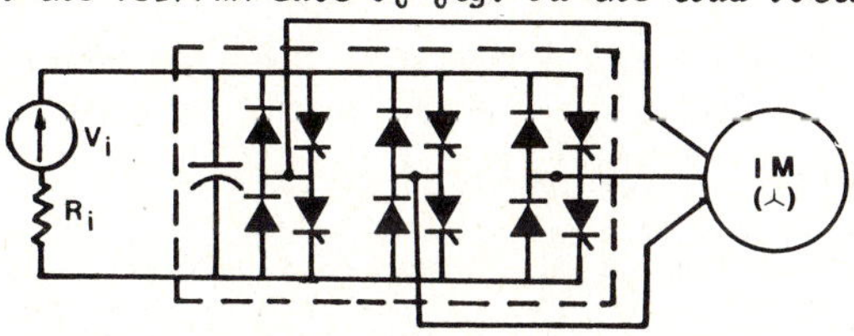

a) VSI WITH DC VOLTAGE SUPPLY

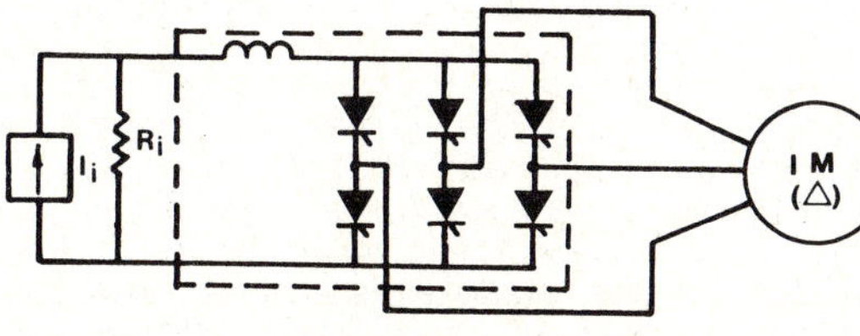

b) CSI WITH DC CURRENT SUPPLY

Figure 2 : VSI AND CSI INVERTER REPRESENTATIONS

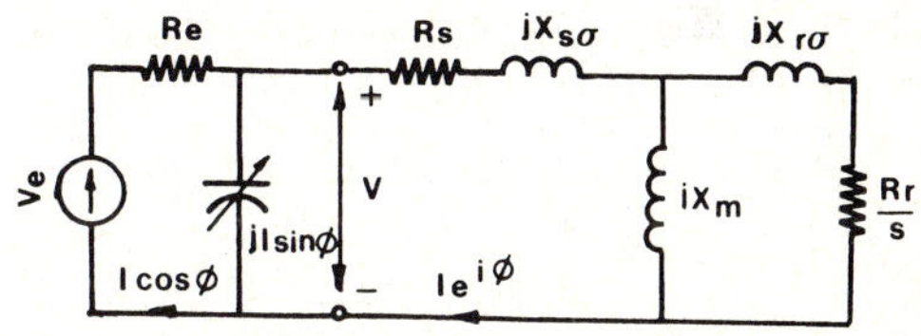

a) VSI WITH DC VOLTAGE SUPPLY
$V_e = (2a/\pi)V_i$; $I\cos\phi = (\pi/3a)I_i$ $(=I_e)$
$R_e = (6a^2/\pi^2)R_i$

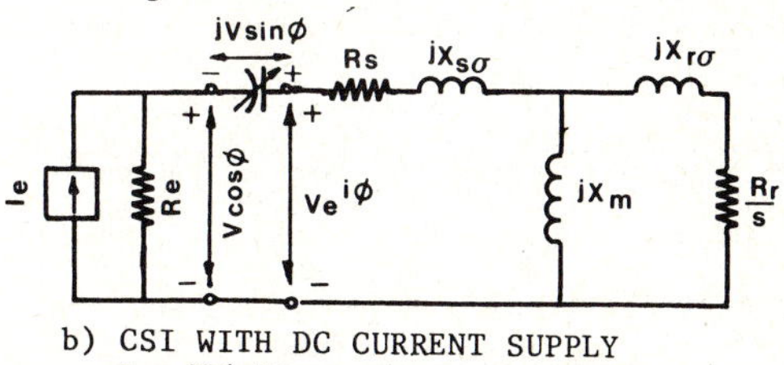

b) CSI WITH DC CURRENT SUPPLY
$I_e = (2/\pi)I_i$; $V\cos\phi = (\pi/3)V_i$ $(=V_e)$
$R_e = (6/\pi^2)R_i$

Figure 3 : EQUIVALENT CIRCUITS FOR VSI AND CSI INVERTERS

been succesfully employed for the analysis of inverter-induction machine self-excitation [12], as will be reported on in the next section.

For the VSI the small signal equations representing the DC link filter and the inverter are (see fig. 3a):

$$\Delta V_{re}+(R_e+pL_e)\Delta I_e = \Delta V_e \tag{1}$$

$$pC_e\Delta V_{re}+ \Delta I_{re} - \Delta I_e = 0 \tag{2}$$

$$\Delta V_{im} = 0 \tag{3}$$

The combination with the small signal saturated machine equations requires however a rotation of the real and imaginary axes of the inverter model to new axes x and y, such that the steady-state magnetising current is aligned with the negative y-axis. If δ is the rotation angle in the positive direction, a vector **U** transforms according to **U'**=**U** $\exp(-j\,\delta)$. For an induction machine $\delta=-\beta$ with $\beta=\arg$ **V** - arg **E** (with **E** the air-gap voltage, to be calculated from the steady-state equations). Eqns. 1,2,3 expressed in the new coordinate frame read:

$$\Delta V_{re}+(R_e+pL_e)\Delta I_e = \Delta V_e \tag{4}$$

$$pC_e\Delta V_{re}+\cos\beta\,\Delta I_{sx}+\sin\beta\,\Delta I_{sy} = \Delta I_e \tag{5}$$

$$\Delta V_{sx}\sin\beta = \Delta V_{sy}\cos\beta \quad (= \Delta V_{re}\cos\beta\sin\beta) \tag{6}$$

The combination with the 5th order induction machine model results then in a 7th order system with as state variables the 5 induction machine state variables (ΔI_{sx}, ΔI_{sy}, ΔI_{rx}, ΔI_{ry}, $\Delta\omega_r$), the DC current ΔI_e and the voltage ΔV_{re} (ΔV_{sx} and ΔV_{sy} in the machine equations are expressed as $\Delta V_{re}\cos\beta$ and $\Delta V_{re}\sin\beta$ respectively). ΔV_e is considered as an input variable.
If the effect of the DC link filter may be neglected, i.e. $L_e=C_e=0$, then the system reduces to a 5th order system with

$$\begin{pmatrix}\Delta V_{sx}\\ \Delta V_{sy}\end{pmatrix} = -R_e\begin{pmatrix}\cos^2\beta & \sin\beta\cos\beta\\ \sin\beta\cos\beta & \sin^2\beta\end{pmatrix}\begin{pmatrix}\Delta I_{sx}\\ \Delta I_{sy}\end{pmatrix} + \begin{pmatrix}\Delta V_e\cos\beta\\ \Delta V_e\sin\beta\end{pmatrix} \tag{7}$$

inserted in the stator equations. From eqn. 7 the effect of the DC resistance R_e is clearly not equivalent to that of a similar increase of the stator resistance; this follows also from the equivalent circuit, fig. 3a.

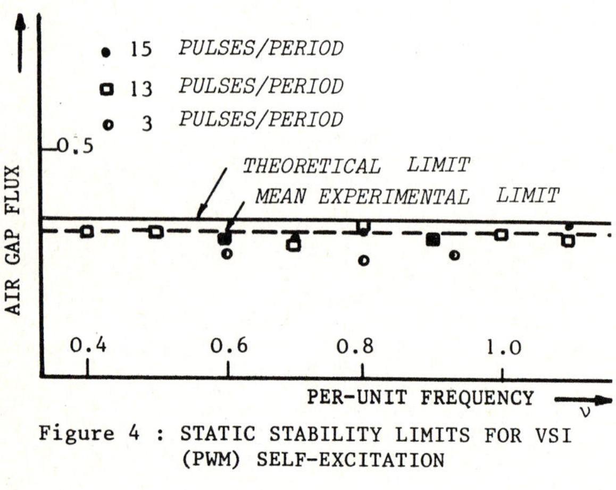

Figure 4 : STATIC STABILITY LIMITS FOR VSI (PWM) SELF-EXCITATION

For the CSI inverter with DC current supply the input equations are

$$\Delta V_e = R_e(\Delta I_e-\Delta I_{re})-pL_e\Delta I_{re} \tag{8}$$

$$\Delta I_{im} = 0 \tag{9}$$

Expressed in terms of an equivalent Thevenin voltage source, eqn. 8 becomes

$$\Delta V_{re} = \Delta V_e - (R_e+pL_e)\,\Delta I_{re} \tag{10}$$

Rotation about an angle $\delta=+\beta=\arg$ **E** - arg **I** and combination with the induction machine equations results in a 4th order system with state variables ΔI_{rx}, ΔI_{ry}, $\Delta\omega_r$ and ΔV_{re} (ΔI_{sx} and ΔI_{sy} are expressed as $\Delta I_{re}\cos\beta$, $\Delta I_{re}\sin\beta$ respectively); ΔV_e (or ΔI_e) is the input variable. The dynamic analysis is much easier in the case of a CSI since the filter impedance simply adds to the stator impedance. For an ideal current source eqns. 8 and 9 reduce to

$$\Delta I_e = \Delta I_{re} \tag{11}$$

$$\Delta I_{im} = 0 \tag{12}$$

and the system becomes of 3rd order (the remaining equation merely determines the stator voltage).

3.3 Inverter-induction machine dynamics

VSI/PWM case:
Without DC link filter and with a small and negligible DC source impedance the fundamental harmonic dynamic behaviour of the VSI/induction machine is identical with that in case of a polyphase mains supply. Induction machine dynamics for polyphase mains supply have already been treated in narrow detail in literature [4,5,6,7,8].
If the source impedance cannot be neglected a separate dynamic analysis is required as the DC source impedance is not equivalent to an increase of the AC resistance (see eqn. 7). Moreover, it is known that the DC source impedance may cause a significant reduction of the system damping and in some cases may result in dynamic instability [13]. A proper choice of the DC link filter might improve the dynamics, but mostly a feedback control might be preferred to obtain proper steady-state characteristics as well as improved dynamic behaviour [12,13,14].
VSI/induction machine dynamics have already been investigated in some detail using the proposed model, the results of which have been presented in previous literature [12,15]. In general, main field saturation has been shown to improve the damping and the stability quite significantly.
Experimental verification has also proven that the proposed modelling of VSI drives using the saturated model produces quite accurate results [12,15]. As an illustration of the value and the accuracy of the method, two types of experiments will breviously be reported on.
The first experiment deals with the regenerative mode of inverter-induction machine drives for constant speed operation [12]. If the classical machine model is used in combination with the inverter model, dynamic analysis shows a zero eigenvalue for constant speed operation.

Nevertheless, because the inverter imposes both frequency and phase on the AC side this is not an autonomous system and therefore no zero eigenvalue should exist for stable operation. Since, on the other hand, actual drives show a stable behaviour in the regenerative mode the classical model clearly is not adequate as it produces either false or useless predictions. With the saturated model, however, this kind of difficulties does not exist. Moreover, excellent agreement is found between actually observed and predicted (static) stability limits, as illustrated in fig. 4.

In the second experiment the dynamic instability of a PWM-drive fed by a DC voltage source with a considerable internal resistance is investigated. Theoretical dynamic analysis of a particular drive using the saturated machine model leads to the following conclusions:

- the particular drive is stable without DC source resistance
- with a DC source resistance of 2 to 15 times the machine stator resistance, dynamic instability at low slip occurs for supply frequencies between 22 Hz and 40 Hz if the machine flux is kept below 0.8; in this case the speed oscillates with frequencies between 4 and 9 Hz, dependent on the DC source resistance
- for higher flux levels, increased main field saturation results in both an increase of the system damping and an increase of the machine gain τ_r / τ_m [8]; the resulting shift of the root locus (for variable gain) to the left half plane together with the change of the gain moves the eigenvalues to the left half plane, although they are still low damped.

These theoretical findings have been confirmed about completely by experiments. Only a small difference in the frequency of oscillation has been noticed (observed frequencies are about 20% higher). However, this is caused by the neglect of the internal resistance of the power transistors which can be considerable at low currents. It has been observed also that the dynamic behaviour is about independent of the switching frequency of the PWM, which justifies the fundamental harmonic approach used in the model.

For CSI drives general results regarding dynamic behaviour are difficult to obtain as the system is not stable without external control. Dynamic behaviour will therefore depend on the particular controller.

4. VOLTAGE CONVERTER FED INDUCTION MACHINE MODELLING

4.1 Steady state model for SCR voltage reduction

Although output voltages of SCR voltage converters are far from purely sinusoidal, in many applications a fundamental harmonic model turns out to result in a sufficiently accurate description. Moreover, for a fairly general load type it is possible to derive a straightforward analytical expression for the fundamental harmonic behaviour of the SCR voltage reduction [16,17]. Indeed, if the 3-phase symmetrical load consists of a sinusoidal emf in series with a reactance (see figure 5) and assuming an ideal sinusoidal power supply and three pairs of completely identical thyristors it can be shown that the SCR voltage reduction comes to an increase of the effective series reactance. This additional reactance is a function of the hold-off angle γ and is proportional to the total series reactance X_σ:

$$X_t = X_\sigma \cdot f(\gamma)/[1-f(\gamma)] \tag{13}$$

with

$$f(\gamma) = \frac{3}{2\pi}[\gamma+\sin\gamma] \quad ; \; 0 \leqslant \gamma \leqslant \frac{\pi}{3} \tag{14}$$

$$f(\gamma) = \frac{3}{2\pi}[\gamma+\sin(\frac{2\pi}{3}-\gamma)]; \frac{\pi}{3} \leqslant \gamma \leqslant \frac{2\pi}{3} \tag{15}$$

The increase of the series reactance is thus completely determined by the hold-off angle γ. To the contrary, the expression

$$\gamma = 2(\alpha - \phi) = 2(\phi - \beta) \tag{16}$$

(see figure 5) shows that the use of either the delay angle α or the extinction angle β as control force implies an inherent internal feedback of the power factor angle. As will be discussed later on, the use of the delay angle as control force therefore results in a dynamically unstable open loop operation.

It is feasible to account for series resistances as well as for harmonic losses. Nevertheless, despite much more involved calculations only a slight modification of the model will result from the inclusion of the series resistance. Moreover, experimental tests have pointed out that the above simplified model is quite satisfactory.

4.2 SCR voltage converter-fed induction machine model

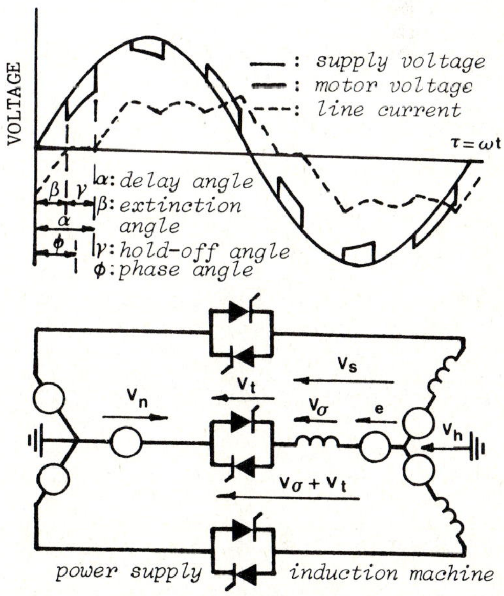

Figure 5 : SCR VOLTAGE CONVERSION

The fundamental harmonic SCR model can also be employed for the modelling of SCR voltage reduction of induction machines, assuming - temporarily at least - a subtransient machine representation consisting of a back emf in series with the machine leakage. The proposed SCR model has indeed been derived on the assumption of a negligible series (stator) resistance. The use of a subtransient machine scheme in view of the voltage reduction model can certainly be justified as the periods of time that are concerned are less than 3.33 ms whereas the rotor time constant of typical 50 Hz machines ranges between 80 ms and 1 s.

Nevertheless, if the approximate SCR model is then combined with a more complete induction machine representation, a fairly reliable description of both the steady-state characteristics and the dynamic behaviour turns out to result.
The SCR voltage converter supply is implemented in the machine equations by way of an input equation for the machine voltage:

$$\mathbf{v}_n = \mathbf{v}_s + \frac{d}{d\tau}\{\ell_t \mathbf{i}_s \exp(j\nu\tau)\}\exp(-j\nu\tau) \qquad (17)$$

with

$$\ell_t = f(\gamma)/\{1-f(\gamma)\} \qquad (18)$$

the per-unit SCR reactance, and $\mathbf{v}_n, \mathbf{v}_s$ the p.u. mains and machine voltages in a synchronously rotating reference frame (p.u. frequency ν).
In steady state eqn. 17 reduces to

$$\mathbf{v}_{no} = \mathbf{v}_{so} + j\nu_o \ell_{to} \mathbf{i}_{so} \qquad (19)$$

The input equation for small displacements about steady state on the other hand reads:

$$\Delta\mathbf{v}_n = \Delta\mathbf{v}_s + \{\frac{d\ell_t}{d\tau} + j\nu_o \Delta\ell_t\}\mathbf{i}_{so} + \ell_{to}\{\frac{d\mathbf{i}_s}{d\tau} + j\nu_o \Delta\mathbf{i}_s\} \qquad (20)$$

with

$$\Delta\ell_t = (d\ell_t/d\gamma)\ \Delta\gamma \qquad (21)$$

$$(d\ell_t/d\tau) = (d\ell_t/d\gamma)(d\gamma/d\tau) \triangleq \ell_t'(d\Delta\gamma/d\tau) \qquad (22)$$

If either the delay angle α or the extinction angle β are to be used as control variable, γ must be expressed in terms of α or β respectively using

$$\Delta\gamma = 2(\Delta\alpha - \Delta\phi) = 2(\Delta\phi - \Delta\beta) \qquad (23)$$

with

$$\Delta\phi = |\mathbf{i}_{so}|^{-2}(i_{syo}\Delta i_{sx} - i_{sxo}\Delta i_{sy}) \qquad (24)$$

In these latter cases, therefore, the internal feedback of the power factor angle will clearly result in totally different steady-state and dynamic drive characteristics. The additional reactance ℓ_t becomes indeed load-dependent for α or β control. For γ control, on the other hand, the drive behaves in steady state as well as dynamically as if the leakage was increased by a constant factor.

4.3 Steady-state characteristics for constant α and γ

Figure 6 shows some calculated steady-state

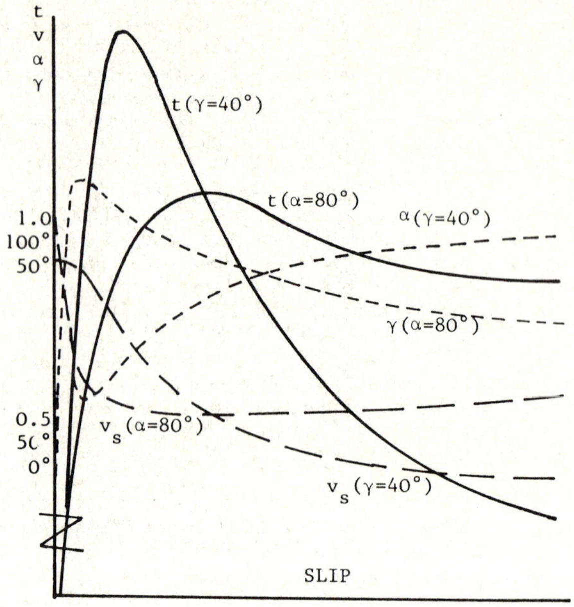

Figure 6 : STEADY-STATE CURVES FOR CONSTANT α AND CONSTANT γ

characteristics for constant α and for constant γ. The typical and well-known differences between the characteristics for both types of open loop control are clearly illustrated. Comparison with actually recorded characteristics demonstrates in most cases a satisfactory correspondence. However, an accurate prediction over the whole slip range may be complicated by considerable leakage saturation, if any.

4.4 Dynamic characteristics for open loop voltage control

Not only the steady state, but in particular the dynamic behaviour differs quite significantly for α and for γ open loop control. As is well known, open loop α control is inherently unstable. Typically, for a small inertia and for low load (small slip), severe huntings with a low oscillation frequency may occur. Open loop γ control on the other hand does not possess such an inherent unstable behaviour.
This fundamental difference regarding dynamic behaviour is quite easily explained by means of the SCR model of this contribution. As explained in more detail in Reference 16, the internal feedback of the power angle in case of α control causes the zeros of the transfer function from the speed to the motor torque to become right half plane, that is the system is not minimum phase any more. A small inertia will therefore result in right half plane eigenvalues (i.e. poles of the transfer function from load torque to speed). For γ control on the other hand, the constant increase of the leakage reactance causes an even more damped behaviour, compared to the machine without voltage reduction [7,8].

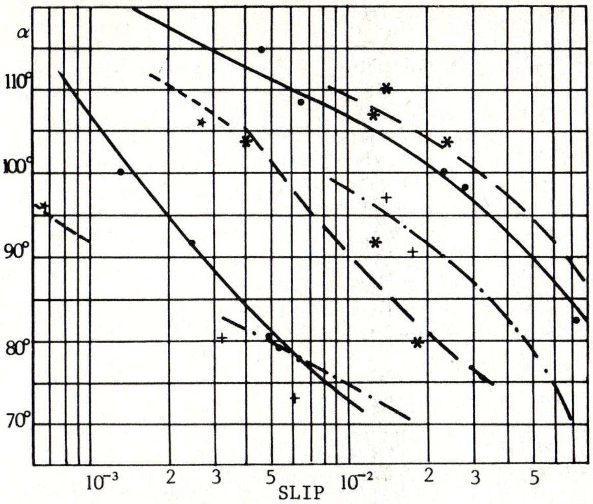

Figure 7: THEORETICAL AND EXPERIMENTAL STABILITY LIMITS OF CONSTANT DELAY ANGLE VOLTAGE REDUCTION

- • *small slip machine*
- ✱ *high slip machine*
- ⁎ *high stator resistance*
- + *very high stator resistance*

As a check on the validity of the model, theoretically predicted stability boundaries for open loop α control have been compared with corresponding actually observed points on the edge of instability. In the investigation, a 2.2 kW motor loaded with a small-inertia Foucault brake has been used. For the same motor, two different rotors have been considered, a high slip one and a low slip one. Further, using series resistances, two additional equivalent stator resistances have been considered for the small slip rotor. The voltage controller on the other hand is micro-processor based and can be used with either α or γ as control force and with either open loop control or closed loop control.

Figure 7 presents some results in the α versus slip plane. Clearly, calculated and experimental stability boundaries match fairly satisfactorily. For a high stator resistance, however, the correspondence is somewhat inferior but this is obviously due to the fact that the SCR model basically assumes a negligible series resistance. Nevertheless, it can be concluded that the SCR model, despite the seemingly drastic simplifying assumptions, produces acceptable results.

4.5 Modelling of a micro-processor based closed loop control

In the micro-processor based voltage converter mentioned above a closed-loop near optimal efficiency voltage control with the hold-off angle as control force has been implemented. Detailled information regarding the hardware is given in [17].

The hold-off angle has been chosen as control force to avoid the inherent instability and the

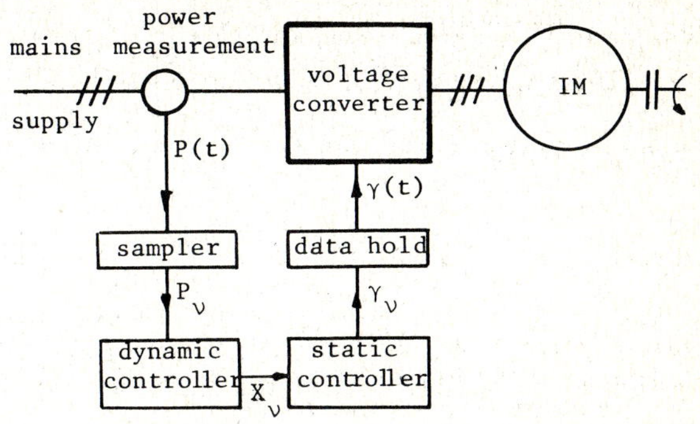

Figure 8 : SCHEMATIC CONTROLLER LAYOUT

strongly load-dependent inherent feedback with which a delay angle control is attended. The static controller algorithm is a linear law relating the hold-off angle and the electrical input power [17]:

$$\gamma = \gamma_m(1-P/P_s) \tag{25}$$

As has been proven by careful experimental investigation, this linear law turns out to result in near optimal efficiency.

As to the dynamic controller characteristics, consider the feedback equation, eqn. 20. Also with γ as control force, the internal feedback of the control angle includes a differentiating action, that is the control variable enters the machine equations via a derivative term. This may explain why the actual controller includes an integrating term. The integrating action is implemented in the actual micro-processor based controller by way of a relaxation.

The actual layout of the controller is represented schematically in figure 8. This control scheme has been implemented in the SCR model as follows:

1- the power measurement, which is a sampling method, generates the discrete data sequence $\{P_\nu\}$ with Z-transform P(z)

2- the relaxation, i.e. the dynamic controller, yields the sequence $\{X_\nu\}$ according to

$$X_{\nu+1} = X_\nu + \frac{T_s}{\tau} P_\nu - \frac{T_s}{\tau} X_\nu \tag{26}$$

with T_s the sampling period and τ the relaxation time. Therefore

$$X(z) = P(z)\frac{T_s/\tau}{z-(1-T_s/\tau)} \tag{27}$$

3- the hold-off angle is then calculated from eqn. 25; the resulting sequence $\{\gamma_\nu\}$ enters the data hold (1st order) yielding the voltage converter input γ(t)

The resulting signal γ(t) can be written as the inverse Laplace transform of

$$\tilde{G}_{dh}(p)\cdot\tilde{\gamma}(p) = \gamma_h(p) \tag{28}$$

with

$$\tilde{G}_{dh}(p) = \frac{1-\exp(-pT_s)}{p} \tag{29}$$

the Laplace transform of the data hold and $\tilde{\gamma}(p)$ the Laplace transform of a series of impulses with intensity $\gamma_\nu = \gamma(\nu T_s)$ (i.e. a mathematical sampler, [18]). Therefore

$$\gamma_h(p) = [\sum_0^\infty \gamma_\nu \exp(-p\nu T_s)]\tilde{G}_{dh}(p)$$
$$= -(\gamma_m/P_s)\,[\sum_0^\infty X_\nu \exp(-p\nu T_s)]\tilde{G}_{dh}(p) \qquad (30)$$

Because $\tilde{\gamma}(p)=\gamma(z)$ *with* $z=\exp(pT_s)$*, this is equivalent to*

$$\gamma_h(p) = -\frac{\gamma_m}{P_s}\cdot\frac{zT_s/\tau}{z-(1-T_s/\tau)}\cdot\frac{1-1/z}{\ln(z/T_s)}\cdot P(z) \qquad (31)$$

with $z=\exp(pT\ )$*. Using*

$$P(z)\Big|_{z=\exp(pT_s)} = \frac{1}{T_s}\sum_{-\infty}^{+\infty} P(p-j\omega_\nu) \approx \frac{1}{T_s}P(p) \qquad (32)$$

(neglecting higher order terms because of the low pass characteristics) we thus obtain

$$\gamma_h(p) \approx -\frac{\gamma_m}{P_s}\left\{\frac{1+pT_s}{(1+p\tau)(1+pT_s/2)}\right\}P(p) \qquad (33)$$

The implementation of the feedback law eqn. 33 in the SCR-induction machine model turns out to produce a model for the closed loop voltage controlled drive whose dynamic characteristics resemble those of the actual drive quite well. This is illustrated in figure 9. Calculated and experimentally measured transfer functions from load torque to speed (for the 2.2 kW machine mentioned above, with total effective inertia = twice the motor inertia) are compared with each other. Except for a relatively small shift in the resonance frequency - which may be due to the effect of the not-exactly rigorous shaft - theoretical and recorded transfer functions match reasonably well.

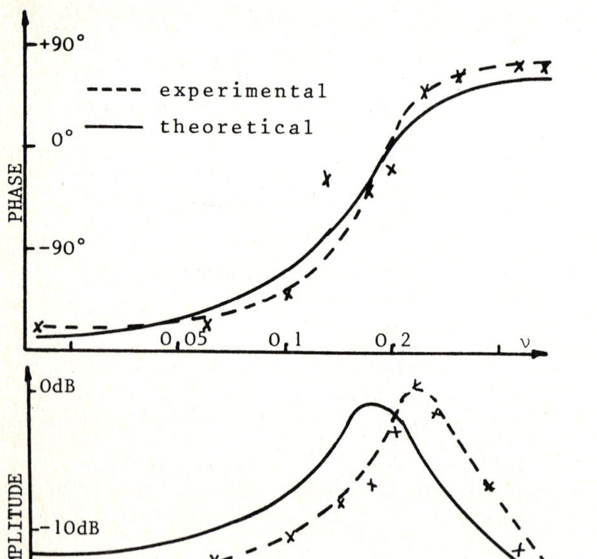

Figure 9 : CALCULATED OUT RECORDED SPEED/TORQUE TRANSFER FUNCTION (CLOSED LOOP γ-CONTROL)

5. REFERENCES

[1] Rogers,G.J., *Linearised analysis of induction motor transients*, Proc. IEE, Vol. 112, 10 (1965), 1917-1925.

[2] Stern,R. and Novotny,D.W., *A simplified approach to the determination of induction machine dynamic response*, IEEE/PAS-97, 4 (1978), 1430-1439.

[3] Cupsa,A., Geysen,W.,et al., *Stability analysis of oscillating induction machines*, Acta Technica Csav (1978), 521-548.

[4] Melkebeek,J.A.A., *Influence of saturation on the stability limits of a voltage-fed induction motor*, 15th UPEC, Leicester (1980), 2B2/1-8.

[5] Melkebeek,J.A.A., *Some effects of magnetic field saturation in induction machines*, 16th UPEC, Sheffield (1981), B23/1-8.

[6] Melkebeek,J.A.A., *Magnetising field saturation and dynamic behaviour of induction machines, Part I: An improved calculation method for induction machine dynamics*, IEE Proc., Vol.130, Pt. B, 1 (1983), 1-9.

[7] Melkebeek,J.A.A., id., *Pt.II: Stability limits of a voltage-fed induction motor and of a self-excited induction generator*, ibid.,10-17.

[8] Melkebeek,J.A.A. and Novotny,D.W., *The influence of saturation on induction machine drive dynamics*, IEEE/IAS, 6, (1983).

[9] Abbas,M. and Novotny,D.W., *Stator referred equivalent circuits for inverter driven electric machines*, IEEE/IAS Meeting, 1978.

[10] Novotny,D.W. and King,T.L., *Equivalent circuit representation of current inverter driven synchronous machines*, IEEE/PAS-100, 6 (1981), 2920-2926.

[11] Melkebeek,J.A.A. and Novotny,D.W., *Steady state modelling of regeneration and self-excitation in induction machines*, IEEE/PAS-102, 8 (1983), 2725-2733.

[12] Melkebeek,J.A.A. and Novotny,D.W., *Small signal dynamic analysis of regeneration and self-excitation in induction machines*, El. Mach. and Power Systems, 5 (1983).

[13] Lipo,T. and Krause,P.C., *Stability analysis of a rectifier-inverter induction motor drive*, IEEE/PAS-88, 1 (1969), 55-65.

[14] Fukazawa,K. and Novotny,D.W., *The influence of volts/hertz control on filter impedance effects in VSI-fed induction machines*, IEEE/IAS, 3 (1982), 230-239.

[15] Y.-K. He and Lipo,T.A., *Saturation effects in stability analysis of a VSI induction motor drive*, Report 82-13, Wempec-ECE, Univ. of Wisconsin-Madison, 1982.

[16] Melkebeek,J.A.A., *Stability analysis of induction motors with SCR voltage control*, 18th UPEC, Guildford (1983), 2C1/1-6.

[17] Van Den Bossche,A. and Melkebeek,J.A.A., *Induction machine SCR voltage reduction: optimized control and dynamic modelling*, Int. Conf. on Power Electr. and Variable Speed Drives, London, May 1-4, 1984.

[18] Elgerd,O.I., *Control Systems Theory*, (McGraw-Hill, 1967).

ELECTRICAL MACHINES AND CONVERTERS - Modelling and Simulation
H. Buyse and J. Robert (editors)
Elsevier Science Publishers B.V. (North-Holland)/© IMACS, 1984

CONTROL LOOP STUDY OF THE INDUCTION MOTOR DRIVEN BY CURRENT SOURCE INVERTER IMPOSING STATOR FREQUENCY IN FLUX CONTROL

J. Santana

Centro de Automática da Universidade Técnica de Lisboa
Secção de Máquinas Eléctricas (I.S.T.)
Av. Prof. Gama Pinto, 2 1699 LISBOA-Portugal

The induction motor driven by current source inverter presents instabilities and uncontrolled flux when operated in an open loop scheme. To overcome these problems feedback controls must be used.
A new flux control using the stator frequency as input variable is studied and comparisons with classical solutions are given.

0. INTRODUCTION

A dynamic model for the induction motor driven by current source inverter by applying the sliding contacts equation |1| was developed. For the conventional three-phase induction motor that model is represented by a set of equations similar to the one obtained from Park's transformation |2|. Both models are a support for the studying of the control system.

Elimination of the open-loop instability was suggested by using a direct current control with a superimposed slip-frequency control in order to ensure constant flux |3| |4|. However, the input variable to be used in the flux control is not defined. In the relevant literature are found systems where the input variable is the direct current or the slip--frequency. These two possible ways were systematically analysed in reference |5|, showing no significant difference between them. Both systems are pratically defined by a pole close to the origine which results in a bad dynamic response. However, by choosing the stator frequency as input variable in order to control the flux a new system can be obtained.

Analysis of this system is given in the paper and a good dynamic response is shown.

1. DESCRIPTION OF THE SYSTEM

Figure 1 shows the induction motor driven by current source inverter, a well known variable speed drive nowadays.

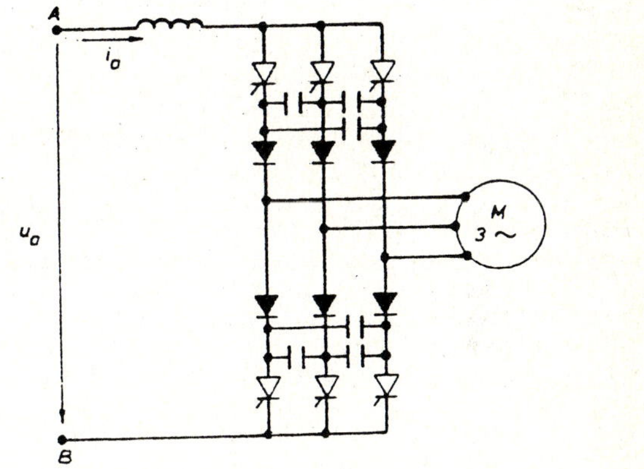

Fig.1: Induction motor driven by current source inverter

From the terminals A B the system can be regarded as a commutator machine and figure 2 shows, clearly, its mechanical analog.

The three stator phases are connected to the mechanical commutator on which one pair of brushes rotates with speed $\dot\theta_r$.

This electromechanical system resulting from the association of Power Electronics and conventional induction motor can be regarded as a commutation machine. Hence its dynamic model is obtained by using the sliding contacts equations. In reference |1|, it was developed a mathematical model governed by the relationships:

$$(1)\quad \begin{bmatrix} U_a \\ 0 \\ 0 \end{bmatrix} = \begin{bmatrix} R_a+L_a p & M_p & \dot\theta_s M \\ Mp & R_r+L_r p & \dot\theta_{sr} L_r \\ -\dot\theta_{sr} M & -\dot\theta_{sr} L_r & R_r+L_r p \end{bmatrix} \begin{bmatrix} i_a \\ i_{dr} \\ i_{qr} \end{bmatrix}$$

$$(2)\quad M i_a i_{qr} - T_L = J\, p\, \dot\theta_r$$

where $\dot\theta_{sr} = \dot\theta_s - \dot\theta_r$

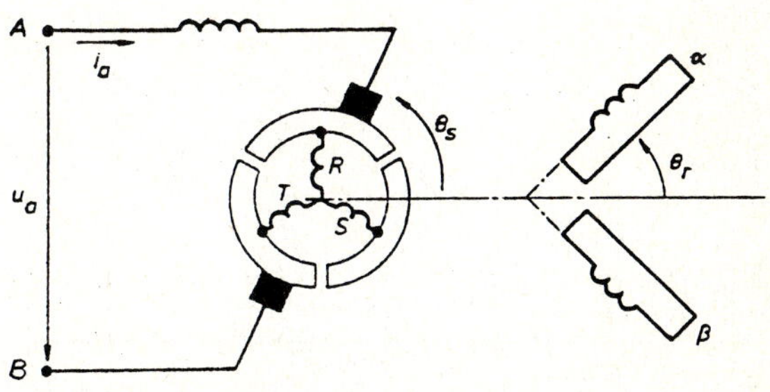

Fig.2: Induction motor mechanical analog driven by current source inverter

These equations are the same given in reference |2|, which were obtained by applying the Park's transformation to the induction motor and choosing a reference frame which is synchronous with the fundamental stator m.m.f..

As represented in Figure 2 for the supply conditions given by the inverter there are only two contacts and so one equation is enough for the stator representation. The electromagnetic behaviour of the induction motor driven by current source inverter can appear as a simplification of the induction motor when connected to the usual three phase supply. In Figure 2, the lack of another stator connection is responsible for some problems existing in the open-loop operation as instability and uncontrolled flux.

It can be shown that system stabilization is not easily achieved by changing the machine parameters. The solutions of the previous problems can be solved by the introduction of a suitable close-loop control.

In Figure 3 the control of direct current i_a plays an important role to improve stability of the system. However the flux is not determined by the variables i_a and $\dot{\theta}_s$ and so in the system shown in Figure 3, with a simple current control the flux is not fixed for all operating points.

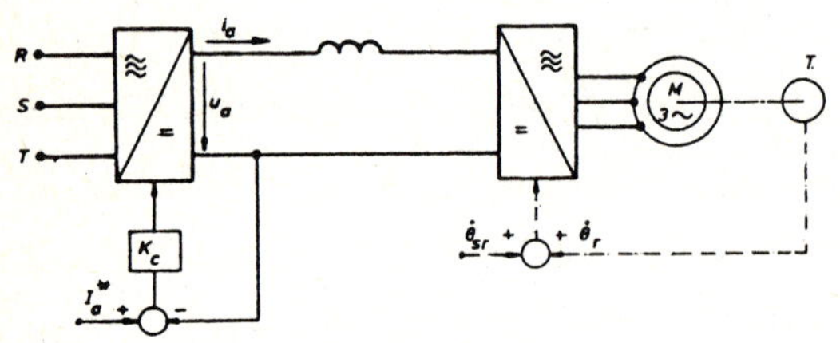

Fig.3: Induction motor direct current control driven by current source inverter

In order to control the flux, the values of direct current and slip frequency must be related in such a way that the magnetization current I_m be constant shown in the following relation

$$(3) \qquad i_a \simeq \sqrt{1+(\dot{\theta}_{sr}\frac{L_r}{R_r})^2}\; I_m$$

For that a second control for $\dot{\theta}_{sr}$ is introduced as represented in Figure 3.

In (3), the relationship between the variables i_a and $\dot{\theta}_{sr}$ it is not imposed which is the input (independent) variable. In the literature are found sistems in which the independent variable is either direct current or the slip frequency. Reference |5| shows that these two possible ways are quite similar when its dynamic behaviour is analysed. Both systems have the same electromechanical characteristic with $dTe/d\dot{\theta}_r \simeq 0$ and therefore they have both a pole related with mechanical time constant which is near the origine. This fact attenuates the drive speed response.

Since the input variable is not imposed in the constant flux operation, other configurations for the system can be developed. Figure 4 shows the configuration which is analysed in this paper where the independent variable is the stator frequency.

The system represented in Figure 4 is described with the equations (1), (2) and (3) together with the following equations

$$(4) \qquad U_a = k_c\,(I_a^* - i_a)$$

$$(5) \qquad \dot{\theta}_{sr} = \dot{\theta}_s - \dot{\theta}_r$$

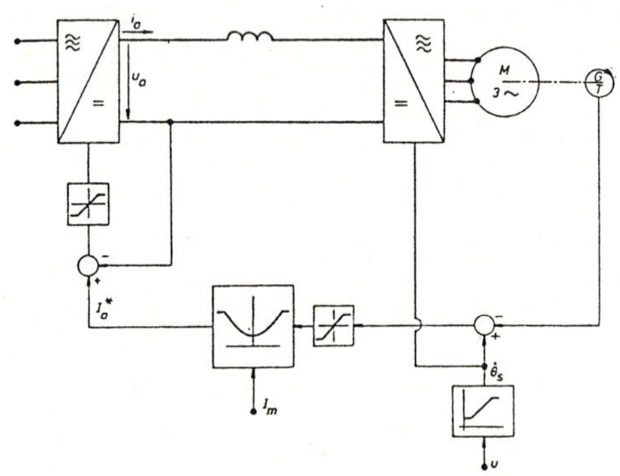

Fig.4: Induction motor driven by current source inverter imposing stator frequency with flux control

2. STEADY STATE BEHAVIOUR

In the steady state the equations (1) and (4) give

$$(6) \qquad \begin{bmatrix} k_c I_a^* \\ 0 \\ 0 \end{bmatrix} = \begin{bmatrix} R'_a & 0 & \dot{\theta}_s M \\ 0 & R_r & \dot{\theta}_{sr} L_r \\ -\dot{\theta}_{sr} M & -\dot{\theta}_{sr} L_r & R_r \end{bmatrix} \begin{bmatrix} i_a \\ i_{dr} \\ i_{qr} \end{bmatrix}$$

The proportional action of the direct current regulator is equivalent to increase the stator resistance.

For high values of k_c, it is possible to identify I_a^* with I_a and in this situation it can be written:

(7)

a) $I_a \simeq I_a^* = \frac{\sqrt{R_r^2 + (\dot{\theta}_{sr} L_r)^2}}{R_r} \cdot I_m$

b) $I_{dr} \simeq - \frac{\dot{\theta}_{sr}^2 \, ML_r}{R_r \sqrt{R_r^2 + (\dot{\theta}_{sr} L_r)^2}} \cdot I_m$

c) $I_{qr} \simeq \frac{-\dot{\theta}_{sr} M}{\sqrt{R_r^2 + (\dot{\theta}_{sr} L_r)^2}} \cdot I_m$

with that condition, the electromagnetic torque is given by

(8) $T_e = \frac{M^2 I_m^2}{R_r} \dot{\theta}_{sr}$

In opposition to the classical solutions, where both i_a and $\dot{\theta}_{sr}$ are constant for a constant input variable |5|, in the new development i_a and $\dot{\theta}_{sr}$ are not constant for a constant input $\dot{\theta}_s$. For a given value of $\dot{\theta}_s$ the torque changes with $\dot{\theta}_{sr}$, as represented in Figure 5.

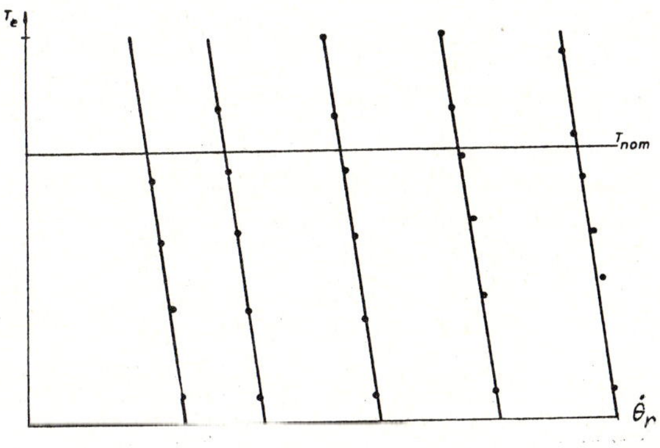

Fig.5: Electromechanic characteristics

It can be seen from Figure 5 that the electromagnetic torque presents a high $dT_e/d\dot{\theta}_r$.

3. DYNAMIC BEHAVIOUR

To illustrate the transient behaviour of the system some experimental results are now presented.

i) Load torque variation

In Figure 6 (i) is firstly considered a two-step load torque variation with constant stator frequency. In first step the system is changed from the no-load to overload condition. After stabilization, a second step is considered in which the system settle to no-load condition.

ii) Step on stator frequency

Another transient is considered in Figure 6 (ii), obtained for constant load torque by applying steps on stator frequency

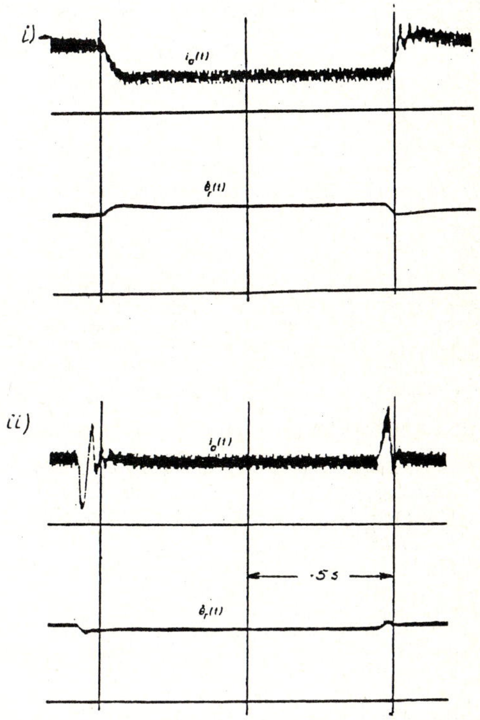

Fig.6: Transient behaviours

The transients in Figure 6 show that the induction motor driven by current source inverter with the control represented in Figure 4 is a system with a good speed response.

The equations (1) to (5) are not linear. For studying the small-signal behaviour a linearization technique must be used.

The electromagnetic torque perturbation is given by:

(9) $T_e = G_1(p) \, \Delta\dot{\theta}_{sr} + G_2(p) \, \Delta\dot{\theta}_s$

This relation with the following equation

(10) $\Delta T_e - \Delta T_L = (Jp + D) \, \Delta\dot{\theta}_r$

establishes the block diagram represented in Figure 7.

It is interesting to observe that the variation of the stator frequency has two effects on

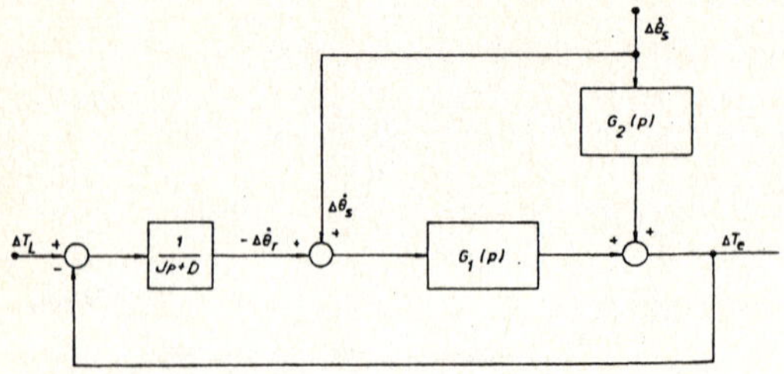

Fig.7: Block diagram of the system

the behaviour of the system.

a) It acts directly on the slip frequency forcing the action of the direct current control through $G_1(p)$.

b) It changes the parameters of the impedance matrix through $G_2(p)$.

On the electromagnetic torque perturbation the first action is the most important. Neglecting $G_2(p)$ effect, the block diagram is simplified as shown in Figure 8.

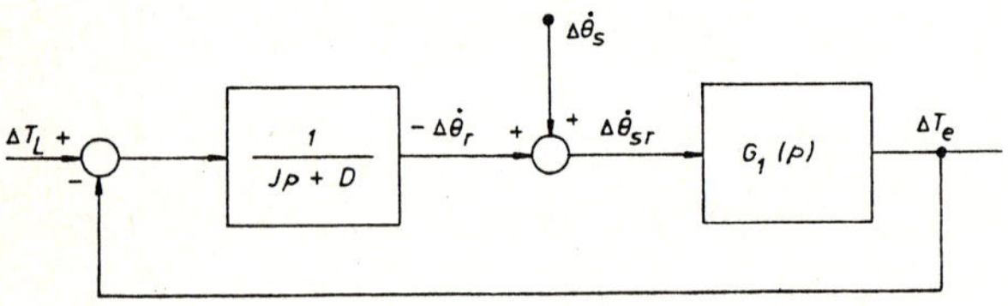

Fig.8: Simplified block diagram of the system

For $k_c \gg \dot\theta_s M$ $G_1(p)$ is given by

$$(12)\quad G_1(p) = \frac{ap^2+bp+c}{L_r^2 p^2 + 2R_r L_r p + R_r^2 + \dot\theta_{sr}^2 L_r^2}$$

where

$$a = \frac{\dot\theta_{sr}^2 M^2 L_r^4 I_m^2}{R_r^2 + (R_r^2 + \dot\theta_{sr}^2 L_r^2)} I_M^2$$

$$b = \frac{M^2 L_r (R_r^2 + 3\dot\theta_{sr}^2 L_r^2)}{R_r^2 + \dot\theta_{sr}^2 L_r^2} I_m^2$$

$$c = \frac{M^2 (R_r^2 + \dot\theta_{sr}^2 L_r^2)}{R_r} I_m^2$$

Table I shows the poles of the block diagrams represented in Figure 7 and Figure 8, for two different operating points with $k_c = 500\Omega$ and $\dot\theta_s = 314$ rad s^{-1}.

Slip	Poles	
	Complete model	Simplified model
.02	$-9.9 \pm j11.8$	$-8.4 \pm j12.9$
	-1630	
	-6.5	-10
.10	$-11.9 \pm j34$	$-12 \pm j33.8$
	-1623	
	-10	-11.3

TABLE I

The simplified block diagram is a good assumption for the system analysis. In fact by the quick action in the direct current control, the stator m.m.f. adapts itself instantaneously to the conditions imposed by the external feedback. Hence the dynamic behaviour of the system is mainly determined by the rotor parameters and mechanical time constant.

If $(\dot\theta_{sr} L_r)^2 \gg R_r^2$, another simplification can be introduced and the function G_1 is given by:

$$(13)\quad G_1(p) = \frac{M^2 I_m^2}{R_r} \cdot \frac{L_r^2 p^2 + 3R_r L_r p + \dot\theta_{sr}^2 L_r^2}{L_r^2 p^2 + 2R_r L_r p + \dot\theta_{sr}^2 L_r^2}$$

The poles and zeros of $G_1(p)$ are

$$p_{1,2} = -\frac{R_r}{L_r} \pm j\dot\theta_{sr} \quad \text{and} \quad Z_{1,2} = -\frac{3R_r}{2L_r} \pm j\dot\theta_{sr}$$

The zeros and the poles are very closed. Then by assuming a complete compensation, $G_1(p)$ has a just proportional action given by

$$(14)\quad G_1(p) \simeq \frac{M^2 I_m^2}{R_r} = -\frac{dT_e}{d\dot\theta_r}$$

With this simplification, the system presents one single pole and assuming D=0, the pole becomes:

$$(15)\quad p = \left(-\frac{M^2 I_m^2}{R_r}\right)/J$$

For the actual system from the equation (15) results:

$$p = -12.9\ s^{-1}$$

This pole which is related to the mechanical time constant is far from origine. As compared to the results given in reference |5| this result shows that a better speed response is obtained. In fact, in reference |5| the pole related with mechanical time constant is pratically zero, because $\frac{dT_e}{d\dot{\theta}_r} \simeq 0$.

4. CONCLUSIONS

The system previously described is a new electromechanical device in which the instability and uncontrolled flux found in an open-loop operation are eliminated using feedback controls.

The flux control considered in this paper is quite similar to the one normaly proposed in the relevant literature, but a different input variable - the stator frequency- is now used. This choice improves the speed response because it establishes an electromechanical characteristic with $dT_e/d\dot{\theta}_r \ll 0$, hence decreasing the mechanical time constant.

REFERENCES

|1| J. Santana: "Modelling and stability analysis of induction motor driven by current source inverter" Proc. of ICEM, Athens 1980.

|2| N. Sawaki and N. Sato: "Steady-state stability analysis of induction motor driven by current source inverter" IEEE Trans. on Industry Applications, Vol. IA - 13 May/June 1977.

|3| B. de Fornel and D. Noyses: "Transient characteristics of an asynchronous machine current feed from a static converter" Proc. IEE, Vol. 124, 10, October 1977.

|4| E. Cornell and T. Lipo: "Modelling and design of controlled current motor drive system" IEEE Trans. on Industry Applications, Vol. IA-43, July/August 1977.

|5| M.L. MacDonald and P.C. Sen: "Control loop study of induction motor driver using D.Q. model" Conf. Rec. IEEE/IAS 1978.

|6| A. Abbondanti and M.B. Brennen: "Variable speed induction motor drive use electronic slip calculator" IEEE Trans. on Industry Applications, Vol. IA-11, 5, September/October 1975.

ELECTRICAL MACHINES AND CONVERTERS - Modelling and Simulation
H. Buyse and J. Robert (editors)
Elsevier Science Publishers B.V. (North-Holland)/ ©IMACS, 1984

NUMERICAL SIMULATION OF A VOLTAGE-FED INVERTER FOR INDUCTION MOTOR SUPPLY

V.P. Nascimento, J. Santana and M.S. Garrido

INIC/IST, 2 Av. Gama Pinto, 1699 Lisboa Codex, Portugal
and UCL, 3 Place du Levant, 1348 Louvain-la-Neuve, Belgium

In a polyphase voltage-fed inverter connected to an active inductive load there is a superposition of the natural behaviour of diodes and the programmed action of switching elements.
In this paper, an asymptotic model of a diode bride rectifier, previously developed, is completed in order to account for the action of the switching elements. A very simple model of the voltage-fed inverter is so obtained, which is suitable for any number of phases and any type of firing control.
Results of simulation will be compared to experimental data obtained with a thyristor inverter.

1. Introduction

A numerical simulation of static converters is necessary if one whishes to calculate the instantaneous constraints on the convertor or on the load.

If the load is an induction motor, it is important to know the harmonic content of currents, the pulsating torques, the behaviour during failure conditions and so on.

There exist two classical models for simulating static converters.

The former represents the semiconductors by ideal logic elements and accounts for the changes in the topology of the system introduced by these elements.

The latter represents each semiconductor by an equivalent non-linear R,L,E circuit and keeps a fixed topology system.

Those models are not very interesting for complex converters, as the voltage-fed inverter, because they require either complicated simulation programs or long computation times.

This paper presents a new model for simulating a voltage-fed inverter supplying an active inductive load, as an induction motor.

The model makes use of the fact that in a voltage-fed inverter there is simply a superposition of the natural behaviour of the diodes and the programmed action of some switching elements, as the forced-commutation thyristors.

Based on this fact, we start from a very simple asymptotic model of a diode bridge rectifier, previously developed |1|, and we modify it in order to account for the action of the switching elements.

A very simple model of the voltage-fed inverter is thus obtained, suitable for any number of phases and any type of firing control.

2. Asymptotic model of a voltage-fed inverter

The figure 1 represents, schematically, an induction motor driven by a voltage-fed inverter.

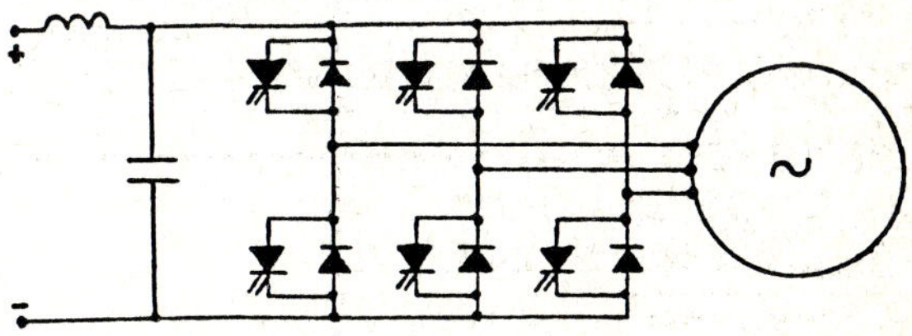

Fig. 1 : Induction motor supplied by a voltage-fed inverter.

For power flow studies, we may ignore the forced-commutation circuitry and replace the thyristors by unidirectional switches.

Further, we are not interested, here, in the analysis of the DC source or of the induction motor, but only in the simulation of the voltage-fed inverter.

So, we will consider in our study the simpler system represented in figure 2.

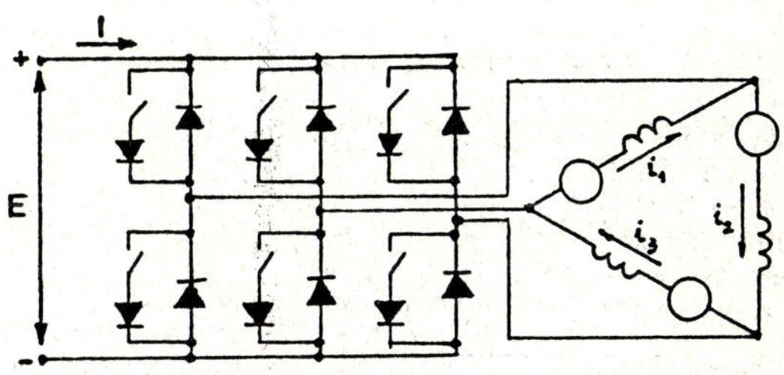

Fig. 2 : Simplified voltage-fed inverter system

If we ignore the switching elements, the system in figure 2 is a simple diode bridge rectifier supplying a DC voltage E.

We know from previous work |1| that the voltage E, seen from the AC side, is equivalent to a polyphase voltage

(1) $$u_k = \gamma_k E$$

with

(2) $$\gamma_k = \lim_{\alpha \to \infty}\left(\frac{A_{k-1}}{1+B_{k-1}\, e^{\alpha(i_{k-1}-i_k)}} - \frac{A_k}{1+B_k\, e^{\alpha(i_k-i_{k+1})}}\right)$$

where A_j, B_j are equal to 1 for non-defective diodes

(3.a) $$A_j, B_j = 1$$

If there is a short-circuit in an up (down) diode, the corresponding coefficient A_j (B_j) becomes zero.

This is just the situation when one of the switches represented in figure 2 is closed. So, we can use for the voltage-fed inverter the same equations (1), (2), but now with

(3.b) $$A_j, B_j = 1 \text{ or } 0$$

according to the state of the thyristors, which is supposed to exclude the short-circuit $A_j = B_j = 0$.

Then, the AC model of the voltage-fed inverter becomes the one represented in figure 3, where the voltages u_k are known in function of the currents i_j by equations (1).

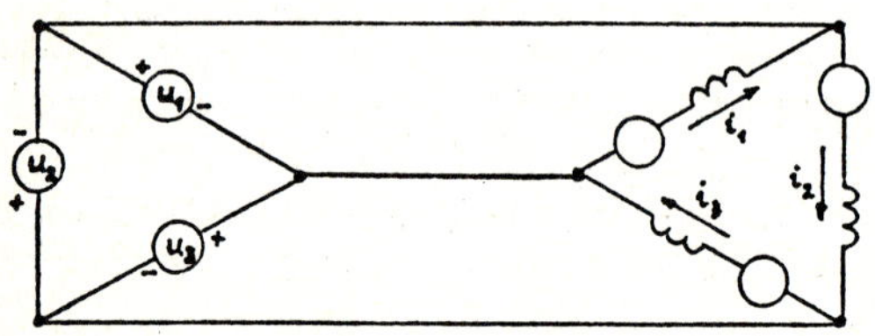

Fig. 3 : Equivalent AC network.

For a numerical simulation of this network it is easier to keep a fixed topology. This is obtained by giving the constant α in the functions γ_k a finite value, which will be fixed by the precision requirements.

The DC current of the inverter is directly obtained as a function of the currents i_k by the node law and we have |1| :

(4) $$I = -\sum_k \gamma_k i_k$$

So, the DC model of the inverter is the one represented in figure 4.

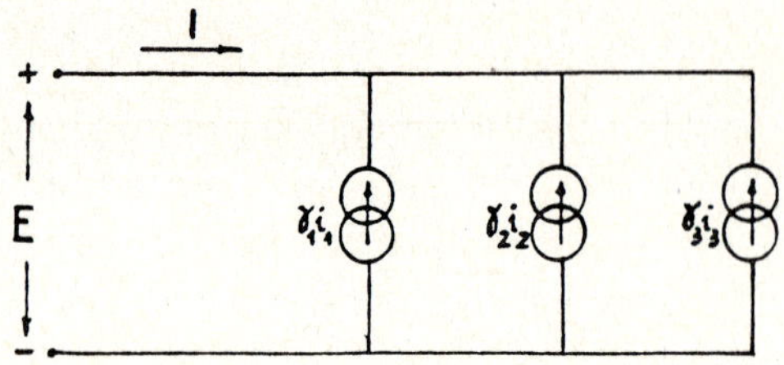

Fig. 4 : Equivalent DC network.

Let us now present some results obtained with the previous model and compare them with experimental data.

3. Examples of simulations

The first case to be considered is the trivial one of a three-phase inverter with the classical 180° wave, feeding a passive R,L load.

Of course we should'nt need a complete model for this case, because the voltages u are defined a priori and have the well known form represented in figure 5.

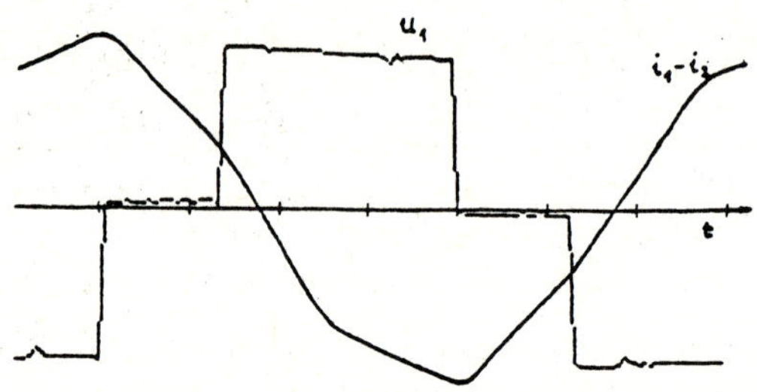

Fig. 5 : Experimental AC line voltage and current.

From equation (2) we can conclude that the γ_k automatically take the rectangular form represented in this figure, just by virtue of the values 0,1 given to coefficients A_j, B_j and without intervention of the exponentials.

Simulation by the model represented in figure 3 gives practically the same result as figure 5.

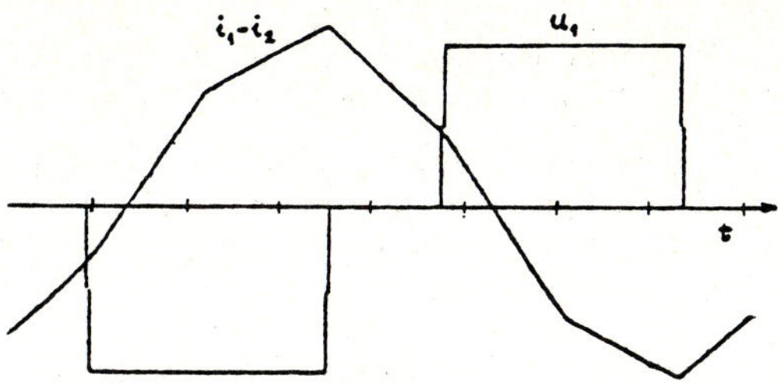

Fig. 6 : Simulated AC line voltage and current.

There are some detail differences introduced by the forced-commutation circuitry, the direct voltage of semiconductors and the small variation of experimental source voltage E.

The computation time using Euler's method with a step of 10^{-4} sec., has been 10 s for a period of 40 ms (25 Hz), on a 16 bits microcomputer.

Let us now consider a less trivial situation.

We will suppose that there is a permanent failure in the firing of one thyristor and this means that a certain A_j, B_j is always 1.

The voltages u_k cannot be known a priori and a complete model is now necessary for describing the behaviour of the system.

The simulated line voltages and currents are represented in figure 7; except for some small details, as in the previous case, the experimental results are practically identical.

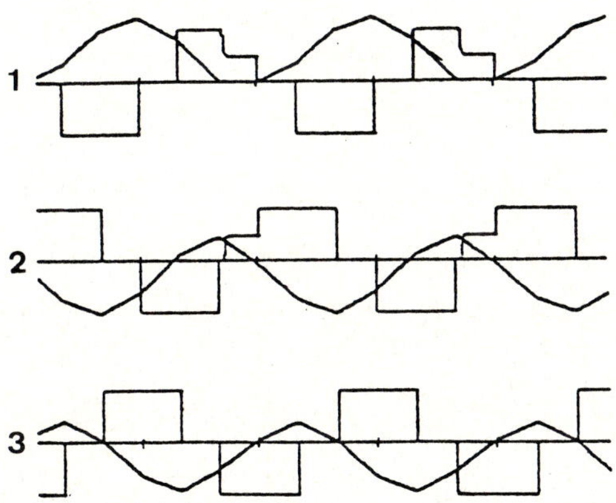

Fig. 7 : Line currents and voltages during failure conditions.

The results have been obtained with α such that $\alpha I \simeq 100$, which gives for a three-phase converter a fairly good precision (of the order of 1%).

There is now an effective intervention of the exponential functions and the numerical computation step has to be reduced by 2.

The form of the voltages u_k depends on the load time constant, as can be seen in figure 8, obtained for a smaller time constant.

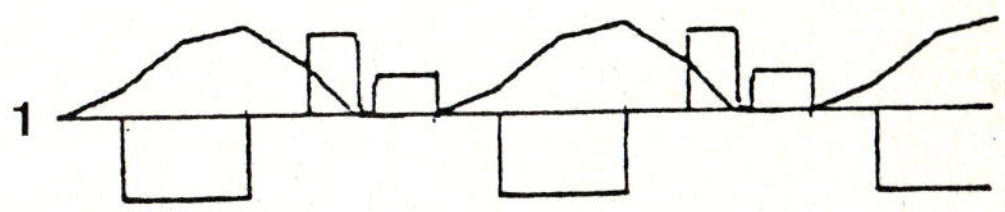

Fig. 8 : Line voltage and current during fault conditions.

Other situations can be simulated and, by changing the laws of A_j, B_j, it is possible to represent different kinds of firing controls : 120° waves, PWM and so on.

The model is, of course, suitable for any number of phases.

4. Conclusion

A general asymptotic model of a voltage-fed inverter was presented, which leads to very simple numerical simulations.

In this model, the AC line voltages of the inverter are directly given by a simple non-linear function of the line currents containing the information about the inverter firing control.

This makes any simulation practically even simple as for the classical case of a three-phase, 180° wave, inverter, where the voltages are a priori known.

References

|1| Garrido M.S., Nouvelle modélisation d'un redresseur idéal polyphasé, to appear in Mesnard G. (ed.), Modelling and Simulation, vol. 2 (AMSE Press, Tassin-la-Demi-Lune, France, 1983); see also the Appendix in the present paper.

Appendix

The asymptotic model of a polyphase diode rectifier is based on a certain representation of the diode voltages in function of the source currents and the output rectified voltage.

Let us consider one phase of the source

and the associated diodes.

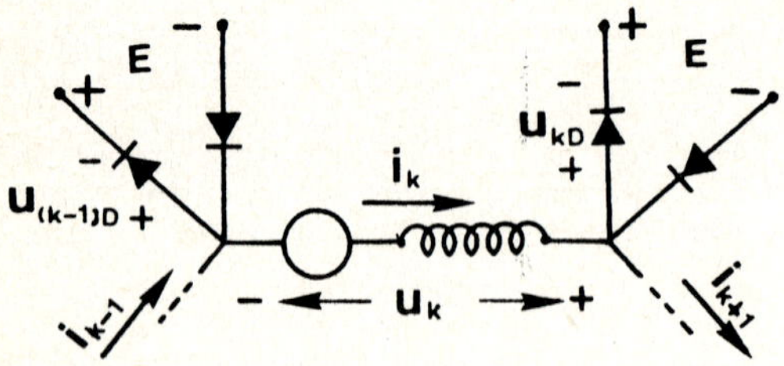

The total voltage u_k in the phase is

$$u_k = u_{kD} - u_{(k-1)D}$$

where u_{kD}, $u_{(k-1)D}$ are the direct voltages in the up diodes.

For perfect diodes, these voltages can be expressed, asymptotically, in function of the phase currents i_j and the output rectified voltage E.

Effectively, for a perfect diode model we can put that voltage u_{kD} is such that

If $i_k - i_{k+1} > 0$ then $u_{kD} = 0$

If $i_k - i_{k+1} < 0$ then $u_{kD} = -E$

If $i_k - i_{k+1} = 0$ then $0 \geqslant u_{kD} \geqslant -E$

This gives the Heaviside function represented in next figure.

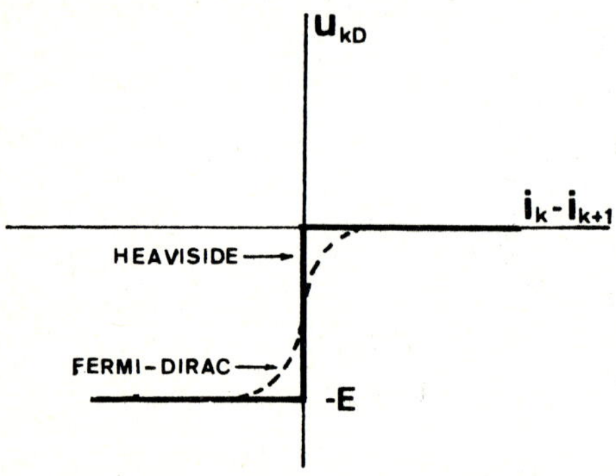

We can, now, replace the Heaviside function by an asymptotic Fermi-Dirac function and put

$$u_{kD} = \lim_{\alpha\to\infty} \frac{-E}{1 + e^{\alpha(i_k - i_{k+1})}}$$

So we have

$$u_k = \lim_{\alpha\to\infty} \left(\frac{1}{1 + e^{\alpha(i_k - i_{k+1})}} - \frac{1}{1 + e^{(i_k - i_{k+1})}} \right) E$$

which is the basic expression used in this paper.

The representation of a short-circuited diode is easily obtained with the same kind of arguments.

The output rectified current, represented by -I in this paper, is obtained by summing all the currents in the up diodes and we have immediatly :

$$-I = \sum_k \lim_{\alpha\to\infty} \frac{-1}{1 + e^{\alpha(i_k - i_{k+1})}} (i_k - i_{k+1})$$

That can also be written,

$$-I = \sum_k \lim_{\alpha\to\infty} \left(\frac{1}{1 + e^{\alpha(i_{k-1} - i_k)}} - \frac{1}{1 + e^{\alpha(i_k - i_{k+1})}} \right) . i_k$$

which was the expression adopted in this paper.

Session 3

MODELLING OF INDUCTION MACHINES AND CONVERTERS (II)

Chairman: G. MAGGETTO

ELECTRICAL MACHINES AND CONVERTERS - Modelling and Simulation
H. Buyse and J. Robert (editors)
Elsevier Science Publishers B.V. (North-Holland)/ © IMACS, 1984

MODELLING IN VIEW OF DIGITAL SIMULATION OF THE THREE PHASED MACHINES ASSOCIATED TO STATIC CONVERTERS

R. LE DOEUFF and E.J. GUDEFIN

Groupe de Recherches en Electrotechnique et Electronique de Nancy (ERA 995)

E.N.S.E.M. 2, rue de la Citadelle 54011-NANCY CEDEX FRANCE

In this paper we present a modelling of the three-phased machines associated to static converters. This modelling has been developped in view of the digital simulation and in order to reduce the computation time. This modelling is based on two différent concepts : a choice of the stator variables depending on the mode of conduction of the machine on one hand and, on the other hand a treatment of the equations which is specific of the structure of the machine (induction or synchronous).

NOMENCLATURE

a) General notation :

$$[T_{32}] = \sqrt{\frac{2}{3}}\begin{bmatrix} 1 & 0 \\ -\frac{1}{2} & \frac{\sqrt{3}}{2} \\ -\frac{1}{2} & -\frac{\sqrt{3}}{2} \end{bmatrix} ; P(\theta) = \begin{bmatrix} \cos\theta & -\sin\theta \\ \sin\theta & \cos\theta \end{bmatrix}$$

p = number of poles pairs.

θ' = angle between the phase a of the stator and the rotor.

θ = angle between the phase 1 of the stator and the rotor.

$\dot{\theta} = \frac{d\theta}{dt}$

b) Induction machine :

L_s, L_r : self inductances of the stator and the rotor phases.

M_s, M_r : mutual inductances between two stator phases and between two rotor phases respectively.

M_o : Maximum value of the mutual inductance between one stator and one rotor phases.

R_1, R_2 : Resistances of one stator and of one rotor phases.

$$L_1 = L_s - M_s \; ; \; L_2 = L_r - M_r \; ; \; M = \frac{3}{2} M_o ; \; l_h = L_s + 2M_s \; ;$$

$$\sigma = 1 - \frac{M^2}{L_1 L_2} \; ; M_1 = M\sqrt{\frac{2}{3}} ; \; M_2 = \frac{M}{\sqrt{3}} ; L_3 = L_s + M_s ; M_3 = \sqrt{2}\, M_s$$

$$M_4 = M\sqrt{\frac{3}{2}} ; \quad \sigma_1 = 1 - \frac{M_2^2}{L_2 L_3} \; ; \; \sigma_3 = 1 - \frac{M_1^2}{L_2 L_s}$$

c) Synchronous machine :

l_h : stator leakage inductance

L_d : stator inductance in the d-axis

L_q : stator inductance in the q-axis

L_f : field inductance

L_d : damper inductance in the d-axis

L_Q : damper inductance in the q-axis

M_{fD} : mutual inductance between the field winding and damper in the d-axis

M_f : mutual inductance between the field winding and d axis equivalent winding

M_d : mutual inductance between the D damper and d axis equivalent winding.

M_q : mutual inductance between the Q damper and q axis equivalent winding

R_1 : résistance of one stator phase

R_f : resistance of field winding

R_D, R_Q : resistance of the damping windings

$$\sigma_{fD} = 1 - \frac{M_{fD}^2}{L_f L_D} ; a = \frac{1}{L_f \sigma_{fD}} ; b = -\frac{M_{fD}}{L_f L_D \sigma_{fD}} ; c = \frac{1}{L_D \sigma_{fD}} ; d = \frac{1}{L_Q}$$

$$k_f = aM_f + b\,M_d \; ; \; k_d = b\,M_f + c\,M_d \; ; \; k_q = dM_q$$

$$L''_d = L_d - k_f M_f - k_d M_d \; ; \; L''_q = L_q - k_q M_q$$

$$l_1 = \frac{L_d + L_q}{2} ; \; l_2 = \frac{L_d - L_q}{2} ; l'_1 = \frac{L''_d + L''_q}{2} \; ; \; l'_2 = \frac{L''_d - L''_q}{2}$$

$$a' = a + \frac{k_f}{L''_d} \; ; \; b' = b + \frac{k_f k_d}{L''_d} ; \; c' = c + \frac{k_d}{L''_d} \; ; \; d' = d + \frac{k_q^2}{L''_q}$$

1. INTRODUCTION

The analytic studies on electrical machines associated to static converters are very difficult and complicated ; thus the digital simulation is a very powerful tool to study these systems. However a simulation programme has to reach two objectives : rapidity and precision. That is not aloways easy because of some specific problems bound to the numerical solving of the differencial equations of the machines which may be written under the general matricial form :

$$[v] = [\mathcal{L}]\frac{d}{dt}[i] + \{[R] + \dot{\theta}' \frac{d}{d\theta}[\mathcal{L}]\}[i] \quad (1)$$

To solve (1) in a numerical way we have to put it under the following form :

$$\frac{d}{dt}[i] = [\mathcal{L}]^{-1}\{[v] - ([R] + \dot{\theta}' \frac{d}{d\theta}[\mathcal{L}])[i]\} \quad (2)$$

When the currents and voltages under consideration are the actual ones in the natural axis, the matrix $[\mathcal{L}]$ depends on θ' and $[\mathcal{L}]^{-1}$ has to be computed in a repetitive way (three times each step when using a fourth order Runge-Kutta method). That inversion has two main disadvantages : first it is largely time-consuming and secondly it may affect the precision of the computation, $[\mathcal{L}]$ being sometimes "ill-conditioned".

A few changes of variables (e.g. the Park transformation) allow to avoid the problem by giving inductance matrices independent of θ'. Unfortunately these transformations are not always easy to use when the machines are associated to static converters. In this paper we present the components which, in our opinion, are the most convenient according to the different conduction modes of the three phased machines

2. THREE PHASED MACHINE HAVING A THREE-PHASED SUPPLY.

When a three-phased machine has a three-phased supply there is no problem and a straight forward use of Park's transformation leads to constant induction matrices for both induction and synchronous machines. We can notice that we have to choose a referencial frame which is linked to the rotor when considering a synchronous machine, but that for the induction machine any referencial frame is convenient.

3. GENERAL CONSIDERATIONS ABOUT THE THREE-PHASED MACHINES HAVING OPEN-CIRCUITED PHASES

When a machine has open-circuited phases, the use of Park's transformation is not very easy. Let us consider that one phase, e.g. the phase a, is open-circuited. In that case the voltage v_a is not determined by the source and it depends only on the fluxes created by the currents in the other phases of the machine. If we use a step by step method for the computation, we do not know v_a at the beginning and in the middle of the step. Thus it is impossible to compute Park's voltages v_d and v_q in terms of v_a, v_b and v_c. Of course it is possible to express the conditions like $i_a = 0$ in terms of Parks variables, but that method leads to very complicated equations and involves a lot of computation (/2/). That's why we prefer to introduce specific machine modellings which allow to avoid numerical matrix inversion and take into account the specific problems of the open circuited phases.

First we can obtain a unified formulation of the problem by denoting "1" the "unsymmetrical phase" (i.e. the only one supplied on the only one open-circuited).

Thus the (1, 2, 3)three-phased system is deduced from the "real"(a, b, c)system by setting :

$$(g_1, g_2, g_3)^t = F^k . (g_a, g_b, g_c)^t \quad (3)$$

with $F = \begin{bmatrix} 0 & 1 & 0 \\ 0 & 0 & 1 \\ 1 & 0 & 0 \end{bmatrix}$, k depending on the name of the unsymmetrical phase (k = 0, 1 or -1), And :

$$\theta = \theta' + \xi \quad (4)$$

with $\xi = 0, \frac{2\pi}{3}$ or $-\frac{2\pi}{3}$

We can then consider that seven modes of conduction may exist for a three phased machine star or delta connected. They are presented on table 1 with thick lines to represent the phases in which the current exists and large points to represent the nodes having their voltages determined by the source.

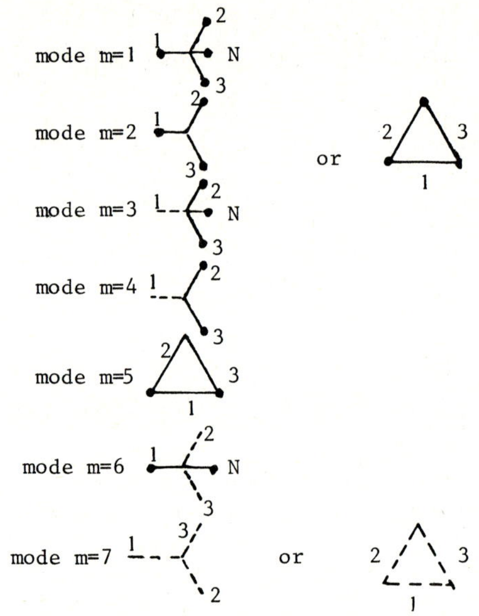

Table 1 definition of the various modes of conduction of a three phased machine.

The modes m = 1, m = 2 and m = 7 being symmetrical we shall only consider in the following the modes m = 3 to 6.

4. MODELLING OF THE INDUCTION MACHINE :

4.1 Generalities

We consider an induction machine having its three rotor phases short-circuited and its stator phases connected to a static converter. The rotor phases are denoted a_r, b_r, c_r ; the stator phases being, 1, 2 and 3 as we said before. Thus the machine equations are :

$$[V] = \{[R] + \dot{\theta} \frac{d}{d\theta}[\mathcal{L}]\}[i] + [\mathcal{L}] \frac{d}{dt}[i] \quad (5)$$

with $[v] = [v_1, v_2, v_3, 0, 0, 0]^t$

$[i] \quad [i_1, i_2, i_3, i_{ar}, i_{br}, i_{cr}]^t$

$$[\mathcal{L}] = \begin{bmatrix} [\mathcal{L}_s] & [\mathcal{L}_{sr}] \\ [\mathcal{L}_{rs}] & [\mathcal{L}_r] \end{bmatrix}$$

$$[\mathcal{L}_s] = \begin{bmatrix} L_s & M_s & M_s \\ M_s & L_s & M_s \\ M_s & M_s & L_s \end{bmatrix} \qquad [\mathcal{L}_r] = \begin{bmatrix} L_r & M_r & M_r \\ M_r & L_r & M_r \\ M_r & M_r & L_r \end{bmatrix}$$

$$[\mathcal{L}_{sr}] = M_o \begin{bmatrix} \cos\theta & \cos(\theta - \frac{2\pi}{3}) & \cos(\theta + \frac{2\pi}{3}) \\ \cos(\theta + \frac{2\pi}{3}) & \cos\theta & \cos(-\frac{2\pi}{3}) \\ \cos(\theta - \frac{2\pi}{3}) & \cos(\theta + \frac{2\pi}{3}) & \cos\theta \end{bmatrix}$$

4.2 Rotor variables in a statoric referencial frome :

In order to obtain a constant inductance matrix we can express the rotor variables in a statoric referencial frame (fig. 1).

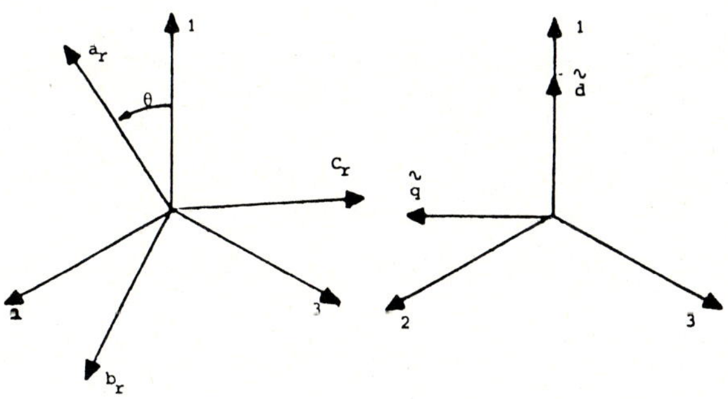

Figure 1

with :

$$(g_{ar}, g_{br}, g_{cr})^t = [T_{32}][P(-\theta)]\,(\tilde{g}_d, \tilde{g}_q)^t \qquad (6)$$

The inductance matrix becomes :

$$[\mathcal{L}'] = \begin{bmatrix} [\mathcal{L}_s] & M\,[T_{32}] \\ M\,[T_{32}]^t & [\mathcal{L}'_r] \end{bmatrix}$$

with $[\mathcal{L}'_r] = \begin{bmatrix} L_2 & 0 \\ 0 & L_2 \end{bmatrix}$

$\mathcal{L}'$ being a constant matrix we don't have to invert it in a repetitive way. However some changes in the statoric variables allow a greater simplification of this matrix. These transformations depend on the conduction mode of the machine.

4.3 Mode m = 3 :

In this mode the phase 1 is not supplied. We can use the $(1, \gamma, \beta)$ components defined by (/3/, /4/) :

$$(g_1, g_\gamma, g_\beta)^t = [D_3]\,(g_1, g_2, g_3)^t \qquad (7)$$

$$\text{with } [D_3] = \frac{1}{\sqrt{2}} \begin{bmatrix} \sqrt{2} & 0 & 0 \\ 0 & 1 & 1 \\ 0 & 1 & -1 \end{bmatrix} \qquad (8)$$

This change of variables leads to the equations:

$$\begin{bmatrix} v_\gamma \\ v_\beta \\ 0 \\ 0 \end{bmatrix} = \begin{bmatrix} L_3 & 0 & -M_2 & 0 \\ 0 & L_1 & 0 & M \\ -M_2 & 0 & L_2 & 0 \\ 0 & M & 0 & L_2 \end{bmatrix} \frac{d}{dt} \begin{bmatrix} i_\gamma \\ i_\beta \\ \tilde{i}_d \\ \tilde{i}_q \end{bmatrix} + \begin{bmatrix} R_1 & 0 & 0 & 0 \\ 0 & R_1 & 0 & 0 \\ 0 & M\dot{\theta} & R_2 & L_2\dot{\theta} \\ M_2\dot{\theta} & 0 & -L_2\dot{\theta} & R_2 \end{bmatrix} \begin{bmatrix} i_\gamma \\ i_\beta \\ \tilde{i}_d \\ \tilde{i}_q \end{bmatrix} \qquad (9)$$

So the inversion of the inductance matrix is very easy and we get :

$$\frac{d}{dt} \begin{bmatrix} i_\gamma \\ i_\beta \\ \tilde{i}_d \\ \tilde{i}_q \end{bmatrix} = \begin{bmatrix} \frac{1}{\sigma_1 L_3} & 0 & \frac{M_2}{\sigma_1 L_2 L_3} & 0 \\ 0 & \frac{1}{\sigma L_1} & 0 & \frac{-M}{\sigma L_1 L_2} \\ \frac{M_2}{\sigma_1 L_2 L_3} & 0 & \frac{1}{\sigma_1 L_2} & 0 \\ 0 & \frac{-M}{\sigma L_1 L_2} & 0 & \frac{1}{\sigma L_2} \end{bmatrix} \times \left[\begin{bmatrix} v_\gamma \\ v_\beta \\ 0 \\ 0 \end{bmatrix} - \begin{bmatrix} R_1 & 0 & 0 & 0 \\ 0 & R_1 & 0 & 0 \\ 0 & M\dot{\theta} & R_2 & L_2\dot{\theta} \\ M_2\dot{\theta} & 0 & -L_2\dot{\theta} & R_2 \end{bmatrix} \begin{bmatrix} i_\gamma \\ i_\beta \\ \tilde{i}_d \\ \tilde{i}_q \end{bmatrix} \right] \qquad (10)$$

The voltage of the non supplied phase being computed by :

$$v_1 = \frac{M_3 L_2 + M_1 M_2}{\sigma_1 L_2 L_3}(v_\gamma - R_1 i_\gamma) - \frac{M_2 M_3 + L_1 L_3}{\sigma_1 L_2 L_3}[R_2 \tilde{i}_d + \dot{\theta}(M i_\beta + L_2 \tilde{i}_q)] \qquad (11)$$

the torque being :

$$\Gamma = p\,(M\,\tilde{i}_d\, i_\beta + M_2\,\tilde{i}_q\, i_\gamma) \qquad (12)$$

4.4 Mode m = 4

In this mode the phase 1 is also in open circuit but there is no neutral connection. So we have $i_1 = o$ and $i\gamma = o$ and the equation (10) becomes :

$$\frac{d}{dt}\begin{bmatrix} i_\beta \\ \tilde{i}_d \\ \tilde{i}_q \end{bmatrix} = \begin{bmatrix} \frac{1}{\sigma L_1} & 0 & \frac{-M}{\sigma L_1 L_2} \\ 0 & \frac{1}{L_2} & 0 \\ \frac{-M}{\sigma L_1 L_2} & 0 & \frac{1}{\sigma L_2} \end{bmatrix} \times \left[\begin{bmatrix} v_\beta \\ 0 \\ 0 \end{bmatrix} - \begin{bmatrix} R_1 & 0 & 0 \\ M\dot{\theta} & R_2 & L_2\dot{\theta} \\ 0 & -L_2\dot{\theta} & R_2 \end{bmatrix} \begin{bmatrix} i_\beta \\ \tilde{i}_d \\ \tilde{i}_d \end{bmatrix} \right] \qquad (13)$$

the voltages of the non supplied phase and of the neutral may be computed from :

$$\begin{bmatrix} v_1 \\ v_\gamma \end{bmatrix} = \left[\begin{bmatrix} \frac{M\dot{\theta}}{L_2}, \frac{R_2}{\sigma_1 L_2}, \frac{L_2\dot{\theta}}{\sigma_1 L_2} \end{bmatrix} \begin{bmatrix} i_\beta \\ \tilde{i}_d \\ \tilde{i}_q \end{bmatrix} \right] \begin{bmatrix} -M_1 \\ M_2 \end{bmatrix} \qquad (14)$$

and the torque is

$$\Gamma = p\, M\, \tilde{i}_d\, i_\beta \qquad (15)$$

4.5 Mode m = 6

In that case phase 1 is the only statoric supplied phase. We can use the $1, \gamma, \beta$ components to get

$$\begin{bmatrix} v_1 \\ 0 \\ 0 \end{bmatrix} = \begin{bmatrix} L_s & M_1 & 0 \\ M_1 & L_2 & 0 \\ 0 & 0 & L_2 \end{bmatrix} \frac{d}{dt}\begin{bmatrix} i_1 \\ \tilde{i}_d \\ \tilde{i}_q \end{bmatrix} + \begin{bmatrix} R_1 & 0 & 0 \\ 0 & R_2 & L_2\dot{\theta} \\ -M_1\dot{\theta} & -L_2\dot{\theta} & R_2 \end{bmatrix} \begin{bmatrix} i_1 \\ \tilde{i}_d \\ \tilde{i}_q \end{bmatrix} \qquad (16)$$

The inversion of the inductance matrix leads to:

$$\frac{d}{dt}\begin{bmatrix} i_1 \\ \tilde{i}_d \\ \tilde{i}_q \end{bmatrix} = \begin{bmatrix} \frac{1}{\sigma_3 L_s} & -\frac{M_1}{\sigma_3 L_2 L_s} & 0 \\ -\frac{M_1}{\sigma_3 L_2 L_s} & \frac{1}{\sigma_3 L_2} & 0 \\ 0 & 0 & \frac{1}{L_2} \end{bmatrix} \left[\begin{bmatrix} v_1 \\ 0 \\ 0 \end{bmatrix} - \begin{bmatrix} R_1 & 0 & 0 \\ 0 & R_2 & L_2\dot{\theta} \\ -M_1\dot{\theta} & -L_2\dot{\theta} & R_2 \end{bmatrix} \begin{bmatrix} i_1 \\ \tilde{i}_d \\ \tilde{i}_q \end{bmatrix} \right] \qquad (17)$$

the voltages of the non supplied phases being deduced from

$$\begin{bmatrix} v_\gamma \\ v_\beta \end{bmatrix} = \begin{bmatrix} \frac{M_3 L_2 + M_2 M_1}{\sigma_3 L_2 L_s} & \frac{-M_1 M_3 - M_2 L_s}{\sigma_3 L_2 L_s} & 0 \\ 0 & 0 & \frac{M}{L_2} \end{bmatrix} \times \left[\begin{bmatrix} v_1 \\ 0 \\ 0 \end{bmatrix} - \begin{bmatrix} R_1 & 0 & 0 \\ 0 & R_2 & L_2\dot{\theta} \\ -M_1\dot{\theta} & -L_2\dot{\theta} & R_2 \end{bmatrix} \begin{bmatrix} i_1 \\ \tilde{i}_d \\ \tilde{i}_q \end{bmatrix} \right] \qquad (18)$$

and the torque is

$$\Gamma = -\,p\, M_1\, i_1\, \tilde{i}_q \qquad (19)$$

4.6 Mode m = 5 :

In that mode the stator phases are delta connected with a single-phased supply (fig. 2)

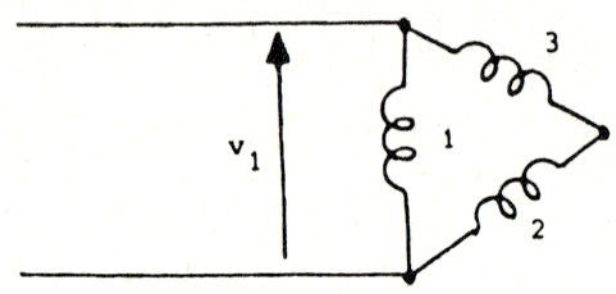

Figure 2

In that case we have no zero-sequence voltage and no zero sequence current so :

$$i_2 = i_3 \qquad (20)$$

Thus we can use either the $(1, \gamma, \beta)$ components or the Concordia components defined by :

$$(g_\alpha, g_\beta)^t = T_{32}{}^t\,(g_1, g_2, g_3)^t \qquad (21)$$

when using these components we get :

$i_\beta = 0$ and :

$$\begin{bmatrix} v_\alpha \\ 0 \\ 0 \end{bmatrix} = \begin{bmatrix} R_1 & 0 & 0 \\ 0 & R_2 & L_2\dot{\theta} \\ -M\dot{\theta} & -L_2\dot{\theta} & R_2 \end{bmatrix} \begin{bmatrix} i_\alpha \\ \tilde{i}_d \\ \tilde{i}_q \end{bmatrix} + \begin{bmatrix} L_1 & M & 0 \\ M & L_2 & 0 \\ 0 & 0 & L_2 \end{bmatrix} \frac{d}{dt}\begin{bmatrix} i_\alpha \\ \tilde{i}_d \\ \tilde{i}_q \end{bmatrix} \qquad (22)$$

after inversion of the inductance matrix we get

$$\frac{d}{dt}\begin{bmatrix} i_\alpha \\ \tilde{i}_d \\ \tilde{i}_q \end{bmatrix} = \begin{bmatrix} \frac{1}{L_1\sigma} & -\frac{M}{L_1 L_2 \sigma} & 0 \\ -\frac{M}{L_1 L_2 \sigma} & \frac{1}{L_2\sigma} & 0 \\ 0 & 0 & \frac{1}{L_2} \end{bmatrix} \left[\begin{bmatrix} v_\alpha \\ 0 \\ 0 \end{bmatrix} - \begin{bmatrix} R_1 & 0 & 0 \\ 0 & R_2 & L_2\dot{\theta} \\ -M\dot{\theta} & -L_2\dot{\theta} & R_2 \end{bmatrix} \begin{bmatrix} i_\alpha \\ \tilde{i}_d \\ \tilde{i}_q \end{bmatrix} \right]$$

and $v_\beta = \frac{1}{L_2}(M\dot{\theta} i_\alpha + L_2 \dot{\theta}\tilde{i}_d - R_2 \tilde{i}_q)$ (24)

If we use the 1, γ, β components we get :

$$\frac{d}{dt}\begin{bmatrix}\tilde{i}_1\\ \tilde{i}_d\\ \tilde{i}_q\end{bmatrix} = \begin{bmatrix}\frac{1}{\sigma L_1} & -\frac{M}{\sigma L_1 L_2} & 0\\ -\frac{M}{\sigma L_1 L_4} & \frac{1}{L_2} & 0\\ 0 & 0 & \frac{1}{L_2}\end{bmatrix}\left[\begin{bmatrix}v_1\\ 0\\ 0\end{bmatrix} - \begin{bmatrix}R_1 & 0 & 0\\ 0 & R_2 & L_2\dot{\theta}\\ -M_4\dot{\theta} & -L_2\dot{\theta} & R_2\end{bmatrix}\begin{bmatrix}i_1\\ \tilde{i}_d\\ \tilde{i}_q\end{bmatrix}\right] \tag{25}$$

The torque being :

$\Gamma = -p\, M\, i_d\, \tilde{i}_q$ (26)

or $\Gamma = -\, p\, M_1\, i_1\, \tilde{i}_q$ (27)

5. SYNCHRONOUS MACHINE MODELLING :

5.1 General considerations :

In the following we shall consider a synchronous machine with three stator phases, a salient rotor with a field winding and two dampers (fig.3)

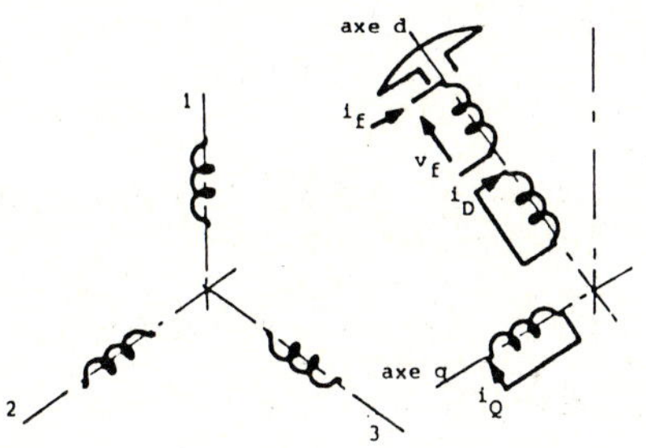

Figure 3

The parameters of such a machine are defined preferably in the Park's d.q axis (fig. 4)

Because of the rotor dissymetries and of the existence of the field winding it is not possible to express the rotor variables in stator axis in order to obtain a constant inductance matrix. Thus we have to keep the rotor variables in the rotor axis and develop an analytical inversion of the inductance matrix using a partition of the stator variables (/3/,/4/).

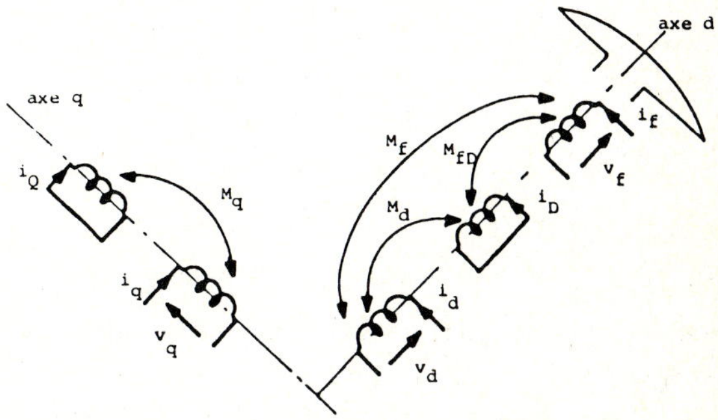

Figure 4

5.2 Partition of the equations - choice of the stator variables.

The stator of the synchronous machine having the same structure as the stator of the induction machine an identical choice of stator variables can be done according to the modes of conduction. Thus we transform the (1, 2, 3) stator variables into the (h, α, β) or the (1, γ, β) variables. When considering the modes m = 3 to 6, there are some real or fictitious open-circuited phases, and the stator voltage vector $[v_s]$ may be divided into two vectors :

- on one hand the vector $[v_{sm}]$ of the supplied "phases" voltages.
- on the other hand the vector $[v'_{sm}]$ of the non supplied "phases".

This partition may be represented by the following table :

Connection mode : m	Supplied phases	vector voltage of supplied phases	vector voltage of unsupplied phases
3		$[V_{s3}] = [V_\gamma\ V_\beta]^t$	$[V'_{s3}] = [V_1]$
4		$[V_{s4}] = [V_\beta]$	$[V'_{s4}] = [V_1 V_\gamma]^t$ or $[V_h = 0,\ V_\alpha]^t$
5		$[V_{s5}] = [V_h=0, V_\alpha]^t$ or $[V_1, V\gamma]^t$	$[V'^{s5}_{s5}] = [V_\beta]$
6		$[V_{s6}] = [V_1]$	$[V'_{s6}] = [V_\gamma\ V_\beta]^t$

The supplied phases are represented by thick lines and the un supplied ones by dotted lines. It will be noted that in mode 5 the currents exist in each phase.

Table 2

After the partition describbd above, the machine equations may be written as follows :

$$\begin{bmatrix} V_{sm} \\ V_r \end{bmatrix} = \begin{bmatrix} A_m & B_m \\ B_m^t & \mathcal{L}_r \end{bmatrix} \frac{d}{dt} \begin{bmatrix} i_{sm} \\ i_r \end{bmatrix} + \left\{ \begin{bmatrix} R_{sm} & [0] \\ [0] & R_r \end{bmatrix} + \dot{\theta} \frac{d}{dt} \begin{bmatrix} A_m & B_m \\ B_m^t & [0] \end{bmatrix} \right\} \begin{bmatrix} i_{sm} \\ i_r \end{bmatrix}$$

and

$$[V'_{sm}] = [G_m \quad H_m] \frac{d}{dt} \begin{bmatrix} i_{sm} \\ i_r \end{bmatrix} + \dot{\theta} \left[\frac{dG_m}{d\theta} \quad \frac{dH_m}{d\theta} \right] \begin{bmatrix} i_{sm} \\ i_r \end{bmatrix}$$

(28) (29)

In these equations the matrices A_m, B_m, R_{sm}, G_m and H_m depend on the conduction mode. They are given in detail in the appendix for the mode m = 3 which is the most complicated one. For the other modes they are easily deduced from the definition of the vectors $[V_{sm}]$ and $[V'_{sm}]$ and of the general equations of the machine.

The submatrix $[\mathcal{L}_r]$always exists in the inductance matrix whatever the conduction mode is. Furthermore $[\mathcal{L}_r]$ is a constant matrix, very easy to invert.

$$[\mathcal{L}_r] = \begin{bmatrix} L_f & M_{fD} & 0 \\ M_{fD} & L_D & 0 \\ 0 & 0 & L_Q \end{bmatrix} \quad \text{leads to :}$$

$$[\mathcal{L}_r]^{-1} = \begin{bmatrix} a & b & 0 \\ b & c & 0 \\ 0 & 0 & d \end{bmatrix}$$

with

$$a = \frac{1}{L_f \sigma_{fD}}, b = \frac{M_{fD}}{L_f L_D \sigma_{fD}}, \quad c = \frac{1}{L_D \sigma_{fD}} \text{ and } d = \frac{1}{L_Q}$$

We can use this easy inversion of $[\mathcal{L}_r]$ to make a block inversion of the partitioned inductance matrix and obtain :

$$\frac{d}{dt} \begin{bmatrix} i_{sm} \\ i_r \end{bmatrix} = \begin{bmatrix} A_m'^{-1} & -(A_m'^{-1}) B_m \mathcal{L}_r^{-1} \\ -\mathcal{L}_r^{-1} B_m^t A_m'^{-1} & \mathcal{L}_r^{-1} + \mathcal{L}_r^{-1} B_m^t A_m'^{-1} B_m \mathcal{L}_r^{-1} \end{bmatrix} \times \left\{ \begin{bmatrix} V_{sm} \\ V_r \end{bmatrix} - \left[\begin{bmatrix} R_{sm} & [0] \\ [0] & \mathcal{L}_r \end{bmatrix} + \dot{\theta} \frac{d}{d\theta} \begin{bmatrix} A_m & B_m \\ B_m^t & [0] \end{bmatrix} \right] \begin{bmatrix} i_{sm} \\ i_r \end{bmatrix} \right\} \quad (30)$$

with $A'_m = A_m - B_m \mathcal{L}_r^{-1} B_m^t$.

$[A'_m]$ is, besides $[\mathcal{L}_r]$,the only matrix to invert. As it is at most a 2 x 2 matrix (for m = 3) it is very easy to invert it in on analytical way. Hence there is no difficulty in putting the equations of the synchronous machine under a form very suitable for the numerical computation. In the appendix we detail A'_3 and the various submatrices appearing in the equation (30) for m = 3.

CONCLUSION.

The modelling presented here for the three phased machines connected to static converters may be summarized as follows :

- The choice of the stator components is specific for the modes of conduction (number of supplied phases and modes of connection)

- The treatment of the equations is specific for the nature of the machine (induction or synchronous).

- The numerical inversion of inductance matrices bay be avoided whatever the conduction mode is :

This modelling allows to save a lot of time for the simulation programmes. For instance, for the synchronous machine the computation time is divided by about three (/5/).

REFERENCES :

/1/ B.A. ADKINS "The general theory of electrical machines" Wiley, New York, U.S.A. 1957.

/2/ R.J. KERKMANN, P.C. KRAUSE and T.A. LIPO "Simulation of a synchronous machine with an open phase" Electric machines and electromechanics, 1, 1977, pp 245-254.

/3/ R. LE DOEUFF "Contribution à la modélisation et à la simulation d'ensembles convertisseurs statiques - machines tournantes" Thèse d'Etat, Nancy, France 1981.

/4/ R. LE DOEUFF et E. GUDEFIN "Modelisation en vue de leur simulation numérique des machines asynchrones associées à des convertisseurs statiques" Comptes-rendus de l'Académie des Sciences, note n° B 215, Janvier 1980.

/5/ R. LE DOEUFF, S. MAHMOUD, A. REZZOUG, "Fast computation of synchronous machine unbalanced transients - application to short circuits". Electric Machines and Power Systems to be published in November 1983.

/6/ R. LE DOEUFF, E. GUDEFIN et J.P. LOUIS "Modélisation en vue de leur simulation numérique des machines synchrones associées à des convertisseurs statiques". Comptes-rendus de l'Académie des Sciences, note n° B 253, mars 1980.

/7/ G. GUILLOUARD and R. LE DOEUFF "Digital simulation of a synchronous motor fed by a rectifier inverter system". IASTED Congrès "Modelling and Simulation", Davos Suisse February 1981.

APPENDIX

Matrices appearing in the synchronous machine modelling for the mode of conduction m = 3.
We use the l, γ, β components and obtain :

$$A_3 = \frac{1}{3}\begin{bmatrix} 2l_h + l_1 + l_2\cos 2\theta & -\sqrt{3}\, l_2 \sin 2\theta \\ -\sqrt{3}\, l_2 \sin 2\theta & 3(l_1 - l_2\cos 2\theta) \end{bmatrix}$$

with $l_1 = \dfrac{L_d + L_q}{2}$ and $l_2 = \dfrac{L_d - L_q}{2}$

$$[B_3] = +\begin{bmatrix} -\frac{1}{\sqrt{3}} M_f \cos\theta & -\frac{1}{\sqrt{3}} M_d \cos\theta & \frac{1}{\sqrt{3}} M_q \sin\theta \\ M_f \sin\theta & M_d \sin\theta & M_q \cos\theta \end{bmatrix}$$

$$[G_3] = \left[\frac{\sqrt{2}}{3}(l_h - l_1 - l_2\cos 2\theta) \quad \sqrt{\frac{2}{3}}\, l_2 \sin 2\theta\right]$$

$$[H_3] = \left[\sqrt{\frac{2}{3}} M_f\cos\theta \;;\; \sqrt{\frac{2}{3}} M_d\cos\theta \;;\; -\sqrt{\frac{2}{3}} M_q \sin\theta\right]$$

$$A'_3 = A_3 - B_3 \mathcal{L}_r^{-1} B_3^t = \frac{1}{3}\begin{bmatrix} 2l_h + l'_1 + l'_2\cos 2\theta & -\sqrt{3}\, l'_2 \sin 2\theta \\ -\sqrt{3}\, l'_2 \sin 2\theta & 3(l'_1 - l'_2\cos 2\theta) \end{bmatrix}$$

$$A_3'^{-1} = \frac{1}{3\Delta_3}\begin{bmatrix} 3(l'_1 - l'_2\cos 2\theta) & \sqrt{3}\, l'_2 \sin 2\theta \\ \sqrt{3}\, l'_2 \sin 2\theta & 2l_h + l'_1 + l'_2\cos 2\theta \end{bmatrix}$$

with $3\Delta_3 = l_1'^2 + 2l_h l'_1 - l_2'^2 - 2\, l_h\, l'_2 \cos 2\theta$

$$(A_3'^{-1}) B_3 \mathcal{L}_r^{-1} = \frac{-1}{3\Delta_3}\begin{pmatrix} \sqrt{3}\, k_f(l'_1 - l'_2)\cos\theta & \sqrt{3}\, k_d(l'_1 - l'_2)\cos\theta & -\sqrt{3}\, k_q(l'_1 + l'_2)\sin\theta \\ -k_f(2l_h + l'_1 - l'_2)\sin\theta & -k_d(2l_h + l'_1 - l'_2)\sin\theta & -k_q(2l_h + l'_1 + l'_2)\cos\theta \end{pmatrix}$$

$$\mathcal{L}_r^{-1} B_3^t A_3'^{-1} B_3 \mathcal{L}_r^{-1} = \frac{1}{3\Delta_3}\begin{pmatrix} k_f^2(l'_1 - l'_2 + 2l_h\sin^2\theta) & k_f k_d(l'_1 - l'_2 + 2l_h\sin^2\theta) & 2k_f k_q l_h \sin\theta\cos\theta \\ k_f k_d(l'_1 - l'_2 + 2l_h\sin^2\theta) & k_d^2(l'_1 - l'_2 + 2l_h\sin^2\theta) & 2k_d k_q l_h \sin\theta\cos\theta \\ 2k_f k_q l_h \sin\theta\cos\theta & 2k_d k_q l_h \sin\theta\cos\theta & k_q^2(l'_1 + l'_2 + 2l_h\cos^2\theta) \end{pmatrix}$$

ELECTRICAL MACHINES AND CONVERTERS - Modelling and Simulation
H. Buyse and J. Robert (editors)
Elsevier Science Publishers B.V. (North-Holland)/ © IMACS, 1984

PARAMETERS OF 3-PHASE ASYNCHRONOUS MOTOR FED BY VARIABLE FREQUENCY AND VOLTAGE SUPPLY

BOULET P., NOTELET F., ROMBAUT M.
Laboratoire "Applications des Redresseurs de Puissance"
University of Lille 1 - FRANCE

It is possible to determine the slip of an asynchronous motor without using a tachometric sensor. By using a simple algorithm and by measuring only the statoric quantities (frequency, tension, current in line and phase angle) the slip can be calculated. This paper shows the factors which intervene strongly in the variations of the equivalent circuit parameter's. It indicates under any limits the classical tests, no-load and short circuit tests, permit with an acceptable precision the knowledge of the values of these parameters and their variations. It shows how we can take into account the more important variations in the indirect determination of the slip.

1. INTRODUCTION

With the cooperation of La Télémécanique Electrique, our laboratory has found and used an indirect method for the determination of the slip of a three phase asynchronous motor fed under variable tension and frequency. So, we utilize only the measured data of the frequency, terminals voltage, current in line and phase angle between the tension and the current.
From the memorized measured values of the resistances and reactances at the preliminary recognition and from measured resistance and reactance of the motor during normal operation, a simple algorithm permits the calculation of the slip.
This algorithm starts from the simplified equivalent circuit of the motor. But, the non constancy of certain elements must be taken into account to get the correct results.

1.1. Used algorithm

We start from the one phase equivalent circuit of the three phase asynchronous motor of figure 1, where the iron losses are neglected. Let :

r_1 : stator phase resistance
l_1 : rotor phase cyclic leakage inductance
r'_2 : rotor phase resistance referred to the stator
l'_2 : rotor phase cyclic leakage inductance referred to the stator
$L\mu$: cyclic magnetizing inductance
V_1 : rms value of the supplied voltages per phase
f : their frequency, $\omega = 2\Pi f$ their pulsation
I_1 : rms value of in line current
s : slip.

It has been shown [1][2] that the slip can be calculated from :

$$s = \frac{1}{\omega\tau_2}\;\frac{X_o - X}{R - X_o}$$

where
R and X , appearent resistance and reactance per phase of the motor seen from the input terminals, for a given slip s.
R_o and X_o, same quantities, but for zero slip.
τ_2 , rotor time constant

$$X_o = (l_1 + L\mu)\omega$$
$$\tau_2 = (L\mu + l'_2)/r'_2$$
$$R_o = r_1$$

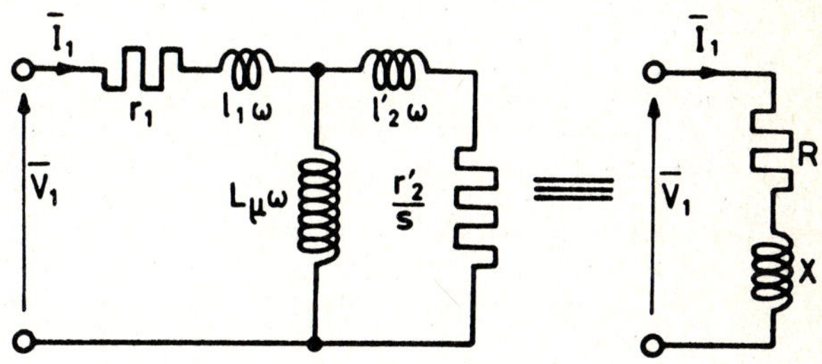

Figure 1

1.2. Direct use of the algorithm

The direct use of the algorithm is not possible unless we assume that l_1, $L\mu$, l'_2, r'_2 and r_1 are constant.
When the rotor is locked, R and X take the values R_c and X_c, and the relation (1) becomes :

$$1 = \frac{1}{\omega\tau_2}\;\frac{X_o - X_c}{R_c - r_1} \qquad (2)$$

divinding (1) by (2), we get :

$$s = a\,\frac{X_o - X}{R - r_1}\ , \text{ where } a = \frac{R_c - r_1}{X_o - X_c} \qquad (3)$$

Furthermore, for evident reasons of easy application, the no-load test is done instead of the one at synchronous speed.

So, we proceed as follows :

- Before putting the equipment on way,
  . We measure, by a direct current, the resistance r_1 ;
  . No loaded the motor, a slow sweeping of the frequency is make up. For each value of ω, we measure and memorize X_o
  . Looked rotor, a new slow sweeping of ω is realised, and now we measure R_c and X_c and, for each ω, we compute and memorize

$$a(\omega) = \frac{R_c(\omega) - r_1}{X_o(\omega) - X_c(\omega)}$$

- at normal operation, when the motor is fed by voltages of pulsation ω,
  . we measure R and X
  . r_1, $X_o(\omega)$ and $a(\omega)$ are called from the memories
  . we compute s from

$$s = a(\omega)\ \frac{X_o(\omega) - X}{R - r_1} \qquad (4)$$

1.3. The posed questions :

As foreseeable, notable differences between the calculated and measured values of the slip are observed. These differences come from the hypotheses of constancy of the parameters and the confusion between the no-load test and the at syncrhonism one.
Whence a certain number of questions posed by the study

- It is evident that r_1 and r'_2 depend on the temperature rise.

The resistance Rμ which represents the iron losses has been neglected. The resistance Rμ and the inductance Lμ depend on the induction in the magnetic circuit, thus on the ratio V_1/ω.
Whence the questions:

1° : How Rμ and Lμ vary as a function of V_1/ω? Are these variations depending only on V_1/ω?

2° : Can the results of no-load test be assimilated to those of the test at synchronism ?

- The secondary resistance r'_2 and leakage reactances depend on the currents, statoric and rotoric frequencies, saturation state of the stator and rotor irons.

Whence the questions :

3° : How the resistance r'_2 and the total leakage inductance $l_1+l'_2$ vary, especially as a function of current and rotoric frequency .

4° : Can the looked rotor test permit to apprehend these variations ?

1.4. Used model. Determination of its elements

To make up this study, the equivalent circuit of figure 2a has been taken, where the total leakage inductances referred to the primary are regrouped in N_1. For r_1 and r'_2/s, it has not been possible to regroup them as in the simplified classical study of the motor fed at the industrial frequency. For very low frequencies, whatever the slip may be, r_1 becomes preponderant in the appearent impedance.

The addition of Rμ permits the representation of the iron losses.
The advantage of this circuit come from the facility to determine the elements from the measured tension, current and power absorbed by the motor.
For easy presentation of calculations and results, the circuit of figure 2b is associated to that of figure 2a. R and l'ω represents the equivalent to the impedances situated after r_1.

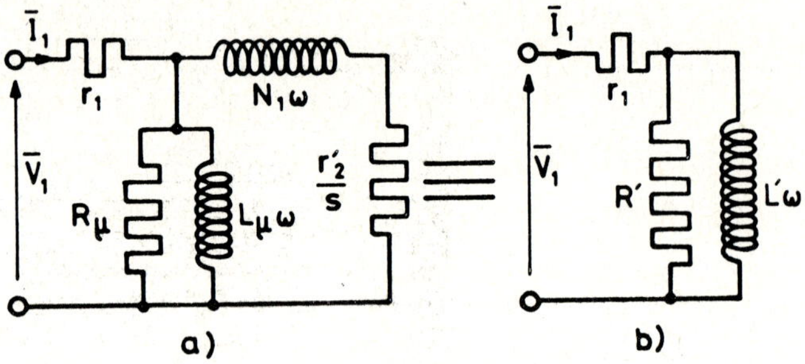

Figure 2

- For each point of operation, we measure the rms value U_1 of the supplied sinusoïdal three phase tensions, their pulsation ω, the rms value I_1, of the currents in line and the active power absorbed by the motor.

We deduce immediatly the appearent resistance and reactance :

$$\left.\begin{aligned} R &= \frac{P_1}{3I_1^2} \\ X &= \frac{1}{3I_1^2}\sqrt{3\,U_1^2\,I_1^2 - P_1^2} \end{aligned}\right\} \qquad (5)$$

Because $R' = \dfrac{(R - r_1)^2 + X^2}{R - r_1}$ and $L'\omega = \dfrac{(R - r_1)^2 + X^2}{X}$,

the resistance and the inductance of the equivalent circuit of figure 2b are given from the measured quantities by :

$$\left.\begin{aligned} R' &= \frac{3r_1\,I_1^2 + U_1^2 - 2r_1\,P_1}{P_1 - 3\,r_1\,I_1^2} \\ L' &= \frac{3r_1\,I_1^2 + U_1^2 - 2r_1\,P_1}{\omega\sqrt{3U_1^2\,I_1^2 - P_1^2}} \end{aligned}\right\} \qquad (6)$$

On the other hand, R' and L' are given as a function of the elements of the circuit of figure 2a, by

$$\left.\begin{aligned} R' &= R\mu\,\frac{R\mu^2\,A + L\mu^2\,.\,\omega^2\,.\,B}{(r_2'^2 + N_1^2\,\omega^2 s^2 + R\mu\,r'_2\,s)\,C} \\ L' &= L\mu\,\frac{R\mu^2\,A + L\mu^2\,\omega^2\,.\,B}{(r_2'^2 + N_1^2\,\omega^2\,s^2 + L\mu\,N_1\omega^2\,s^2)C} \end{aligned}\right\} \qquad (7)$$

where

$$A = (r_2'^2 + N_1^2\,\omega^2\,s^2 + l\mu\,N_1\,\omega^2\,s^2)^2$$

$$B = (r_2'^2 + N_1^2\,\omega^2\,s^2 + R\mu\,r_2'\,s)^2$$

$$C = (R\mu r_2' - L\mu N_1\,\omega^2 s)^2 + (r_2' L\mu\omega + R\,N_1\,\omega s + R\,L\mu\omega s)^2$$

The tested motor has the following characteristics :

G HP, 4 poles, 50 Hz, 380 V, 10 A

$r_1 = 0{,}903\Omega$ at ambiant temperature

2. STUDY OF THE MAGNETIZING IMPEDANCE

It is the synchronous speed test which permits the determination of $R\mu$ and $L\mu$. For $s = 0$, the relations (7) give :

$$R' = R\mu \quad ; \quad L' = L\mu$$

After we see how $R\mu$ and $L\mu$ vary as a function of V_1/ω for various values of frequency, we will see in which conditions the no-load test gives acceptable values for the two parameters.

2.1. Test at the synchronous speed

The rotor of the asynchronous motor is carried at the synchronous speed in the sens of its statoric rotating field, with the help of an auxiliar motor. In these tests, the temperature rise of stator is neglected ; which is verified by the constancy of r_1. The stator has been fed at 10,20,30,40 and 50 Hz successively. Figure 3 gives the variations of $R\mu$ and $L\mu$, releaved as function of V_1/ω, ofr different values of the frequency f.

- **Resistance** $R\mu$ - The resistance $R\mu$ (figure 3a) depends on the induction, so, on V_1/ω and the frequency f. For a given value of V_1/ω , $R\mu$ increases with f and is sensibly proportional to it. For a given value of f, $R\mu$ increases with

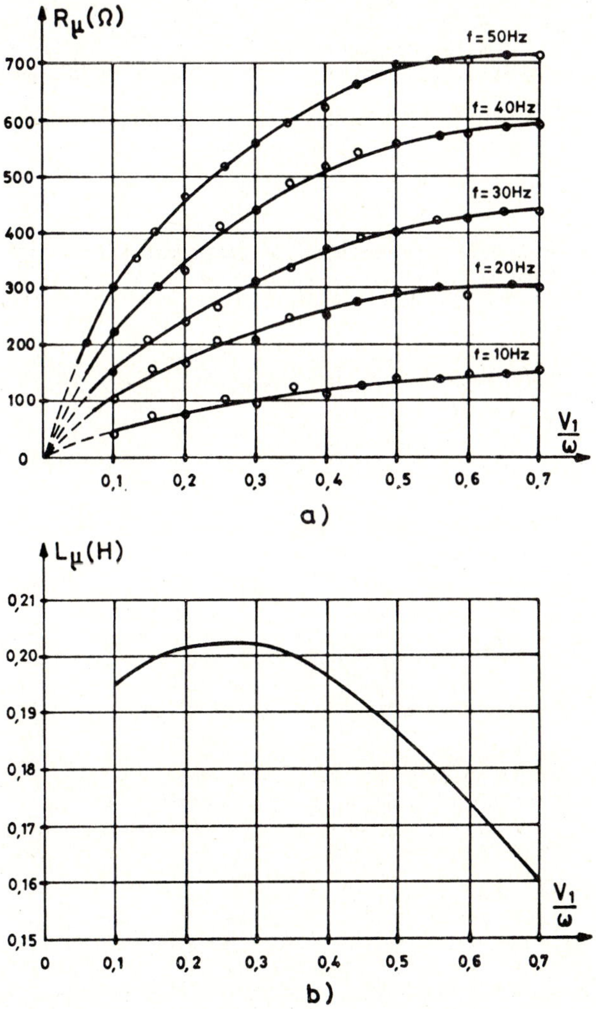

Figure 3

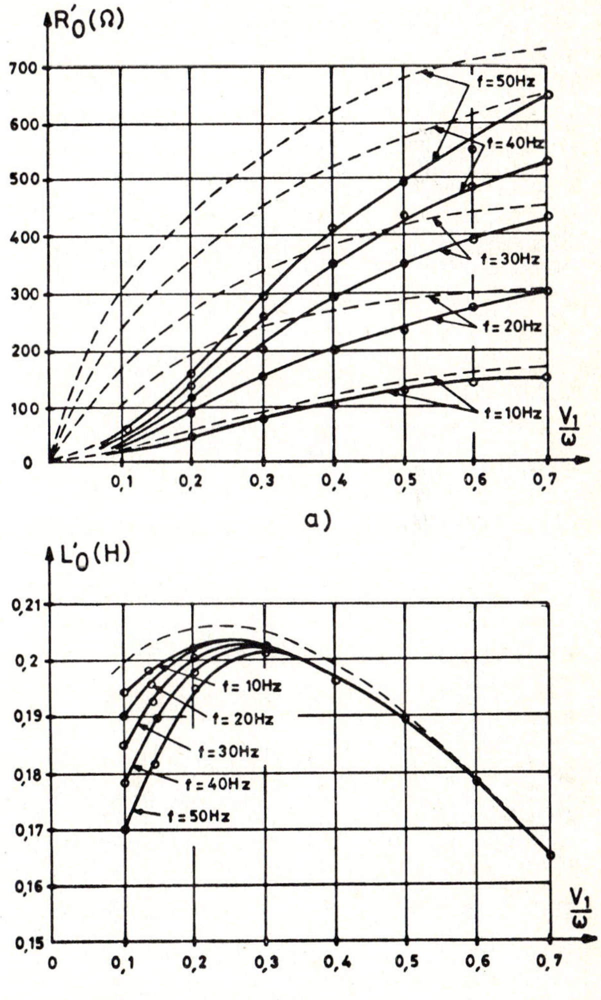

Figure 4

V_1/ω , but this increase, firstly fast, becomes slow for the values of V_1/ω greater than the nominal (here 0,7 wb).
In this test, the precision in the determination of $R\mu$ is weak, because the power absorbed P_1, is very reduced.

– **Inductance** $L\mu$ – The inductance $L\mu$ (figure 3b) increases firstly with V_1/ω, pass through a maximum which corresponds to the maximum of the magnetic circuit's permeability, then decreases slowly because of saturation. For a given value of V_1/ω, the variation of $L\mu$ with the frequency is very small. An expanded scale is requiered to be abble to separate the correspondent curves for various values of frequency, because the deviations are always less than 3%, figure 3b gives the mean curve.

– **shunting of** $L\mu$ **by** $R\mu$ – More the frequency and the flux V_1/ω are low, more the resistance $R\mu$ shunts effectively $L\mu$. So, $R\mu$ is eliminated from the equivalent circuit, only when the flux and the frequency are sufficiently high.

2.2. No-load tests

It is the no-load test which is usually used to determine $R\mu$ and $L\mu$. We will design by R'_o and L'_o, the values of R' and L' deduced from the measurement realised in this test.
Firstly, we note that at no-load, the slip being low but not zero, it will result an unevitable deviation between $R\mu$ and R', $L\mu$ and L'_o. These deviations have been calculated from the relations (7) :

$$R\mu - R'_o = \frac{R\mu^2 \cdot r'_2 \cdot s}{N_1^2 (L + N_1)^2 \omega^2 s^2 + r'^2_2}$$

$$L\mu - L'_o = \frac{L\mu^2 N_1 \omega^2 s}{N_1 (L\mu + N_1) \omega^2 s^2 + r'^2_2}$$

These relations show simply that R'_o and L'_o are respectively lower than $R\mu$ and $L\mu$ and the deviations tend to zero when the slip itself tends to zero. It is not possible to use more these relations because $R\mu$, $L\mu$, r'_2 and N_1 depend also on the frequency and the slip.

The no-load test has been done, feeding the motor by tensions of variable values and of frequency equal to 10,20,30,40 and 50 Hz successively. The figure 4 shows the evolutions of R'_o and L'_o as function of V_1/ω by curves in continous lines (the interrupted curves are those deduced from the test at the synchronous peed).

The resistance R'_o is very less than $R\mu$. But, for high values of V_1/ω, the difference between them becomes more little. That is due to the fact that no-load slip is more high when V_1/ω is low.

Also, **the inductance** L'_o is less than $L\mu$ for the low values of V_1/ω, more the frequency is high more L'_o is less than $L\mu$ but this deviation becomes fastly neglected when V_1/ω increases.
It must be to note that the difference between the appearent inductance X_o/ω calculated directly from the relation $(5)^o$ and L'_o is neglected whatever the frequency.

As conclusion, the no-load test allows to determine $L\mu$ and $R\mu$ with a good approximation from the values of V_1/ω which must be sufficient to neglect the slip. The precision in the determination of the inductance is, therefore, much well than that for the resistance.

3. STUDY OF THE ROTORIC RESISTANCE AND TOTAL LEAKAGE INDUCTANCE

It is classical locked rotor test used, under reduced tension, to determine the total leakage inductance N_1 and secondary resistance r'_2 referred to the primary. But, in this test, the rotoric frequency f_R is the same as one f of the stator, even in the most applications f_R, equal to sf_1, is very lower than f. We propose to determine N_1 and r'_2 under the conditions when f_R and f are different and to compare them with the obtained values from the locked rotor test. We will see also the influence of the current on r'_2 and N_1.

To deduce r'_2 and N_1 from the measured values of U_1, I_1, P_1, f and s, we use the relations (7) which give :

$$\left.\begin{aligned} r'_2 &= \frac{R\mu^2 L\mu^2 L' (R\mu - R') \omega^2 s}{(R\mu - R')^2 L\mu^4 \omega^2 + R\mu^4 (L\mu - L')^2} \\ N_1 &= \frac{R\mu^4 (L\mu - L') \; L\mu L'}{(R\mu - R')^2 L\mu^4 \omega^2 + R\mu^4 (L\mu - L')^2} \end{aligned}\right\} \quad (8)$$

3.1. Realized tests

In same times, we give the frequency f of the statoric currents and that f_R of those of the rotor.
To get f_R equal to 50 Hz for exemple, we fed the motor by tensions of frequency equal to 50, 40,30,20 and 10 Hz successively.
The motor carried by an auxiliar one, is forced to run at a speed such that f_R be always equal to 50 Hz. We proceed by the same manner for f_R equal to 40,30,20 and 10 Hz.
For each serie of tests, the releaved characteristics, for f equal to f_R, correspond to the classical locked rotor test.
Moreover, we adjust the supplied voltage to get always the same current I_1.
Each point must be releaved very quickly to evit the heating and the correspondent variations of the resistances. We verify that the measured resistance r_1 by a direct current has not varied before and after the measurement ; we keep an interval of time between the successive readings.

To use the relations (8), we take the values of $R\mu$ and $L\mu$ given by the tests at synchronism. We take the correspondent values of V_1/ω to the tension between the terminals of the magnetizing impedance of the circuit of figure 2a, Let :

$$\frac{V'}{\omega} = \frac{1}{\omega}\sqrt{\frac{U_1^2}{3} + r_1^2 \cdot I_1^2 - \frac{2}{3} r_1 P_1}$$

3.2. Obtained results

The figure 5 shows the variations of r'_2 and N_1 as function of f_R thus releaved for constant current and frequency f.

- the **resistance** r'_2 increases notably with the rotoric frequency, the variation is sensibly linear. For a given value of f_R, r'_2 decreases when f increases (fig. 5a)

- the **leakage inductance** N_1, decreases slowly when f_R increases (fig. 5b). Also, for a given f_R, we note that N_1 decreases slowly as function of f.

The mixed line curves correspond to f equal to f_r, otherwise, those which correspond to the locked rotor test or for a slip equal to unity.

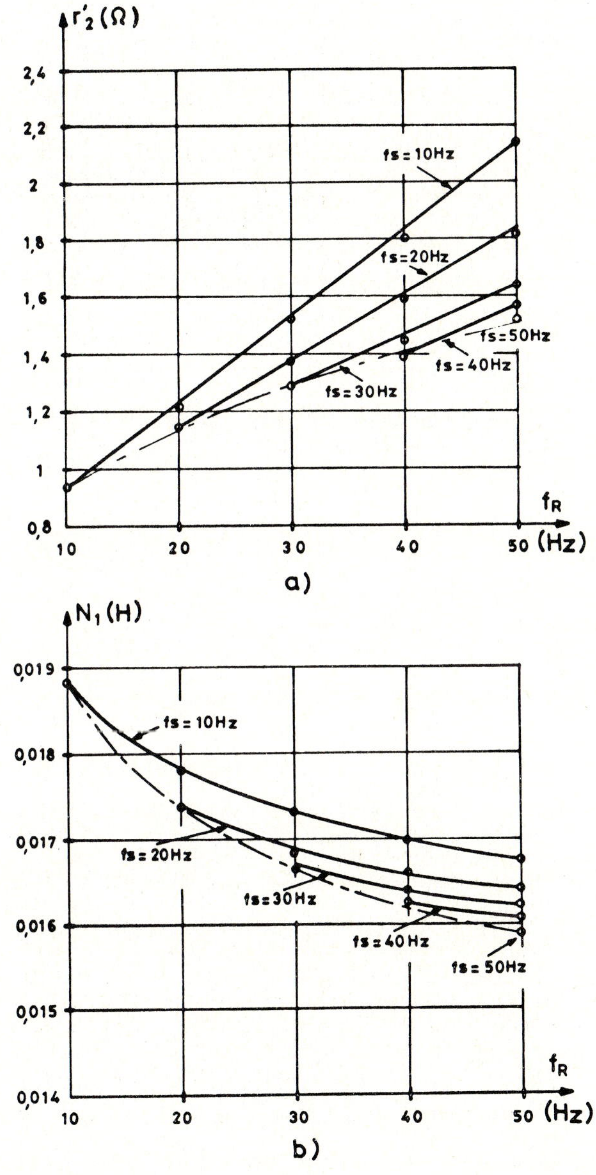

Figure 5

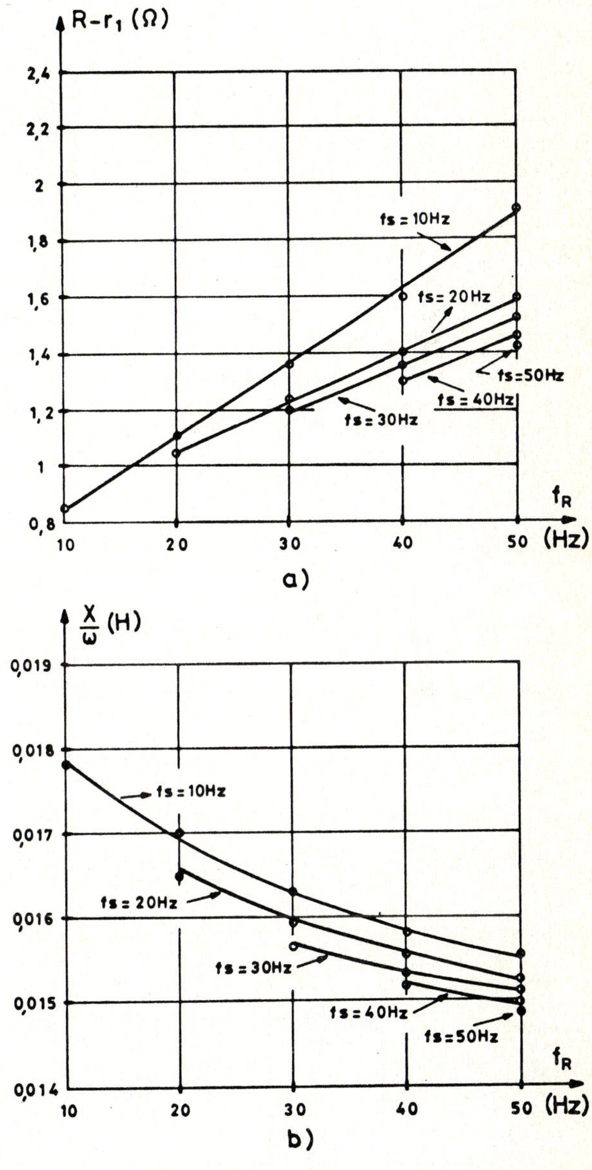

Figure 6

For a given f_R, when s is equal to 1, N_1 and r'_2 are less than those when s is greater than 1. This means that the resistance and reactance for normal operations, for s < 1, are less than, those deduced from the locked rotor test.

The tests and calculations done for different values of primary current, therfore, of secondary current, show a very little difference between the values of r'_2 and N_1.

3.3. Effects of the calculation's simplifications

For nominal values of V_1 and f, for the locked rotor test, we say always that the magnetizing impedance is very much greater than the impedance formed by r'_2 and $N_1\omega$. So, we write :

$$N_1\omega \simeq X \ , \ r'_2 \simeq R - r_1$$

Figure 6 shows the variations of $R - r_1$ and X/ω as function of f_R, for various values of f. The comparison between the curves of figure 5 shows that $R - r_1$ and X/ω are, respectively, less than r'_2 and N_1 ; the influence of the magnetizing impedance is not neglected.

4. EXEMPLE OF APPLICATION

In the prototype equipment built up to verify the possibility of an indirect determination of slip, we have taken into account the variation of r'_2 and N_1 with the rotoric frequency. The first tests done, assuming the constancy of these two parameters, have shown effectively that this hypothese has been asource of important errors.

4.1. Presentation of the equipment and the chain of measurement

The asynchronous motor A.M. is fed (fig. 7) from power supply, by a rectifier bridge and current inverter of six thyristors which furnish to the machine, a pulsated current, where the length of each pulse is equal to 120°, and of variable frequency. The appearents resistances and reactances of the motor are calculated from U_d, I_d and U_1. The inverter bridge is piloted by a clock H which gives consequently the image of the supply signals of the motor.

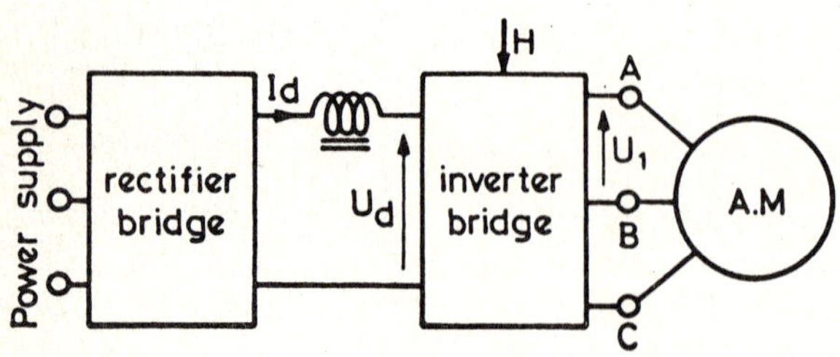

Figure 7

- The chain of measurement and treatment (fig. 8) is realized about a microprocessor which can generate a clock signal H of choiced periodicity, and read a period of a signal F which is an external clock.
The interface sheet translate the useful quantities of V_d, I_d, U_1 into others comprehensible by the microprocessor which commands the multiplexer and the A/D, D/A, converters.

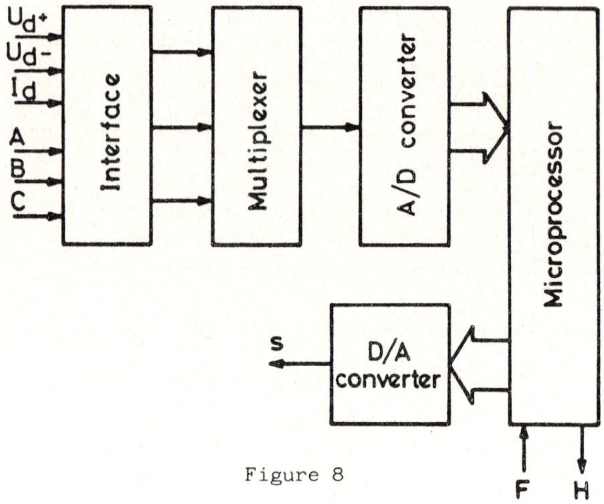

Figure 8

4.2. Putting in hand of the algorithm

To take the variations of τ_2 and f_R into account, we use the relation (1) by the following manner :

$$s\omega\,\tau_2 = \omega_R\,\tau_2 = \frac{X_o - X}{R - r_1}$$

where ω_R is the pulsation of the rotoric quantities. From the readings, we deduce $\omega_R\tau_2$, therefore, we deduce ω_R, and s from $s = \omega_R/\omega$.

We have two steps to calculate the slip, they are :

recognition of the motor

slip calculation of the machine under operation

Recognition of the motor

This preliminary step decompose itself into three phases corresponding to the classical tests of the motor :

- Phase 1 ; we measure, by a direct current, the resistance r_1
- Phase 2 : the motor at no-load, a slow sweeping of the frequency is done. For each value of ω, we measure X_o. These values of X_o are memorized in a table as function of ω :
- Phase 3 : the rotor is locked, again a slow sweeping of the frequency is done. For each value of the rotoric pulsation ω_R (here, equal to ω), we measure the appearents resistance R_c and inductance X_c of the rotor, we calculate $\omega_R\,\tau_2 = [X_o(\omega_R) - X_c(\omega_R) \ / \ R_c(\omega_R) - r_1]$.

The calculated values are memorized in a table as function of ω_R. During these two last phases, it is the microprocessor which pilote, by its clock H, the current inverter by a manner to proceed to a methodical sweeping of the frequency.

Normal operation

At normal operation of the motor fed at a pulsation

- we measure R and X
- we calculate $\omega'_R\tau_2 = [X_o(\omega)-X]/[R-r_1]$
- we search in the table the value of ω_R such that $\omega'_R\tau_2 = \omega_R\tau_2$.
- we calculate the slip from $s = \omega_R/\omega$.

During this second step, the inverter is piloted by the clock F which is commanded directely by the user. The microprocessor compute, by an interrupt program, the period of the clock signal. Figure 9 shows the flowshart of this step.

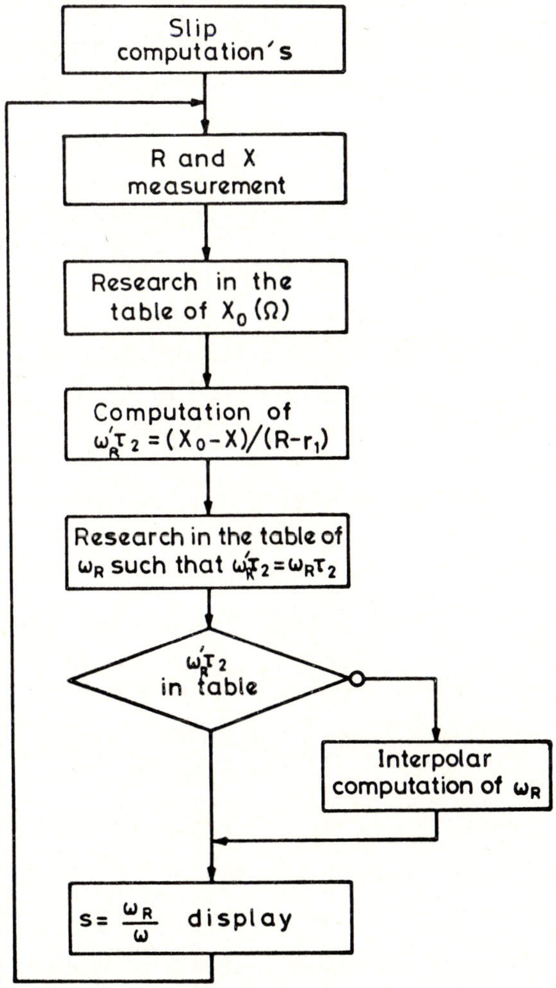

Figure 9

For each phase, the operator indicates the microprocessor if its indications on the motor states are wright, to make up the following measurements.

The realized equipment utilize a simple algorithm which take into account the variation of the parameter which give the greatest variation of the rotoric resistance. It has gived place to several patent depositions, and allowed to obtention of the slip with sufficient precision for the driven and protection functions which has been assigned to it.

REFERENCES

-o-o-o-o-

[1] : SEGUIER G., NOTELET F., PILLAIS J. "Détermination indirecte du glissement du moteur asynchrone triphasé pour la commande des équipements à vitesse variable." MECO'82 - Tunis, Sept. 1982, pp1-1,1-4.

[2] : NOTELET F., BOULET P., SEGUIER G. "Identification des paramètres du moteur asynchrone triphasé permettant de calculer le glissement à des fins de commande et de protection." 1rst IASTED Symp. on Applied Informatics, Lille, Mars 1983, Vol 2, pp141-144.

ELECTRICAL MACHINES AND CONVERTERS - Modelling and Simulation
H. Buyse and J. Robert (editors)
Elsevier Science Publishers B.V. (North-Holland)/ © IMACS, 1984

THE DOUBLE CAGE INDUCTION MOTOR MODEL FOR THE SIMULATION OF INVERTER FED MACHINES.

R.DE DONCKER,W.GEYSEN

LABORATORY OF ELECTRICAL MACHINES AND DRIVES
KATHOLIEKE UNIVERSITEIT LEUVEN,BELGIUM

1. ABSTRACT.

An improved induction motor model which is automatically self adaptive for changes in frequency of the rotor parameters is proposed. A comparison between this model and the commonly used single cage model of the asynchronous machine is made both for the simulation of the machine and design of field oriented control circuits .

2. INTRODUCTION

Speed control of asynchronous machines is done by aid of frequency inverters. During this control it should be possible to maintain the total flux in the machine constant at its rated value. The torque delivered by the machine must follow dynamical laws. Both quantities, flux and torque, are straightly connected to the currents in the machine. These currents are driven by the frequency inverter.

The goal of the field-oriented control is to control torque and flux independently. This gives a very sophisticated dynamical control of the induction machine, which can be compared with a dc-motor.

The design of the field-oriented control is strongly related to the chosen model of the induction motor. The present paper covers the comparison between two models and the theoretical background of the improved model.

3. FIELD-ORIENTED CONTROL OF AN INDUCTION MACHINE BASED UPON THE SINGLE CAGE MODEL

Neglecting skin effect, finite permeability of the iron, saturation, spatial harmonics in the air gap, slot effects, eddy current, iron losses, unsymmetry, eccentricity of the rotor, one can derive the following differential equations for the transient behaviour of the induction motor:

$$\underline{u}_S = R_S\underline{i}_S + \frac{d\underline{\psi}_S}{dt} + j\omega_\varphi\,\underline{\psi}_S \quad (1)$$

$$0 = R_R\underline{i}_R + \frac{d\underline{\psi}_R}{dt} - j\omega_{m\varphi}\,\underline{\psi}_R \quad (2)$$

$$\frac{J}{p}\frac{d\omega_m}{dt} = \frac{3p}{2}\,\mathrm{Im}(\underline{i}_S\cdot\underline{\psi}_S^*) - M_L \quad (3)$$

$$\omega_\varphi = \frac{d\gamma_\varphi}{dt} \quad \text{(speed of the field axis),} \quad (4)$$

$$\omega_{m\varphi} = \omega_m - \omega_\varphi = \frac{d\gamma_m}{dt} - \frac{d\gamma_\varphi}{dt} \quad (5)$$

(ω_m : mechanical speed)

$$\underline{u}_S = u_{SD} + ju_{SQ} \text{ (etc.),} \quad (6)$$

$$\underline{\psi}_R = L_{R\sigma}\underline{i}_R + L_h(\underline{i}_S+\underline{i}_R), \quad (7)$$

$$\underline{\psi}_S = L_{S\sigma}\underline{i}_S + L_h(\underline{i}_R+\underline{i}_S), \quad (8)$$

and

$$i^\varphi_{SQ} + i^\varphi_{RQ} = 0. \quad (9)$$

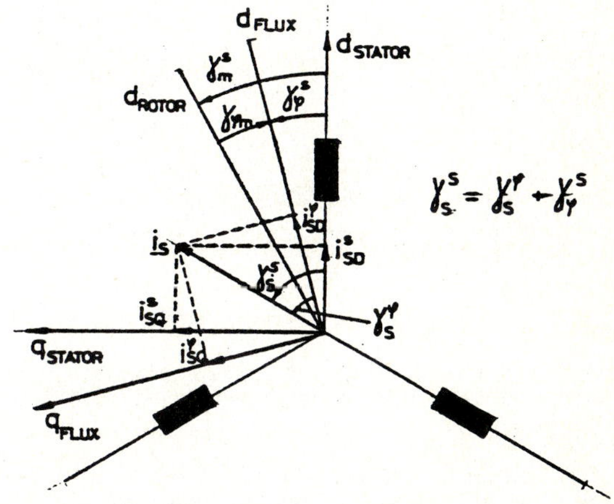

Fig.1 Different mainframes, equations.

All equations and space vectors are written in a mainframe which is fixed with the airgap flux of the machine.(Fig.1)

From these equations one is able to derive a current-model (Fig.2) of the machine, as it was principally explained by F.Blaschke[1].

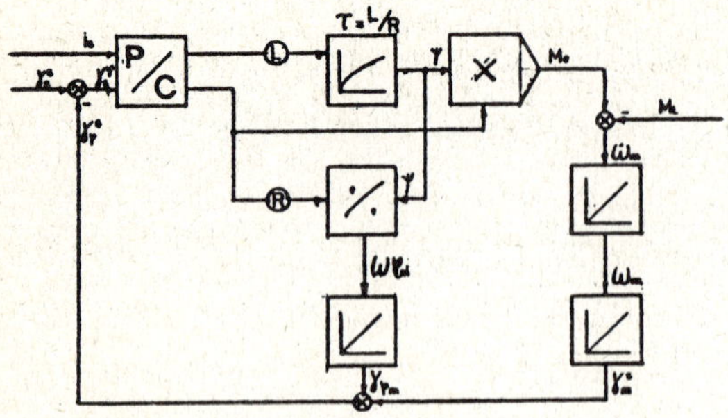

Fig.2 Model of single-cage induction motor with impressed currents.

When the leakage inductance of the rotor is neglected, the direct component of the stator current i^{φ}_{SD} is proportional to the airgap flux of the machine. Once i^{φ}_{SD} is fixed, the quadrature component i^{φ}_{SQ} determines the torque in the machine. The desired components i^{s}_{SD}, i^{s}_{SQ} of the stator current in stator-coordinates are also calculated from this current-model. These components have to be amplified by the inverter (after transformation to a 3 phase system). The principal circuit necessary to calculate the stator current components i^{s}_{SD}, i^{s}_{SQ} is shown in figure 3.

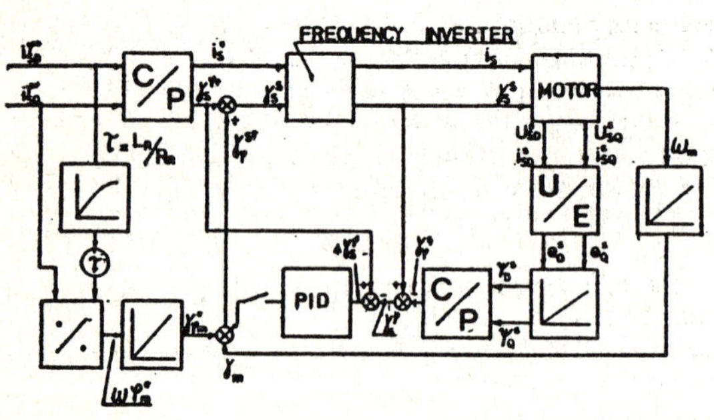

Fig.3 Block diagram of field oriented control based upon a single-cage inductionmotor model.

The angle γ_{φ} between stator axis and flux axis must be calculated as accurate as possible.

In practice the real machine parameters dependent on temperature, frequency and saturation. Especially the rotor parameters have to be adapted to these changes during operation. This will eliminate large steady state errors once the system has achieved a stationary operation. Also, the response of the system is improved by using a closed loop control, which consists of a measurement of the line currents and line voltages of the machine, followed by a calculation of the flux vector and a PID-controller (Fig.3).

Concluding:
- the parameters of the flux model are few in number (rotor time constant)
- the model and the calculation are only correct in a small operation area. They suffer mostly from saturation, temperature rises and frequency variations of the rotor parameters.

4. DOUBLE CAGE INDUCTION MOTOR MODEL.

In order to improve the behaviour of the regulation system for frequency variations of the rotor parameters it is necessary to refine the model of the induction machine.

To attain latter goal, the double cage model is used. Rotor parameters are mostly influenced by the "deep bar" effect generated by eddy currents in the rotor cage, especially when using a real double cage induction motor. The model is based on the following equations:

$$\underline{u}_S = R_S \underline{i}_S + \frac{d\,\underline{\psi}_S}{dt} + j\,\omega_{\varphi}\,\underline{\psi}_S \tag{10}$$

$$0 = R_1 \underline{i}_{R1} + \frac{d\,\underline{\psi}_{R1}}{dt} - j\omega_{m\varphi}\,\underline{\psi}_{R1} \tag{11}$$

(upper cage)

$$0 = R_2 \underline{i}_{R2} + \frac{d\,\underline{\psi}_{R2}}{dt} - j\omega_{m\varphi}\,\underline{\psi}_{R2} \tag{12}$$

(lower cage)

$$\frac{J}{p}\frac{d\,\omega_m}{dt} = \frac{3p}{2}\,\mathrm{Im}(\underline{i}_S \cdot \underline{\psi}^{*}_S) - M_L \tag{13}$$

$$\underline{\psi}_S = L_S \underline{i}_S + L_{h1}\underline{i}_{R1} + L_{h2}\underline{i}_{R2} \tag{14}$$

$$\underline{\psi}_{R1} = (L_{h1}+L_{\sigma 1})\underline{i}_{R1} + L_{\sigma 12}\underline{i}_{R1} + L_{h1}\underline{i}_S + L_{\sigma 12}\underline{i}_{R2} \tag{15}$$

$$\underline{\psi}_{R2} = (L_{h2}+L_{\sigma 2})\underline{i}_{R2} + L_{\sigma 12}\underline{i}_{R1} + L_{h1}\underline{i}_S + L_{\sigma 12}\underline{i}_{R2} \tag{16}$$

and

$$i^{\varphi}_{SQ} + i^{\varphi}_{RQ1} + i^{\varphi}_{RQ2} = 0 \tag{17}$$

In stationary state the equivalent scheme, shown in figure 4, describes the induction machine in the same manner. ($L_{h1}=L_{h2}=L_h$, $L_{\sigma 1}=0$)

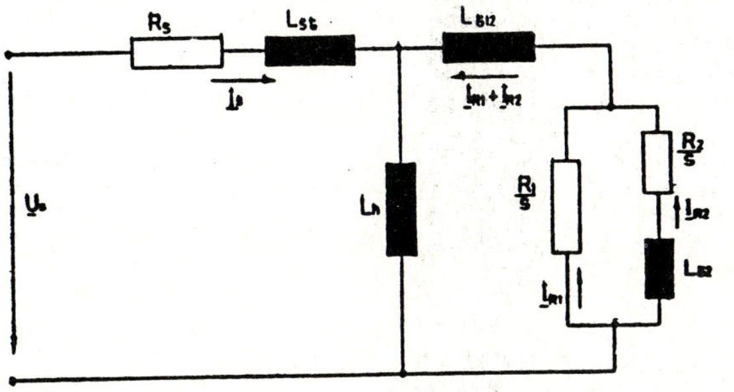

Fig.4 Double-cage model, equivalent scheme.

This scheme is simply recalculated to fig.5 which shows the dependency of the rotor parameters on the rotor frequency.

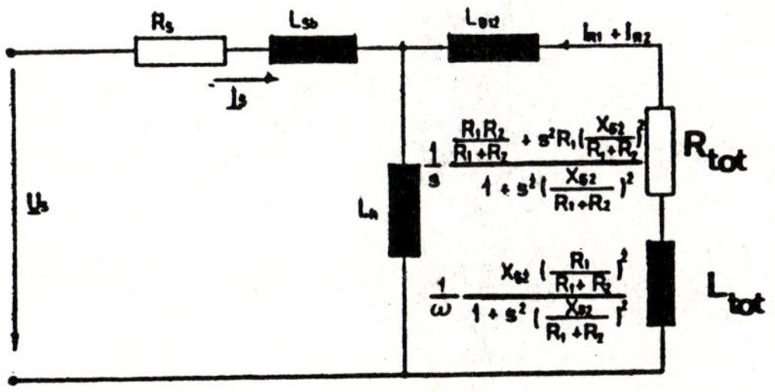

Fig.5 Double-cage model, recalculated parameters.

The model adapts itself to frequency variations.

5. COMPUTATION OF THE ROTOR PARAMETERS.

More parameters have to be calculated. This seems a great disadvantage of the double cage model. On the other hand more degrees of freedom are allowed to compute those parameters. The four rotor parameters $R_1, R_2, L_{\sigma 12}, L_{\sigma 2}$ are determined by four requirements. These requirements can differ depending on the problems that have to be calculated. Four possible evaluations are :

1. rated torque at rated speed (slip)
2. rated stator current at rated speed.
3. at slip s=1 (50 Hz); $R_{tot}(1)$ equals the resistance calculated from measurements.
4. at slip s=1 (50 Hz); $L_{tot}(1)$ equals the leakage inductance calculated from measurements.

Completely different requirements are possible (e.g. four working points at different frequencies and small slip). The requirements named above are determined in a more simple manner (rated values, short circuit test) and present good results for the calculations.

6. VERIFICATION OF THE DOUBLE CAGE MODEL.

A verification of the model was worked out on a 11 kW machine fed by a six step voltage source frequency inverter. The rated values and the calculated stator current, compared with the measurements are shown in figure 6.

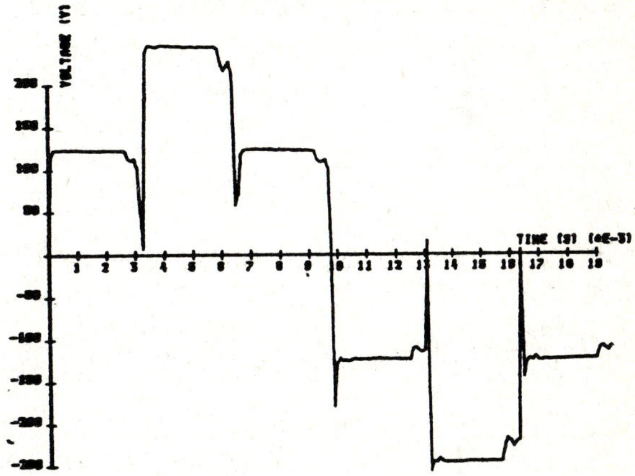

Fig.6.1 Measured phase voltage.

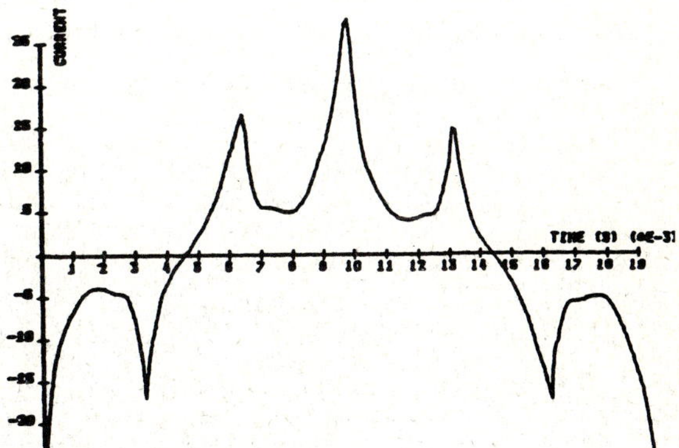

Fig.6.2 Measured current.

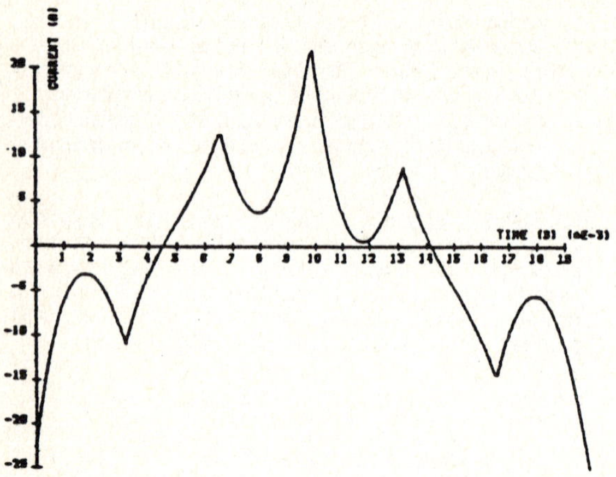

Fig.6.3 Calculated current, single cage model.

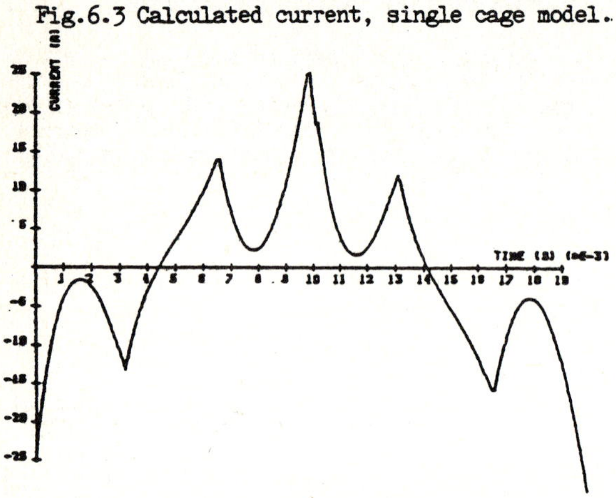

Fig.6.4 Calculated current, double cage model.

The calculation and measurements conform very well for all frequencies (10 to 150 Hz.) Even for the fundamental component of the waveforms at rated slip, the frequency has an influence on the rotor parameters form 10 to 100 Hz that is in the same order of a temperature rise during operation.

Conclusion:
- calculation of parameters is very simple
- the model can be fitted very well for each problem by choosing the four requirements properly
- the double cage model adapts itself to frequency variations in the rotor.
- transient phenomena are accurately calculated (for every type of induction machine).

7. FIELD ORIENTED CONTROL BASED UPON THE DOUBLE CAGE MODEL.

The advantage of the double cage model are implemented in the field oriented control of the induction motor. In a similar way one can derive a currentmodel of the machine. With this model one reaches a fourth order equation for $\omega_\varphi - \omega_m$. Neglecting the higher order terms $L^2_{\sigma1}$, $L^2_{\sigma12}$, $L_{\sigma12} \cdot L_{\sigma2}$ next formula is obtained:

$$\omega_\varphi - \omega_m = - \frac{i^\varphi_{SQ}}{\dfrac{i^\varphi_{SD} L_h (R_1 + R_2 + pL_{\sigma2})}{p^2 L_1 L_{\sigma2} + p(R_2 L_1 + R_1 L_2) + R_1 R_2}} \quad (18)$$

$$L_1 = L_h + L_{\sigma12};\ L_2 = L_1 + L_{\sigma2} \quad (19)$$

The principal block diagram is shown in fig.7.

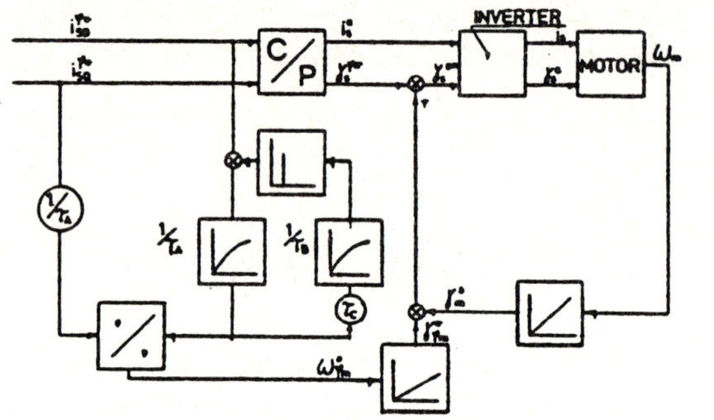

Fig.7 Field-oriented control based upon the double cage model.

A control circuit based on this diagram will be developed. The current-model is calculated by the use of microprocessors. A better dynamical behaviour and stability of the system is expected.

8. ACKNOWLEDGEMENT :

This research is financed by the Belgian National Science Foundation (N.F.W.O.).

Ir. R. De Doncker, prof. dr. ir. W. Geysen, dr. ir. A. Vandenput and ir. R. Belmans are with the Laboratory of Electrical Machines and Drives of the Electrotechnical Department of the Katholieke Universiteit of Leuven, Kardinaal Mercierlaan 94, B-3030 Leuven.

9. REFERENCES.

1. Blaschke F., Das Verfahren der Feldorientierung zur Regelung der Asynchronmachine, Siemens Forschungs- und Entwicklungsberichte I, 1972

2. Blaschke F., Regelverfahren für Drehfeldmaschinen, VDI-Bildungswerk BW3232, 1982

3. Jordan H., Seinsch H.-O., Ermittlung der Daten des zu einem Keilstabläufer equivalenten Doppelkäfigläufer, Technisch Bericht st-2, K.U. Leuven, 1979

4. Jordan H., et al., Asynchronmaschinen, p.234-295, 1975

5. Schumacher W., Microprocessor controlled ac-servo drive, ETG-Fachberichte 11, 1982, Darmstadt.

ELECTRICAL MACHINES AND CONVERTERS - Modelling and Simulation
H. Buyse and J. Robert (editors)
Elsevier Science Publishers B.V. (North-Holland)/ ©IMACS, 1984

RADIAL VIBRATIONS OF INDUCTION MOTORS WITH SQUIRREL CAGE.

R. Belmans, A. Vandenput, G. Zeebroek, W. Geysen.

Laboratory for Electrical Machines and Drives
of the Katholieke Universiteit of Leuven, Belgium.

As the speed of an asynchronous motor fed by a frequency-inverter, is not constant, it is necessary to predict the shaft bending when passing the critical speed in a very accurate way.

Furthermore the inverter does not always allow to accelerate swiftly enough through the resonance zone, in order to avoid inadmissible shaft bending and even collision of stator and rotor.

In this paper a vibrational equivalent scheme, consisting of a mechanical spring constant, a mechanical damping coefficient and a concentrated mass is derived from calculated and experimental values.

Furthermore the radial bending when accelerating through the critical speed is calculated and experimentally verified. If the bending as function of the motor speed is known, the fatal collision between stator and rotor can be avoided by adapting the machine dimensions.

INTRODUCTION

Most authors use the Laval rotor model when analysing specific phenomena influencing the radial vibrational behaviour of induction motors. They consider the coefficients appearing in this model (mass m, spring constant c_m, damping coefficient d_m) as known and start their analysis at that point.

In this paper we will show how the model parameters (mass m and spring constant c_m) can be derived from measured or calculated critical speed values and how the modal analysis technique can yield the value of the damping coefficient d_m. All dynamic effects in the bearings and their supports are neglected.

The rotor model obtained in this way is used to simulate the shaft bending when accelerating through the first critical speed. This phenomenon becomes more and more important as induction motors, fed by frequency converters pass their critical speed during the regulation cycli. This passing of the critical speed is sometimes not swiftly enough to avoid inadmissible shaft bending and even collision of stator and rotor. The results of the simulation are compared with experimentally obtained values.

CALCULATION AND MEASUREMENT OF THE CRITICAL SPEED

In [1] three methods for the calculation of the critical speed of an induction motor with aluminium cast squirrel cage are compared. All three are used on a machine model with shaft dimensions given in figure 1. The dimensions of the cross section of the squirrel cage bars and the lay-out of the rotor cage are represented on figure 2 and figure 3.

The first and very elementary method neglects the shaft mass and concentrates the rotor mass in a disc in the middle of the shaft. The shaft is the only flexible element of the system. The critical speed obtained from this calculation is n_{cr}=3607 r.p.m..

In order to obtain a better approximation, some authors take the real stiffness and mass distribution into account [2]. Doing this yields n_{cr}=2461 r.p.m..

However both theoretical values were too far away from the experimental value obtained by the modal analysis technique which gives n_{cr}=2863 r.p.m..

Therefore an other calculation scheme has beendeveloped taking into account not only the shaft mass and stiffness distribution, but also the rotor length

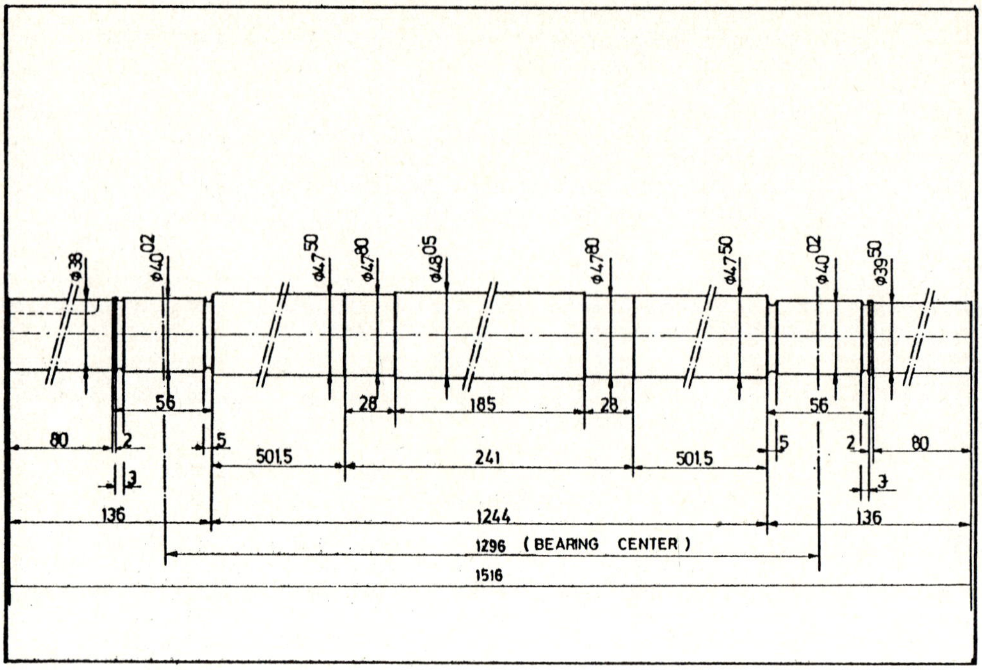

Fig. 1 Dimensions of the machine shaft.

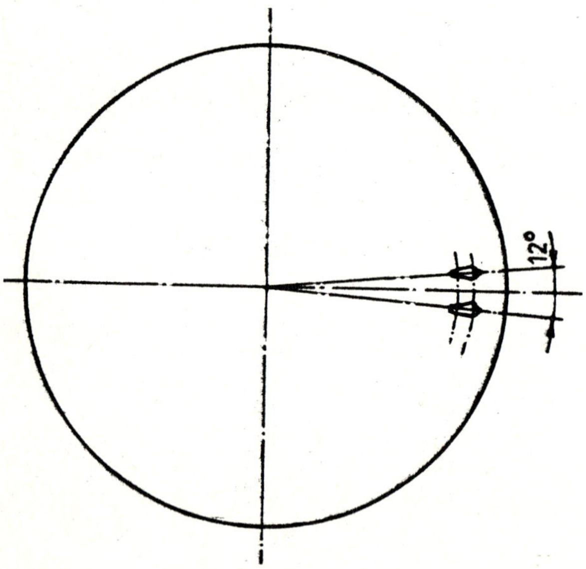

Fig. 2 Lay-out of the rotor cage.

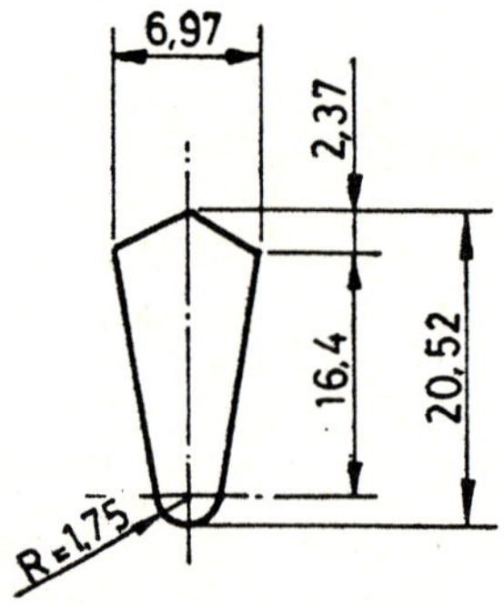

Fig. 3 Dimensions of the rotor bars.

and the supplementary stiffness due to the rotor bars, which is very high compared to the shaft stiffness. This method gives n_{cr}=2821 r.p.m. which result is very near to the experimentally obtained value.

DERIVATION OF THE PARAMETERS OF THE MACHINE MODEL

Undamped system

As said above, most authors dealing with radial vibrations of induction motors, use the Laval rotor model. This model, shown in figure 4, consists of a massless, flexible shaft with spring constant c_m and a rotor disc with mass m in the middle of the two supporting bearings.

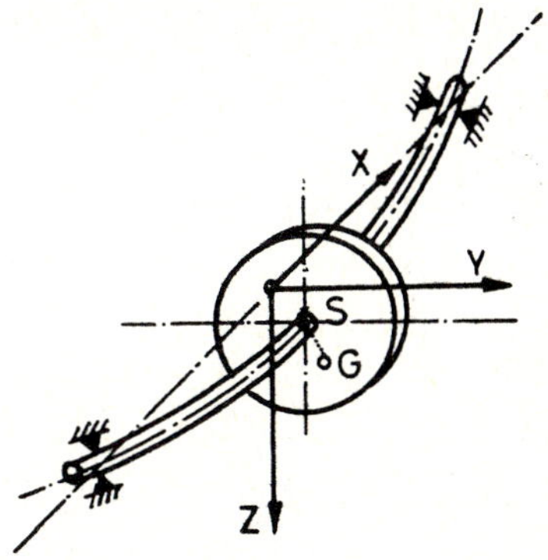

Fig. 4 Laval rotor model.

If the parameters are known, special problems are analysed using this machine model. The unbalanced magnetic pull e.g. can be represented by a negative spring constant c_e [3, 4] as this force tends to enlarge the existing eccentricity, yielding a lower critical speed of the machine:

$$\omega_{ne} = \sqrt{\frac{c_m - c_e}{m}} \quad . \tag{1}$$

We will now show how the parameters of the model can be derived from the machine parameters.

As the model and the real machine must have the same critical speed, the knowledge (experimental or theoretical) of this speed gives a first relationship between the spring constant and the mass of the model:

$$\omega_n = \sqrt{\frac{c_m}{m}} \quad . \tag{2}$$

A second relationship between the model parameter c_m and the real rotor can be obtained if we state that the maximum bending, due to an uniformely distributed force on the real rotor, must be the same as the maximum bending of the model under influence of the same force concentrated on the disc rotor.

The maximum bending of the model is:

$$w_{max} = \frac{F}{c_m} \quad . \tag{3}$$

An elementary calculation yields for the real machine

$$w_{max} = \frac{F}{(EI)_5}\left(\left(\frac{l_r}{4}\right)\left(\frac{5}{24}\left(\frac{l_r}{2}\right)^2 + \frac{5}{6} l_4 \left(\frac{l_r}{2}\right) + l_4^2\right) + \frac{1}{6}\left(k_1 l_1^3 + k_2 l_2^3 + k_3 l_3^3 + k_4^3\right)\right) \tag{4}$$

with

$$\begin{aligned} k_1 &= \left(\frac{1}{(EI)_1} - \frac{1}{(EI)_2}\right)(EI)_5 \\ k_2 &= \left(\frac{1}{(EI)_2} - \frac{1}{(EI)_3}\right)(EI)_5 \\ k_3 &= \left(\frac{1}{(EI)_3} - \frac{1}{(EI)_4}\right)(EI)_5 \\ k_4 &= \frac{(EI)_5}{(EI)_4} \end{aligned} \tag{5}$$

The combination of equations (3) and (4) taking into account the dimensions given in figure 4 yields:

$$c_m = 1.526 \; 10^6 \text{ N/m} \; .$$

In this way the machine model gets the same bending as the real motor, when submitted to the same stationary force.

The uniform force distribution along the rotor is a realistic representation for the electromagnetically induced forces (unbalanced magnetic pull). These forces are function of the distance between stator and rotor, which is practically constant along the rotor length due to the supplementary stiffness of the rotor cage.

On figure 5 the bending of the real shaft due to a uniform load of 1000 N is shown.

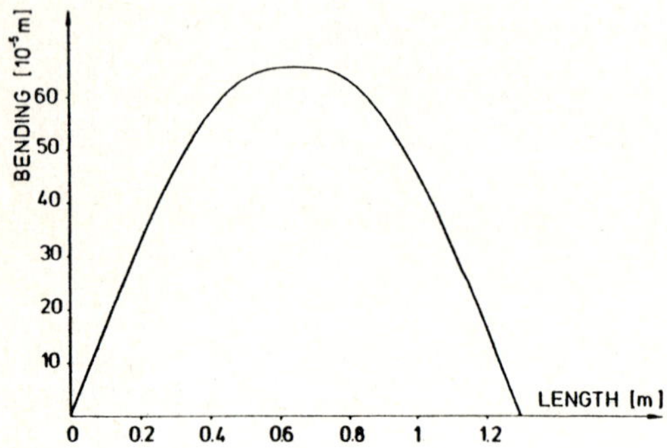

Fig. 5 Shaft bending due to a uniform load on the rotor.

If the spring constant c_m and the critical speed $_n$ are known, equation (2) gives us the equivalent mass of the model, yielding for our example:

m = 16.91 kg.

Damping

Using the modal analysis technique a relative damping coefficient for the different modes is obtained [5]. The tested structure is excited in 6 points (figure 6) and the transfert function is measured and evaluated. The results for the first mode (first critical speed) are shown in table. Accounting for the coherence the mean relative damping coefficient of the system is:

σ_1 = 0.805% .

In [5] we find:

$d_m = \sigma . 2m\omega$,

yielding:

d_m = 0.272 N s/m .

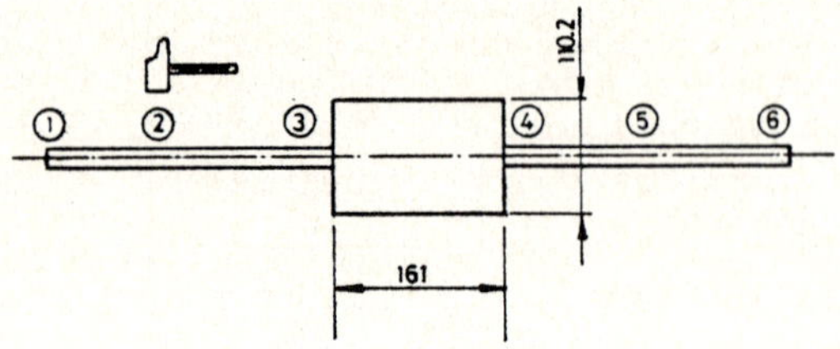

Fig.6 Excitation points.

measuring point	frequency (Hz)	damping coefficient	Ψ	coherence
1	47.92	1.17	- 3.81	.8
2	47.99	0.135	- 14.06	.97
3	47.58	1.14	- 17.14	.995
4	47.75	1.13	- 17.75	.995
5	47.40	.482	- 11.6	.997
6	47.73	.832	- 2.45	.93

We will now use our machine model to simulate the shaft bending during an acceleration through the critical speed.

SIMULATION OF THE BENDING DURING ACCELERATION AND EXPERIMENTAL VERIFICATION

If the machine is represented by the Laval rotor, the equilibrium of forces acting in horizontal and vertical direction is (figure 7):

$$m\ddot{y}_G + d_m\dot{y}_S + c_m y_S = 0$$
$$m\ddot{z}_G + d_m\dot{z}_S + c_m z_S = 0 \qquad (6)$$

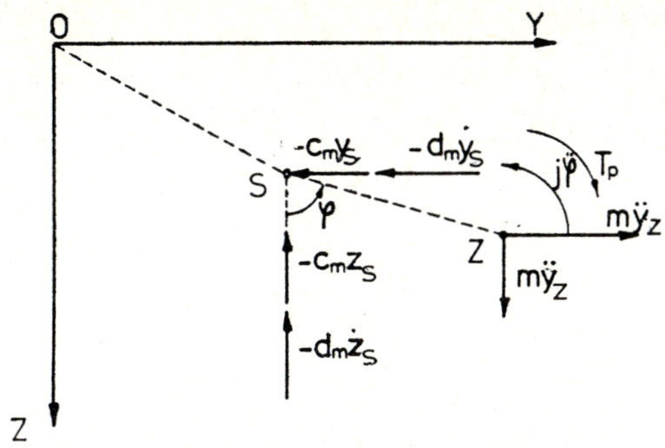

Fig. 7 Forces acting on the machine rotor.

With e the eccentricity, the relationship between the coordinates of the centre of gravity G and the centre of the shaft S are:

$$y_G = y_S + e \sin\varphi$$
$$z_G = z_S + e \cos\varphi \qquad (7)$$

Substituting eq. (7) in (6) yields:

$$m\ddot{y}_S + d_m\dot{y}_S + c_m y_S = me\dot{\varphi}^2 \sin\varphi - me\ddot{\varphi}\cos\varphi$$
$$\qquad (8)$$
$$m\ddot{z}_S + d_m\dot{z}_S + c_m z_S = me\dot{\varphi}^2 \cos\varphi + me\ddot{\varphi}\sin\varphi$$

This set of two differential equations contains three unknown variables y_S, z_S and φ.

Therefore we have to find a third relationship between these quantities, which is given by the torque equation.

If the torque used to accelerate the system is T, the torque equation is:

$$J\ddot{\varphi} = T - c_m e(z_S \sin\varphi - y_S \cos\varphi) - d_m e(\dot{z}_S \sin\varphi - \dot{y}_S \cos\varphi). \qquad (9)$$

The constant accelerating torque is the difference of the driving torque and the load torque (e.g. friction, ventilation, ...).

Equations (8) together with equation (9) form a set of simultaneous non-linear differential equations, which can only be solved exactly by using digital simulation techniques. For our system, we used a fourth order Runge-Kutta integration method.

The system used for the experimental verification consists of a d.c. motor, driving the rotor with parameters given above. The mass moment of inertia appearing in equation (9) is the sum of the mass moments of inertia of the d.c. motor and the rotor of the asynchronous motor and is:

$$J = 0{,}98 \text{ kg m}^2$$

The eccentricity of the system is:

$$e = 2{,}7 \text{ mm}.$$

In figure 8, the results of this simulation are shown, the driving torque is:

$$T = 9{,}4 \text{ Nm}.$$

In figure 9, the measured shaft bending is plotted. The shaft vibrations are measured using inductive displacement transducers and are recorded on a digital scope.

The subharmonic vibrations appearing in the measured curve are not found in the simulation, as the model does not account for the phenomena, causing them (e.g. misaligment).

However the vibrations at the motor critical speed are quite good represented by the model.

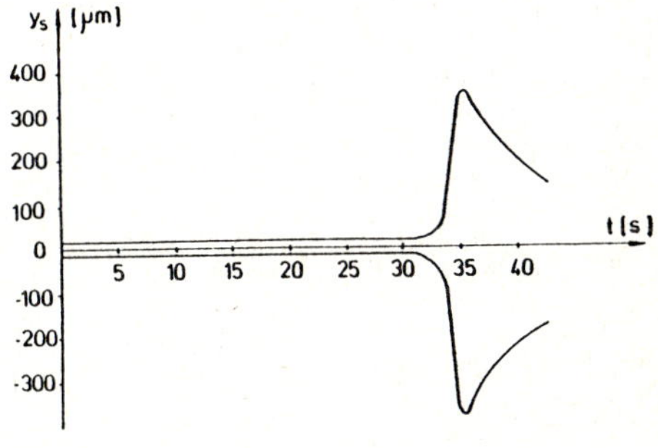

Fig. 8 Bending of the shaft during acceleration through the first critical speed.(theory).

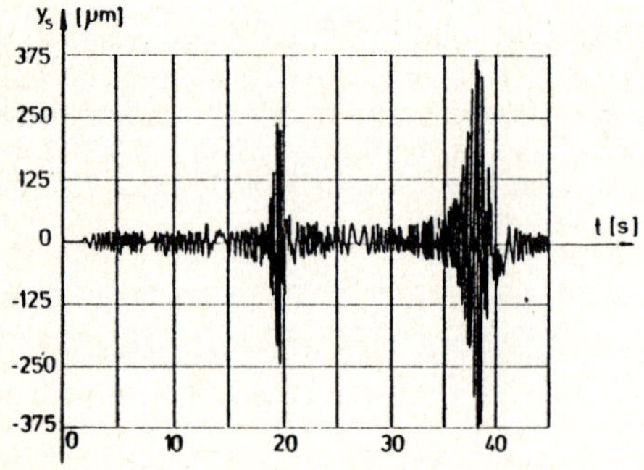

Fig. 9 Bending of the shaft (experiment)

CONCLUSION

The model derived here describes quite good the mechanical vibrations of a shaft accelerating through its critical speed.

The parameters of the model can be derived from measurements (modal analysis) or calculation (except damping coefficient).

ACKNOWLEDGMENT

The authors are mostly gratefull to the Council of the Joint Fundamental Research Fund (F.K.F.O.) of the Belgian National Science Foundation (N.F.W.O.) for granting a project that financed this research.

REFERENCES

[1] R.Belmans, W.Heylen, W.Geysen, A. Vandenput, R.De Doncker
Calculation and measurement of the critical speed of induction motors with cage rotor. To be published IEE Proc.-B, Electric Power Applications.

[2] Freise W.
"Ueber die kritischen Drehzahlen elektrischer Maschinen". Ph.D.Thesis Techn.Hochsch.Hannover 1961.

[3] Freise W., Jordan H.
Einseitige magnetische Zugkräfte in Drehstrommaschinen. ETZ-A 83 (1962) p.298-303.

[4] Jaenicke P., Jordan H.
Berechnung der elektromagnetischen Radialkraft bei Drehstromasynchronmaschinen mit exzentrischer Verlagerung des Läufers. Siemens Forschungs- und Entwicklungsberichte, Bd.5 (1976), p.249-256.

[5] Formenti D.
Analytical and experimental model testing Seminar on Modal Analysis, K.U.Leuven, 1977.

ELECTRICAL MACHINES AND CONVERTERS - Modelling and Simulation
H. Buyse and J. Robert (editors)
Elsevier Science Publishers B.V. (North-Holland)/ © IMACS, 1984

SIMULATION OF TRANSIENT BEHAVIOUR OF CAPACITOR-RUN INDUCTION MOTOR FED BY POWER CONVERTERS.

G.A. CAPOLINO

ENSET 5,Avenue Taha Hussein 1008 TUNIS-TUNISIA.
on leave from:
ESIM 28,Rue des Electriciens 13012 MARSEILLE-FRANCE.

A non-linear state space model of a capacitor-run induction motor is presented in order to provide transient analysis of the machine fed by power converters. The non-linear model is used for investigations of two types of converters which permitts to control both the speed and the stator current of the machine.
The first type of converter is a series connected thyristor switches for for the main winding of the motor. The analysis of the problem gives two modes of feeding with the main winding openned or directly connected to the AC line. Then, the firing angle permitts to limit the starting current and to control the steady state speed.
The second type of converter is a DC-AC inverter with a particular modulation process which provides a voltage and frequency control.
The starting performances of the machine fed by both converters are simulated and compared to the plugging on an AC line. Some experimental results are given and compared to the simulation showing the accuracy of the non-linear model.

1. NOTATIONS.

dq	stationnary reference frame.
$\alpha r \beta r$	axes related to the rotor reference frame.
$m_s a_s$	axes related to the stator reference frame.
i_{ms}	current for the main stator winding.
i_{as}	current for the auxiliary stator winding.
$i_{\alpha r}, i_{\beta r}$	equivalent currents for the rotor squirrel-cage.
e_{ms}	voltage for the main winding.
e_{as}	voltage for the auxiliary winding.
v_{ca}	voltage to the terminals of the capacitor-run.
C	capacitor.
q	charge of the capacitor.
$\emptyset$	equivalent fluxes for a winding (with convenient index).
R_m	resistance of the main stator winding.
R_a	resistance of the auxiliary stator winding.
R_r	equivalent resistance of a rotor winding.
L	total inductance for a winding (with convenient index).
M	mutual inductance between two windings (with convenient index).
θ_s	relative position of the stator reference frame.
θ_r	relative position of the rotor reference frame.
θ	relative position stator-rotor.
w	rotor speed (electrical).
J,f	mechanical characteristics of the motor (inertia,friction).
m_L	load torque.
a_1	firing angle of the thyristor switches (series connected).
	phase-shift angle (voltage source control).
$\hat{V}$	maximum value of the input-voltage.
w_s	pulsation of the voltage supply.

2. INTRODUCTION.

The widespread interest in applying static converters to induction motor speed and/or current control has created means of predicting performances related to the non-linear characteristic of this association. Today, numerous papers have been published concerning the modelling and simulation of three-phase machines fed with voltage or current sources.

Unfortunatly, the capacitor-run induction motor has not yet suscited such great interest as the three-phase machine for speed reducing controls. While both the cross-field and the revolving-field theories have been treated in the litterature before 1930, no satisfactory model for transient investigations have been available. Kron in his tensor theory of electrical machines provided a model suitable for state-space analysis in 1938. The first paper /1/ which uses this theory related to a capacitor-run induction motor fed with series connected thyristor switches appeared in 1968.

The method is unique in that it achieves solution both for transient and steady-state analysis and provides model for modern control theory. For this reason, we present in this paper a non-linear state-space model for a capacitor-run induction motor suitable for different types of feeding semiconductor devices. The model is presented in a previous paper /10/ where conditions to obtain a time-invariant system are performed. The state model of this motor has linear and non-linear terms which can be reduced into a quadratic form. Then, the behaviour of the machine is described over a wide speed and current range and it is very powerful compared with linearized models.

Several authors have shown the possibility to fed a capacitor-run induction motor with voltage sources such as DC-AC converters /3/,/5/,/7/,/9/. Our first papers /7/,/9/ related for steady state analysis both from complex analysis with first harmonic method and time-varying analysis associated to Newton algorithm to find steady-state vector. An other method to calculate the performances of a capacitor-run induction motor fed with non-sinusoidal voltage consists in applying Fourier series transformation to the voltage waveforms /3/.

In the second part of the paper, the capacitor-run induction motor controlled by series connected thyristor switches is related. Most of the papers published on the subject /1/,/2/,/4/,/6/,/8/ presented only the steady-state analysis of the machine for speed variations. The different theories related from sinusoidal analysis in steady-state running using only the first harmonic /2/ or the Fourier series /4/. An other approach consists in solving linear differential equations describing the operation by Laplace transform method /6/.

It is obvious that all these methods are not convenient to perform transient analysis of the machine; so we have tried to develop an uniform modelisation for all types of non-sinusoidal waveform supply. The starting formulation is derived from Kron's point of view in his tensor theory of electrical machines and networks.

3. BASIC CONSIDERATIONS.

Let us consider the generalized equivalent machine for a capacitor-run induction motor (figure 1.) where the stator windings are in quadrature (axes m_s ans a_s) so as the rotor windings (axes α_r and β_r). The main winding of the stator armature is fed by a voltage source e_{ms} while the auxiliary winding is fed by a similar source e_{as} connected in series with a capacitor.

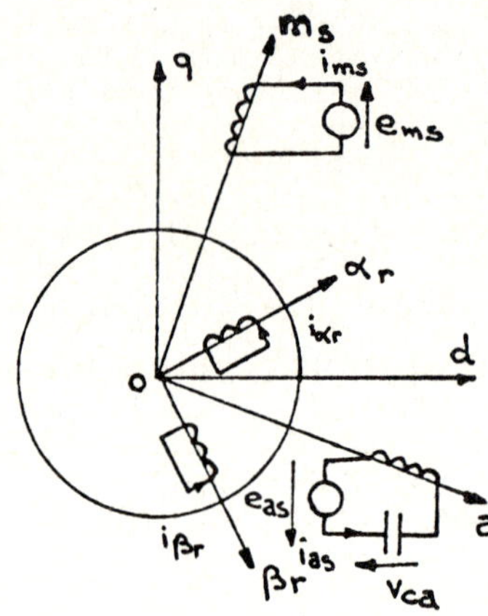

Figure 1. Equivalent two-phase machine for capacitor-run motor.

The modified Maxwell equation for the described system becomes :

$$e = R.q + \dot{\phi} \qquad 1)$$

with:

$e = (e_{ms} \; e_{as} \; 0 \; 0 \; 0)^t$ input vector.

$q = (i_{ms} \; i_{as} \; i_{\alpha r} \; i_{\beta r} \; v_{ca})^t$ state vector.

$\phi = (\phi_{ms} \; \phi_{as} \; \phi_{\alpha r} \; \phi_{\beta r} \; q)^t$ fluxes and charge vector.

R is a partition of the evolution matrix:

$\text{diag } R = (R_m \; R_a \; R_r \; R_r \; 0)$ and $r_{25} = +1.$ $r_{52} = -1.$

The machine is supposed to be linear for the fluxes and then:

$$\phi = L.q$$
$$\dot{\phi} = \dot{L}.q + L.\dot{q} \quad \text{with:}$$

$\text{diag } L = (L_{ms} \; L_{as} \; L_{\alpha r} \; L_{\beta r} \; C)$

$l_{13}=l_{31}= M_d.\cos(\theta_r-\theta_s)$

$l_{23}=l_{32}=-M_q.\sin(\theta_r-\theta_s)$

$$l_{14}=l_{41}= M_d.\sin(\theta_r-\theta_s)$$
$$l_{24}=l_{42}= M_q.\cos(\theta_r-\theta_s)$$

The aim of the vector transformation is to give the system 1) the form of a state time-invariant equation. It is well known that a rotation tensor permitts to provide such references for three-phase systems in a dq frame. For a two-phase system the choice of Hancock provides an energy conservation between model 1) and the modified one.

$$P = \begin{matrix} P_1 & 0 & 0 \\ 0 & P_2 & 0 \\ 0 & 0 & 1 \end{matrix} \qquad 2)$$

with the submatrices P_1 and P_2:

$$P_1 = \begin{bmatrix} \cos\theta_s & \sin\theta_s \\ \sin\theta_s & -\cos\theta_s \end{bmatrix}$$

$$P_2 = \begin{bmatrix} \cos\theta_r & \sin\theta_r \\ \sin\theta_r & -\cos\theta_r \end{bmatrix}$$

This transformation leads to a new state vector:

$$\mathfrak{q} = P.q \qquad 3)$$

with:

$$\mathfrak{q} = (\, i_{ds} \; i_{qs} \; i_{dr} \; i_{qr} \; v_{ca} \,)^t$$

The new formulation of equation 1) gives:

$$\mathfrak{e} = (P.R.P^{-1}+P.(\dot{L}.P^{-1}+L.\dot{P}^{-1})).\mathfrak{q} + P.L.P^{-1}.\dot{\mathfrak{q}} \qquad 4)$$

with $\mathfrak{e}$ the control vector in the dq reference frame.
The equation 4) permitts to write a singular form of the electrical state equation for the capacitor-run induction motor:

$$\dot{\mathfrak{q}} = -(P.L.P^{-1})^{-1}.(P.R.P^{-1}+P.(\dot{L}.P^{-1}+ L.\dot{P}^{-1})).\mathfrak{q} + (P.L.P^{-1})^{-1}.\mathfrak{e} \qquad 5)$$

The electrical state equation have to be complete with the mechanical one. The electromechanical torque is derived from equation 4) multiplying left by $\mathfrak{q}^t$ in order to obtain the power equation. The classical torque expression of White and Woodson becomes:

$$T_e = \mathfrak{q}^t.G.\mathfrak{q} \qquad 6)$$

with $G.w = P.(\dot{L}.P^{-1}+L.\dot{P}^{-1})$.

In a previous paper /10/, we have shown that the only mean to obtain a time-invariant equation is to impose the condition $\theta_s = 0$ and then $w = \dot{\theta}_r$. The important result is that the non-linear state equation of a capacitor-run induction motor is time-invariant only if the system is observed from a fixed reference frame related to the stator. This condition leads to a new state equation which contains only linear and non-linear terms:

$$\begin{bmatrix} \dot{\mathfrak{q}} \\ \dot{w} \end{bmatrix} = \begin{bmatrix} -(P.L.P^{-1})^{-1}.(P.R.P^{-1}) & 0 \\ 0 & -f/J \end{bmatrix} . \begin{bmatrix} \mathfrak{q} \\ w \end{bmatrix} + \begin{bmatrix} -(P.L.P^{-1})^{-1}.P.(\dot{L}.P^{-1} + L.\dot{P}^{-1}) & 0 \\ \mathfrak{q}^t.(G/J) & 0 \end{bmatrix} . \begin{bmatrix} \mathfrak{q} \\ w \end{bmatrix} + \begin{bmatrix} (P.L.P^{-1})^{-1} & 0 \\ 0 & -1/J \end{bmatrix} . \begin{bmatrix} \mathfrak{e} \\ m_L \end{bmatrix} \qquad (7)$$

The first term of the state equation is linear, the second is non-linear (state affine) and the control matrix is also linear. The different submatrices of the state equation 7) will be given in Appendix.

4. PHASE CONTROL MODE.

The technique of phase-angle control of a capacitor-run induction motor consists in connecting series semiconductor switches with the main stator winding while the auxiliary winding is directly fed by the AC source. Refered to the equivalent two-phase machine, we can give a model of this converter where e_{ms} and e_{as} become the same AC source and a switch is connected in series with e_{ms}. If the principle of this switch is examined, it can be noted that a firing angle a_1 refered to the AC line is provided and then the conduction angle c_1 depends on the nature of the load (figure 2.).

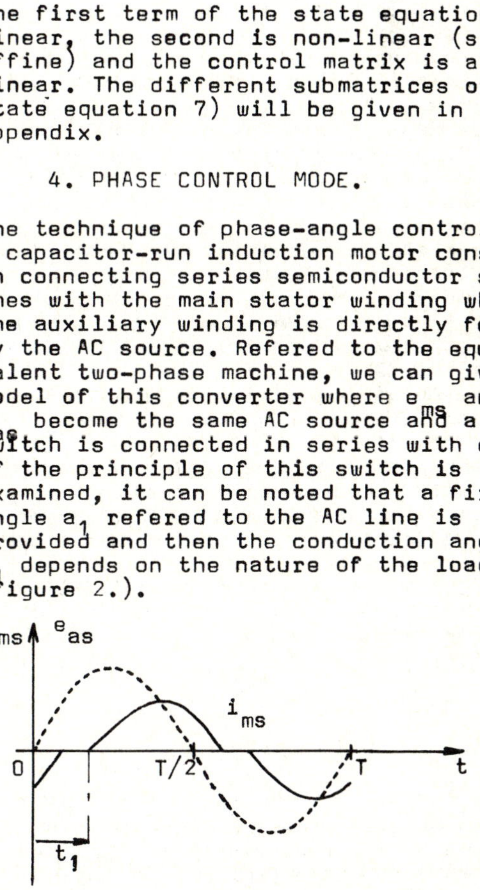

Figure 2. AC line and main winding current in phase control mode.

$t_{mul} = t - (k-1).T$ 8)

where k is the number of the period and:

$t_1 = a_1 . T/2.\Pi$

$t_2 = (a_1+c_1).T/2.\Pi$

Then five intervals will be defined for t_{mul}:

I_1 for $t_{mul} \in [0, t_2-0.5.T]$

I_2 for $t_{mul} \in [t_2-0.5.T, t_1]$

I_3 for $t_{mul} \in [t_1, t_2]$

I_4 for $t_{mul} \in [t_2, t_1+0.5.T]$

I_5 for $t_{mul} \in [t_1+0.5.T, T]$

The conditions of openning and closing the series connected thyristor switch lead to two different sub-models of the machine.

Mode 1. Two stator windings connected to the AC line.

For I_1, I_3 and I_5, the switch is closed and then, the stator windings are fed by the original voltage source. The model can be simulated by means of equation 7) as it was defined in the non-linear analysis.

Mode 2. Auxiliary winding connected to the AC line, main winding openned.

For I_2 and I_4, the switch is openned and then, no more current feeds the machine from the main stator winding. In equation 4), the influence of the main current i_{ds} has to be removed and the voltage at the terminals of this winding is only influenced by the rotor fluxes. For practical means of simulation, the same dimension of the state equation may be conserved with $i_{ds}=0$. In this mode, it is obvious that the conducting angle c_1 is computed and not constant if the transient behaviour is examined. The firing angle has to be greater than a bounded value which coincide with the loss of control.

The model can be simulated from a modified equation 7) with new state variables:

$q_1 = (\, i_{ds} \; i_{qs} \; i_{dr} \; i_{qr} \; v_{ca} \,)^t$

for mode 1.

$q_2 = (\, 0 \; i_{qs} \; i_{dr} \; i_{qr} \; v_{ca} \,)^t$ 9)

for mode 2.

5. INVERTER CONTROL MODE.

The inverter control mode of a capacitor run induction motor consists in connecting both main and auxiliary windings to a square-wave source. Then, e_{ms} and e_{as} are the same voltage as defined in figure 3.

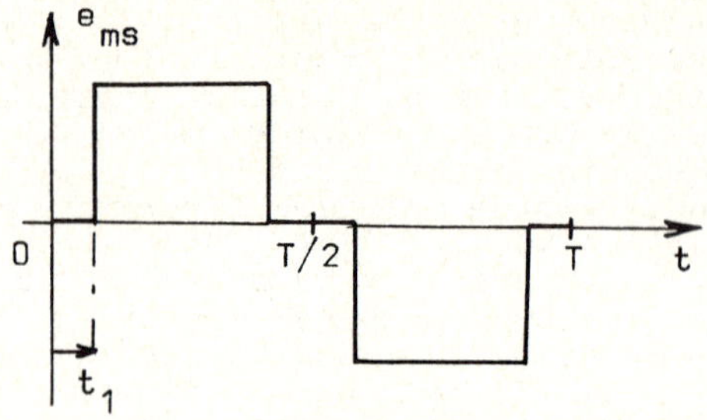

Figure 3. AC voltage in inverter control mode.

In the inverter control mode, a phase-shift is also defined in order to modify the r.m.s. value of the common AC voltage. It is obvious that a_1 is bounded by the angles 0 and $0.5.\Pi$ and corresponds to a time $t_1=a_1.T/2.\Pi$. The three-level control voltage is related to one period by the same time reduction method and a level multiplier X_{mul}.

$e_{ms}(t_{mul}) = X_{mul} \cdot \hat{V}$ 10)

with t_{mul} bounded by 0 and T and

$t_{mul} = t - (k-1) . T$

where k is the number of the period

$X_{mul} = +1$ for $t_{mul} \in [t_1, 0.5.T-t_1]$

$X_{mul} = -1$ for $t_{mul} \in [0.5.T-t_1, T-t_1]$

$X_{mul} = 0$ elsewhere.

The equation 10) provides a describing control voltage function for a period and the simulation of the state vector is given from equation 7).

6. SIMULATION RESULTS.

The previous investigations are applied to simulation of the transient behaviour of the capacitor-run induction motor plugging on the feeding source with both phase control and inverter control mode.

Phase control mode PCM (graphic table I).

The starting performances with constant phase control is presented with an output vector containing main stator voltage, total source current, capacitor voltage, electromagnetic torque, rotor speed and source power.

For a convenient phase control ($a_1 = 0.5 . \Pi$), the capacitor-run induction motor starts approximatively with constant r.m.s. current. It can be seen that the conduction angle vanished when the motor reaches the steady-state. The r.m.s. value of the capacitor voltage is also constant during the transient.

The rotor speed grows very slowly while the electromagnetic torque is always positive with smooth double frequency oscillations. The peak source power is practically constant during the starting period.

Inverter control mode ICM (graphic table II).

The same parameters are simulated for plugging the motor on a voltage inverter with constant voltage and frequency output. The results are approximatively the same as plugging the machine on an infinite power AC line bus. The source current has a short transient mode where its peak value decreases to the steady-state while the capacitor voltage increases till its high final value.
The electromagnetic torque has very important oscillations and becomes negative so as decreasing speed can be observed during the transient.
The comparison between all simulation results and the experiments can be performed for the peak values plugging on AC line and both with the two control modes.

		AC	PCM	ICM
source current A.	sim.	8.1	4.9	10.68
	cal.	8	5	11
capacitor voltage V.	sim.	462	375	537
	cal.	460	370	540
em torque mN.	sim.	2.2	1.2	2.7
starting time ms.	sim.	140	460	130
source power VA.	sim.	2300	1110	3330

7. CONCLUSIONS.

A versatile simulation of transient behaviour of the capacitor-run induction motor fed by power converters has been presented. This simulation has been performed from a non-linear model of the machine with state-space analysis which is very powerful compared with the classical method found in the litterature. The model permitts to simulate the behaviour of the motor fed both with series connected switches (PCM) and voltage inverter (ICM). In the first case the starting transient with constant angle control has been simulated with two sub-models relative to the feeding of the main stator winding. In the second case the plugging on the inverter with constant voltage and frequency output has been investigated with only one non-linear model.
The simulation permitts to calculate the output vector with different electrical and mechanical parameters of the machine. The results are presented on graphic tables and compared with the experiments. This comparison shows the great accuracy of the non-linear model and the versatily of the program allows simulation of other problems.

The simulation permitts to investigate transient, running around a working point, disturbances ,stability and other control problems for a capacitor-run induction motor fed with non-conventional electrical sources.

REFERENCES.

/1/ D.W. NOVOTNY and F. FATH: 'Analysis of induction machines controlled by series connected semiconductor switches',IEEE Trans. PAS,vol.PAS-87, Feb.1968,pp.597-605.

/2/ A.K. WALLACE and al.: 'The design optimisation of a split-phase fan motor with triac voltage control', IEEE Trans. PAS,vol.PAS-94,May 1975, pp.778-785.

/3/ S. NONAKA and K. TSUBOI: 'Effects of distorded power source voltages on condenser motor performances', Elect. Eng. Jap.,vol.97,n°2,1977, pp.49-56.

/4/ T. FUJII and S. YAMADA: 'A simplified analysis of steady-state characteristics of single-phase induction motors controlled by thyristor switches',Elect. Eng. Jap.,vol.97, n°5,1977,pp.64-70.

/5/ C.V. ABROAL and V.B. MURTY: 'Equilent hunting networks of an invertor fed capacitor-run induction motor', Proc. of ICEM,Bruxels 1978,E2/7.

/6/ D.R. KOHLI and P.J. PUROHIT: ' Steady-state analysis of phase-controlled single phase induction motor' ,Elect.Mach.Electromech.,vol.5,1980, pp.473-484.

/7/ G. CAPOLINO and M. HAND BAHIOL: 'Steady state analysis of single-phase induction motor with variable frequency and voltage supply',Proc. of ICEM,Athens 1980,T.3.

/8/ E.H.T. EL-SHIRBEENY: 'Control configured design of single phase induction motor',Proc. of ICEM,Budapest 1982,T.2.

/9/ G. CAPOLINO and Z. SOILIHI: 'Analysis of capacitor-run induction motor controlled with voltage inverter' ,Proc. of ICEM,Budapest 1982,T.2.

/10/ G. CAPOLINO et M. GAUTIER: 'Quelques aspects de la modelisation non-lineaire d'un systeme electromecanique-le moteur asynchrone monophase', Proc. of AMSE, Session 2c, Nice 1983.

APPENDIX.

A1. Parameters of the simulated machine.

$V = 311$ Volts.
$f = 50$ Hz.
Bipolar motor.
$R_m = 14.85$ Ohms.
$R_a = 14.85$ Ohms.
$R_r = 17.82$ Ohms.
$L_d = .591$ H.
$L_q = .591$ H.
$M_d = .549$ H.
$M_q = .549$ H.
$L_r = .591$ H.
$C = 10$ uF.
$J = 3.E-04$ kg.m^2.
$f = 1.4E-04$ USI.
$m_L = 0.$

A2. Matrix identification for different modes.

Mode I.

$-(P.L.P^{-1})^{-1}.(P.R.P^{-1}) = AL$

with:

$AL_{11} = -R_m.L_r/D_d$
$AL_{13} = R_r.M_d/D_d$
$AL_{22} = -R_a.L_r/D_q$
$AL_{24} = R_r.M_q/D_q$
$AL_{25} = L_r/D_q$
$AL_{31} = R_m.M_d/D_d$
$AL_{33} = -R_r.L_d/D_d$
$AL_{42} = R_a.M_q/D_q$
$AL_{44} = -R_r.L_q/D_q$
$AL_{45} = -M_q/D_q$
$AL_{52} = -1/C$ and elsewhere 0.

$-(P.L.P^{-1})^{-1}.P.(L.P^{-1}+L.P^{-1}) = ANL$

with:

$ANL_{12} = M_d.M_q/D_d$
$ANL_{14} = L_r.M_d/D_d$
$ANL_{21} = -M_d.M_q/D_q$
$ANL_{23} = -L_r.M_q/D_q$
$ANL_{32} = -L_d.M_q/D_d$
$ANL_{34} = -L_r.L_d/D_d$
$ANL_{41} = L_q.M_d/D_q$
$ANL_{43} = L_r.L_q/D_q$ and elsewhere 0.

all the terms ANL_{ij} have to be multiplied by the state variable w.

$\dot{q}^t.G/J = AN$ line matrix with:

$AN_1 = -M_d.i_{qr}/J$
$AN_2 = M_q.i_{dr}/J$
$AN_3 = -L_r.i_{qr}/J$
$AN_4 = L_r.i_{dr}/J$ and elsewhere 0.

$(P.L.P^{-1})^{-1} = B$ with:

$B_{11} = L_r/D_d$
$B_{13} = -M_d/D_d = B_{31}$
$B_{22} = L_r/D_q$
$B_{24} = -M_q/D_q = B_{42}$
$B_{33} = L_d/D_d$
$B_{44} = L_q/D_q$
$B_{55} = 1./C$ and elsewhere 0.

Mode 2.

The structure of the different matrices are the same, only the terms which change are given.

$AL_{11} = 0.$
$AL_{13} = 0. = AL_{31}$
$AL_{33} = -R_r/L_r$
$ANL_{12} = 0.$
$ANL_{14} = 0.$
$ANL_{32} = -M_q/L_r$
$ANL_{34} = -1.$
$B_{11} = 0. = B_{33}$
$B_{13} = 0. = B_{31}$

with the parameters:

$D_d = L_r.L_d-M_d^2$
$D_q = L_r.L_q-M_q^2$.

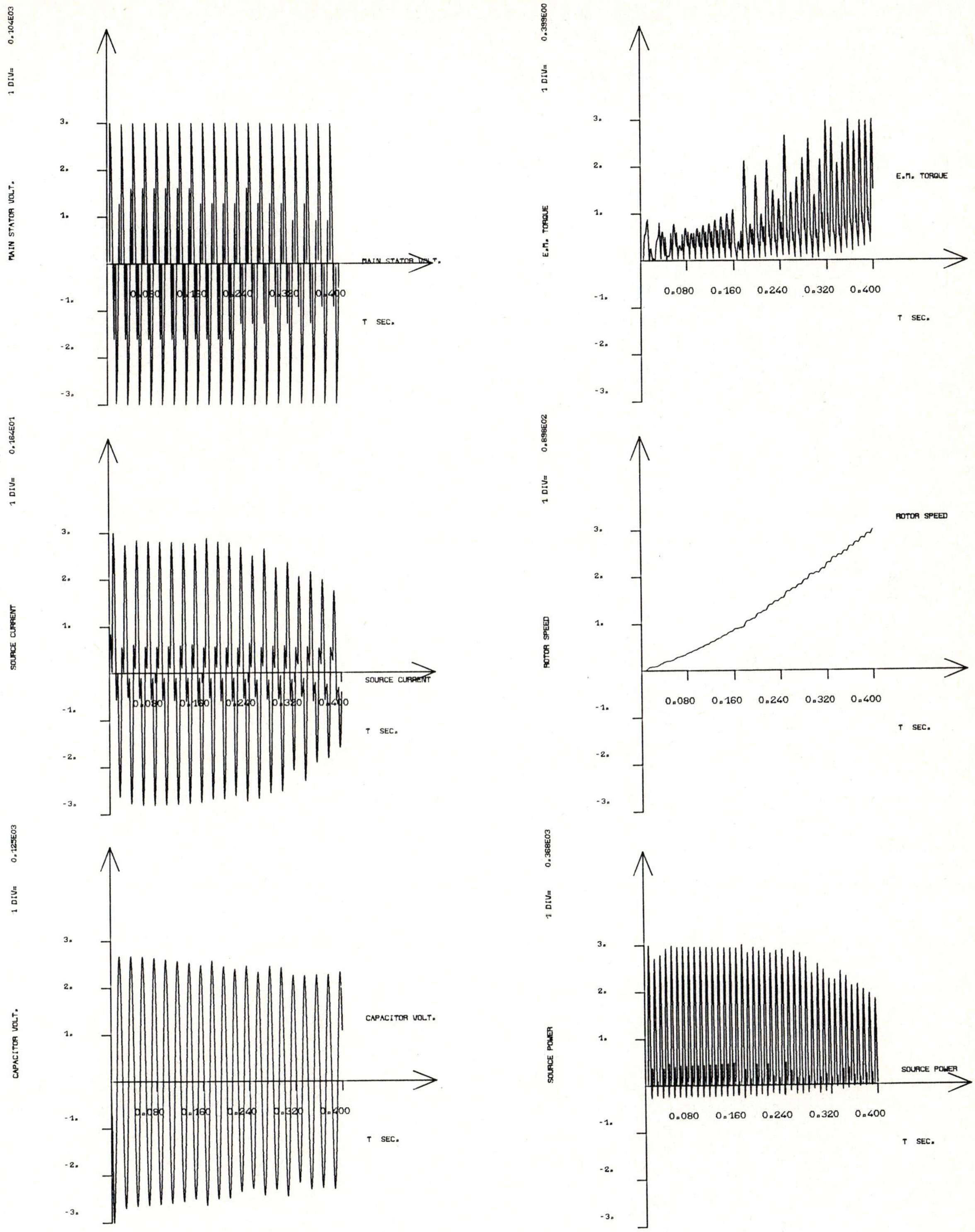

Graphic table I.

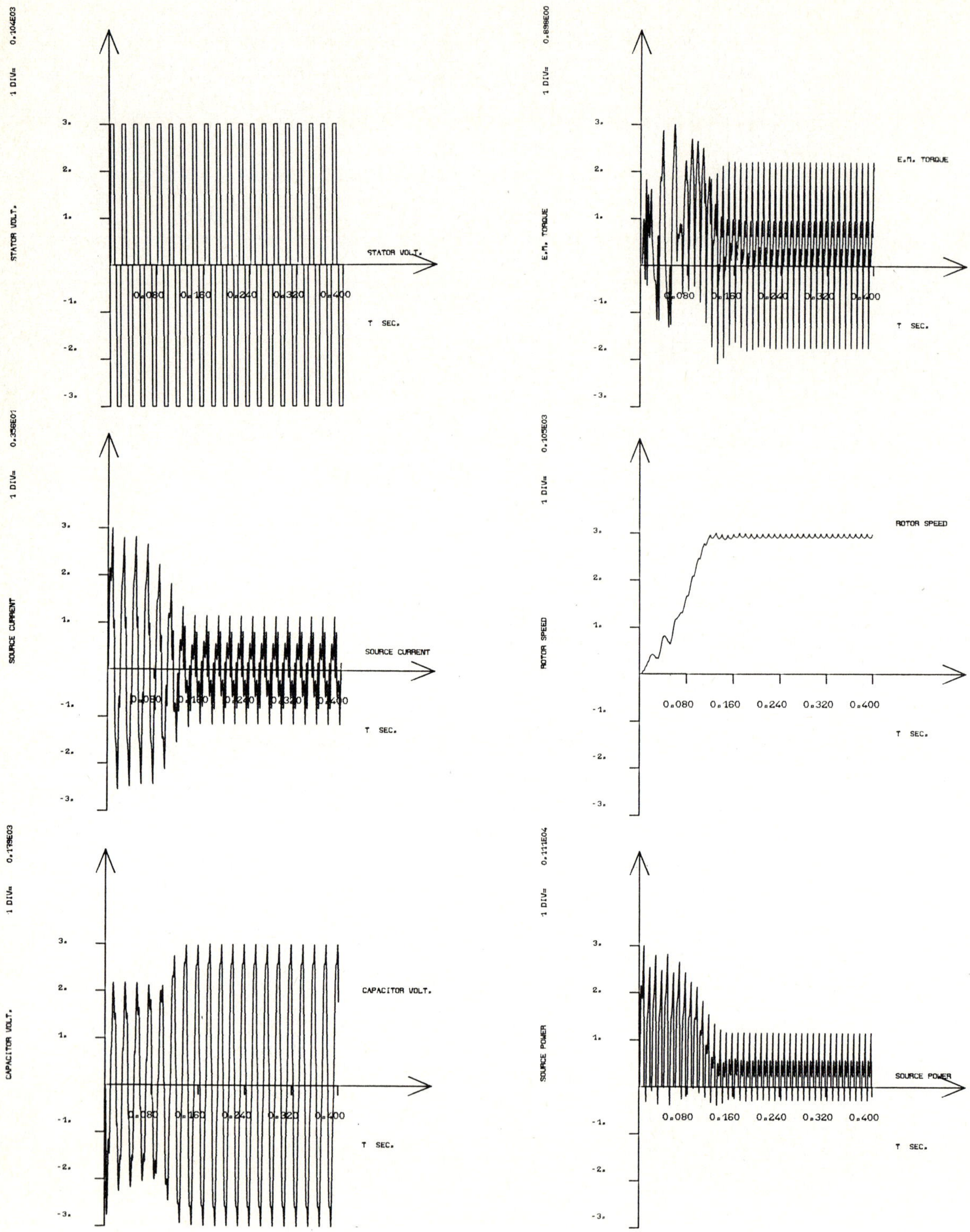

Graphic table II.

Session 4 A

FIELD COMPUTATION IN ELECTRICAL MACHINES (I)

Chairman: W. GEYSEN

ELECTRICAL MACHINES AND CONVERTERS - Modelling and Simulation
H. Buyse and J. Robert (editors)
Elsevier Science Publishers B.V. (North-Holland)/ © IMACS, 1984

THE FINITE-ELEMENT COMPUTATION OF THE FIELD IN A PERMANENT MAGNET MACHINE

A. Hameed, K.J. Binns, A. Vandenput, W. Geysen

Prof. dr. K.J. Binns is with the Dept. of Electrical Engineering & Electronics, University of Liverpool, England.
Prof. dr. ir. W. Geysen, Dipl-Ing. A. Hameed and dr.ir.A. Vandenput are with the Dept. of Elect. Engr., Power Division, Catholic University of Leuven, Belgium.

A particular configuration of synchronous permanent-magnet machine is studied using a finite element method of computation of the magnetic field. Machines of this type have important uses as inverter-fed drives. A model representing the machine is carefully chosen for the efficient evaluation of torque for imposed current and load angle.
From the range of parameters investigated a rotor design can be selected for a given magnet material. Results are presented for ferrite and rare-earth magnets. The finite element program shows good convergence for non-linear magnetic materials.
The boundary-value problem is treated as two dimensional. The region is discretised with a careful choice of elements which is made in the light of an investigation of the error due to the choice of the number of finite-elements. Convergence is studied for two types of magnet material and a convergence factor is proposed. The machine torque is evaluated from the flux density in the air gap.
The results are processed so that the torque can be expressed as a function of load and imposed voltage. Operation as a motor supplied from a voltage source or current-source inverter is briefly summarised.

1. INTRODUCTION

Permanent magnet synchronous machines have undergone rapid development in the last ten years (1). The main reason for this is the development of new configurations coupled with the production of new magnetic materials. Some of these materials make use of the rare-earths and have a relatively high energy density. Others such as the anisotropic ferrites, which have been available for quite some time, have the merit of cheapness coupled with a reasonably high coercivity. Another factor in the development of synchronous machines is the steep reduction in cost and size of inverters. Current source inverters, many using transistors as output devices, are likely to play an increasing part in machine drive systems in the future.

There are several types and configurations (2) of permanent magnet machine and this paper is concerned with one of them which lends itself to a requirement for low speed stability and conveniently has a relatively high pole number, that is 8 pole or higher. Some designs involve salient magnets but, apart from the danger of demagnetisation, they suffer from the problem that the average flux per pole pitch is, of necessity, less than that in the magnet itself. The configuration discussed here leads to a gap density on load which is higher than that in the magnet. The machine is not self starting so that for motoring operation an inverter is required. The combination of inverter and machine has all the characteristics of an efficient drive system.

The performance of the machine in terms of the synchronous torque and the flux density wave form in the air gap are evaluated by a two-dimensional finite element analysis. The calculation was carried out for a prescribed range of constant current operation. The performance under constant fundamental gap density was obtained using linear interpolation. Care is taken to ensure that the discretisation is sufficiently precise to give useful results without excessive use of processor time. The program was developed at the University of Leuven with guidance and advice from Professor K.J. Binns, of the University of Liverpool.

2. MACHINE CONFIGURATION AND FLUX DISTRIBUTION

A repeatable section of the permanent magnet excited rotor is shown in Figure 1 together with flux lines on no-load. It is seen that rectangular block magnets tangentially magnetised are recessed into rectangular slots in a rotor made of laminated or solid steel. A multipolar field is established across the air gap and it is evident that the air-gap field can be higher than that in the magnet. The rotor dimensions are critical and can be broadly established regardless of the precise stator geometry. For this reason the stator is represented without slotting and the stator current is disposed over the elements adjacent to the stator surface.

Figure 2 shows a field plot on load for a rare earth magnet. The discontinuity in the flux lines due to the use of first order elements can be seen though it is of no practical significance. The flux density distribution in the air gap is also shown, the 'dashed' line being the no-load distribution. Figure 3 shows the field map for the same load but with excitation from an anisotropic ferrite magnet of the same width but having a different depth. The 'air-gap density' for a pole pitch is also shown.

Though lower in magnitude than for the rare earth magnet, the shape of the flux density distribution is not very different.

The shaft is normally non-magnetic as shown in the previous field maps, but for some rotor configurations this is not essential. Figure 4 shows a field plot for a magnetic shaft and it is apparent that the leakage flux below the magnet is not excessive.

3. THE VARIATION OF TORQUE WITH LOAD

The torque variation for a given geometry as the load varies is quite interesting. The load angle of the machine is most conveniently defined as the angle between the peak of the fundamental gap density and the centre of a rotor pole. This angle is, of course, zero on no-load. The length of the machine was assumed to be one meter. Figure 5 shows the variation of torque with load angle, the torque being obtained by summing BIL for each element. Figure 6 shows the variation of torque with the current loading for different values of the fundamental gap density. The characteristic 'C' is a useful one. The variation of torque with power factor is shown in Figure 7. The relation between the pull out torque and the stator current loading is shown in Figure 8. The curve, at low stator current, is very close to a straight line. At higher values of the stator current, however, the slope increases since the operating point of the magnet is now significantly changed.

4. CONCLUSIONS

It has been demonstrated that the electro magnetic design of the rotor can be achieved using a simplified model. Optimisation of torque can be achieved for a given rotor diameter and magnet material. Extensive computation has shown for a given magnet material the relation between pull out torque and rotor geometry. The rotor dimensions giving the highest pull-out torque per unit volume of magnet for a given rotor diameter is clearly of interest particularly for rare-earth magnets. The work of Mr Hameed will be submitted quite soon as a doctoral thesis.

ACKNOWLEDGEMENTS

The computational work was carried out as research for a doctorate by Mr Hameed of the Catholic University of Leuven. His work was directed by Professor K.J. Binns of the University of Liverpool and Dr A. Vandenput and Professor W.J. Geysen of the K.U. Leuven, under an arrangement for collaboration between the Universities concerned. The authors wish to thank ABOS for offering a scholarship to Mr Hameed to carry out this work and also Professor R. Coussement, Director of the Institute for Nuclear Physics of K.U. Leuven for offering them the computer facilities needed for the extensive numerical analysis.

REFERENCES

[1] Binns, K.J., Barnard, W.R. and Jabbar, M.A., Hybrid permanent magnet synchronous motors, Proc. I.E.E. Vol. 125 No. 3 (1978) 203-208.

[2] Binns, K.J. and Jabbar, M.A., High field self starting permanent magnet synchronous motor, Proc. I.E.E. Vol. 128 (1981) 157-160.

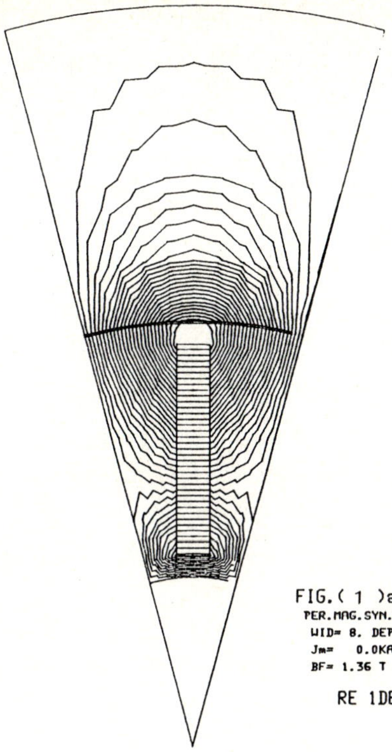

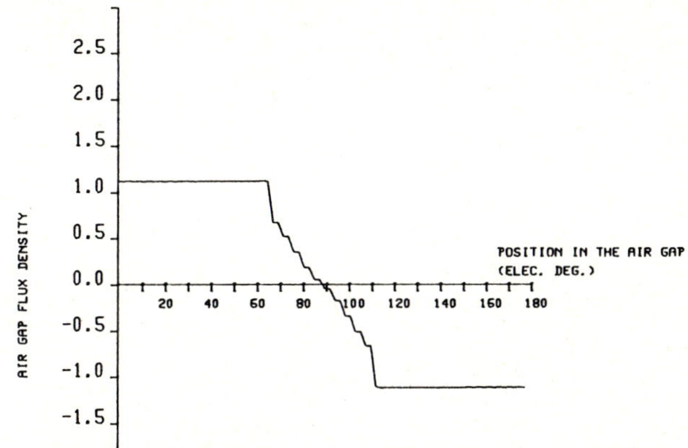

FIG.(1)b
DISTRIBUTION OF B IN THE AIR GAP.

FIG.(1)a FIELD PLOT
PER.MAG.SYN.MOTOR
WID= 8. DEP=50.0 RAD= 40. DM=5.0 DG=.500
Jm= 0.0KA/m, THETAI= 90.00 DEG
BF= 1.36 T ALFA= -1.13DEG. ITER= 12

RE 1DEL. VP=.001

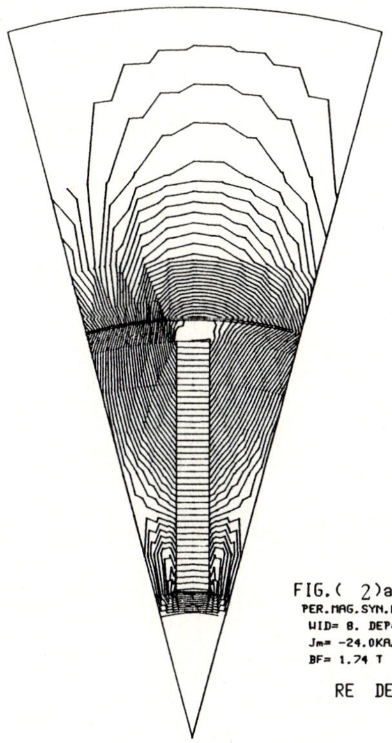

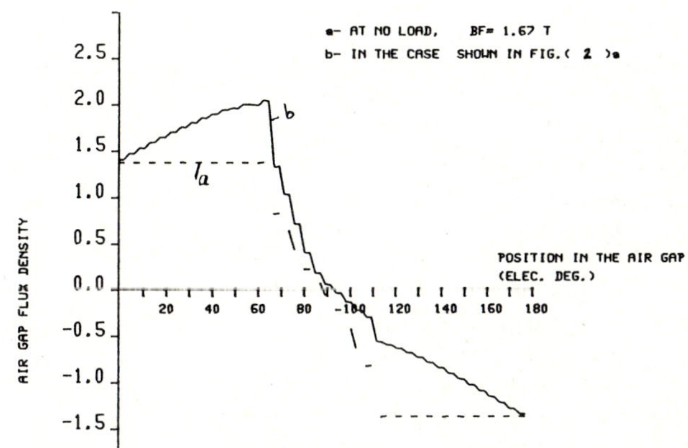

FIG.(2)b
DISTRIBUTION OF B IN THE AIR GAP.

FIG.(2)a FIELD PLOT
PER.MAG.SYN.MOTOR
WID= 8. DEP=60.0 RAD= 30. DM=5.0 DG=.500
Jm= -24.0KA/m, THETAI= 90.00 DEG
BF= 1.74 T ALFA= 17.73DEG. ITER= 15

RE DEL. VP=.001

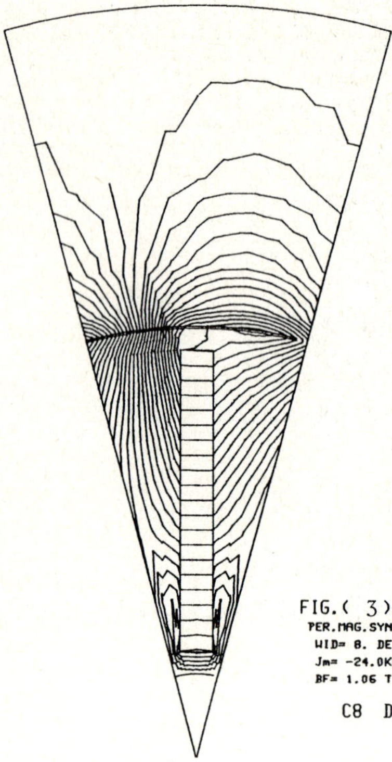

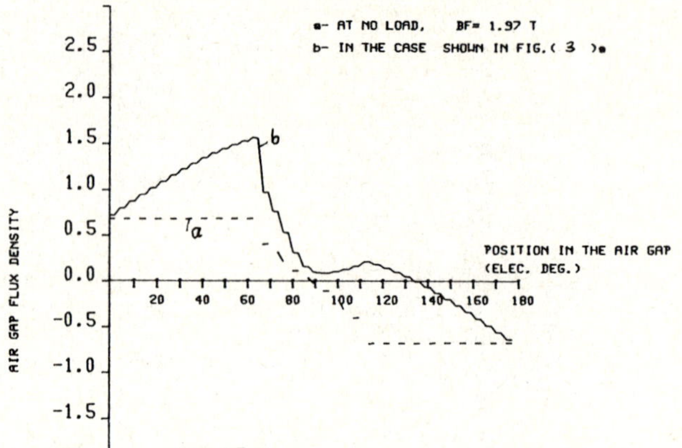

FIG.(3)b
DISTRIBUTION OF B IN THE AIR GAP.

FIG.(3)a FIELD PLOT
PER.MAG.SYN.MOTOR
WID= 8. DEP=70.0 RAD= 20. DM=5.0 DG=.500
Jm= -24.0KA/m, THETAI= 90.00 DEG
BF= 1.06 T ALFA= 37.79DEG. ITER= 15

C8 DEL. VP=.001

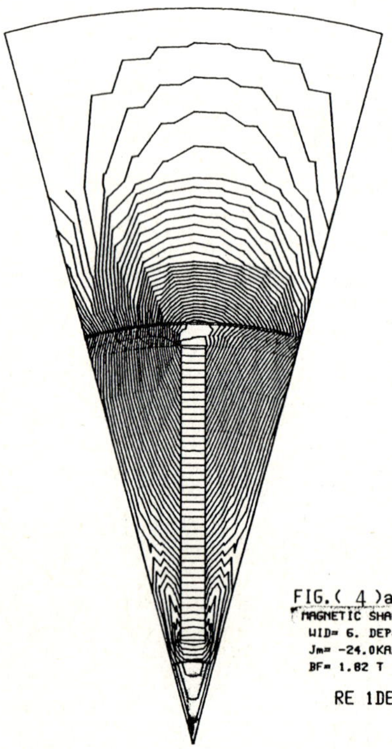

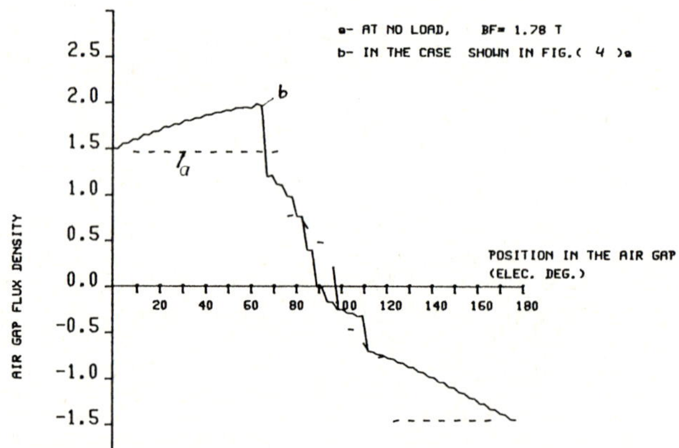

FIG.(4)b
DISTRIBUTION OF B IN THE AIR GAP.

FIG.(4)a FIELD PLOT
MAGNETIC SHAFT MOTOR
WID= 6. DEP=70.0 RAD= 20. DM=5.0 DG=.500
Jm= -24.0KA/m, THETAI= 90.00 DEG
BF= 1.82 T ALFA= 15.39DEG. ITER= 12

RE 1DEL. VP=.001

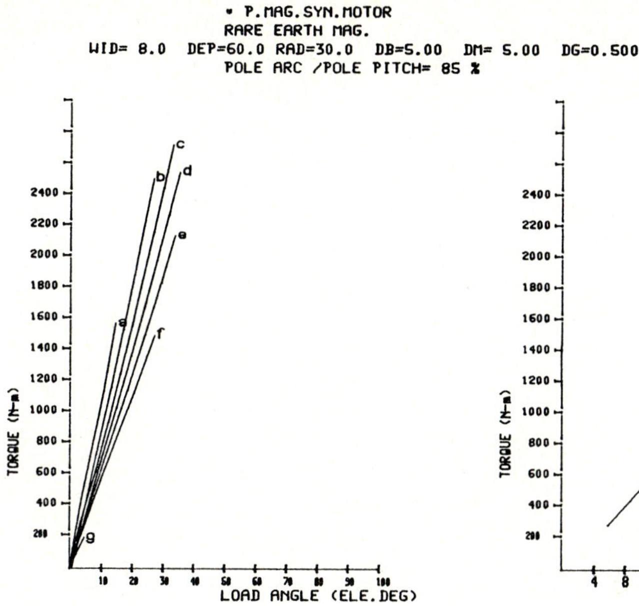

Fig. 5. The variation of the torque with load angle for different values of fundamental gap density(BF).

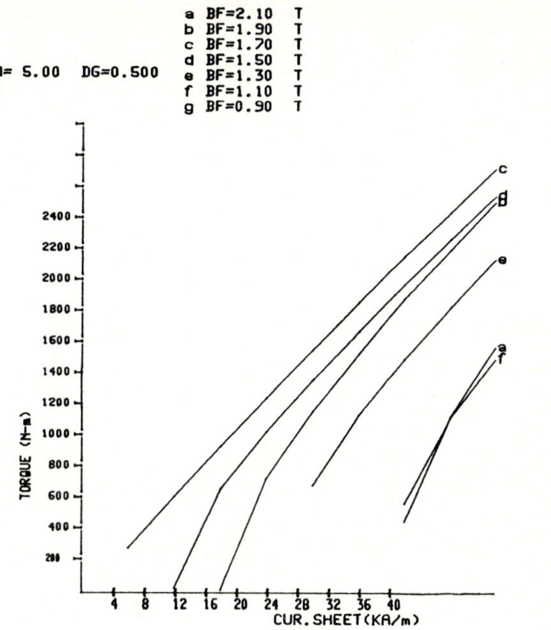

Fig. 6. The variation of the torque with current sheet density for different values of fundamental gap density.

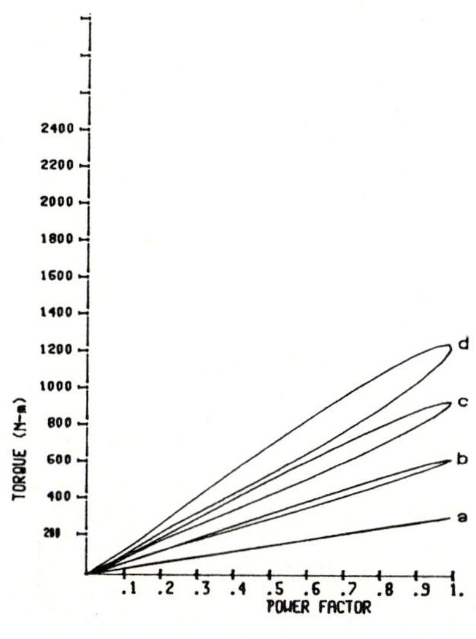

Fig. 7. The variation of torque with power factor of the machine for different values of stator current sheet density (Jm) in KA/m.
a) Jm=6.0 KA/m b) Jm=12.0 KA/m
c) Jm=18.0 KA/m d) Jm=24.0 KA/m

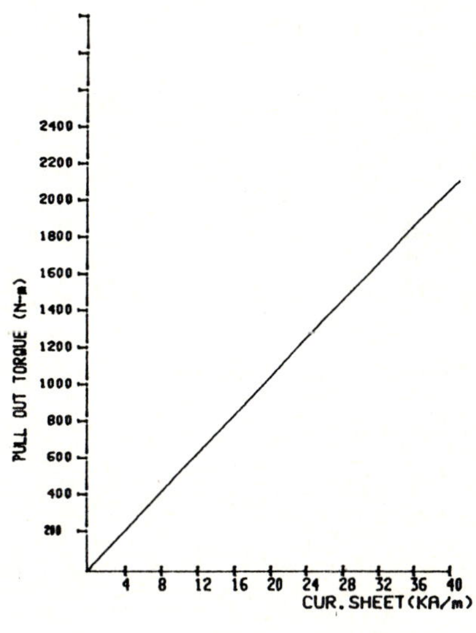

Fig. 8. The relation between the pull out torque and stator current sheet density (Jm).

ELECTRICAL MACHINES AND CONVERTERS - Modelling and Simulation
H. Buyse and J. Robert (editors)
Elsevier Science Publishers B.V. (North-Holland)/ © IMACS, 1984

FIELD INVESTIGATION ON A COMPENSATED D.C. TRACTION MOTOR BY FINITE ELEMENT METHOD

M. Rizzo[+], A. Savini[++], and C. Zimaglia[+++]

+ Università di Palermo, Palermo, Italy
++ Università di Pavia, Pavia, Italy
+++ Politecnico di Torino, Torino, Italy

The finite element method is applied to field analysis and construction of flux density diagrams in a d.c. compensated motor. The machine is supposed operating at rated load and low excitation; the flux density distribution in interpole airgap is taken into particular account, and compared with an ideal one, coming from a detailed analysis of inductive effects on all coil sides in a commutating slot. Different pole arc compensation degrees and their effects on such a distribution, being pole pitch compensation degree kept constant, are also considered.

1. RESEARCH PURPOSES AND LIMITS

The fundamental task of providing a d.c. machine with a satisfactory commutative behaviour involves, in the various cases, to fulfill specific functional conditions which today applications are making more and more severe as concerns values and change rate of surcharges, speed regulations by field weakening, and so on. However, in normal cases, design and arrangement processes develop through an usual succession of steps:
a) choice of a rotor winding having, with a prefixed value of peripherical current density, the maximum spreading as concerns the number of commutator bars; the consistency of such a condition with technological opportunities implies a condition on the choice of rotor diameter, and must then be considered as startpoint for the whole design;
b) development of rotor winding and commutator design such as to allow all known improvement tools for commutation to be adopted, as concerns slot number for a given bar number, slot form and size, commutation bar covering by brushes, etc.;
c) design of pole face winding, when surcharges, field regulations and any cause apt to emphasize bar-to-bar voltages make it necessary. Two conditions need then to be fulfilled: - pole arc compensation degree of armature reaction (to be, from now on, shortly indicated by ACD) to approach 100%; - stator teeth pitch to be prime with the rotor one, to avoid running irregularities;
d) evaluation of a mean "reactance voltage", which, as resulting from a schematic simplification, describes commutation in a widely approximate way, and, as based on particular assumptions among the several possible ones, has a purely comparative validness;
e) assignment to interpole of presumably proper shape, size, airgap;
f) approximate evaluation of pole pitch compensation degree of armature reaction (to be, from now on, shortly indicated by PCD), according to the previously evaluated reactance voltage and the selected interpole airgap;
g) design and construction of devices, apt to retain, in dynamic conditions, the maximum possible amount of commutation guarantees, as pursued for permanent operation (interpole and yoke laminations);
h) test adjustment of interpole gap, if required, PCD being normally kept fixed.
Being test results on similar and similarly loaded machines not available, or particularly critical situations being expected, the factors which enter the commutation and are neglected in the above mentioned approach call for a deeper insight. In particular, high speed operation emphasizes the differences between the real commutation process and its usual linearized model, so that no role but a roughly indicative one can be recognized to the latter. A thorough treatment of the commutation problem (old in its general formulation, but getting new contents in new operation frameworks) should take into account all geometric

and electromagnetic, time depending conditions, which characterize the space wholly interested by the active parts of the machine; such elements as number and distribution of cooling holes, local saturation conditions, permeance cyclic variations and others may then get importance. Advanced methods for field analysis can help the approach, allowing simplifying assumptions to be abandoned, or partial and local effects to be more exactly considered /1/.

The present paper is oriented towards such a direction. For a given machine geometry (including shape and size of airgaps) and for a given PCD (ratio between statoric and rotoric interpole linked m.m.f.'s), its purposes are:

- to plot resulting flux density distribution along a pole pitch;
- to compare flux density values and distribution in the interpole airgap with an ideal situation, as could be determined through the individuation, as detailed as possible, of e.m.f.'s to be compensated during commutation;
- to check, on such a comparison, the previously and conventionally established design data;
- to investigate possible consequences on flux density distribution in interpole airgap, owing to ACD changes, i.e. to different distributions of statoric m.m.f.

The investigation applies to a traction motor supposed operating with maximum series field shuntage, at rated armature voltage and current. Such a condition, in practical cases, is the most critical for commutation, owing to the corresponding high speed. However, the choice is mainly due to the special questions to be examined, independently on: - conventional calculation methods being presumably sufficient in the particular case for a good commutation result, to be reached if necessary through reasonable test room airgap settlements; - ACD changes being extended far beyond practically occurring ones, and without considering other negative effects on commutation, which so large changes, through resulting bar-to-bar voltages, might introduce. On the contrary, the emphasizing of causes can give more accurate conclusions on effects to be drawn within practical limits.

2. THE MOTOR UNDER EXAMINATION

The following one-hour performances of the motor have been assumed: 550 kW; 1500 V; 400 A; 800/1800 rpm. Correspondingly, a design of active parts was laid down, according to the cross section reported in Figure 1.

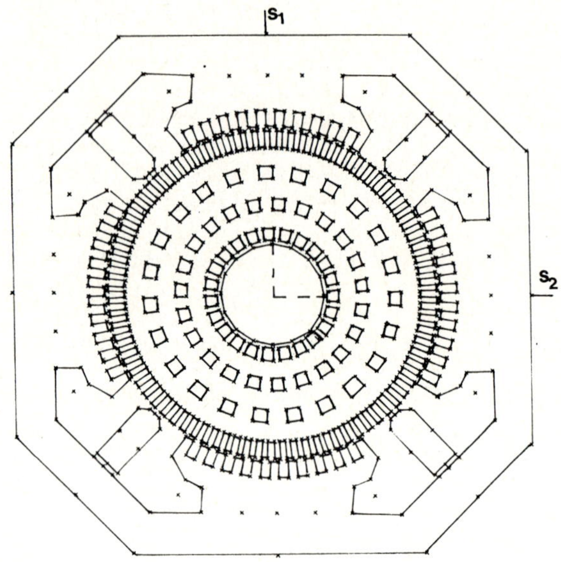

Figure 1: Cross section of the motor

Some basic data for commutation calculations are here detailed.

Pole number: 4.

Rotor core: external diameter 550 mm; internal diameter 170 mm; axial length 470 mm; 3 concentric hole sets; 82 open slots, width 10,4 mm, height 32 mm.

Rotor winding: simple lap type; 10 conductors per slot, each formed by two 1.32x5 (mm) parallel wires; slot pitch 20; commutator pitch 1; equalization pitch 205.

Commutator: diameter 450 mm; bar number 410; brush tangential width 20 mm.

Main pole: arc 310 mm; airgap 6 mm; 10 semiopen slots, constant width 14.8 mm, total height 27.5 mm.

Main pole winding: 40 = 14 + 14 + 12 turns per pole, copper size 4.5 x 10 (mm); pole connection in two parallel paths. In the assumed operating condition with maximum field weakening at rated load, only the winding section having 12 turns per pole is supposed to be active.

Pole face reference winding: 7 conductors per slot, each formed by two 1.5 x

10 parallel wires; pole connection in two parallel paths.
Commutating pole: core width 48/20 mm; bevel angle 30° against pole axis direction; central airgap 9 mm; amagnetic additional gap between pole and stator yoke 3 mm.
Stator yoke: octagonal shape; height 68 mm; axial length 740 mm.
With the established pole face reference winding the usual calculations of reactance voltage suggest for reference interpole winding 28 turns per pole, when pole connection in two parallel paths is adopted; copper size is then assumed to be 2.5 x 16 (mm).
The corresponding reference compensation degrees are then: PCD 123%; ACD 97.3%. The former shall be kept constant from now on. Eventual required modifications should concern: in final design the pole shape; in final design or/and in test settlement the pole total gap.
A step towards a closer, although still approximate evaluation of inductive effects associated with commutation is allowed by the method of "partial variations"; it gives the diagram of e.m.f.'s being induced in a "coil side" apt to represent all real coil sides in a slot according to their commutation sequence /2/. The basis of the diagram is the displacement, normally expressed in commutator bar number, of coil during overall commutation of slot conductors; it depends on winding constitutive features and commutator bar covering. The ordinate values are determined time by time by self or mutual inductances (between coil sides in the same slot and the same layer; in the same slot and different layers; in different slots magnetically coupled through teeths and interpole body; in end connections); ordinate changes occur when anyone among coil sides in the considered slot or in the next ones, magnetically coupled, begins or ends to commutate. Single contributing e.m.f.'s are however evaluated on the assumption of current linear change. The improvement in the commutation analysis consists then in replacing a constant linearization of commutation for a single real coil along its actual duration, by a piecewise linearization for a "mean" coil along the whole slot commutation time.

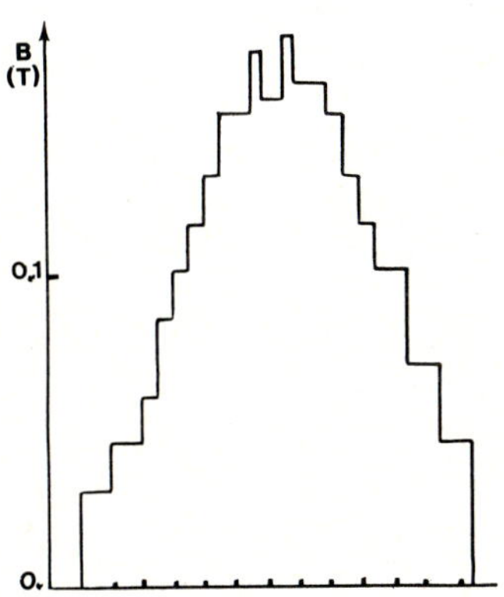

Figure 2: Ideal flux density distribution in interpole airgap

Figure 2 gives the diagram for the above rotor winding; the ordinates are directly expressed as flux density values. A few simplifications were brought in, by averaging some narrow or small steps. The diagram should, as strictly as possible, be covered by the flux density distribution set up by the resulting interpole linked m.m.f..
First of all, interpole shape, dimensions and gaps, with the above indicated reference turn assignments, shall be checked on diagram covering , and possible design modifications shall be individuated. Pole face conductor number shall then be increased or decreased, while opposite changes occur in the interpole winding, and corresponding situations compared with the ideal diagram.

3. FIELD ANALYSIS

The finite element method /3/ is here employed to obtain an accurate field simulation. For the sake of simplicity, axial end effects are neglected, and so a two-dimensional simulation of the cross section of the motor is performed. Other assumptions are listed below.
All magnetic materials are taken to be isotropic and homogeneous. Hysteresis effects are neglected. Saturation is expected to play in general an important role in field configuration. However, owing to the assumptions made in Sec. 1 about the operating conditions to be studied (for which the main pole m.m.f. is not for itself sufficient to establish

saturation levels in magnetic media, and the armature reaction is at least partially compensated), the iron reluctivity may be supposed constant to a first,not arbitrary,approximation. Being the rotor slot number per pole pair kept odd to allow half slot pitch shortening, and also the total cooling hole number being odd, the whole cross section needs to be considered. A complete analysis should take into account several positions of rotor against stator, from alignment of interpole axis with a rotor slot axis, up to alignment with a rotor tooth axis. The consideration of the extreme relevant positions (to be respectively denoted A and B, and to be approached in successive pole pitchs) is thought to be sufficient for a consistent methodological exposition and for the validness of conclusions.
An exact picture of rotor m.m.f. distribution could, in principle, be obtained only through a convergent sequence of trials, in any subsequent two of which an interpole flux density distribution and a current space distribution in commutating intervals should correspond to each other. The first step of such an iterative process assumes current to vary linearly between two equal and opposite values in each "commutation zone" (i.e. in the already mentioned interval corresponding to the whole commutation of a moving slot, and forming the basis of the partial variation diagram reported in Figure 2). Flux density diagram have been drawn on such an assumption.
For discretizing the field region, first-order triangular element are chosen, as easy to treat mathematically and suitable to accomodate complex profiles. The initial mesh of elements is so drawn as to allow easy adjustments when geometrical variations have to be introduced. Particular care is taken on numbering vertices as to assure the matrix bandwidth of the final system of equations to be as narrow as possible. Such a system,as resulting, in terms of vector potential, from the finite element formulation of the problem, is solved by the frontal method /4/, making effective use of the properties of sparsity, symmetry,positive definiteness of the coefficient matrix. The only boundary condition is that vector potential is zero along the external stator profile, i.e. no flux fringing is supposed to occur in the external region around the stator.
The finite element mesh resulting from discretization comprises 920 nodes and

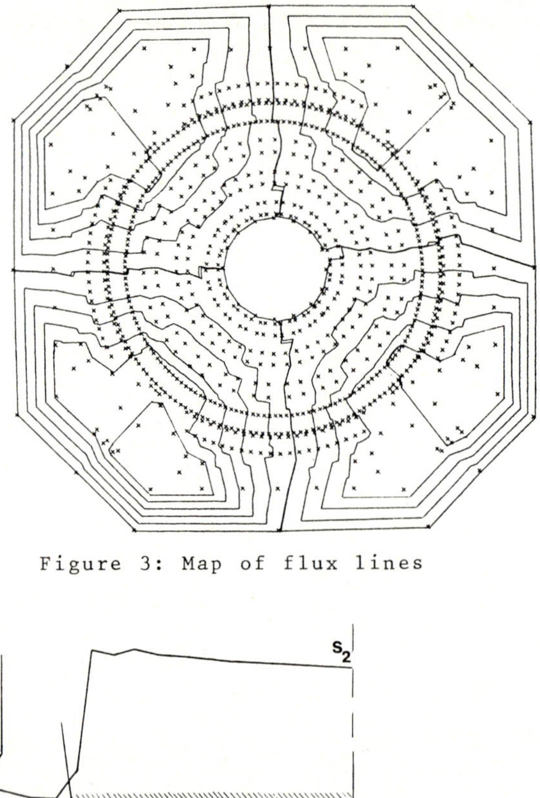

Figure 3: Map of flux lines

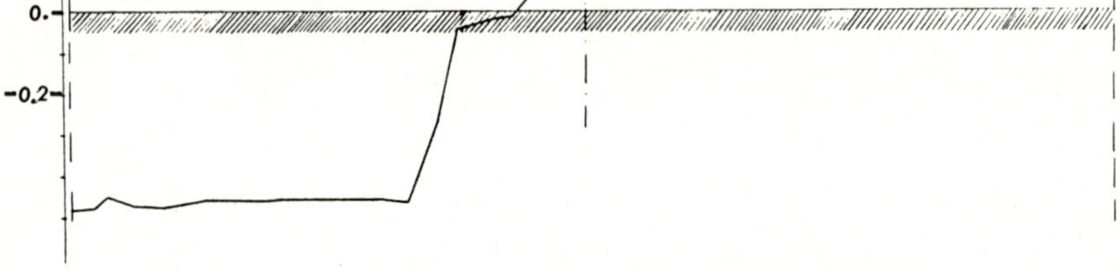

Figure 4: Flux density diagram for unslotted machine; pos. A

1808 elements. Computations were performed on a Honeywell DPS 8/44 digital computer at Pavia University; each field calculation took about 300 s CPU time.

4. COMPUTED RESULTS IN THE REFERENCE CONDITION

The analysis in the reference condition assumes as absolute values of the involved m.m.f.'s:

- total m.m.f. contribution for each rotor slot outside the commutating zone: 1000 At;
- total m.m.f. contribution for each pole face slot: 1400 At;
- interpole m.m.f.: 5600 At;
- main field m.m.f.: 2400 At.

Based on such values the following dia-

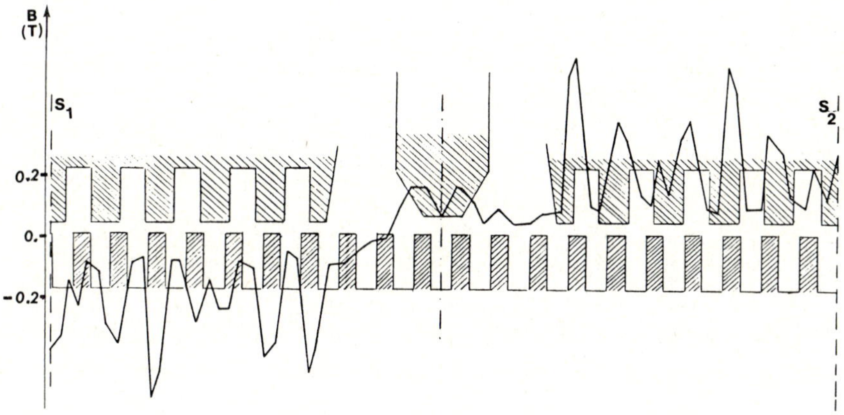

Figure 5: Flux density diagram for real machine; pos. A

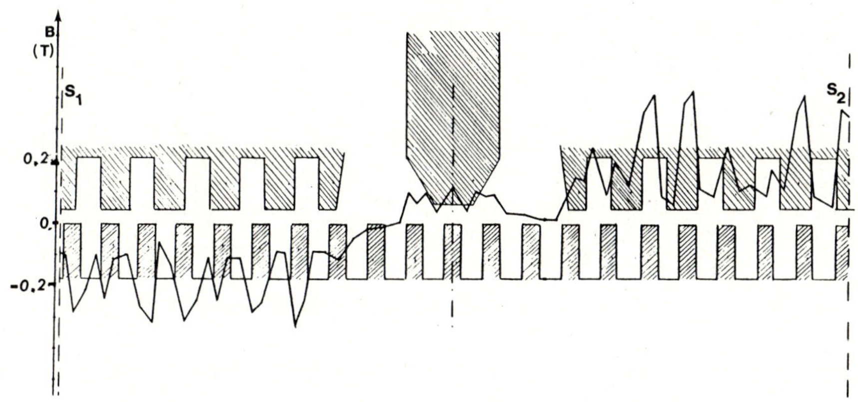

Figure 6: Flux density diagram for real machine; pos. B

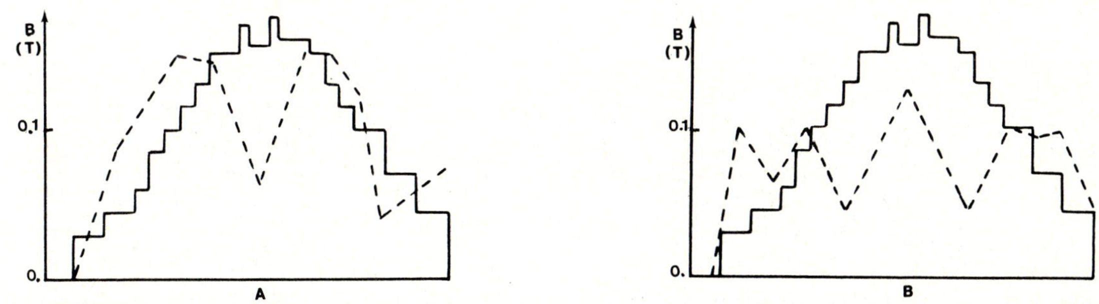

Figure 7: Flux density in interpole airgap (full line:ideal diagram; broken line:computed diagram)

grams were laid down:
a) map of flux lines for the whole real machine (Figure 3);
b) flux density diagram along a pole pitch (position A) for the unslotted machine (Figure 4);
c) id. for the real machine (Figure 5);
d) flux density diagram along a pole pitch (position B) for the real machine (Figure 6);
e) detail for interpole airgap region, desumed from Figures 5,6, and comparison with ideal diagrams (Figure 7 A,B).

Through a proper mean consideration of differently positioned effects of slotting, Figure 7 shows that the obtained diagrams are not too far from the ideal ones, at least as concerns their forms and positions; therefore the established commutating pole shape can be retained, and the design of the machine, as concerns commutation, gets a substantial confirmation by the performed analysis. Clearly improvements are still expected; however they may reasonably be sent to adjustments of sheets in test room, and

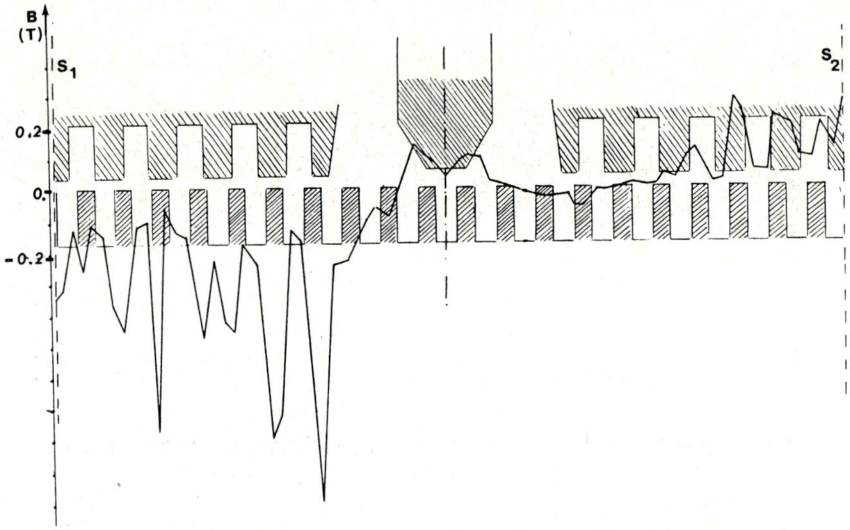

Figure 8: Flux density diagram (hypocompensated case)

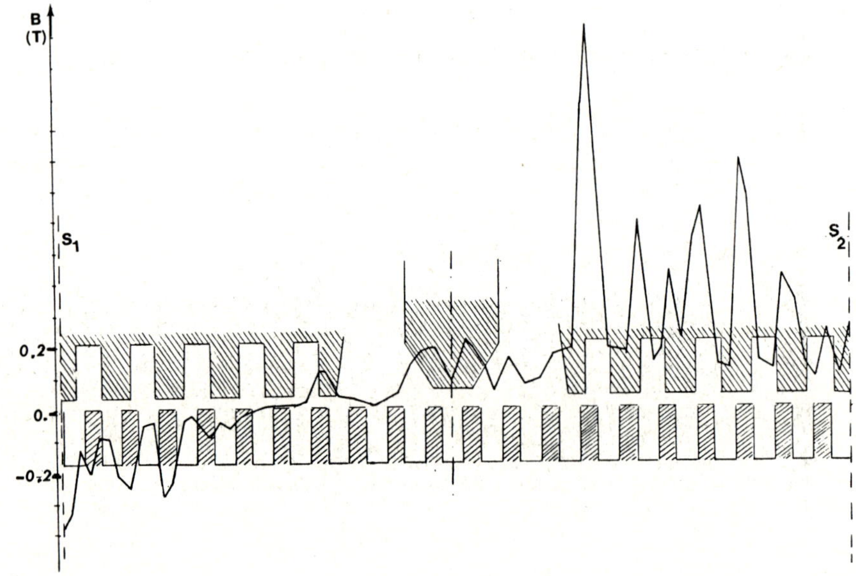

Figure 9: Flux density diagram (hypercompensated case)

coordinated with proof results in all other significant operation conditions.

5. COMPUTED RESULTS IN OTHER CONDITIONS

In the hypocompensated and respectively hypercompensated conditions, the following values differ from the reference ones:

- total m.m.f. contribution for each pole face slot: 800/2000 At;
- interpole m.m.f.: 8600/2600 At;

The corresponding bar numbers per slot are 4 and 10; turn numbers per pole 43 and 13. ACD values are 55.6% and 139%. Figures 8,9 show the flux density diagrams along a pole pitch, both for position A only; Figures 10,11 detail interpole airgap situations.

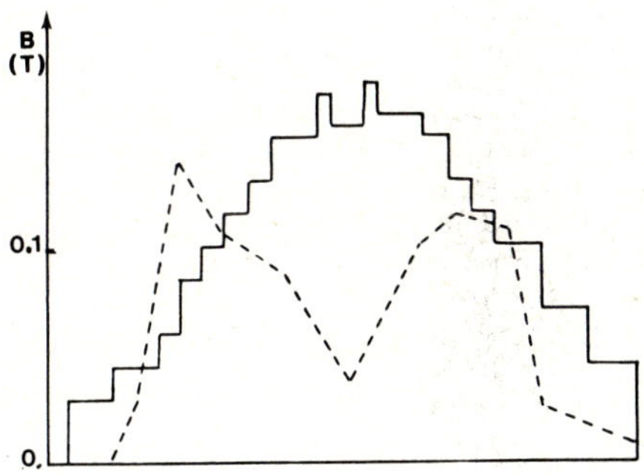

Figure 10: Flux density in interpole gap in the hypocompensated case (full line:ideal diagram;broken line:computed diagram)

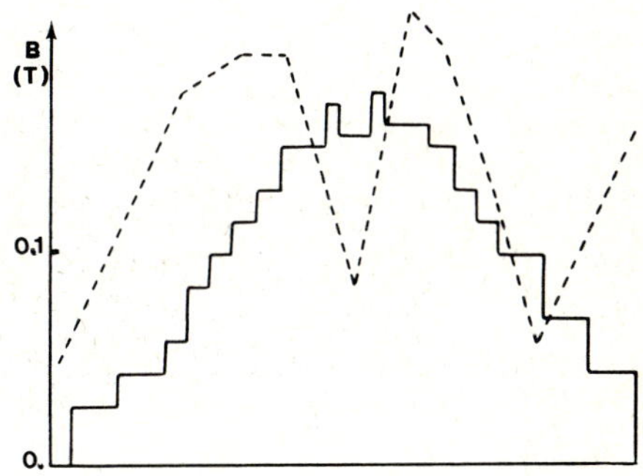

Figure 11: Flux density in interpole gap in the hypercompensated case (full line:ideal diagram;broken line:computed diagram)

The hypocompensated case, even though accepting the pole shape, suggests a reduction of the total gap, or an increase in the interpole turns, i.e. in PCD. The hypercompensated case appears less adjustable in either way, owing to the remarkable advancement of flux density distribution on the required one. This fact can justify difficulties in test room settlement of commutation, sometimes occurring even with ACD values of about 110%. In effect (also leaving out any consideration concerning the dynamic effect of a positive internal feedback) hypercompensation, when not imposed as migth be in large a.c. commutator motors, should be in general avoided even for permanent operation reasons.

6. REFERENCES

/1/ Ahamed,S.V. and Erdelyi,E.A., Flux distribution in DC machines on-load and overloads, IEEE Trans,PAS 85 (1966)

/2/ Loutzky,S., Calcul pratique des machines électriques à courant continu, (Editions Eyrolles, Paris, 1963)

/3/ Zienkiewicz,O.C., The Finite Element Method in Engineering Science (Mc Graw Hill, New York, 1971)

/4/ Irons,B.M., A Frontal Solution Program for Finite Element Analysis, I.J.N.M.E., 2 (1970)

ELECTRICAL MACHINES AND CONVERTERS - Modelling and Simulation
H. Buyse and J. Robert (editors)
Elsevier Science Publishers B.V. (North-Holland)/© IMACS, 1984

MODELLING OF LINEAR INDUCTION MOTOR: ANALYTICAL-NUMERICAL METHOD

Aurio Gilberto Falcone and Jorge Mieczyslaw Janiszewski

Escola Politécnica da Universidade de São Paulo

São Paulo - Brazil

Resistance and slot leakage inductance of equivalent circuit for a linear motor are initially calculated by analytical method. Comparison is carried out with measurements made in a real-size model of slot. Then, for better accuracy, finite element method is used in association with analytical values.

1. INTRODUCTION

The aim of this analysis was the search for data in order to design a linear induction motor with single primary [1], [2], [3] and where the secondary was a continuous sheet of non-magnetic metal, at high temperature. The motor was intended to be used in sheet transportation.

Owing to the rigorous thermal conditions, the only viable way of construction, is with a water cooled primary or, at least, with conductors at a certain distance from the iron, allowing a clearance between the conductors and the iron. Besides, in order to avoid problems with insulation, the motor will operate with low voltage (some tens of volts) and consequently with high currents (some thousands of amperes). In this way, the proposed winding was a single layer one, concentric, and with full-pitch.

The mentioned conditions have resulted in slot shapes and conductor position inside the slots very different from those used in traditional cylindrical motors.

The shape and large cross-section of the conductors give rise to skin effect, and non-uniform current distribution will exist even at low frequencies (50 or 60 Hz), modifying the values of slot leakage inductance and conductor apparent resistance.

Among several proposed slots for the winding, two of them can be seen in Fig. 1(a) and (b). The last is the most appropriate for this application: it presents at the top a non-ferromagnetic cylinder with high mechanical and thermal resistance, used as support and sliding element.

Being the motor linear and with open gap, the current is very high, mainly due to the magnetizing component. Then, series parameters of equivalent circuit are very important (high voltage drop) and must be accurately calculated. Since the slot shapes are complicated, the traditional method of calculating the inductance by integration of flux [4], [5], [6] with uniform current distribution is very inaccurate. So, two methods for the determination of model parameters are presented: the first is analytical, starting with Maxwell equations and the second is a numerical one, based on finite elements.

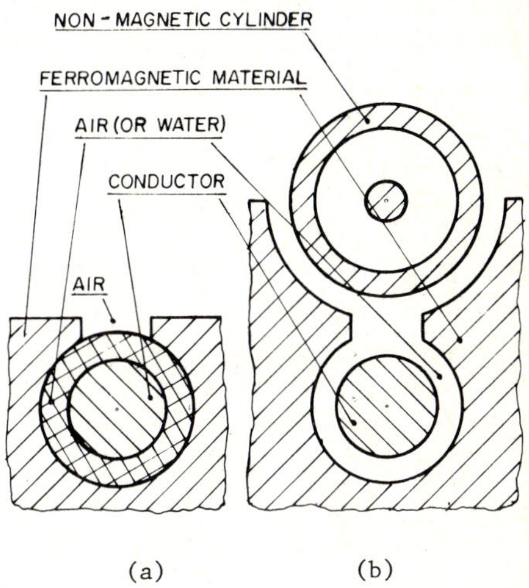

Fig. 1 - Proposed slot shapes and conductor position

2. LEAKAGE INDUCTANCE AND APPARENT RESISTANCE

2.1 - Analytical Method

The following method is an adaptation of Annell's work [7]. Consider the slot of Fig. 1(a) with solid conductor, whose dimensions and cylindrical coordinates z, ρ, ψ are shown in Fig. 2, the z axis directed along the conductor axis. Medium 1 is non-magnetic ($\mu = \mu_0$) with high conductivity ($\sigma = 58 \times 10^6$ S/m) while media 2 and 3 have $\mu = \mu_0$ and $\sigma = 0$. Medium 4 is ferromagnetic with low carbon steel sheets whose magnetization curve $B = f(H)$ is known.

For analytical solution the approximation of $\mu = \infty$ can be made in the core, because its flux density is low.

Taking into consideration slowly varying fields,

$$\text{rot}\ \vec{E} = -\frac{\partial \vec{B}}{\partial t} \tag{1}$$

$$\text{rot}\ \vec{H} = \vec{J} \tag{2}$$

$$\vec{B} = \mu_r \mu_0\ \vec{H} \tag{3}$$

$$\vec{J} = \sigma \vec{E} \tag{4}$$

from which

$$\text{rot rot}\ \vec{E} = -\mu_r\ \mu_0\ \sigma \frac{\partial \vec{E}}{\partial t} \tag{5}$$

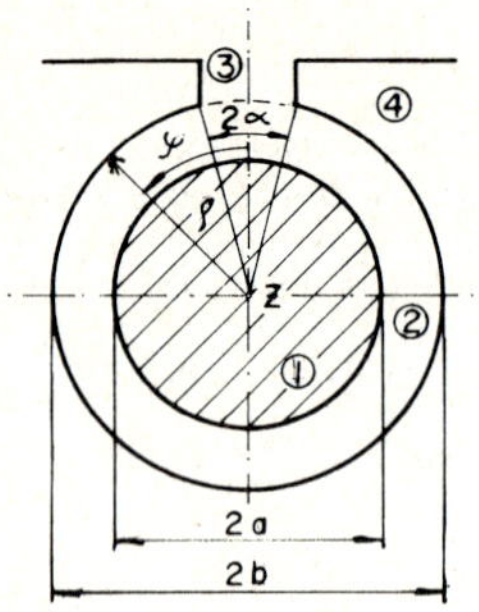

Fig. 2 - Slot dimensions and coordinate system

Neglecting end effects at the bar, $\vec{E}$ will not depend on coordinate z and, after some work, the z component of (5) results

$$\frac{\partial^2 E_Z}{\partial \rho^2} + \frac{1}{\rho}\frac{\partial E_Z}{\partial \rho} + \frac{1}{\rho^2}\frac{\partial^2 E_Z}{\partial \psi^2} - \mu_r \mu_0 \sigma \frac{\partial E_Z}{\partial t} = 0 \tag{6}$$

Omitting index z and considering phasor notation, it follows, for media 1 and 2,

$$\frac{\partial^2 \dot{E}_1}{\partial \rho^2} + \frac{1}{\rho}\frac{\partial \dot{E}_1}{\partial \rho} + \frac{1}{\rho^2}\frac{\partial^2 \dot{E}_1}{\partial \psi^2} - j\omega\mu_r\mu_0\sigma\dot{E}_1 = 0 \tag{7}$$

$$\frac{\partial^2 \dot{E}_2}{\partial \rho^2} + \frac{1}{\rho}\frac{\partial \dot{E}_2}{\partial \rho} + \frac{1}{\rho^2} + \frac{\partial^2 \dot{E}_2}{\partial \psi^2} = 0 \tag{8}$$

For equation (7) the solution can be obtained by variable separation, and noting the symmetry of the slot about the $\psi = 0$ plane,

$$\dot{E}_1 = A_0 J_0(k\rho) + \sum_{n=1}^{\infty} A_n J_n\ (k\rho)\ \cos(n\psi) \tag{9}$$

where

$$k^2 = -j\omega\mu_r\mu_0\sigma$$

A_0 and A_n - constants imposed by boundary conditions.

J_0 and J_n - Bessel functions of first kind, order 0 and n.

For equation (8), by the same procedure,

$$\dot{E}_2 = C_0 + C_0' \ln\frac{\rho}{a} + \sum_{n=1}^{\infty} (C_n\rho^n + C_{-n}\rho^{-n})\ \cos n\psi \tag{10}$$

where

C_0, C_0', C_n, C_{-n} - constants imposed by boundary conditions, i.e.:

I. $\dot{E}_1(\psi) = \dot{E}_2(\psi)$ at $\rho = a$

II. $\dot{H}_1(\psi) = \dot{H}_2(\psi)$ at $\rho = a$

III. $\oint \vec{H} \times d\vec{\ell}$ in region 3 is constant and equal to $\dot{I}$ in the conductor.

Final solution for $\dot{E}_1$ and $\dot{E}_2$ is obtained with a lot of work and results in complicated expressions. Considering that we are interested on the field in region 2, it can be written:

$$\dot{E}_2 = -\dot{I}R_0\left\{\frac{ka\ J_0(ka)}{2J_0'(ka)} + \frac{(ka)^2\ \ln \rho/a}{2\mu_r} + \frac{(ka)^2}{\alpha}\sum_{n=1}^{\infty}\frac{\left[\left(\frac{\rho^n}{a^n} + \frac{a^n}{\rho^n}\right) J_n(ka) + \left(\frac{\rho^n}{a^n} - \frac{a^n}{\rho^n}\right)\frac{ka}{\mu_r n} J_n'(ka)\right] \text{sen}\ n\alpha\ \cos n\psi}{n\left[\left(\frac{b^n}{a^n} - \frac{a^n}{b^n}\right)\mu_r n J_n(ka) + \left(\frac{b^n}{a^n} + \frac{a^n}{b^n}\right) ka\ J_n'(ka)\right]}\right\} \tag{11}$$

with

α = angle of slot opening

$R_0 = \dfrac{1}{\sigma\pi a^2}$ - D C resistance of conductor

J_0' and J_n' - derivatives of Bessel functions

Now, the AC impedance of the conductor bar, for unity length, can be obtained by

$$\dot{Z} = \frac{\dot{E}_2}{\dot{I}} + j\omega \frac{\dot{\emptyset}_e}{\dot{I}} \qquad (12)$$

where

$\dot{E}_2$ - calculated for $\psi = 0$ and $\rho = b$, corresponding to internal flux of the slot. The surface limited by $\rho = b$ is the boundary between internal and external flux of the slot.

$\dot{\emptyset}_e$ - external flux of the slot, including its opening (slot neck).

The first term of equation (12) will give the AC internal resistance and reactance of the slot while the second corresponds to external reactance. Since external reactance can be accurately calculated by traditional method of flux integration, the analysis is limited to the internal part, and then,

$$R_i = R_e \{\frac{\dot{E}_2}{\dot{I}}\} \qquad (13)$$

$$X_i = |j\omega L_i| = I_m \{\frac{\dot{E}_2}{\dot{I}}\} \qquad (14)$$

Defining the factors:

$\frac{R_i}{R_0}$, resistance factor

$\frac{\omega L_i}{R_0}$, reactance factor

and dividing the second by $\pi(ka)^2$, there results the inductance factor, L_i/μ_0, wich corresponds exactly to permeance coefficient usually used in electrical machine design.

Now, consider a slot like that of Fig. 2, with dimensions:

$$2a = 25,4 \times 10^{-3} \text{ m}$$

$$|ka| = a\sqrt{\omega\mu_r\mu_0\sigma} = 0,272\sqrt{f}$$

$$2b = 30 \times 10^{-3} \text{ m}$$

$$2\alpha = 40^o$$

$$\frac{a}{b} = 0,85$$

Preparing conveniently equation (11), substituting the above values, adapting the curves presented by Annell [7] and changing the variable ka to frequency, result the curves for R_i/R_0 and L_i/μ_0 presented in Fig. 5 (labeled "analytical values").

2.2 - Experimental Results

A laminated core with insulated sheets was constructed (Fig. 3), corresponding to the dimensions previously considered. The conductor was maintained apart from the core by means of an insulating tube.

The voltage between bar ends was obtained by a suitable probe, in order to measure only the contribution of internal flux. A variable power supply was used and currents up to 200 A could be injected into the bar, with frequencies varying from 10 to 100 Hz. Conductor impedance was calculated from $\dot{V}$ and $\dot{I}$ and real and imaginary parts were separated by measurement of the phase angle between $\dot{V}$ and $\dot{I}$ using osciloscope in association with digital time counter.

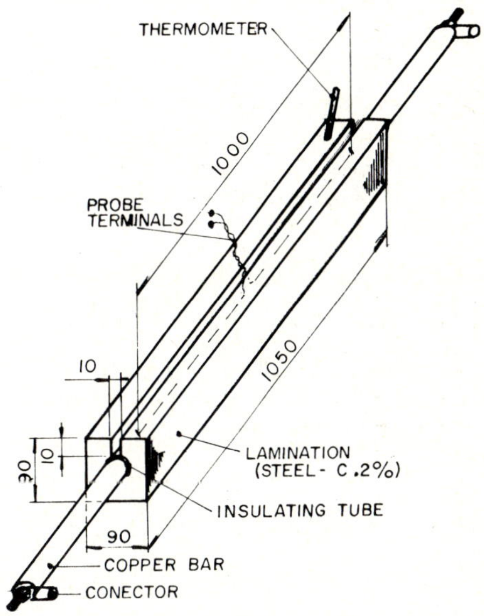

Fig. 3 - Constructed slot model for measurements of L_i and R_i

Experimental values appear in Fig. 5 where it can be noticed that inductance factors present a large deviation from analytical values (about 25% - 30%), probably due to approximations in analytical method. For resistance a reasonable agreement is observed, with deviation of 6% to 9%.

2.3 - Finite Element Evaluation

Finite elements have been largely used in problems dealing with electrical machines [8], [9], [10].

For application of the method, the region to be analysed was divided into triangles (Fig. 4). In this case, current density has only the z component and the potential vector $\vec{A}$ satisfies the following Poisson equation:

$$\frac{\partial}{\partial x}\left(\gamma \frac{\partial A_z}{\partial x}\right) + \frac{\partial}{\partial y}\left(\gamma \frac{\partial A_z}{\partial y}\right) = -J_z \qquad (15)$$

Where γ is the reluctivity of the media under consideration.

Boundary condition corresponds to zero values of A_z on the contour of discretized region and symmetry of the system allows to consider only the right side of the slot.

For variational approach, the functional

$$F(A) = \iint_R \left(\int_0^B \gamma\, b db\right) dxdy - \iint_R JA\, dxdy \qquad (16)$$

was defined and linear interpolation for potential A was used in each element [10].

Imposing extremum condition for functional F(A), a system of equations was obtained, written in matrix form:

$$[S]\ [A] = [J] \qquad (17)$$

This system was constructed and solved by computer [11] and since the problem is non-linear, magnetization curves were introduced into the computation and, by iterative procedure [8], flux density, B, was calculated.

Finally, inductance was calculated from magnetic energy. In this way,

$$\frac{L_i}{\mu_0} = \frac{1}{\mu_0} \frac{2}{I^2} \Sigma W_i \qquad (18)$$

W_i being the magnetic energy associated with a triangle of area Δ_i,

$$W_i = \frac{1}{2\mu} B_i^2 \Delta_i \qquad (19)$$

The value of L_i/μ_0 was computed initially for DC, with uniform distribution of current in the conductor, and resulted

$$\frac{L_i}{\mu_0} = 0{,}5608$$

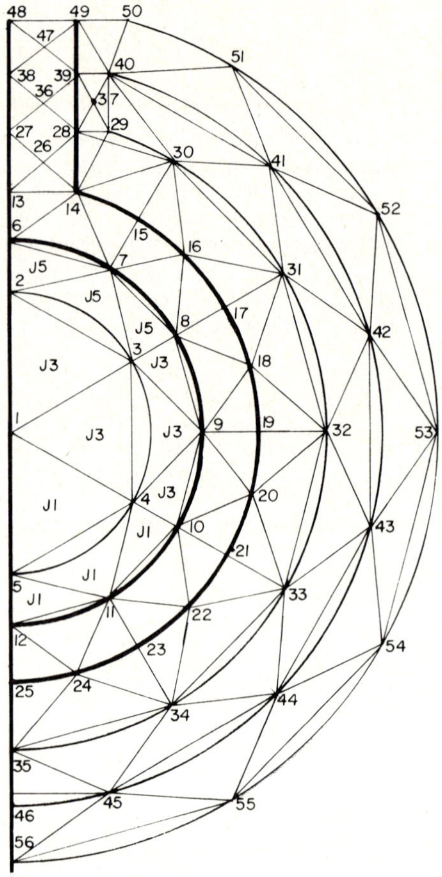

Fig. 4 - Finite elements for right half of the slot of Fig. 1(a)

wich can be compared with the extrapolated experimental value,

$$\frac{L_i}{\mu_0} = 0{,}60$$

obtained from Fig. 5.

For frequencies of 35 Hz, 60 Hz and 90 Hz the non-uniform current distribution was simulated in such a way that it will produce, for the same current, the apparent resistances obtained by analytical method. This was based on the fact that the values of R_i/R_0 obtained by analytical method were in good agreement with measured values.

The results of finite element evaluation were plotted in Fig. 5 for comparison with measured values. We can notice that now the deviation in the inductance is small (about 1% - 5%).

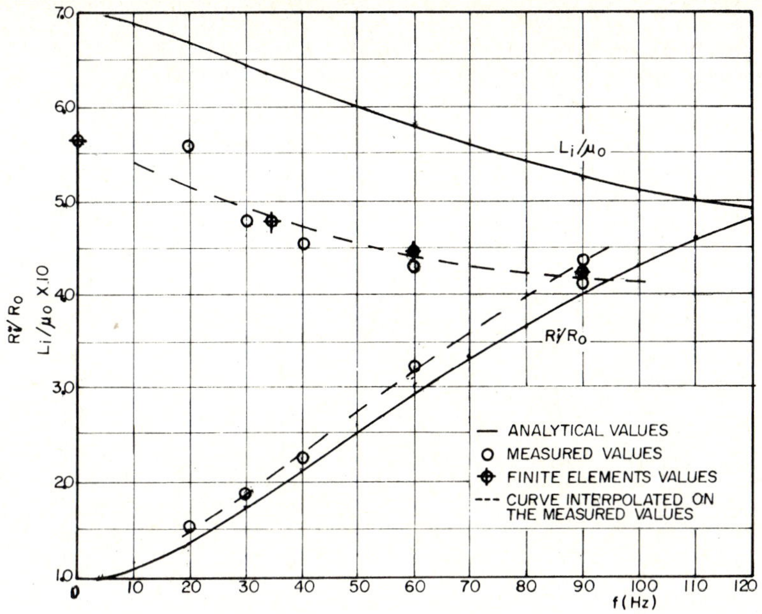

Fig. 5 - Calculated and experimental values for R_i/R_0 and L_i/μ_0

3. CONCLUSIONS

According to the obtained results, some conclusions can be carried out:

I. The numerical method gives satisfactory results for the stationary value of L_i/μ_0. This can be seen by comparing it with the extrapolated measured value (Fig. 5).

II. The analytical-numerical method for the evaluation of L_i/μ_0 is accurate. This method consists on the simple application of the finite elements supposing magnetostatic fields, but with non-uniform conductor current density. The current distribution was obtained from the analytical calculated resistance factor, R_i/R_0. As has been pointed out, this can be done because the analytical method has given little errors in the apparent resistance values.

III. The analytical-numerical method can be used for complicated slot shapes like that of Fig. 1(b) wich will be used on the design of the linear motor [12] for plates and sheets transportation. In this case the value obtained for the inductance factor was L_i/μ_0 = 1,025 wich includes the internal inductance factor (corresponding to the conductor region) and the external factor (corresponding to the non-magnetic cylinder region).

4. REFERENCES

[1] Yamamura, S., Theory of Linear Induction Motors (University of Tokio Press / John Wiley & Sons, Japan, 1979).

[2] Nasar, A., Linear Motion Electric Machines, (John Wiley & Sons, USA, 1972).

[3] Poloujadoff, M., The Theory of Linear Induction Machines (Oxford University Press, USA/Lowe & Brydone Ltd, England, 1980).

[4] Carrer, A., Costruzione di Machine Elettriche (Ed. Libreria Editrice Universitaria Levrotto & Bella, Italy, 1951).

[5] Puchstein, A.F., Calculation of Slot Constants, AIEE Transaction, vol. 66 (1947), 1315-23.

[6] Alger, P.L., The Nature of Induction Machines (Ed. Gordon and Breach Science Publishers, USA, 1965).

[7] Annell, H., The Alternating Current Impedance of a Solid Conductor of Circular Cross Section Placed in a Semi-Closed Circular Slot, Transaction of the RIT, nº 200(1963), Sweden.

[8] Demerdash, N.A. and Nehel, T.W., An Evaluation of the Method of Finite Elements and Finite Diferences in the Solution of Non-Linear Electromagnetic Fields in Electrical Machines, IEEE Transactions on PAS, vol. PAS-98, nº 1(1979), 74-87.

[9] Chiampi, M. and Cozza, T., Il metodo degli elementi finiti nella risoluzione di campi magnetostatici piani, L'Energia Elettrica, nº 9 (1977), Italy.

[10] Silvester, P. and Chari, M.V.K., Finite Element Solution of Saturable Magnetic Field Problems, IEEE Transactions on PAS, vol. PAS-89 (1970).

[11] Janiszewski, J.M., Método dos Elementos Finitos Aplicado a Problemas de Campos Eletromagnéticos Estáticos, Thesis, Escola Politécnica da Universidade de São Paulo, Brazil, (1978).

[12] Falcone, A.G., Proposta para uma Nova Aplicação dos Motores Planos de Indução, Thesis, Escola Politécnica da Universidade de São Paulo, Brazil, (1983).

ELECTRICAL MACHINES AND CONVERTERS - Modelling and Simulation
H. Buyse and J. Robert (editors)
Elsevier Science Publishers B.V. (North-Holland)/ © IMACS, 1984

IMPROVEMENT OF EFFECTIVENESS OF NUMERICAL FIELD DETERMINATION IN ELECTRICAL MACHINE BY EXTRAPOLATION

Andrzej L. Biczel, Paweł Staszewski

Computing Centre CYFRONET, Otwock - Świerk, Poland
Institut of Electrical Machines, Technical University of Varsav, Poland

This paper is devoted to the presentation and the discussion of the extrapolation algorithms as applied to the numerical determination of the magnetic field distribution in the non - linear, heterogenous medium of electrical machine. In the number of tests the different extrapolation applicability conditions are compared. Finally, the most effective algorithm one was choosen. The practical calculations were done for the case of the inductor generator tooth - zone.

1. INTRODUCTION

The iterative methods of large equations systems solution, which may occur e.g. in the finite difference or finite element techniques applied to the magnetic field determination in complicated, nonlinear and heterogenous media, are time consuming and expensive. Thus it is desirable to undertake the works concerning improvement of their effectivity, that is the reduction of the calculation time and the improvement of the solution physical accuracy.
Until now a lot of attention was paid to the problem of the iterative computation convergence accelaration. One have to mention as the examples the works concerning line succesive relaxation [1], alternating directions method [2, 3], bloc relaxation [4, 5] or relaxation coefficients application [1, 6, 7, 8]. In the nonlinear problems arising in the electrical machines field calculations the effectivness of the iterative process can be also increased by the non-stationary reluctivity underrelaxation [8, 9].
This research on the further iterative process acceleration with the extrapolation procedures is based on the particular iterative method of the vector potential calculations with the forward -backward line sweeping in the alternating directions and one - step relaxation of vector potential and reluctivity. This method is discussed in details in other papers [1,3,10]. The work presented here is devoted to the presentation and the discussion of the extrapolation algorithms as applied to the numerical magnetic field calculation in the tooth - zone of the inductor generator.

2. EXTRAPOLATION ALGORITHMS

The theory of the extrapolation algorithms is well developped for linear problems [11,12]. The goal of this work is the application of the extrapolation methods for the nonlinear problems. We stresse that we do not attempt to develop the extrapolation theory for the non-linear problems; the algorithm well known to be effective for the linear problems was applied to nonlinear case and then modified. The modifications were introducend as the result of the test problem calculations.
In this paper the extrapolation algorithms are applied to the potential and to the magnetic reluctivity as well. The current values of the vector potential solution are extrapolated in the whole region while the reluctivity is extrapolated in the nonlinear subregions.

2.1 Basic extrapolation applicability condition

The solution extrapolation is done independently in a mesh point provided the following basic condition is fulfilled:

$$\frac{RM_i}{RM_{i-4}} > \frac{RM_{i-4}}{RM_{i-8}}$$

where:

$$RM_i = \max_{j,k} \left| A_i^{j,k} - A_{i-1}^{j,k} \right|$$

i - iteration number
j,k - mesh point coordinates
A - vector potential

This condition shall be called the global condition in opposition to the local extrapolation conditions which shall be introduced later. The condition /1/ is verified at each fourth iteration because of the four-sweep

nature of the iteration method, what was mentioned earlier. The global condition for the extrapolation applicability is postulated as the replacement of such condition in the linear problems. In the linear case the condition consist in reaching the asymptotic convergence rate [12, 13]. In nonlinear case it has to be replaced because the asymptotic convergence rate estimation is not available. Accordingly to the global condition /1/ the extrapolation procedure is initiated when the iterative process convergence rate decreases.
If the global condition /1/ is met the current solution values of vector potential A and magnetic reluctivity ν are saved as the first elements to be used in the extrapolation algorithm. Next the two subsequent four-sweep iteration cycles are performed. So, the sequence $\{x_n\}$ of the three successive solution approximations is formed for mesh point:

$$x_1 = A_i^{j,k} \quad \text{or} \quad \nu_i^{j,k}$$

$$x_2 = A_{i+4}^{j,k} \quad \text{or} \quad \nu_{i+4}^{j,k}$$

$$x_3 = A_{i+8}^{j,k} \quad \text{or} \quad \nu_{i+8}^{j,k}$$

This sequence is used in the three-step extrapolation formulae.

2.2 Extrapolation formulae

The research on the extrapolation algorithms was started from application of the extrapolation formula known as the δ - Aitken algorithm [12]:

$$x = \frac{x_2 - x_1 \cdot q}{1 - q} \qquad /2/$$

$$q = \frac{x_3 - x_2}{x_2 - x_1} \qquad /3/$$

The formula /2/ gives the value of extrapolated solution x. The equations /2/, /3/ can be derived from the assumption that the sequence $\{x_n\}$ form the geometrical sequence with the quotient q.
The direct application of formula /2/ turned - out not to be succesfull enough. It was necessary to consider the sequence with the weaker convergence properties than properties of the geometrical sequence. It seems, that simplest way to obtain such a sequence is to replace the geometrical sequence by geometrical series.
Let S_n be the partial sums of unknown geometrical series. It is required that given sequence $\{x_n\}$ approximate those partial sums. Then, if the series is convergent, the extrapolated solution value is defined by the sum of this series. To obtain the best approximation the following function is minimised:

$$F = \sum_{n=1}^{N} (x_n - S_n)^2 \qquad /4/$$

It is well known that the all terms S_n are uniquely determined by the first element of series a_o and the quotient q. In order to get the linear equations for a_o and q the known relation for geometrical series is used:

$$S_n = a_o + q\, S_{n-1} \qquad /5/$$

As the terms $\{x_n\}$ are considered as the approximations of S_n one can to postulate:

$$x_n = a_o + q\, x_{n-1} \qquad /6/$$

Substituing this relation to the equation /4/ one obtains:

$$F(a_o,q) = \sum_{n=2}^{N} (x_n - a_o - q\, x_{n-1})^2 \qquad /7/$$

The minimisation of the function $F(a_o,q)$ in respect to a_o, q leads to the following solution for N=3:

$$q = \frac{x_3 - x_2}{x_2 - x_1} \qquad /8/$$

$$a_o = \frac{x_1 + x_2}{2}\left[(1 - q) + \frac{x_3 - x_1}{x_1 + x_2}\right] \qquad /9/$$

Since the sum of the convergent ($|q|<1$) geometrical series is given by

$$S = \lim_{n\to\infty} S_n = \frac{a_o}{1 - q} \qquad /10/$$

the extrapolated solution value x=S is calculated from the following equation:

$$x = \frac{x_1 + x_2}{2}\left[1 + \frac{1}{1 - q} \cdot \frac{x_3 - x_1}{x_1 + x_2}\right] \qquad /11/$$

2.3 Local conditions for extrapolation

The extrapolation defined by the equations /2/ or /11/ is performed for the given mesh element if the local conditions are fulfilled. The introduction of those conditions was necessary for the following reasons:

a/ the extrapolated sequence $\{x_n\}$ must obey the basic assumption of the extrapolation formulae, that is the convergence condition,

b/ it seems that the extrapolation application should depend on the relative solution change compared with the average relative solution stability measure $\mathcal{E}$.

c/ the extrapolated solution value must

retain the physical validity.
The following different local conditions will be considered and tested in order to meet the above requirements
- in respect to a/:

$$|q| < 1 \qquad /12a/$$

or

$$|q| < 1 - 4 \qquad /12b/$$

- in respect to b/:

$$\frac{|x_3 - x_2|}{x_2} > \varepsilon \qquad /13a/$$

or

$$\frac{|x_3 - x_2|}{x_2} < 10\,\varepsilon \qquad /13b/$$

- in respect to c/:

$$x_{min} < x < x_{max} \qquad /14/$$

Case Nº	Algorithm equations	i	RM_{i+4}	RM_i/RM_{i+4}	ε_{i+4}	$\varepsilon_i/\varepsilon_{i+4}$
		-	Wb/m	-	-	-
1.	/2/,/12b/,/13b/	24	4.84 E-4	1,21	3.035 E-3	2,53
		40	3.74 E-4	0,59	1.164 E-3	1,39
2.	/11/,/12b/,/13b/	24	4.84 E-4	1,21	3.035 E-3	2,53
		40	0.74 E-4	2,63	0,637 E-3	2,56
3.	/11/,/12b/,/13a/	24	7.29 E-4	0,80	3.040 E-3	2,53
		40	4.20 E-4	1,32	2.064 E-3	1,31
4.	/11/,/12a/,/13a/	24	65.11E-4	0,09	8.741 E-3	0,88
		36	8.10 E-4	1,00	4.896 E-3	0,75

The condition /12a/ was cosidered as the usual convergence condition of the geometrical sequence and series. This condition was used at the begining of test calculations. Notice it, that the sequence quotient is calculated with the error coming from the error ε of the sequence terms $\{x_n\}$. One can prove that in this case the convergence condition should take the stronger form /12b/.
The condition /13a/ was introduced because the extrapolation seemed to be reasonable if the relative solution change is graeter then the solution instability ε . The condition /13b/ was introduced as the opposition of the condition /13a/. The relation /13b/ excludes the extrapolation application if the relative solution changes are too large.
One cane verify the physical validity of the extrapolated solution value /the condition/14// provided the estimates of the lower and upper limits of the solution value are available.
If the local conditions /12/, /13/, /14/ are not fulfilled than the actual solution value for the given mesh element is taken accordingly to the following formulae:

$$x = \begin{cases} x_3 & \text{if} \quad q > 0 \\ 0{,}5/x_2 + x_3/ & \text{if} \quad q < 0 \end{cases} \qquad /15/$$

3. TEST CALCULATIONS RESULTS AND CONCLUSIONS

To verify the previously described algorithms the magnetic field calculations were performed for the tooth - zone of the homopolar inductor generator. The disceretistion mesh, boundary conditions, initial guess for potential and reluctivity and convergence criteria values were taken the same as for the case without extrapolation [10].
The most important results of the test calculations are presented in the table 1.

Table 1. Extrapolation algorithms comparsion

This table specifies for the different extrapolation algorithms the effect of their application expressed by selected measures of the solution convergence. In the table the four different cases are compared. The case specification is given by the numbers of the equations describing the extrapolation algorithm together with the local conditions for is applications. The notation used in the table headline is follwing:

- i - iteration number after which extrapolation is performed,
- RM - maximum of the solution stability error taken from all the mesh points /eqn./1//,
- ε - average relative stability error of the solution.

The quotiens RM_i/RM_{i+4} and $\varepsilon_i/\varepsilon_{i+4}$ express the main effect of the extrapolation application i.e. the reduction of the solution stability errors.
The analysis of the table 1 easily leads to the conclusion, that from the algorithms tested, the best results are obtained with the algorithm described by equations /11/,/12/,/13b/ - /table 1, case Nº 2/. The convergence and efficiency application of the extrapolation parameters for this case are many times better in comparison with those parameters for another cases. It means that:
- the geometrical series application leads to the better approximation

of the extrapolated sequence$\{x_n\}$
- the extrapolation procedure shoud not be applied if the solution changes are too large
- it is necessary to take into account the error ε of the extrapolated sequence in the convergence criterion.

The final effect of the chosen extrapolation algorithm application is given in table 2.

Table 2. The net effect of the extrapolation application

Case specific.	IT	$\delta\Theta$	F
	-	%	A
without extrap.	152	0,20	-0,8
with extrap.	120	0,11	0,6

In this table the calculation results according to extrapolation algorithm proposed here are compared with the results for the algorithm without extrapolation [9]. The notation IT denotes the total number of iterations. The quantities $\delta\Theta$ and F represent the physical accuracy criteria /thery are equal to zero for the exact solution/.
One can see that the proposed extrapolation algorithm guarantee the 20% reduction of the iterations number and approximately the same reduction of the computing time. At the same time the physical accuracy of the solution is improved.

4. REFERENCES

[1] Erdelyi,E.A.,Fuchs,E.F.,Binkley,D.H., Nonlinear magnetic field analysis of DC machines pt I,II,III, IEEE Trans. on PAS 7 /1970/

[2] Peaceman,D.W., Rachford,H.H., The numerical solution of parabolic and elliptic differential equations, J. Soc. Indus. Appl. Math. 3 /1955/

[3] Freundl,H., Reichert,K., Erdelyi,E.A., On the computation of the magnetic field of betatron magnets with finite-difference method, in Proc. of the Third Int. Conf. on Magnet Technology A-21, /Hamburg, 1970/

[4] De la Vallée Poussin,F., Lion,A., Calcul iterative de l'induction magnétique das les machines électriques, RGE 4 /1967/

[5] Mastero,S., Empirical convergence algorithm for successive overrelaxation, Proc. IEE 5/1974/

[6] Varga,R.S., Matrix iterative analysis /N.J., 1962/

[7] Kreisinger,V., Numerical analysis of nonlineare magnetic fields in electrical machines pt.I-Application of block-iterative methods, Acta Technica CSAV 3 /1971/

[8] Sarma,M.S., Wilson,J.C., Accelerating the magnetic field iterative solutions IEEE Trans. on MAG 6/1976/

[9] Biczel,A.L., Staszewski,P., New approach to the non-stationary reluctivity underrelaxation in the iterative magnetic field calculations, in II Krajowa Konferencja Teorii Pola Elektromagnetycznego, /Poland, Rydzyna, 1983/

[10] Staszewski,P., Numeryczna metoda rozwiązania nieliniowego obwodu magnetycznego maszyny elektrycznej, Rozprawy Elektrotechniczne 2 /1979/

[11] Smith,P.D., Linear extrapolation algorithms for accelerating the convergence of scalar and vector iterations, Report GA-A13090, General Atomic Company /San Diego, 1974/

[12] Guziński,W., Padé aproximation. General properties and applications for accelerating the convergence of the sequences by means of the ε - algorithm, Report INR 1966, Cyfronet/PM/A /Poland, Warszawa-Świerk, 1978/

[13] Marczuk,G.I., Lebiedev,V.I., Numerical methods in the neutron transport theory /in russian, SU, Moskwa 1981/.

Session 4 B

SIMULATION AND MODELLING OF CONVERTERS (I)

Chairman: V. RAJAGOPALAN

ELECTRICAL MACHINES AND CONVERTERS - Modelling and Simulation
H. Buyse and J. Robert (editors)
Elsevier Science Publishers B.V. (North-Holland)/ © IMACS, 1984

COMPUTATION OF ALMOST PERIODIC STEADY-STATE RESPONSE OF POWER ELECTRONIC CONVERTER SYSTEMS

V. Rajagopalan, A. Jacob, A. Sévigny
Université du Québec à Trois-Rivières
Trois-Rivières, Québec, Canada

K. Sankara Rao
North Dakota State University
Fargo, N.D., U.S.A.

The results obtained from the implementation of a general method for the computation of steady-state from an arbitrary initial state are discussed, in the case of any power electronic converter system. The proposed method makes use of the state variables techniques for the formulation of system equations; the Euler Backward Algorithm for the integration of the state equations and the computation of the Jacobian; the Newton-Raphson method for the computation of the iterated values for the initial conditions for the successive integration periods. Results obtained in the case of two pratical Power Electronic Converters are discussed.

1. INTRODUCTION

General purpose simulation programs such as ATOSEC I, SACSO are available for the computation of transient response of any power electronic converter systems containing semi-conductor switching elements. (1, 2) The system is described in the form of a set of first order differential equations (state equations). Presently, the method used to compute the steady-state is by solving the state equations using numerical integration starting from a given initial state until steady-state is reached. (3) Although this method is quite general, it is prohibitively expensive for highly damped circuits, where it takes a very long time for the transient component to die out. Therefore, the method is named "brute force method" in a recent publication, wherein efficient algorithms for computing almost periodic steady-state of non-linear systems to multiple input frequencies are described. (4) The problem of computation of the steady-state solution in an efficient manner for specific cases of power electronic converter-fed motor systems and for H.V.D.C. transmission system has been reported. (5, 6, 7, 8, 9) However, a generalized approach to an efficient computation of steady-state solution starting from network topology description and implementation of such a method in power electronic global simulation programs such as ATOSEC I will be useful in the computer-aided analysis and design of power electronic systems. In this paper, the authors discuss the results obtained from the implementation of a general method for the computation of steady-state from an arbitrary initial state. (10) The method has been implemented for the ATOSEC I simulation program and results obtained in the case of two power electronic converter systems are discussed.

2. METHOD OF COMPUTATION OF STEADY-STATE

Consider a system described by a set of first order differential equations (state equations):

$$[\dot{X}] = [A]\ [X] + [B]\ [U] \qquad \ldots(1)$$

Let the initial state vector be

$$[X]_{t=0} = [X_0] \qquad \ldots(2)$$

Let the state and input matrices $[A]$ and $[B]$ have time-varying coefficients. Assuming that the Euler's Backward Algorithm has been used for the step-by-step computation of the solution at the end of first step-length "h_1":

$$[X_1] = [X_0] + h_1 \{[A_1]\,[X_1] + [B_1]\ [U_1]\} \ldots(3)$$

where

$[X_1]$ is the state variable value at the end of first step-length, h_1

$[A_1]$ is the state matrix at the end of first step-length with constant elements during this step-length

$[B_1]$ is the input matrix at the end of first step-length; constant elements during this step-length

We can write the solution for X at the end of nth step-length (h_n)

as follows:

Equation (3) is simplified as:

$$X_1 = \phi_1^{-1} X_0 + h_1\phi_1^{-1} B_1 U_1 \qquad \ldots(4)$$

where

$$X_1 = [X_1]\ ;\ \phi_1 = \{[I] - h_1 [A_1]\}$$
$$B_1 = [B_1]\ ;\ U_1 = [U_1]$$

similarly,

$$X_2 = \phi_2^{-1} X_1 + h_2\phi_2^{-1} B_2 U_2$$

or

$$X_2 = \phi_2^{-1} \cdot \phi_1^{-1} X_0 + h_1\phi_2^{-1} \phi_1^{-1} B_1 U_1 + h_2\phi_2^{-1} B_2 U_2$$

The solution at the nth step length is:

$$X_n = \phi_n^{-1} X_{n-1} + h_n\phi_n^{-1} B_n U_n$$

or

$$X_n = (\phi_n^{-1} \cdot \phi_{n-1}^{-1} \; .... \; \phi_1^{-1})\, X_o$$
$$+ h_1\, (\phi_n^{-1} \cdot \phi_{n-1}^{-1} \; .... \; \phi_1^{-1})\, B_1\, U_1 \; +$$
$$+ h_2\, (\phi_n^{-1} \cdot \phi_{n-1}^{-1} \; .... \; \phi_2^{-1})\, B_2\, U_2 + ...$$
$$+ h_n \phi_n^{-1} \cdot B_n\, U_n \qquad ...(5)$$

Examination of the nth term X_n in equation 5 reveals that the solution is dependent on two major terms:

- one term depends directly on the initial vector X_0
- another term depends on the input matrices and sources

We, therefore, define a Jacobian Matrix so that it describes the partial variations of the solution X_n at the nth step-length: (10)

$$J(X_0) = \frac{\partial X_n}{\partial X_0} = (\phi_n^{-1} \cdot \phi_{n-1}^{-1} \cdot\cdot \phi_1^{-1}) \qquad ...(6)$$

It is interesting to note that the evaluation of the Jacobian is done as and when the numerical solution is obtained by using the Backward Euler Algorithm.

In case, a different numerical integration method is used for the solution of the state equations, it is easy to evaluate the element of the Jacobian Matrix at each step length using equation (6).

Assume that the period of the "Almost periodic" solution is given by

$$T = nh \qquad ...(7)$$

then $X_n = X_T$ represents the values of the state variables at the end of the period.

If the steady-state has been assumed to be reached, then,

X_T is almost equal to X_0

that is:

$$X(T, X_0) = X_0 \qquad ...(8)$$

where $X(T, X_0)$ signifies the values at the end of a period, T, starting from the initial vector X_0. We have to solve equation (8) by an iterative method based on the Newton-Raphson algorithm. (3)

$$X_0^{(j+1)} = X_0^{(j)} - (1 - J(X_0^{(j)}))^{-1}\, (X_0^{(j)} - X(T, X_0^{(j)}))$$

where $X_0^{(j+1)}$ represents the initial vector to be used for the

(j+1)th iteration step

$X_0^{(j)}$ is the initial vector at the jth iteration step

$X(T, X_o^{(j)})$ is the value of X at time T(period) and at the jth iteration

The iterations are continued until all the values for the final state vector at the end of a period are such, that they differ from the initial values by a small value.

Therefore, the test for the periodicity condition is given by:

$$\frac{|X_m(T, X_0^{(j)}) - X_{m0}^{(j)}|}{|X_m(T, X_0^{(j)})|} \leq \varepsilon \text{ for } m = 1 \text{ to } n \qquad ...(10)$$

Where ε is a suitably chosen small positive value, representing the relative error between the final value at time T of any state variable and its initial value.

Now the algorithm for the calculation of initial state vector under steady-state can be stated:

Step 1:

Assume arbitrary values for the initial vector (zero values, for example); compute, at each step length, the Jacobian matrix until time T, the period of the "Almost periodic solution". Compute the final values at time T.

Step 2:

Test, if periodicity condition as given by equation (10) is satisfied: if not satisfied, use equation (9) to compute the initial vector for the next iteration. Go to step 1, if the periodicity condition is satisfied. Print the initial values. STOP.

We will illustrate the method with the aid of a numerical example.

3. A NUMERICAL EXAMPLE

Let the state equation for a system (Figure 1) be given as follows:

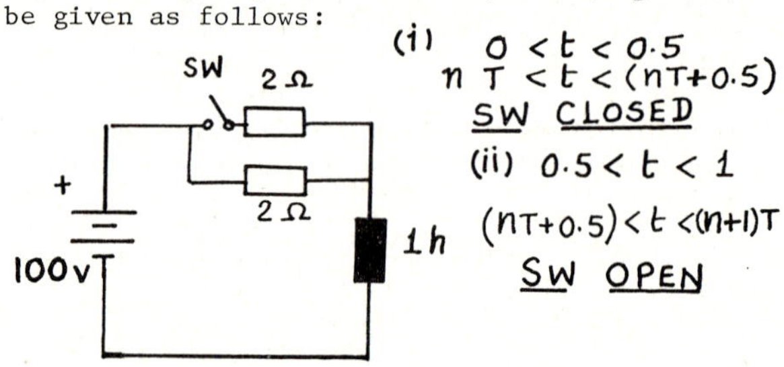

FIGURE 1

$$\frac{di}{dt} + 1.i = 100 \; : \text{ in the interval } 0 < t \leq 0.5;$$

$$\frac{di}{dt} + 2.i = 100 \; : \text{ in the interval } 0.5 < t < 1$$

Let the period T = 1 and let the initial value of current under steady-state condition be assumed to be i_0.

Case 1

Step 1:

The step length is assumed to be 0.1. The Jacobian at each step length for the interval $0<t\leq 0.5$ is:

$$\phi_1^{-1} = (1 - 0.1(-1))^{-1} = (1.1)^{-1}$$

similarly,

$$\phi_5^{-1} = (1.1)^{-1} \cdot (1.1)^{-1} \cdot (1.1)^{-1} \cdot (1.1)^{-1} \cdot (1.1)^{-1} = (1.1)^{-5} = 0.62092$$

The Jacobian at each step length for the interval $0.5<t<1.0$: $\phi^{-1} = (1.2)^{-1}$

similarly,

$$\phi_{10}^{-1} = (1.2)^{-1} \cdot (1.2)^{-1} \cdot (1.2)^{-1} \cdot (1.2)^{-1} \cdot (1.2)^{-1} \cdot \phi_5^{-1} = (1.2)^{-5} \cdot (1.1)^{-5} = 0.24953$$

The final value of i can be calculated from equation (10);

$$i_{10}^{(1)} = i(T,\ i_0^{(1)}) = 38.63 \text{ assuming } i_0^{(1)} = 0$$

Step 2:

$$\frac{|i(T,\ i_0^{(1)}) - i_0^{(1)}|}{|i\ (T,\ i_0^{(1)})|} = 1 \geq \varepsilon \ : \text{Let } \varepsilon = 0.01$$

Periodicity condition is not satisfied. Apply Newton-Raphson algorithm to compute the next iterated value for the initial value $i_0^{(2)}$ from equation (9):

$$i_0^{(2)} = 0 - (1 - 0.24953)^{-1}\ (0-38.63) = 51.43$$

Repeat Step 1:

The Jacobian remains the same for this problem and it is equal to 0.24953

The final value of $i\ (T,\ i_0^{(2)})$ is calculated from equation (10):

$$i_{10}^{(2)} = i\ (T,\ i_0^{(2)}) = 51.46$$

Step 2:

Periodicity condition is satisfied:

$$\frac{51.46 - 51.43}{57.46} = 0.0005 < \varepsilon \text{ chosen}$$

STOP

The initial value of "i" under steady-state condition is 51.46. The initial value of "i" for this simple problem can be evaluated exactly as 59.3.

Case 2

We shall now demonstrate, for the same numerical example, the usefulness of calculating precisely the Jacobian and the $X(T,\ X_0^{(i)})$ at any iteration.

Let us assume that the step length be decreased to 0.01 and the solution be calculated more precisely by (better numerical algorithm or) the Euler Backward algorithm.

Step 1:

Jacobian is calculated as:

$$= (1.02)^{-50} \cdot (1.01)^{-50} = 0.2259$$

Final value of i

$$= i_{10}^{(1)} = i\ (T,\ i_0^{(1)}) = 45.39$$

Step 2:

Periodicity condition is not satisfied. The iterated value for i is:

$$i_0^{(2)} = 0-(1 - 0.2259)^{-1}(0 - 45.39) = 58.63$$

Repeat Step 1:

Jacobian is the same the one calculated earlier.

Final value $i_{10}^{(2)} = 58.63$

Step 2:

Periodicity condition is satisfied. The initial value is calculated as 58.63 which is very close to exact value 59.3. The advantage of decreasing the step length by a factor of 10 greatly improves the accuracy of the solution.

4. IMPLEMENTATION OF STEADY-STATE COMPUTATION ALGORITHM

The flow chart is shown in figure 2. In the beginning of the computation, arbitary initial values are assumed for all the state variables. Using Euler Backward Algorithm, the computation of the system of first order State equations is done. A constant step length has been chosen for the solution of the state equations. The Backward Euler Algorithm has been chosen in order to ensure numerical stability of the solution of the system of stiff differential equations. At each step length, the Jacobian required for the application of Newton-Raphson iteration formula is computed. The inverse of the required matrix is computed only in the case when the state matrix A_t changes. The computation of the solution continues until the required period is reached. At the end of the period, the condition for the periodicity is verified; if this condition is not fulfilled, then the initial values for the next period of integration are computed from the Newton-Raphson formula.

In the next section, we discuss the results obtained from ATOSEC 2 simulation program in the case of two practical power electronic converter systems.

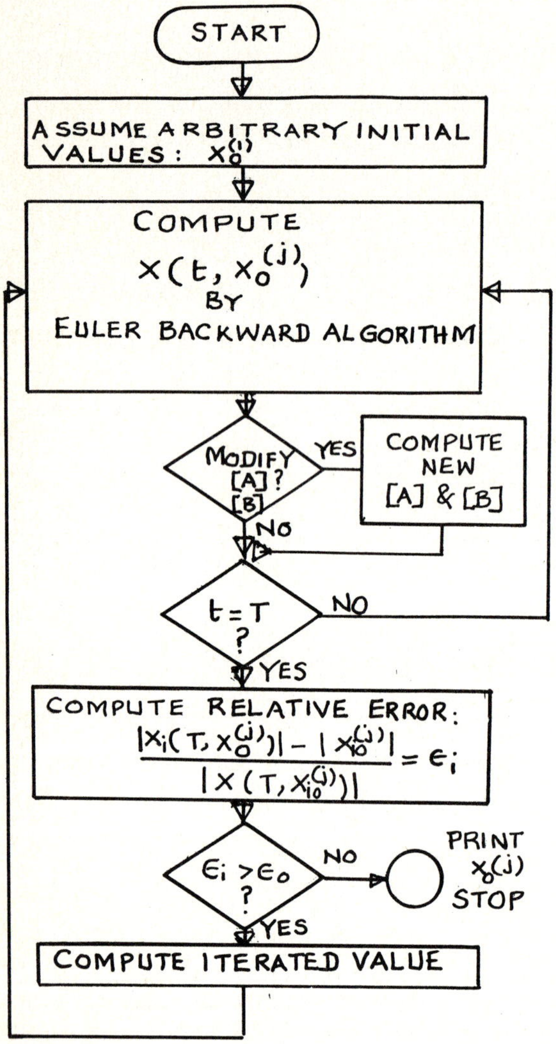

FIGURE 2

5. SAMPLE PROBLEMS

5.1 A transistor chopper-fed R-L circuit problem

Figure 3.1.1 shows a transistor chopper-fed d.c. motor:

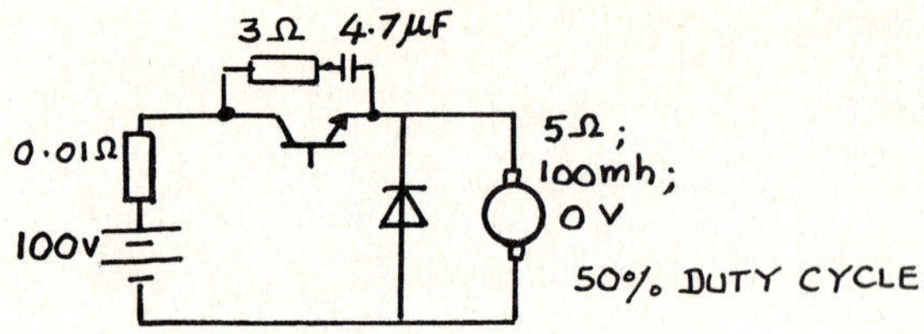

FIGURE 3.1.1

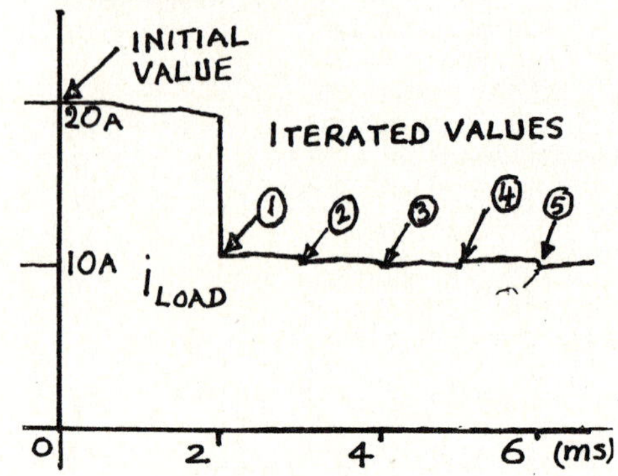

FIGURE 3.1.2

Figure 3.1.2 shows the evolution of the load current transient from zero initial conditions as well as the iteration steps. Figure 3.1.2 shows the evolution of the load current for the first two periods. The iterative values for the initial values of the variables are computed only at the second period. In power electronic circuits, there are widely varying components of the transients: some involving fast decaying transients and some other involving very slowly decaying ones. Since the initially assumed values for the variables could be theoretically any arbitrary initial state, we wanted to calculate the transient solution at least for two periods so that periodicity in the operation of the switches is properly ensured in the simulation program. This is a particular feature of our simulation program and has no relation to the steady-state calculation method used.

Steady-state initial values have been calculated to a desired accuracy of 1 percent after 6 iteration steps.

Figure 3.1.3 shows the evolution of the transient state from zero initial conditions.

It can be concluded from figures 3.1.2 and 3.1.3, that the initial condition under steady-state as well as the steady-state solution are computed requiring less than 18 seconds computer processing time. If the transient state solution had been computed until the steady-state was reached (within 1 percent relative error for the periodicity condition), the computer processing time for this problem would be about five times more than the time required for the calculation of the same by using the algorithm proposed in the paper.

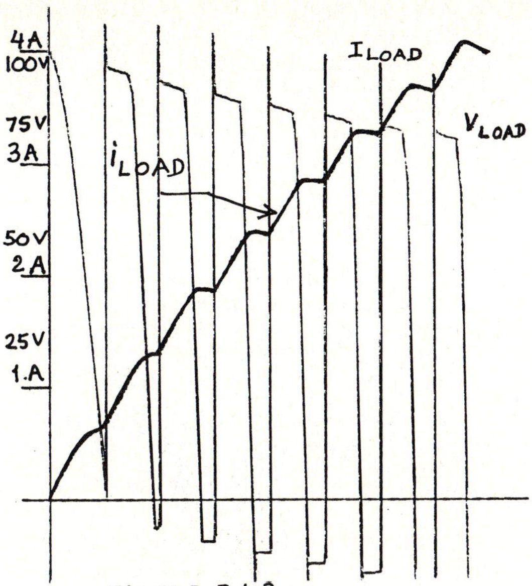

FIGURE 3.1.3

5.2 A three phase thyristor converter-fed R-L Load:

Figure 4.1.1 shows the three phase thyristor converter system. The number of state variables for this problem is 16.

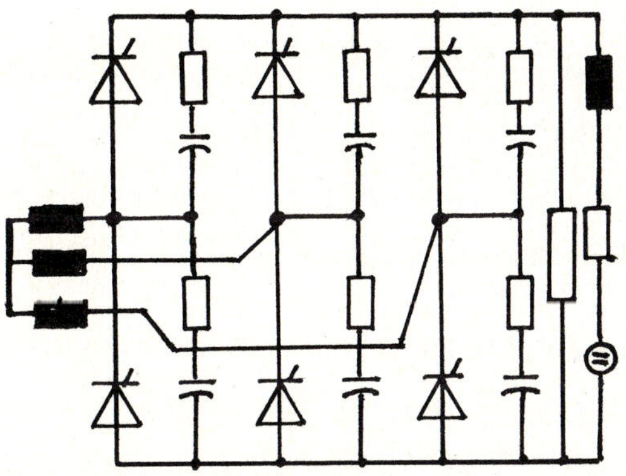

FIGURE 4.1.1.

Figure 4.1.2 shows the steady-state solution as well as the iterative values computed for the initial values.

The computer processing time requirement for obtaining the initial values to compute the steady-state solution within 2% accuracy is 80 seconds. The number of iterations required is 4.

Figure 4.1.3 shows the evolution of the transient state computation of steady-state within 2% accuracy would require at least 500 seconds of CP time and it is prohibitively long.

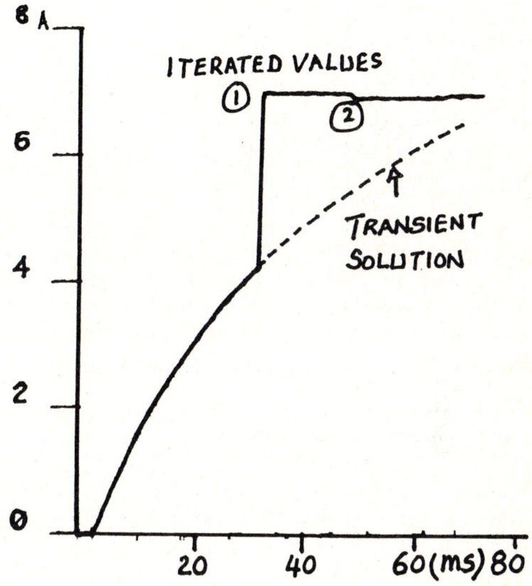

FIGURES 4.1.2 & 4.1.3

6. DISCUSSION OF THE PRESENT METHOD WITH RESPECT TO EXISTING METHODS:

The method described in this paper follows closely the development reported by Aprille and Trick. (10) A general steady-state analysis program ANASP for thyristor circuits based on adjoint-network approach is described by Y. Kuroe et al. (11) Stig Skelboe describes the different techniques available for time-domain steady-state analysis of non-linear systems.(12) The Extrapolation algorithm described by Stig Skelboe and N. Rhumin seem to be very promising since this algorithm simply processes the results of the time-domain network analysis using any of the general purpose circuit analysis programs and returns a new set of initial conditions. (13, 14) Further work is in progress with a view to implementing this algorithm for the efficient evaluation of the steady-state solution.

7. ACKNOWLEDGEMENTS

This work was possible thanks to the grants received from the Natural Sciences and Engineering Council of Canada. The authors are grateful to Dr Skelboe of the Institute of Datalogy, Denmark and to Professor Rhumin of McGill University, Canada for helpful suggestions during the progress of this work.

8. REFERENCES

1 V. Rajagopalan et al., "Simulateur digital des convertisseurs de puissance à thyristors" Revue canadienne de génie électrique, vol. 3, no 1, (1978) pp. 5-10.

2 H. Foch et al., "A general digital computer simulation program for thyristor static converters (program SACSO) Application examples", IFAC Symposium on control in power electronics and electrical drives, Dusseldorf, (1977) Pergamon Press Book, pp. 1-5.

3 Leon O. Chua and Pen-Min Lin, "Computer-aided analysis of electronic circuits: algorithms and computational techniques" chapter 17, section 17-5, Prentice-Hall Inc. (1975) pp. 687-691.

4 Leon O. Chua and A. Ushida, "Algorithms for computing almost periodic steady-state response of non linear systems to multiple input frequencies", IEEE Trans. on circuits and systems, vol. CAS-28, no 10, October (1981) pp. 953-971.

5 V.V. Sastry and K.R. Rao, "A numerical approach for the analysis of inverter-fed induction motor schemes", Electric Machines and Electromechanics, vol. 3, no 2, Jan.-March (1979) pp. 157-170.

6 V. V. Sastry and N. Prabhakaran, "Digital simulation of voltage-controlled A.C. motor drives", Electric Machines and Electromechanics, vol. 3, no 4, Oct-Dec. (1979) pp. 233-253.

7 K.Y.G. Li, "Analysis and operation of inverter-fed variable speed induction motor", Proc. IEE, London, vol. 116, no 9, Sept. (1969) pp 1571-1580.

8 C. Chellamuthu and V.V. Sastry, "Simulation of a 1-phase static converter - DC motor drive system", Electric Machines and Electromechanics, vol. 5, no 6, Nov.-Dec. (1981) pp. 263-279.

9 B.T. Ooi, M. Menemenlis and H.L. Nakra, "Fast steady-state solution for HVDC analysis", IEEE Transaction on Power Apparatus and Systems, Pas-99, Nov.-Dec. (1980) pp. 2453-2459.

10 T.J. Aprille and T.N. Trick, "Steady-state analysis of non-linear circuits with periodic inputs", Proceedings of the IEEE, vol. 60, no 1, January (1972) pp. 108-114.

11 Y. Kuroe, H. Haneda and T. Maruhashi, "A general steady-state analysis program ANASP for thyristor circuits based on adjoint-network approach", PESC'82 Conference Record (1982) pp. 180-189.

12 S. Skelboe, "Time-domain steady-state analysis of nonlinear electrical systems" proceedings of the IEEE, October (1982) pp. 1210-1228.

13 S. Skelboe, "Computation of the periodic steady-state response of non-linear networks by extrapolation methods", IEEE transactions on circuits and systems, Vol. CAS-27, No 3, March (1980) pp. 161-175.

14 N. Rhumin, "Evaluation of periodic steady-state algorithms for large non-linear dynamic circuits", IEEE circuits and systems symposium, California (1983) 4 pages.

ELECTRICAL MACHINES AND CONVERTERS - Modelling and Simulation
H. Buyse and J. Robert (editors)
Elsevier Science Publishers B.V. (North-Holland)/© IMACS, 1984

APPLICATION OF A SAMPLED-DATA MODELLING OF STATIC CONVERTERS TO THE ANALYSIS AND SYNTHESIS OF CERTAIN REGULATIONS

Jean-Paul LOUIS

Laboratoire de Recherche du Département d'Electricité de l'E.N.S.E.T.
61 Avenue du Président Wilson
94230 CACHAN (FRANCE)

Presentation of applications of a new modelling of control systems with static converters. The method principle is defined on a didactic example : a thyristors chopper with current control where the commutation interval is taken into account. Methods of syntesis are presented on D C motor speed control including current control. The continuous and discontinuous modes are considered.

1. INTRODUCTION

This paper deals with modelling and synthesis of a large class of control systems including static-converters. These systems are non-linear and sampled-data, but often are approximated by linear, continuous modellings (1). In these theories, assumptions are used : the non-linear effects are neglegible and the bandwith is not too large. For the cases where these assumptions are wrong a previous paper (2) has presented a systematic method to establish the rigorous non-linear model and the linearized model around an operating point.

We give here a presentation of this method applied to a didactic example : a thyristors chopper which presents three different intervals of operation during a basic cycle. After that we show how to use the linearized modelling for the synthesis of controllers with the example of a speed control of D C rotor including an internal current loop in continuous and discontinuous mode of operation.

PART I : MODELLING PRINCIPLE

Static converters present usualy a basic cycle which is divided into several intervals. The problem is generaly the following one :

Let us consider the initial state : X(n) at instant tn (end of the $(n-1)^{th}$ cycle). How to definie the final state X(n + 1) at instant tn + 1 (end of the nth cycle) :

$$(\underline{X}(n), t_n) \longrightarrow (X(n+1), t_{n+1})$$

Here we will define linearized models, then we consider the variation of $\underline{X}(n)$ around an operating point $\underline{X}(\infty)$: $\underline{X}(n) = \underline{X}(\infty) + \delta\underline{X}_n$, and we want to write the recurrency matrix M_R such that :

$$\delta\underline{X}_{n+1} = M_R\, \delta\underline{X}_n$$

Then the problem is : how to write the relations between $d\underline{X}_n$ and $d\underline{X}_{n+1}$ for a system which presents various structures during a basic cycle ?

THE E - FUNCTION

The first tool is given by a vectorial function which permit to write easily the sensibility of the trajectory with respect to the variations of the limit states ($\delta\underline{X}(n)$, δt_n).

We suppose that the cycle is constituted by a sequence of various structures or modes. Let us consider a mode indexed m. Its state vector is $\underline{X}_m$, and his duration is $T_m = (t_n, t'_n)$. The limit states are : the initial : $(\underline{X}_n^{(n)}, t_n)$, and the final $(\underline{X}'_n{}^{(n)}, t'_n)$.

We consider its equation under two forms : first the differential equation :

$$F_m\, d\underline{X}_m/dt + G_m\, \underline{X}_m = \underline{V}_m(t)\ ,\ t \in (t_n, t'_n)$$

and, second, its state equation

$$d\underline{X}_m/dt = A_m\, \underline{X}_m + \underline{B}_m(t)$$

with $A_m = -F_m^{-1} G_m$ and $\underline{B}_m = F_m^{-1} \underline{V}_m$.

Then the relation between the limit states is the classical one :

$$\underline{X}'_m(t'_n) = \exp\left[(t'_n - t_n)A_m\right] \underline{X}_m(n) + \int_{t_n}^{t'_n} \exp\left[(t'_n - t)A_m\right] B_m(t)\, dt$$

To have a symetrical form for this recurrency equation, we define the E - function :

$$\underline{E}_m(\underline{X}_m, t) = \exp(-tA_m)\, \underline{X}_m - \int_0^t \exp(-\tau A_m)\, B_m(\tau)\, d\tau$$

Then the recurrency equation becomes

$$\underline{E}_m(\underline{X}'_m(n), t'_n) = \underline{E}_m(\underline{X}_m(n), t_n) \qquad (1)$$

The E -function is an <u>invariant</u> on the trajectory (X(t), t). Its variation verifies :

$$\delta\, \underline{E}_m(n) = d\, \underline{E}'_m(n)$$

$\delta\, \underline{E}_m$ is easy to write with the variation of the state :

$$\delta\, \underline{E} = \exp(-tA_m)\, \left[d\underline{X}_m - \underline{Y}_m(t)\, \delta t\right]$$

Where $\underline{Y}_m(t) = A_m\, \underline{X}_m + \underline{B}_m(t)$

The equation (1) is not sufficient to describe the system. It is necessary to have joining conditions with the modes before and after the m-mode and condition of end of this mode.

THE DIDACTIC EXAMPLE

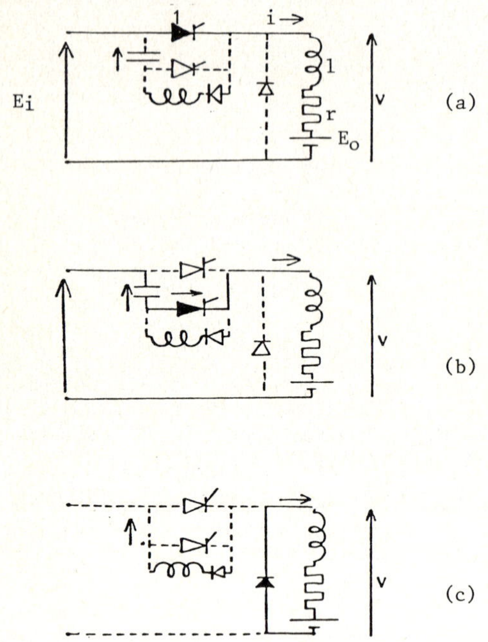

Figure 1

To define the complete recurrency relation it is necessary to write the relation between the variations of $\delta \underline{X}$ when an event provoques a changing of structure. To describe the joining conditions, we will consider as a typical example a thyristors chopper (Fig. 1). We will take into account its commutation interval. Then this systems presents three modes :

- the a-mode : normal conduction of duration ta; we do not consider the oscillation of the Vc voltage which has no effect on the systems. The elements of the differential equation are :

$\underline{X}_a = (i)$; $F_a = (1)$; $G_a = (r)$; $\underline{V}_a = (E_i - E_o)$

- the b-mode : commutation interval ; duration tb. Its elements are :

$\underline{X}_b = \begin{pmatrix} i \\ vc \end{pmatrix}$; $Fb = \begin{pmatrix} 1 & 0 \\ 0 & c \end{pmatrix}$; $G_b = \begin{pmatrix} r & 1 \\ -1 & 0 \end{pmatrix}$; $\underline{V}_b = \begin{pmatrix} E_i - Eo \\ 0 \end{pmatrix}$

- the c-mode : free-wheeling interval ; duration tc. Its elements are :

$\underline{X}_c = (i)$; $F_c = (1)$; $G_c = (r)$; $\underline{V}_c = (-Eo)$

These three modes are separated by three events : two external events, the firing of the thyristors, and an internal event : the turning on of the free-wheel diode (Fig. 2).

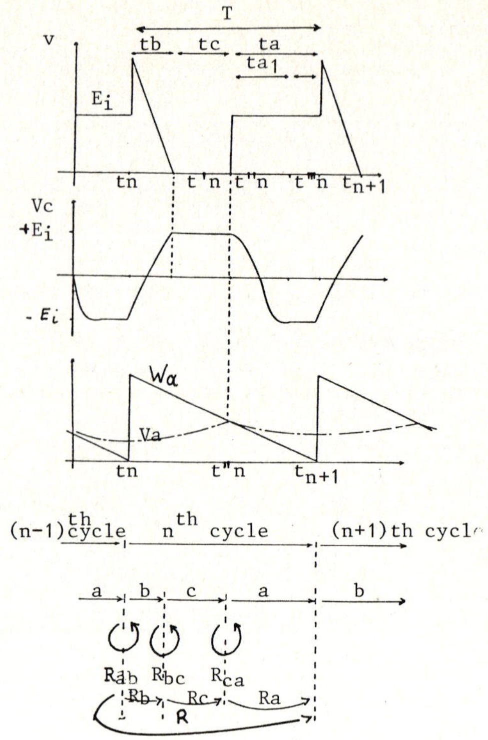

Figure 2

The b-mode begins at instant tn where the Th 2-thyristor is fired by a timer at fixed instants :

$$tn = nT$$

Then the variation of tn is : $dt_n = 0$.

The a-mode begins at instant t"n, when the control voltage Va and the timing voltage Wa are equal :

$$Va\ (t"n) = Wa\ (t"n) \quad (2)$$

The control voltage is generated by a controller and depends on the system variables on the c-mode (t"n is the end of it) : $\underline{X}_c$ and on a set-point : Vao :

$$Va = \underline{P}_r^T\ \underline{X}_c + Vao \quad (3)$$

In the example, we consider a proportionnal current control, then :

$$Va = k\ (I_o - i)$$

or : $\underline{P}_r^T = (-k)$, and : $Vao = k\ I_o$

The turning voltage is created by the pulses generator, and is supposed to be linear :

$$Wa\ (t) = A_1\ \left(1 - \frac{t - tn}{T}\right)$$

its derivative is : $\dot{A}n = dW_a/d_t \,\big|_{t\, =\, t"n}$

Then the variations verify :

$$\underline{P}_r^T\ \delta \underline{X}_c\ (t"n) = \dot{A}_n\ \delta t"n$$

THE JOINING CONDITIONS

We apply here methodology explained in (2). The linearized modelling is given by the following datas which describe a complete cycle with all recurrencies R_k and R_{kl} necessary to describe the global point mapping : $\delta X_n \xrightarrow{R} \delta X_{n+1}$

(i) The initial condition is : ($X_a(tn) = X_a(n)$, tn)
(ii) Rab For the instant tn we admit the existence of two constant matrices H_{ba} and H_b such that :

$$\underline{X}_b\ (tn) = H_{ba}\ \underline{X}_a\ (n) + \underline{H}_b\ E_i$$

In our example :

$$\underline{X}_a\ (n) = (i(tn))\ ;\ \underline{X}_b\ (tn) = \begin{pmatrix} i(tn) \\ vi(tn) \end{pmatrix}\ ;$$

$$H_{ba} = \begin{pmatrix}1\\ o\end{pmatrix}\ ;\ H_b = \begin{pmatrix}0.\\ -1\end{pmatrix}$$

The variations of $\underline{E}_b$ and $\underline{X}_b$ are (note that $\delta tn=0$):

$$\delta\,\underline{E}_b(tn) = \exp(-tnA_b)\ \delta\underline{X}_b(tn) = \exp(-tnA_b)\ Hba\ \delta\underline{X}_a(n)$$

This is the joining condition at instant tn.

(iii) R_b: Between tn and t'n the evolution of the b-mode (duration tb = t'n-tn) is described by a recurrency whose variation is :

$$\delta\underline{E}_b\ (tn) = \delta\underline{E}_b\ (t'n)$$

(iv) R_{bc} : The instant t'n occurs when vc = E_i. After it, this property becomes true during the c-mode. vc is a state-variable for the b-mode, but not for the c-mode where it is constant. Then we may admit the existence of the constant matrices C_{cb} and C_{bc} such that :

$$\underline{X}c = C_{cb}\ \underline{X}_b\ ;\ \underline{V}_c = D_{cb}\ \ \underline{V}b\ ,\ \forall\ t$$

$$\underline{X}_b = C_{bc}\ \underline{X}_c\ ,\ \forall\ t \in T_c$$

In our example :

$$C_{cb} = D_{cb} = C_{bc}^{\ T} = (1\ 0)$$

Furthermore, the relations between the differential equations of mode c and b are :

$$F_c = D_{cb}\ F_b\ C_{bc}\ ;\ G_c = D_{cb}\ G_b\ C_{bc}$$

and it can be proved that the variations of $\underline{E}_b$ and $\underline{E}_c$ at instant t'n verify the joining condition :

$$F_c \exp(t'n\ Ac)\ \delta\underline{E}_c(t'n) =$$

$$C_{cb}\ F_b\ \exp\ (t'n\ A_b)\ \ \delta\underline{E}_b(t'n)$$

(v) Rc : The end of the c-mode occurs at t"n, control instant of the main thyristor Th1. Between t'n and t"n (duration : tc = t"n - t'n) the variation of the reccurency equation is :

$$\delta\underline{E}_c\ (t'n) = \delta\underline{E}_c(t''n)$$

(vi) R_{ca} : We have seen the condition which definies the instant t"n ((2) to (3)). To defin the transition from th c-mode to the a-mode, we admit the existence of a constant matrix Hac such as :

$$\underline{X}_a\ (t''n) = Hac\ \underline{X}_c\ (t''n)$$

In the example : Hac = (1). It can be proved that the joining condition at firing instant (as t"n) is :

$$\delta\underline{E}a\ (t''n) = \exp\ (-t''nAa)\ M_B \exp(t''nA_c)\delta\underline{E}_c(t''n)$$

$$\text{with } M_B = Hac + \frac{\left[\underline{Y}''_a\ (n) - Hac\ Y''_c(n)\right]\ \underline{P}_r{}^T/A'}{1 - \underline{P}_r{}^T\ \ \underline{Y}''_c\ (n)/A'}$$

where $Y''_m\ (n) = Am\underline{X}\ m\ (t''n) + \underline{B}m\ (t''n)$

(vii) $\underline{R}_a$: Between t"n and tn + 1, the evolution of the a-mode (duration $t_a = t_{n+1} - t''n$) is described, for the variations, by :

$$\delta\underline{E}a\ (tn+1) = \delta\underline{E}a\ (t''n)$$

and at tn + 1, the firing of Th2 is provoqued by a timer (tn+1 = (n + 1)T and δtn + 1 = 0), and we have :

$$\delta\underline{E}a\ (tn+1) = \exp(-tn+1\ Aa)\ \delta\underline{X}a\ (tn+1)$$

MODELLING

With this sequence of relations we can write the recurrency relative to small variations as :

$$\delta\underline{X}_a\ (n+1) = M_R\ \delta\underline{X}_a(n)$$

with :

$$M_R = \exp(taAa)\ M_B \exp(tcAc)\ Fc^{-1}$$

$$D_{cb}\ F_b\ \exp\ (tbAb)\ H_{ba}$$

In the example, which is of first order, we have : $F_c{}^{-1} D_{cb}\ F_b = (1\ 0)$; $H_{ba} = \begin{pmatrix}1\\ o\end{pmatrix}$; exp (tAa) = exp (tAc) = (exp(-rt/l)) ; exp (tAb) = (φij (t)), where only φ11(t) is used :
$\varphi\ 11 = e^{-\ 1t} \sin(\beta 1t + \theta 1)/\sin \theta 1$, with $\alpha\ 1 = r/2l$;

$\rho\ 1^2 = 1/lc$; $\beta 1^2 = \rho 1^2 - \alpha 1^2$; $\sin\theta 1 = \beta 1/\rho 1$; $\cos\theta 1 = -\alpha 1/\rho 1$.

Then M_B has the expression :

$$MB = 1 - Ge, \text{ with } Ge = \frac{Ei}{lA_1/k + ri\ (t''n) + Eo}$$

And the linearized model is :

$$\delta i\ (n+1) = zo\ \delta i(n)$$

Where the pole, zo, has the expression :

$$zo = (1-Ge)\exp{=}\frac{T - tb}{Te}\ \varphi 11\ (tb)$$

Numerical Example

We consider the system defined by the following datas : r : 5.27Ω ; l = 72.6 mH ; C = 4.74 x 10^{-6}F; A_1 = 5V ; Ei = 200 V ; Eo = 80 V; Io = 3.8A ; T = 5 ms.

We can compute the controller using the following optimization : the system present a deadbeat response (here : a response in one cycle) if the pole is zero. Then zo = 0 leads to Ge=1. But Ge depends on the steady -state value of i(t"n),and this value is not foreseeable because it depends also on the gain of the corrector which is parameter which we have to determine. Then the simple solution is given by an approximation :

$$i(t''n) \simeq Io$$

which is good if the gain k is sufficiently large. Then the optimal gain, $\hat{k}$, which imposes the deadbeat response is roughly given by :

$$\frac{1}{\hat{k}} = \frac{T}{A_1 l}\ \left[Ei - (rIo + Eo)\right]$$

In the example : $\hat{k}$ = 0.73. The fig. 3 gives the numerical simulation of a response frome the initial condition i(o)=0. The first load ont the caparitor has an important duration (t''_o=1,09ms), but the deadbeat response is almost perfect, and the sequence of the value at the instants $t_n = nT$ are :
$i_o = 0$; $i_1 = 4.061$; $i_2 = 3.974$; $i_3 = 3.979$;
$i_4 = i_5 = 3.977$

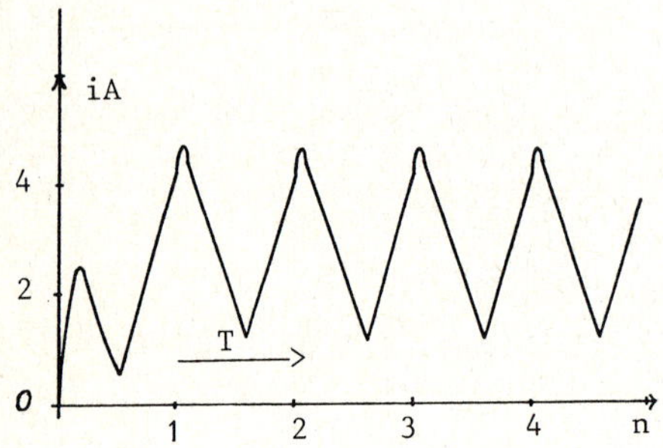

Figure 3

PART II SYNTHESIS OF CONTROL SYSTEMS

INTRODUCTION

In this part we wish to apply the previous modelling for the synthesis of control systems.

We will consider the example of a D C motor bridge fed by an AC-DC full thyristors rectifier with speed and current control (Fig. 4), in two cases : continuous mode without overlap and discontinuous mode. These systems are respectively of 4th and 3rd order. Previous works done on that type of system with discrete models do not exceed usualy the order 2.

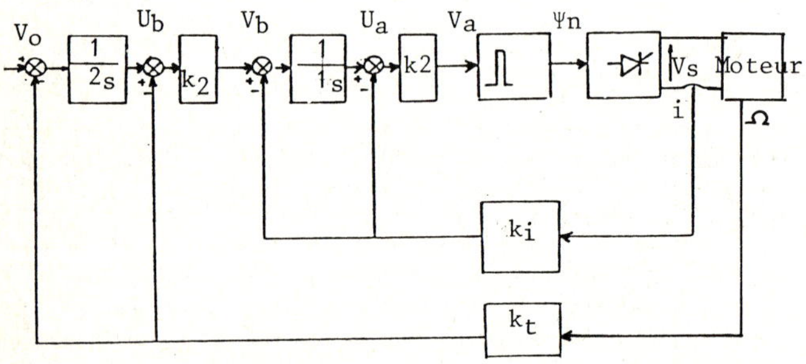

Figure 4

In continuous mode, where 2 thyristors are conducting, the motor equation are :

$$V_m \cos\left(wt - \frac{\pi}{6}\right) = ri + l\frac{di}{dt} + K\Omega ,$$

$$\Psi n < wt < \Psi n + 1 + \frac{\pi}{3}$$

$$K_i = J\frac{d\Omega}{dt} + f\Omega + \Gamma c$$

The controllers equations are :

$$\begin{cases} \dfrac{dU_a}{dt} = \dfrac{1}{\tau_1}(V_b - k_i i) \\ V_a = k_1 (V_a - k_i i) \end{cases}$$ (V_a is the control voltage of the converter)

$$\begin{cases} \dfrac{dUb}{dt} = \dfrac{1}{\tau_2}(V_o - k_t\Omega) \\ V_b = k_2 (Ub - k_t\Omega) \end{cases}$$

The states variables are : i (armature current), Ω (speed), U_a and U_b (associated to the pI controllers). The other coefficients are definied in annexe I. To simplify the calculus we adopt reduced variable, as an angular

$$\theta = wt$$

Then, as the converter is a 6 pulses per period, its period is $wT = \pi/3$. The reduced variables are :
$\xi 1 = Ri/Vm$; $\xi 2 = K\,\Omega/Vm$; $ua = Ua\,R/ki\,Vm$; $ub = Ub\,K/kt\,Vm$; other reduced variables are definied by annexe I which also gives numerical values.

The firing angle is $\Psi n = wt\,n$ and its value is given by the equality of the control voltage V_a and timing voltage W_a :

$$W_a(\psi n) = V_a(\Psi n)$$

W_a is supposed linear and synchronized with the supply :

$$W_a = A_1(-\theta + \delta)$$

Its reduced expression is

$$w_a = -\theta + \delta$$

The synthetic modelling is given by the differential equation

$$F_2 \frac{d\underline{X}_2}{d\theta} + G_2 \underline{X}_2 = \underline{V}_2(\theta)$$

with

$$X_2 = \begin{pmatrix} \xi 1 \\ \xi 2 \\ u a \\ u b \end{pmatrix} ; \quad F_2 = \begin{pmatrix} \Delta e & & & \\ & \Delta m & & \\ & & w\tau 1 & \\ & & & w\tau 2 \end{pmatrix}$$

$$G_2 = \begin{pmatrix} 1 & 1 & 0 & 0 \\ -c & 1 & 0 & 0 \\ 1 & kb & 0 & -kb \\ 0 & 1 & 0 & 0 \end{pmatrix} ; \quad V_2 = \begin{pmatrix} \cos(\theta - \pi/6) \\ -c\,\xi 1c \\ 0 \\ \xi 20 \end{pmatrix}$$

and the control condition

$$V_a = \underline{Pr}^T\,\underline{X2}(n) = Wa(\Psi n)$$

with $\underline{Pt}^T = k_a(-1\ 0\ 1\ 0)$

In the discontinuous mode, there are two intervalls to consider :

- between Ψn (firing angle) and $\Psi' n$ (extinction angle where $i(\Psi'_n) = 0$) that is the normal conduction whose equations are identical to the previous case : this is the 2 mode.

- between $\Psi'n$ and $\Psi n+1+\pi/3$ that is the no-current interval which can be deduced from the precedent by putting

$$i = 0$$

This is the 1-mode. Its state vector is

$\underline{X}_1^T = (\xi 2 \; u_a \; u_b)^T$, and its equation is :

$$F_1 \frac{d\underline{X}_1}{d\theta} + G_1 \underline{X}_1 = \underline{V}_1 (\theta)$$

F_1, G_1, $\underline{V}_1$ are easily deduced from F_2, G_2, $\underline{V}_2$ by the mean of joining matrices C_{12}, D_{12} and C_{21}, defined by : $\underline{V}_1 = D_{12}\underline{V}_2, \forall t$; $\underline{X}_1 = C_{12}\underline{X}_2, \forall t$; $\underline{X}_2 = C_{21}\underline{X}_1$ on the 1-mode. In this example, we have :

$$D_{12} = C_{12} = C_{21}^T = \begin{pmatrix} 0 & 1 & 0 & 0 \\ 0 & 0 & 1 & 0 \\ 0 & 0 & 0 & 1 \end{pmatrix}$$

and the relations between the differential equations are :

$F_1 = D_{12} F_2 C_{21}$; $G_1 = D_{12} G_2 C_{21}$; $\underline{V}_1 = D_{12} \underline{V}_2$

The states equations :

$$\frac{dX_2}{d\theta} = A_2X_2 + B_2(\theta) \text{ and } \frac{d\underline{X}_1}{d\theta} = A_1\underline{X}_1 + B_1(\theta)$$

are also obvious.

The firing instant is definied at the end of the 1-mode :

$$v_a(\Psi n)\, Pr1^T\, \underline{X1}(n) = W_a(\Psi n)$$

with $P_{r1}^T = (0 \; ka \; 0)$

MODELLING IN THE CONTINUOUS-MODE CASE

The linearized discrete model in the continuous mode case is

$$\underline{Z}_{n+1} = M_R \; \underline{Z}_n$$

with :

$$\underline{Z}_n = \left\{ \mathbb{1} - \left[A_2\underline{X}_2(\infty) + \underline{B}_2(\Psi\infty + \pi/3) \right] \underline{P}_r^T/A'\infty \right\} . \delta \underline{X}_2(n)$$

This transformation is regular. $\mathbb{1}$ is the unity matrix. The index ∞ denotes the operating point. $A'\infty$ is the derivative of Wa at instant $\Psi\infty$.

and $M_R = M\phi \; M_B$

ϕ is the transition matrix :

$$M\phi = \exp(-\frac{\pi}{3} A_2)$$

and M_B describes the feed-back and sampling effects :

$$M_B = \mathbb{1} + \frac{\left[B_2(\Psi\infty + \pi/3) - B_2(\Psi\infty x)\right] \underline{P}_r^T}{A'\infty - \underline{P}_r^T \left[A_2 \underline{X}_2(\infty) + B(\Psi\infty + \pi/3)\right]}$$

For analytic computation, $M\phi$ is partitioned as :

$$M\phi = \exp(\theta A_2) = \begin{pmatrix} \phi_2 & 0 \\ W & \pi \end{pmatrix}$$

where ϕ is relative the motor only :

$$\phi_2 = \left| \exp\left[\theta \begin{Bmatrix} -1/\Delta & -1/\Delta_e \\ +c/\Delta_m & -1/\Delta_m \end{Bmatrix}\right] \right| = \begin{pmatrix} \varphi_{11} & \varphi_{12} \\ \varphi_{21} & \varphi_{22} \end{pmatrix}$$

π is a triangular matrix :

$$\pi = \begin{pmatrix} 1 & \frac{\pi}{3} kb/w\tau_1 \\ 0 & 1 \end{pmatrix}$$

W describes the coupling between the motor and the controller. It can be proved that W can be written as :

$$W = \begin{pmatrix} W_{11} & W_{12} \\ W_{21} & W_{22} \end{pmatrix} = \begin{pmatrix} 0 & kb/w\tau_1 w\tau_2 \\ 0 & 0 \end{pmatrix} \begin{pmatrix} \alpha_{11} & \alpha_{12} \\ \alpha_{21} & \alpha_{22} \end{pmatrix} \Big|_{\theta = \pi/3} + \begin{pmatrix} 1/w\tau_1 & kb/w\tau_1 \\ 0 & 1/w\tau_2 \end{pmatrix} \begin{pmatrix} \beta_{11} & \beta_{12} \\ \beta_{21} & \beta_{22} \end{pmatrix} \Big|_{\theta = \pi/3}$$

where the α_{ij} and β_{ij} are independant of the controllers parameters.

M_B is written as :

$$M_B = \begin{pmatrix} 1-G_{ea} & 0 & G_{ea} & 0 \\ 0 & 1 & 0 & 0 \\ 0 & 0 & 1 & 0 \\ 0 & 0 & 0 & 1 \end{pmatrix}$$

G_{ea} has the physical meaning of a dynamic gain. Its expression is :

$$G_{ea} = \frac{h}{\left(\frac{dV_a}{d\theta} - \frac{dW_a}{d\theta}\right)\Big|_{\theta = \Psi\infty + \pi/3}} \; \frac{k_a}{\Delta e}$$

with $h = \cos(\Psi\infty - \pi/6) - \cos(\Psi\infty + \pi/6)$

G_{ea} is the only coefficient which depends on the gain ka.

Then M_R can be written analytically, and so is its characteristic equation whose roots give the poles of the system :

$$p(z) = \mathrm{Det}(M_R - z\,\mathbb{1}) = z^4 + p_3 z^3 + p_2 z^2 + p_1 z + p_0 \quad (4)$$

Where the p_i are easy to compute ; particularly $p_0 = 1 - G_{ea}$.

We must note that the dynamic modelling depends on the operating point which is defined by the references ; $\Psi\infty$ is given by :

$$\xi 1c + \frac{1 + C}{c} \; \xi 20 = \frac{\sin \pi/6}{\pi/6} \cos \Psi\infty$$

MODELLING IN THE DISCONTINUOUS-MODE CASE

In that case, the system has two distinct intervals, but with similar concepts seen in the example of the chopper (or see (2)) it is possible to definie the linearized model :

$$\delta \underline{Z}_{n+1} = M_R \; \delta \underline{Z}_n$$

with $M_R = M\phi \; M_B$

where $M\phi$ is similar to a transition matrix :

$$M\phi : \exp(\theta\infty A_1) \; F_1^{-1} \; D_{12} \; F_2 \exp(\gamma\infty A_2)$$

where $\gamma\infty = \Psi'\infty - \Psi.\infty$ is the duration of the 2-mode and $\theta\infty = \pi/3 - \gamma\infty$ is the duration of the 1-mode.

$M\phi$ is partitionned as :

$$M\phi = \begin{pmatrix} \phi_1(\theta\infty) & \begin{matrix}(o)\\(1)\end{matrix} & \phi_2(\infty) & 0 \\ & \mathcal{W} & & \pi \end{pmatrix}$$

ϕ_2 has the same signification as in the previous paragraph (motor in continuous interval) and ϕ_1 is the exponential associated to the motor when $i = o$: $\phi_1 = \exp(-\theta/\Delta m)$. π has been preniously defined describes the coupling between the motor and the controller and is noted :

$\mathcal{W} = \begin{pmatrix} W11 & W12 \\ W21 & W22 \end{pmatrix}$, it can be written in a similar manner as the one seen in the continuous model of operation :

$$(W_{ij}) = \begin{pmatrix} 1/w_1 & kb/w\tau_1 \\ 0 & 1/w\tau_2 \end{pmatrix} (\beta'_{ij}) + \begin{pmatrix} o & kb/w\tau_1 w\tau_2 \\ 0 & o \end{pmatrix} (\alpha''_{ij})$$

where the β'_{ij} and α'_{ij} are independant of the controllers parameters. MB is defined by

$$M_B = H_{21} + \frac{H_{21}\,[A_1X_1(\infty)+B_1(\Psi_\infty+\pi/3)]\frac{P_r^T}{A_\infty} - [A_2X_2(\infty)+B_2(\psi_\infty)]\frac{P_r^T}{A'_\infty}}{1 - P_r^T\,[A_1X_1(\infty) + B_1(\psi_\infty + \pi/3)]/A'_\infty}$$

where H_{21} is the joining matrix at the firing instant :

$$\underline{X}_2 = H_{21}\,\underline{X}_1$$

$$\text{in our example : } H_{21} = \begin{pmatrix} 0 & 0 & 0 \\ 1 & 0 & 0 \\ 0 & 1 & 0 \\ 0 & 0 & 1 \end{pmatrix}$$

Then M_B may be written :

$$M_B = \begin{pmatrix} 0 & G_{eb} & 0 \\ 1 & 0 & 0 \\ 0 & 1 & 0 \\ 0 & 0 & 1 \end{pmatrix}$$

$$\text{with } G_{eb} = \frac{\cos(\psi_\infty - \pi/6) - \xi 2(\infty)}{\Delta e \left[-\frac{A'_\infty}{ka} + \frac{kb}{w\tau_1}(-\xi 2(\infty) + ub(\infty))\right]}$$

The dynamic model depends on the steady state which is not easy to compute exactly. Then we admit that the speed is constant on a period, then : $\xi 2(\infty) \simeq 20$

$$ub(x)\ \xi 20 + \bar{\xi} 1(\infty)/Ub$$

where $\bar{\xi}1(\infty)$ is the mean value of the (discontinuous) current which can be digitally computed.

Then the eigen-values of M_R are the roots of the characteristic equation, which can be determined by classical calculus with, the following form :

$$p(z) = \mathrm{Det}(M_R - z\,\mathbb{1}) = -z^3 + p'_2 z^2 + p'_1 z + p'_o \quad (5)$$

SYNTHESIS OF THE CONTROL SYSTEM

For synthesis, several strategies can be used :
(i) exact synthesis

We can use the characteristic equation to make a poles placement. As the system is sampled data, we can want to have the poles at the origin (z = o). This is practically realisable for naturaly fast systems, as a current regulation. For mechanical systems, this is irrealisable (the torque will be too large and the results will be very sensitives). Then we propose a hybrid placement which is the following for our example :

- two (for the continuous case) or one (for the discontinuous case) pole(s) are zero : this corresponds approximatively to a deadbeat response for the current loop.
- two poles : z_3 and z_4 choosen as :

$$z_{3,4} = \exp(T\,s_{3,4})$$

where T is the period of the converter (T = 3,3ms in our example), and $s_{3,4}$ is a pair of poles choosen is the s-plane for example :

$$z_{3,4} = \rho(-\sqrt{2}/2 \pm i\,\sqrt{2}/2)$$

which are "optimal" poles in the sense of the optimal control of second order continuous systems.

Then we identify the characteristic polynoms with :

$$z^r\,(z - z_3)\,(z - z_4) = 0$$

(where r = 2 for continuous case, r = 1 for discontinuous case).

In the continuous case, we obtain the following condition

$$G_{ea} = 1$$

and explicit equations to compute ka, kb, τ_1 and τ_2 (5).
Similar results are obtained in discontinuous case.
A precise digital simulation /6/ permits to compare exact response to the desired one. The figure 5 gives the response to a small variation of the speed reference around a steady-state for different values of the modulus ρ of the dominant poles in the s-plane (continuous case)

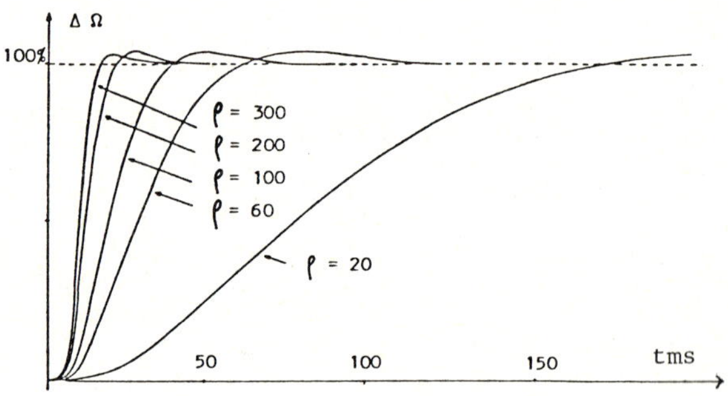

Figure 5

The table I compares the results of the simulation of the 4th order system to the performances associated to the dominant pair of poles.

tr is the time to reach the first maximum, and D is the overshoot. The results are very near. tr given by the simulation is a little longer : this can be explained by the presence of the two supplementary poles (equal to zero) in the complete system.

TABLE I

	SIMULATION		Performances associated to the desired poles	
	tr (ms)	D (%)	tr (ms)	D %
20	230	4	222	4,5
60	80	3,6	74	4,5
100	45 ou 50	3,3	44	4,5
200	25 ou 28	3,7	22	4,5
300	20	3,2	15	4,5

(ii) approximative synthesis
This is the most popular approach, because that class of systems is generaly approximated by linear continuous modelling. For a first example,

we propose to sudy the system in the continuous case where the non-linearities are small. Then, we approximate the converter by a simple gain Go :

$$G_o = \frac{d\,\overline{V}_s}{d\,V_a}$$

Where $\overline{V}_s$ is the mean value of the out put voltage. The modelling is linear and continuous and we can easily compute the controllers parameters (k_1, $\tau 1$, k_2, $\tau 2$) such as the poles have a desired configuration.

The fig. 6 illustrates this approach : the desired places for the four poles are choosen in the s_ plane : two "fast" poles (modulus : $R_1 = 300s^{-1}$; arguments : $\pm 3\ \pi/4$) and two "slow" poles (modulus $R_2 > 20_s{}^{-1}$, arguments : $\pm 3\pi/4$). The fig. compares, in the z-plane, the position of these desired poles for the linear continuous modelling, with the position of the poles obtained with the linearized modelling (digital computation of the roots of (4). We see that for a slow dynamic ($R_2 < 150s^{-1}$) the two modelling methods give similar results. This can be explained : in the case of the continuous mode of operation the effect of the non-linearities and of the sampling are light when the band-width is not too large. Furethermore the inner current loop has also a favourable effect. This property is largely used by engineers in industrial applications. But we observe than for large bandwidth (here for $R_2 > 250s^{-1}$) the precision of the prevision given by the linear continuous modelling method becomes rough. The figure 7 gives simulation results.

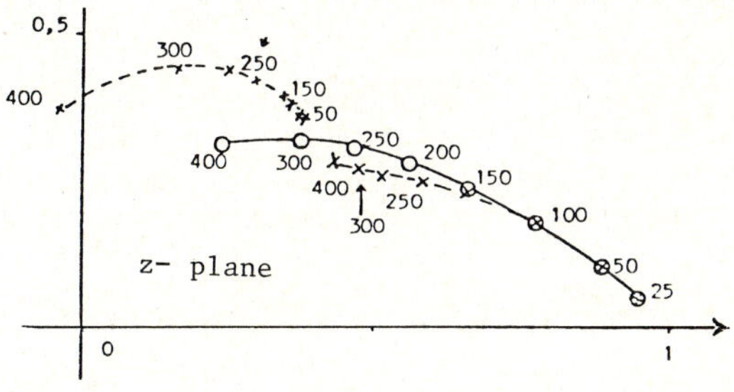

Figure 6

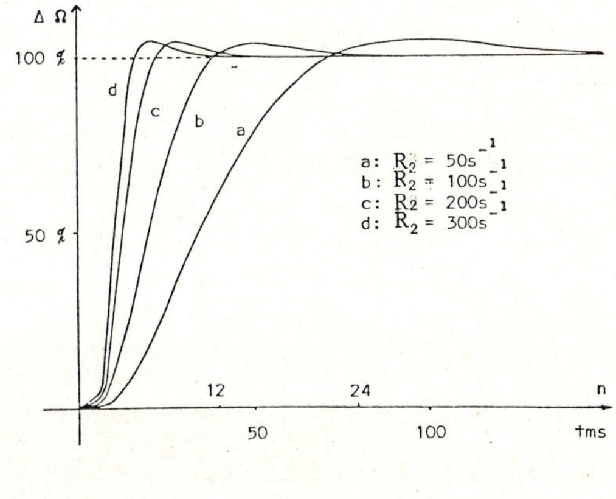

Figure 7

The second approach is also classical : this is the decoupling technic. First the internal current loop is designed ; second the speed loop is computed, assuming that the current loop is perfect. We give here results concerning the discontinuous case, but similar results would be obtained in the continuous mode case with a precise, linearized model, we determine an optimal current control (deadbeat response). Then the speed loop is considered as a two-order system and the speed controller is designed to give to the system two "slow poles" of modulus R and arguments : ± 3 /4. The fig. 8 illustrates the results. For a slow dynamic (here $R < 100\ s^{-1}$) the real poles are near by the desired poles and the simulation confirms these results (Fig.9) For fastes dynamic ($R > 150s^{-1}$), the obtained poles are farer from the desired poles and the system becomes unstable for $R > 250s^{-1}$

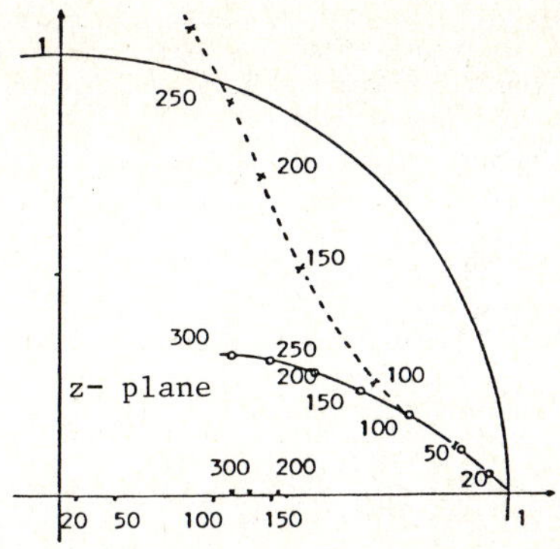

Figure 8

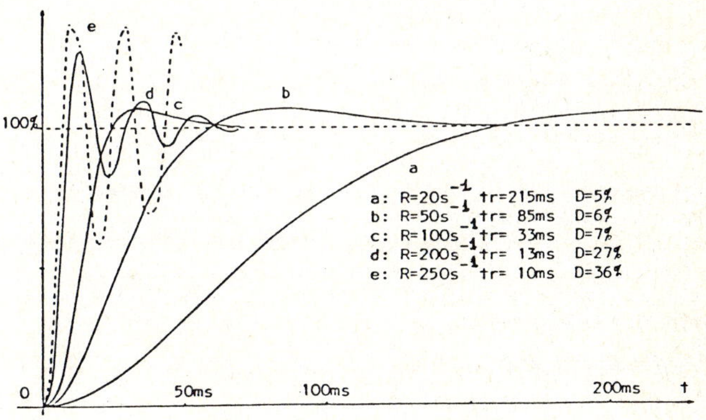

Figure 9

CONCLUSION

This paper deals with the applications of a rigorous modelling for control systems including static converters, as DC-DC converters and AC-DC converters. The use of recurrencies equations and matricial joining conditions at instants of structure changes permits to determine a global linearized model.

Then the characteristic equations and the poles of the system can be computed. So we have a powerful tool for the synthesis if a placement poles strategy is used. When analytical calculus are possible, a rigorous pole placement can be used and we have proposed a hybrid placement : poles at the origin for the fast part of the physical system, and poles associated to a slower dynamic for the part remaining of the system. When these calculus are too complicated, the designer can use approximative, simple modelling and after that can verify if the obtained poles placement is correct.

ANNEXE I

Definitions and numerical values of the system parameters taken as an example in the part II.

For the DC motor : r = 1 Ω and l = 18 mH : resistance and inductance of the armature ; Vm = 312 V; K = 1,6V/rds^{-1} or Nm/A : Torque constante ; J = 0,070 kg m^2 and f = $6,7.10^{-3}$ N/rds^{-1} : inertia and viscous torque coefficient Γc is the load (constant) torque ; Δe = wTe = lw/r = 5,65 ; Δm = Jw/f = 3280 ; c = K^2/rf = 382 : reduced coefficients ξ1c = RΓc/KVm ; ξ20 = KVo/ktVm . The pulsation of the supply is w = 100 π ; the converter receives 6 pulses per period T = 3,3ms or wT = π/3. The turning voltage Wa is defined by A1 = 3,47 and δ = π /2.

The gain transducers are : ki = 1/30 V/A(current); kt = 0,0296 V/rds^{-1} (speed)
The controllers parameters are : τ1, k_1 (time constant and gain of the of the current controllers) ; τ2, k2 (speed controller). The reduced coefficients are : wτ_1 ; ka = k_1 k_i v_m/RA_1
kb = Rkt k_2/kiK.

ANNEXE II

Steady state point (continuous mode case)
mean values : $\bar{I}_\infty$=15A ; Ω_∞=125rd/s ; Γc=23,2 Nm.
Instantaneous values : i(∞) = 11,6A ; Ψ_∞ = 43°7.

The exponentiel associated to the motor in continuous mode case is :

$$\phi\ (T) = \begin{pmatrix} \varphi 11 = 0,821 & \varphi 12 = -0,168 \\ \varphi 21 = 0,109 & \varphi 22 = 0,989 \end{pmatrix}$$

ANNEXE III

Steady state point (discontinuous mode case)
mean values : $\bar{I}_\infty$ = 3 A ; Ω_∞ = 20,94 rd/s ;
angles : firing : Ψ_∞ = 87° ; extinction :
Ψ_∞ = 138°5
exponential associated to the moteur :
$\phi(\gamma_\infty) = [\varphi 11 = 0,847 ; \varphi 12 = 0,145 ;$
$\varphi 21 = 0,094 ; \varphi 22 = 0,892]$

BIBLIOGRAPHY

(1) LEONHARD W.: Regelung in der Elektrischer Autriebstechnik, Teubner, Stuttgart, 1973.

(2) LOUIS J. P. : Control in Power Elect. and and Elec. Drives, 3rd IFAC Symposium, 12-19 sept 1983, Lausanne, R Zwicky ed., Pergamon Press, pp. 9 à 16.

(3) LEONHARD W : Elektrotechnische Zeitschrift tome A 86, n° 16, 1965, pp 513.

(4) BÜHLER H : 2nd symp. (FAC, Dusseldosf 1977, Pergamon Press, pp, 67-77.

(5) LOUIS J. P. : Thèse Doctorat ès-sciences, Nancy, INPL, 1981.

(6) IUNG C. LOUIS J. P. : 8th A/CA Congrès, Delft 1976. I. Dekker ed., North Holland Pub. cy, pp, 805-813.

AKNOWLEDGEMENT

The major part of this work was done when the author was with the Centre de Recherche en Automatique de Nancy (CNRS-ERA 905, Directeur : Prof. R. HUSSON).

ELECTRICAL MACHINES AND CONVERTERS - Modelling and Simulation
H. Buyse and J. Robert (editors)
Elsevier Science Publishers B.V. (North-Holland)/© IMACS, 1984

MODELLING AND SIMULATION OF A PWM-INVERTER-FED LINEAR NETWORK

G. K. Creighton B.Sc., Ph.D., C.Eng., M.I.E.E

Department of Electronic and Electrical Engineering
The University of Technology
Loughborough, Leicestershire, LE11 3TU, England

The paper presents some solutions of linear state-equations, which are valid when the control inputs are combinations of steps and sinusoids, and describes their application to the analysis of a single-phase pwm-inverter-fed linear network. A computational technique for generating the necessary pwm voltage waveform is described. The advantages of the method of analysis used are emphasised, and some computer results for the performance of a 10kVA pwm inverter supplying a loaded passive filter are presented. Some problems involved in the design of such a system are discussed.

PRINCIPAL SYMBOLS

x, z = $(n \times 1)$ state variable vectors
u = $(s \times 1)$ control variable vector
B = $(n \times n)$ matrix of constants
C = $(n \times s)$ matrix of constants
P = $(n \times n)$ matrix of eigenvectors of B
Λ = $(n \times n)$ diagonal matrix of eigenvalues of B
λ = eigenvalue of B
ω = angular frequency
ω_m = angular frequency of the modulating wave
ω_c = angular frequency of the carrier
ϕ = phase angle
t = time
V = direct voltage
M = modulation depth: [amplitude of modulating wave]/[amplitude of carrier]

1. INTRODUCTION

Any linear system can be modelled with a set of first-order differential equations, which are often called state equations. The general solution of these state equations for the system state at any time can be expressed in terms of the initial state, the transition matrix and the control inputs to the system[1]. When the control inputs are combinations of step and sinusoidal functions, some simple algebraic operations on the general solution convert it into an ideal form for system analysis and modelling on a digital computer. For example, this form permits the system state at the beginning of a cycle of steady-state operation to be obtained without iteration, a feature which considerably simplifies the calculation of steady-state system performance. The transient and steady-state components of the system response can be separated, thereby permitting a straightforward calculation of the settling time. The exponential and oscillatory components of the system response can be separately identified, and, further, the system response can be directly calculated at any selected time instants, so that the problems associated with step-by-step integration over corners or high-frequency transients are completely avoided. This ideal form of the solution of the state equations is described and developed in the paper.

Now most inverters produce voltages or currents at the load terminals that have waveforms essentially comprising steps, and, therefore, when these inverters are loaded with linear networks, they are systems that are well suited to modelling and analysis using the method described. Inverters are commonly loaded with ac machines which are nonlinear devices, but they are often assumed to be linear when the speeds are constant[2]. Thus the linear network could comprise a constant speed ac machine, or a passive filter supplying a sinusoidal voltage, for example, to a computer, medical instrumentation or protection circuits.

2. MODELLING AND SIMULATION OF A SINGLE-PHASE PWM INVERTER SYSTEM

2.1 General

An expression is derived in Appendix 5.1 that can be used for the determination of the state of a linear system at any time, when the control inputs comprise combinations of steps and sinusoids. It incorporates the eigenvalues and the eigenvectors of the system which are readily calculated with the aid of a digital computer. Thus, a digital simulation is assumed throughout the paper.

As any periodic control input comprises a train of sinusoidal components it is possible to use superposition of the states, calculated for the dominant input sinusoids, to obtain the system response for any periodic controls. However, as most inverters produce output voltage steps, it is the step control signal which is of interest here. In particular, a single-phase pwm inverter is chosen for illustrating an application of the expression derived in Appendix 5.1 to the determination of a linear system's performance. This type of inverter commonly produces a naturally-sampled pwm waveform, and the computational technique for generating an example of such a waveform is described. The technique is well suited to the inclusion of errors introduced by the electronic circuits used in practice to generate the waveform.

2.2 Single-phase pwm inverter output voltage waveform

Figure 1 shows a typical single-phase inverter network connected to a filter that supplies the load with a voltage having a good sinusoidal waveform. The inverter supplies the filter with voltage steps approximately equal in amplitude to the direct supply voltage acting in either the forward or reverse directions. It is common practice[3] to reverse the direct voltage supply to the filter at the time instants that a sinusoidal modulating wave and a triangular carrier wave intersect, as shown in Figure 2. The pwm wave applied to the filter then has a fundamental component of amplitude proportional to the modulation depth, and its most significant harmonics are at or near to the carrier frequency. An expression showing the harmonic composition of the pwm waveform[4] is given in Appendix 5.2.

The pwm waveform is defined by the direct voltage amplitude and the wave edge positions. The first step in the development of a systematic procedure for determining the edge positions on a digital computer is to convert the sawtooth carrier into a straight line that intersects a series of modulating waves as shown in Figure 3. If the paper of Figure 3 were folded at values of x equal to integral multiples of π, the modulating waves would overlap and the upper part of Figure 2 would appear. The trailing edges of the pwm waveform then occur at the intersections for which

$$x = \frac{\pi}{2}\left[1 + M\cos(x\omega_m/\omega_c)\right] + 2k\pi \qquad (1)$$

and the leading edges occur when

$$x = \frac{\pi}{2}\left[3 - M\cos(x\omega_m/\omega_c)\right] + 2k\pi \qquad (2)$$

where k is any integer. The next step is to apply the Newton Raphson method to determine the values of x that satisfy equations 1 and 2. The values of x used to start the iterations occur at the positions of the dotted lines in Figure 3 and can be obtained from equations 1 and 2 by setting M=0. As $\omega_c >> \omega_m$ in practice, the right-hand sides of equations 1 and 2 are nearly linear functions of x between the initial values and the intersection points. Thus, the Newton Raphson method normally produces the required x to a satisfactory accuracy with only one iteration. Errors in the positions of the pwm waveform edges can be deliberately introduced by a modification of equations 1 and 2. Thus an imperfect sinewave of modulation could be modelled or some switching delay could be incorporated.

2.3 Simulation of a linear network with a pwm supply

Appendix 5.1 shows that the state equations $\dot{x} = Bx + Cu$ characterising a linear system can

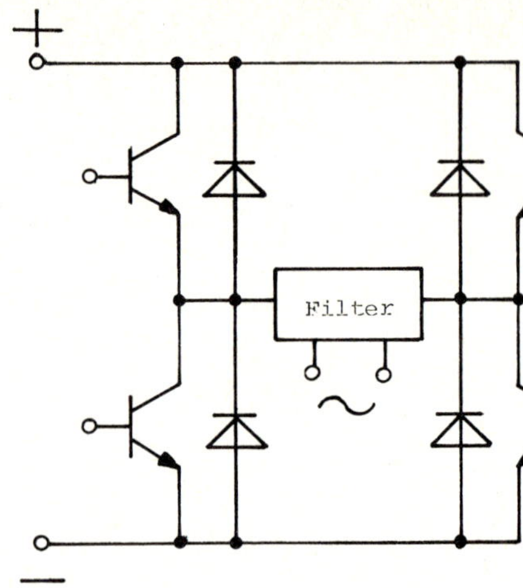

Fig 1: Single-phase pwm inverter

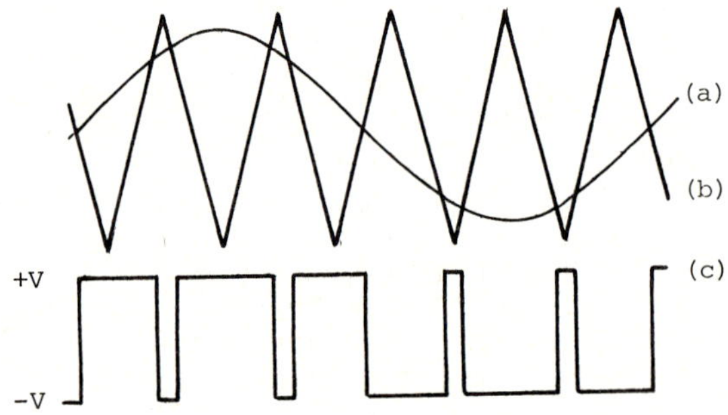

Fig 2: Production of the pwm waveform

(a) Sinusoidal modulating wave
(b) Triangular carrier wave
(c) Pwm voltage wave

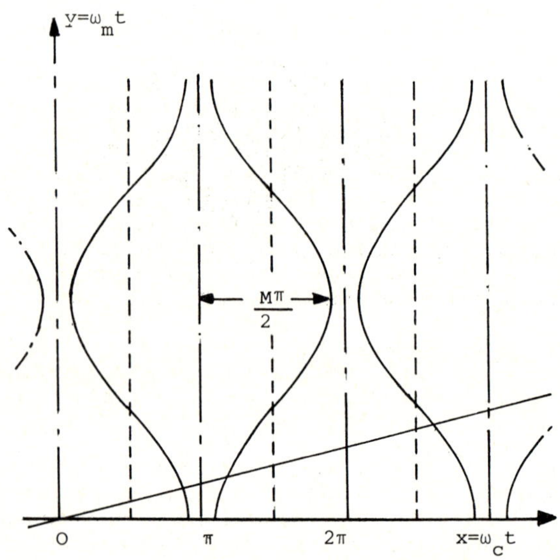

Fig 3: Determination of the crossover points

be solved for the state x(t), with the aid of the transformation x = Pz, when u comprises steps and sinusoids. Expressions are given for z(t) which can be converted to x(t) by using this transformation. The pwm voltage signal produced by the inverter acts like a single bang-bang control input to the linear network loading the inverter. In general, a linear network could be supplied with several controls of this kind, for example if the supply were a 3-phase pwm inverter. A method is required for obtaining x(t) either during a transient or in the steady state when the controls are bang-bang. The results given in the Appendix can be extended to produce x(t) in this case as follows.

Assume that the kth control input u_k comprises steps with edges occurring at times $t_1, t_2 \ldots t_j < t$ so that t is such that j vertical edges in the waveform of u_k have been passed. ω_k and ϕ_k may be set to zero in equation 9, the integration of equation 11 can be performed analytically, and the resulting ith component of z(t) is

$$z_i(t) = \exp(\lambda_i t) z_i(0) + \lambda_i^{-1} \sum_{k=1}^{s} d_{ik} \Big[\exp(\lambda_i t) + 2 \sum_{m=1}^{j} (-1)^m \exp\{\lambda_i (t - t_m)\} + (-1)^{j+1}\Big] \quad (3)$$

In the derivation of this result it is assumed that d_{ik} is obtained using the magnitude of the control input, i.e. $|a_k|$ is used in equation 9. Equation 3 can be used to determine $z_i(t)$ no matter how many edges have been passed. However, as the number of edges passed increases, so does the number of terms to be evaluated. To limit the continual increase in the number of terms to an easily manageable figure, when, for example, two edges have been passed the value of $z_i(t)$ is regarded as a new initial condition $z_i(0)$, the position of the time origin is changed accordingly, and a counter for the number of edges passed is set to zero. The required components of x are of course obtained using x = Pz. In the single-phase inverter simulation there is only one control input, so that s can be set to 1 in equation 3.

It should be noted that equation 3 permits $z_i(t)$ to be determined at any time, and therefore the corners at the pwm wave edge positions can be stepped over if necessary. The restrictions accompanying a step-by-step integration of the state equations are avoided.

2.4 Steady-state simulation

When an inverter is operating in the steady state the values of the state at each end of a cycle of operation are equal. Steady-state performance of the inverter can be computed by solving the system's state equations over one cycle when the initial conditions are known. The initial state may be determined by correcting an estimate of it using Newton Raphson iteration, or by solving the state equations, using a state estimate for the initial conditions, for as long as is necessary for the steady state to be reached. A method of obtaining the initial state which involves less computing time than these possibilities is obtained from equation 3 as follows.

For steady-state operation of an inverter, x(0) = x(T) where T is the period of the pwm voltage waveform. Thus, since it follows that $z_i(0) = z_i(T)$, the substitution of this result into equation 3 gives

$$z_i(0) = \lambda_i^{-1} [1 - \exp(\lambda_i T)]^{-1} \sum_{k=1}^{s} d_{ik} \Big[\exp(\lambda_i T) + 2 \sum_{m=1}^{j} (-1)^m \exp\{\lambda_i (T - t_m)\} + (-1)^{j+1}\Big] \quad (4)$$

where j is the number of edges of the pwm waveform passed during the period T. This expression can be used for the determination of the initial conditions for steady-state operation directly.

For many applications the linear network is a filter which supplies a good sinusoidal voltage to a linear load. In such cases the load voltage, power factor and rating are usually known and the necessary modulation depth of the pwm supply voltage wave has to be determined. This can be done by assuming that the input to the filter is a unit amplitude voltage sinusoid and then using the results given in the Appendix to determine the corresponding output voltage amplitude. Linear amplitude scaling can be used to determine the necessary input fundamental amplitude, and the modulation depth follows. The figure for the modulation depth should not exceed about 0.9 to ensure that the minimum pulse-width is long enough to allow the inverter switching devices to turn fully on and off.

2.5 Transient simulation

A transient occurs in the inverter system when, for example, the steady-state condition is disturbed by a sudden change in the load. The transient state variation with time can be determined by using equation 3, but also of interest is the settling time. As the difference between the transient state and the steady state can be computed using the expressions presented, the determination of the settling time is straightforward if the modulation depth is constant. It becomes complex if a control system is used to continuously vary the modulation depth, perhaps to ensure that the output voltage remains approximately constant. However, much useful information can be obtained from the first few cycles of a transient when the control system usually does not have enough time to alter the modulation depth significantly.

2.6 Structure plan

The solution of the linear state-equations characterising the system depends on the matrices B and C, the initial state, and the control inputs. When the configuration of an inverter-fed linear network is specified, B and C are readily obtainable. If the network is a loaded filter with a specified voltage at the load terminals, and the direct voltage, the modulation frequency and the carrier frequency are specified, the modulation depth can be determined and the pwm input voltage waveform defined. If the steady-state performance is required, the initial state is obtained and the states at any desired time instants can be determined. These states could be used for plots of current and voltage waveforms at the various filter components or, for example, for the calculation of the harmonic distortion in the load voltage waveform. To simulate a transient following a load switching, the B and C matrices are modified so that the filter together with its new load are modelled, and the system's states are obtained as before.

It is evident that a computer program for simulating the inverter-fed linear network may be written such that only minor modifications are required when the network configuration is altered, as only B and C and the information printed out need be changed. Thus, for example, essentially the same computer program could be used for the performance investigation of pwm inverter-fed single-stage filters and two-stage filters.

2.7 Discussion of the results of a simulation

The expressions developed in the paper have been incorporated in a Fortran program which has been run on a Prime 460 computer at Loughborough University, and performance results for single-phase pwm inverters supplying both single and two-stage LC filters have been calculated. A sample of the computer's graphical output following its calculation of the steady-state and transient performances of a single-phase pwm inverter-fed single-stage loaded filter with the data given in Appendix 5.3 is shown in Figures 4 and 5. The steady-state results were obtained for full-load operation, and the transient results for a switch from full load to no load. Points of interest in these results are the peak input currents and the harmonic distortion of the output voltage wave. The peak input current is of value to the designer who has to select the inverter components, and the harmonic distortion must be low enough to meet the customer's specification. The computer's output includes not only the figure for the harmonic distortion but also the volt-ampere ratings and the peak voltages and currents for each component of the filter and load. The rating of the filter inductor is especially important because the inductor is an expensive component and its rating tends to be high. The inductor has to withstand the peak load voltage plus the direct voltage. The designer aims to minimise the inductor rating, to reduce the equipment cost, while maintaining the peak input current and the output harmonic distortion at satisfactory levels. In addition, he must ensure that the transient performance meets the specification. Another design problem concerns the choice of the carrier frequency. A high carrier frequency can be filtered with small values of the inductance and capacitance, but can produce excessive switching losses in the inverter components and high iron-losses in the inductor. Thus, compromises must be made to ensure overall minimum cost while satisfying the performance specifications.

3. CONCLUSIONS

The paper has shown that the general solution of a set of linear state-equations can be expressed in forms which are well suited to a computer simulation when the system inputs comprise combinations of steps and sinusoids. The example of a pwm-inverter-fed linear network is used to show the application of the expressions to the determination of a linear system's performance during both transient and steady-state operation. The generation of the naturally-sampled pwm input voltage waveform has been described, and the special results for the state which are required when the input is of this form have been given. The results of a computer simulation of a pwm-inverter-fed linear filter are presented and some problems in the design of such systems are briefly discussed. It can be concluded that the expressions given provide some worthwhile advantages when they are used for the determination of linear system performance.

4. REFERENCES

[1] Dorf, R.C., Time-Domain Analysis and Design of Control Systems (Addison-Wesley, Reading, Mass., 1965).

[2] Adkins, B., The General Theory of Electrical Machines (Chapman and Hall, London, 1962).

[3] Bowes, S.R., New sinusoidal pulsewidth-modulated inverter, Proc.IEE, Vol.122, No.11, (1975) 1279-1285.

[4] Bowes, S.R. and Bird, B.M., Novel approach to the analysis and synthesis of modulation processes in power converters, Proc.IEE, Vol.122, No.5, (1975) 507-513.

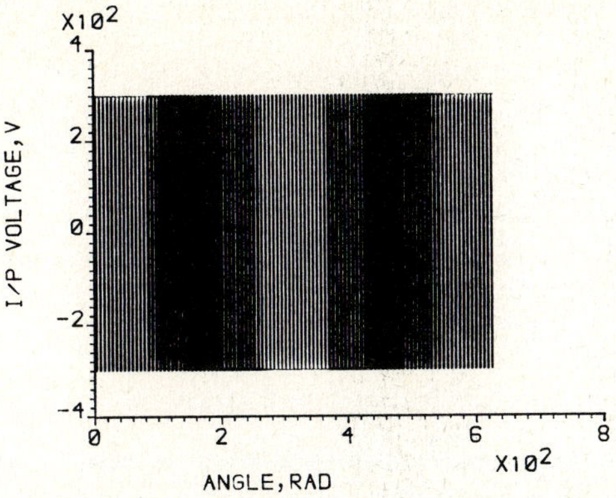

(a)

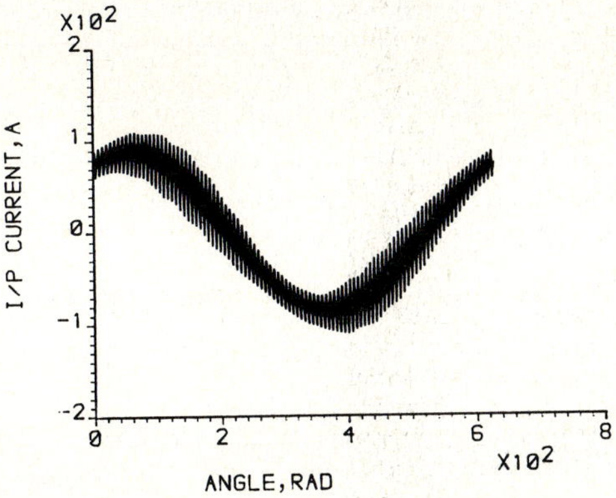

(b)

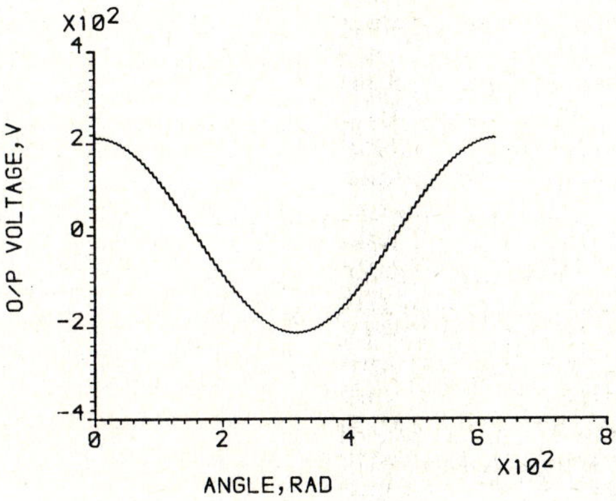

(c)

Fig 4: Steady-state results

(a) Pwm input voltage wave
(b) Input current wave
(c) Output voltage wave

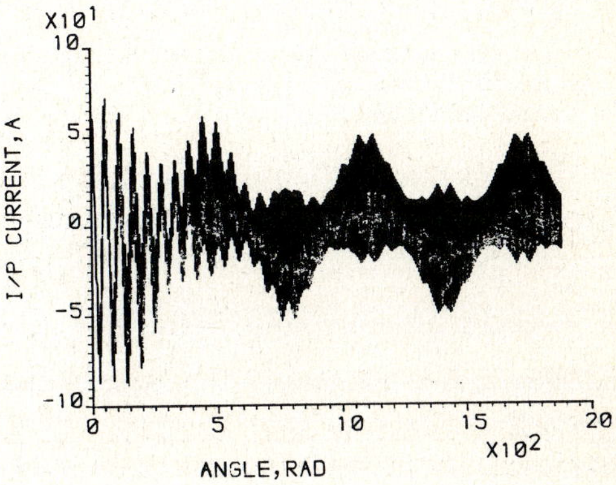

(a)

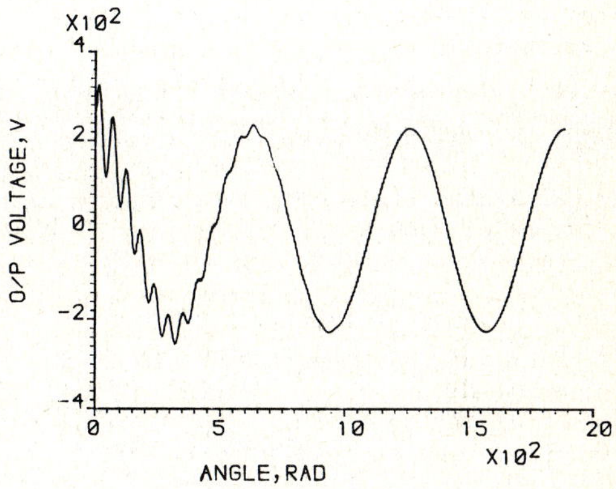

(b)

Fig 5: Transient results

(a) Input current wave
(b) Output voltage wave

5. APPENDICES

5.1 Solution of linear state equations with step and sinusoidal inputs

The state equations of a linear time-invariant system are

$$\dot{x} = Bx + Cu \tag{5}$$

When the eigenvalues of B are distinct, equation 5 may be converted to diagonal form by applying the transformation

$$x = Pz \tag{6}$$

Thus, when $\Lambda = P^{-1}BP$ is a diagonal matrix comprising the eigenvalues of B

$$\dot{z} = \Lambda z + P^{-1}Cu \tag{7}$$

The standard solution of equation 7 is

$$z(t) = \exp(\Lambda t)\, z(0) + \int_0^t \exp[\Lambda(t-\tau)]P^{-1}Cu(\tau)\,d\tau \tag{8}$$

The transition matrix $\exp(\Lambda t)$ may be obtained by inspection, given the eigenvalues, and the integration may be performed analytically when u comprises steps and sinusoids.

Let the kth control input be represented by

$$u_k(t) = a_k(\omega_k t + \phi_k) \tag{9}$$

where $1 \leqslant k \leqslant s$. This expression may also represent a constant control of magnitude a_k when ω_k and ϕ_k are zero.

Now if C_k represents the kth column of C, let

$$P^{-1}C_k a_k = [d_{1k}, d_{2k}, \ldots d_{nk}]^t \tag{10}$$

Then if I represents the integral in equation 8, it can be written as

$$I = \sum_{k=1}^{s} \int_0^t \begin{bmatrix} \exp\{\lambda_1(t-\tau)\}d_{1k} \\ \exp\{\lambda_2(t-\tau)\}d_{2k} \\ \vdots \\ \exp\{\lambda_n(t-\tau)\}d_{nk} \end{bmatrix} \cos(\omega_k\tau+\phi_k)\,d\tau \tag{11}$$

The ith component of z(t) may be obtained from equation 8 after the substitution of the integrated ith row of equation 11. The result is

$$\begin{aligned} z_i(t) = {} & \exp(\lambda_i t)\, z_i(0) \\ & - \sum_{k=1}^{s} d_{ik}[\lambda_i\cos(\omega_k t+\phi_k) - \omega_k\sin(\omega_k t+\phi_k) \\ & - \exp(\lambda_i t)(\lambda_i\cos\phi_k - \omega_k\sin\phi_k)]/(\omega_k^2+\lambda_i^2) \end{aligned} \tag{12}$$

Equation 12 can be used to determine the values of z_i for $1 \leqslant i \leqslant n$ at any time t, and then x(t) can be obtained by using equation 6. If the denominator $(\omega_k^2+\lambda_i^2)$ of a term in the summation of equation 12 is zero, that term should be replaced by $d_{ik}t$ as the integration is different in that particular case.

Not all of x(t) need be determined as the required rows of P may be selected in the product Pz.

For steady-state operation of the system the exponential terms must disappear from equation 12, leaving the steady-state z_i given by

$$\begin{aligned} z_{is}(t) = {} & \sum_{k=1}^{s} d_{ik}[\omega_k\sin(\omega_k t+\phi_k) \\ & - \lambda_i\cos(\omega_k t+\phi_k)]/(\omega_k^2+\lambda_i^2) \end{aligned} \tag{13}$$

There is no need to determine the initial conditions before going on to obtain the steady-state results as equation 13 can be used directly.

The transient component of equation 12 is

$$z_{it}(t) = \left[z_i(0) + \sum_{k=1}^{s} \frac{d_{ik}}{\omega_k^2+\lambda_i^2}(\lambda_i\cos\phi_k - \omega_k\sin\phi_k)\right]\exp(\lambda_i t) \tag{14}$$

Use of this equation permits not only the determination of the transient component for any element of the state vector, but also the determination of the coefficients of every exponential term. Thus, the components of the system response can be calculated as well as the settling time.

5.2 Fourier components of an ideal naturally-sampled pwm waveform

The Fourier series for the pwm waveform of the type shown in Figure 2 is

$$\begin{aligned} F(t) = {} & MV\cos(\omega_m t) \\ & + \frac{4V}{\pi}\sum_{m=1}^{\infty} \frac{J_0(mM\pi/2)}{m}\sin\left(\frac{m\pi}{2}\right)\cos(m\omega_c t) \\ & + \frac{4V}{\pi}\sum_{m=1}^{\infty}\sum_{n=\pm1}^{\pm\infty} \frac{J_n(mM\pi/2)}{m}\sin\left\{(m+n)\frac{\pi}{2}\right\} \\ & \qquad \cos\{(m\omega_c \pm n\omega_m)t\} \end{aligned} \tag{15}$$

where J_n is a Bessel function or order n, and 2V is the peak to peak amplitude of the pwm waveform.

5.3 Inverter and filter data

Rating 10 kVA; direct voltage 300 V; modulation frequency 50 Hz; carrier frequency 5 kHz; output voltage on full load 150 V; full-load power factor 0.8; filter inductor 0.4 mH, 0.1 Ω; filter capacitor 200 μF.

ELECTRICAL MACHINES AND CONVERTERS - Modelling and Simulation
H. Buyse and J. Robert (editors)
Elsevier Science Publishers B.V. (North-Holland)/ ©IMACS, 1984

SIMULATION OF THE FORCED COMMUTATION IN A STATIC FREQUENCY CHANGER

J. Robert G. Comans R. Montfort

Institut Electrotechnique Montefiore
Université de Liège
Rue Saint-Gilles 33, B-4000 Liège - Belgium.

Summary

This survey starts with a short explanation of the unrestricted frequency changer principles. The power circuit and the commutation process are then further developped. Next comes the elaboration of the simulation programme. Assumptions and models are adopted to allow the final simulation process to be introduced.

The simulating method uses a signal flow graph for each model. Those signal flow graphs were united into one general flow graph to simplify the setting of the final programme.

1. Introduction

Before studying simulation, it is necessary to explain briefly the unrestricted frequency changer principle (UFC).

The UFC is an AC-AC converter directly changing the input power at frequency f_{in} to the output power at frequency f_{out}, using a bidirectional switches system connecting the input terminals to the output terminals (Fig. 1).

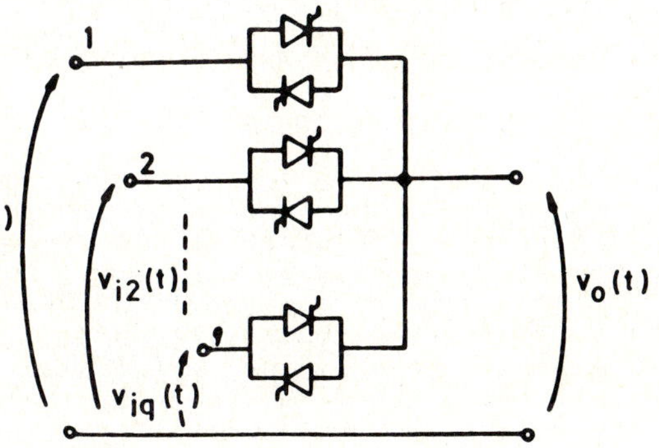

Fig. 1 : q-pulse frequency changer circuit with single-phase output.

The UFC is the only converter in his category providing an output frequency greater than the input frequency. An other advantage is that it inverts the phase angle of the load : it means that a self-inductance is seen as an capacitor. This property may be used in reactive power compensators. In that case the UFC load is a self-inductance and by changing the output frequency and the output voltage, one can continuously change the reactive power supply. An other application is the speed-control of the induction motor with short circuited rotor. It is possible because the UFC permits independently the regulation of the output frequency and the output voltage; so that one can keep the magnetic flux constant (f/v = constant).

The output voltage wave shown in Fig. 2 is independent of the phase of the load so that forced commutation is required. Such an output wave is constructed by connecting in turn the input terminals to an output terminal.

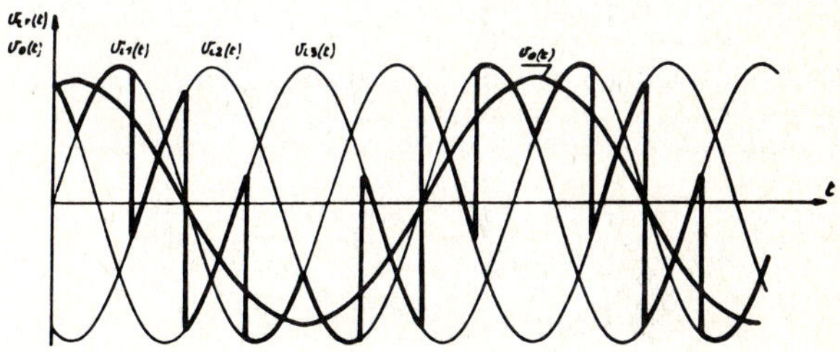

Fig. 2 : Typical three-pulse UFC output voltage waveform.

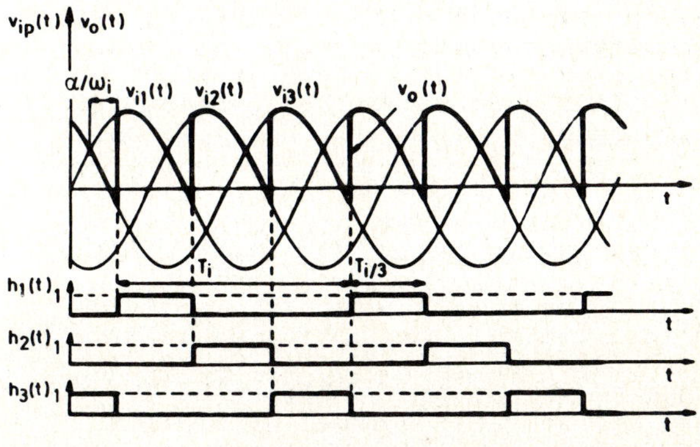

Fig. 3 : Describing functions for a DC output voltage.

The describing function of a bidirectional switch is a binary function whose value is high when the switch is on and low when the switch is off. Fig. 3 shows this function in the case of a DC output voltage and a 3 phase input.

2. Power circuit and commutation process.

The practical problem is the realization of a static bidirectional switch. A silicon controled rectifier (SCR or thyristor) has been adopted to reach the solution of this problem. In fact the thyristor is an unidirectional element so that it was necessary to join it with a rectifier bridge. An other property of the thyristor is that it can't be turned off at any time. So one has to add a mutual inductance in series to the thyristor and a capacitor in parallel with both the thyristor and mutual inductance. The resultant power circuit is presented at Fig. 4.

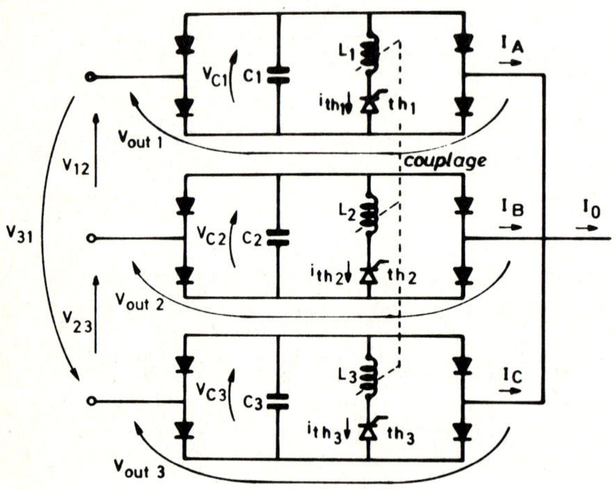

Fig. 4 : Power circuit.

The thyristor will be turned off by turning on the next switch. Proceeding in this manner the command circuits are simplified because it is not necessary to use an auxiliary thyristor to turn off the main thyristor. Next, the commutation process will be explained in a few words, for example, the commutation from switch 1 to switch 2 (Fig. 4). The charge of the capacitor C_2 is at least half the maximum input voltage. In fact C_2 is at first charged with $|V_{23}|$ (rectified line to line voltage) because switch 3 was activated after the conduction period of switch 2. In a second phase, C_2 is charged by $|V_{12}|$. So C_2 is at least charged at minimum of V_{12}, V_{23}. ($V_{max}/2$). The capacitor C_1 is totaly discharged because it was short circuited by thyristor 1. Firing thyristor 2, rapidly applies the voltage of C_2 to the mutual inductance; because it is a quick phenomenon, the mutual inductance behaves like a transformer, that means the voltage of C_2 is directly applied to the mutual inductance of the first switch. So the voltage applied to thyristor 2 is $V_{C1} - V_{C2} = -V_{C2}$.

The current of thyristor 1 is quickly transfered to thyristor 2 and thyristor 1 turns off. While V_{C1} is smaller than V_{C2}, a reverse voltage is applied accross Th 1 so that it regains its forward voltage blocking capability. During the discharge of C_2, the current in Th 2 continuously increases and becames larger than the load current. Assuming the load inductance is great enough so that the load current does not vary during the commutation, this leads to an excedentary current wich flows to the thyristor and the rectifier bridge.

3. Assumptions and model.

For simulation, some assumptions were adopted and a model was elaborated.

The thyristors and the diodes are considered as ideal switches. The leakage capacitance and the wire inductance are neglected. The load has a large inductance so that the output current doesn't change during the commutation. One assumes that the commutation is very quick so that the input voltage remains constant. The internal circulating current has been reduced to zero a long time before commutation has taken place, so that Th 1 only carries the output current. Forced commutation only was studied because it is the most important process of the UFC with thyristors.

Since the circuits are built up with non-linear components, it is not possible to discuss it from one model only. The equations of the system change with the state of the semi-conductors. Thus there are many possible models in this case. The different possibilities of combinations of the state of the semi-conductors were analysed so the practical configurations were found. For example, it is impossible that only one diode conducts in a bridge of diodes. (Kirchhoff's law). A rectifier bridge can only have four possible configurations wich are (Fig. 5) :

B blocked (all diodes are interrupted)
F forward (two diodes are turned on)
R reverse (the other two diodes are turned on)

CC shorted (all diodes are turned on).

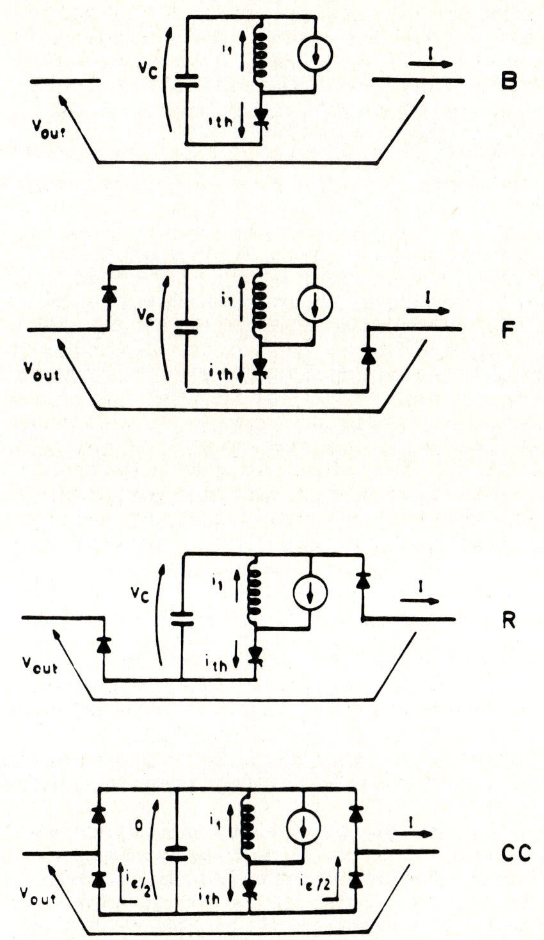

Fig. 5 : Possible configurations of the diode bridges when the thyristor is "ON".

The B,F,R configurations can exist either when the thyristor is turned on or turned off. The CC configuration leads the thyristor to be turned on. In earlier publications there were only sixteen configurations considered before the complete discharge of the incoming capacitor, end of calculations. The following table gives the configurations that have been dealt with :

FBB	FRB	RFB	BFB
FBF	FRF	RFR	BFR

The first character means the configuration of the first bridge, the second the configuration of the second bridge and the third the configuration of the third bridge. All these configurations are possible with the outgoing thyristor turned on or turned off, so that we get a total of sixteen configurations. After simulation, it was found out that there are more possible configurations than firstly assumed. For example, physically impossible results were obtained.

The simulation is realized through a continuous simulation programme (SIMULT). This programme calculates the values of the variables, each variable is developped in Taylor's series in terms of time. The programme uses a resolution method based on signal flow graphs. So that the equations of each configuration may be turned into a flow graph as shown on Fig. 6 for the FBB configuration. The mutual inductances are replaced by an equivalent circuit. In the circuit there is a self inductance and a current source, whose value depends on the current circulating in the self inductances of the other switches.

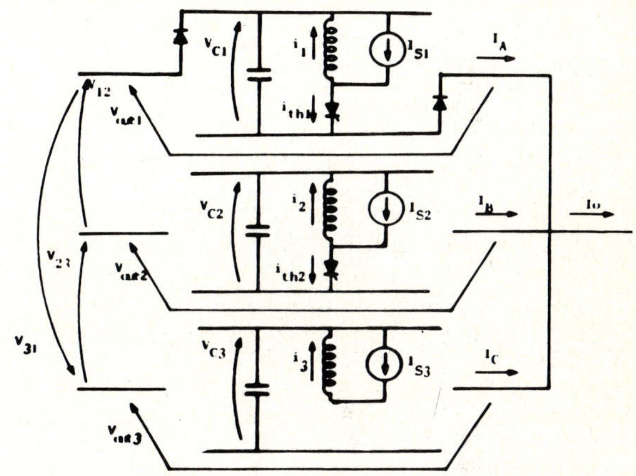

Fig. 6 : FBB equivalent circuits with the incoming end the outgoing thyristor "ON".

These currents are in two particular cases :

Th 1 and Th 2 turned on

$$I_{S3} = I_3 = \frac{k}{k+1}(i_1 + i_2)$$

$$I_{S1} = \left(\frac{k}{k+1}\right)^2 i_1 + \frac{k(2k+1)}{(k+1)^2} i_2$$

$$I_{S2} = \left(\frac{k}{k+1}\right)^2 i_2 + \frac{k(2k+1)}{(k+1)^2} i_1$$

Th 1 only turned on

$$I_{S1} = I_{S3} = i_1 = i_3 = ki_2$$

$$I_{S2} = \frac{2k^2}{k+1}\, i_2$$

where k is the mutual coupling coefficient.

The other equations of the configurations are determined using the equivalent circuit, and we get for the current :

$$I_A + I_B + I_C = I_O \qquad I_A = i_{Th1} + i_{C1}$$

$$I_C + I_{Th2} = 0 \qquad I_B = I_C = 0$$

$$I_{C1} + i_{Th1} = I_A \qquad I_{C1} = C\,\frac{dV_{C1}}{dt}$$

$$I_{S1} - i_1 = i_{Th1} \qquad I_{C2} = C\,\frac{dV_{C2}}{dt}$$

$$I_{S2} - i_2 = i_{Th2}$$

and the voltages :

$$V_{C1} = V_{L1} = L_e\,\frac{di_1}{dt}$$

$$V_{C2} = V_{L2} = L_e\,\frac{di_2}{dt}$$

These equations translated to a flow graph are shown in Fig. 7.

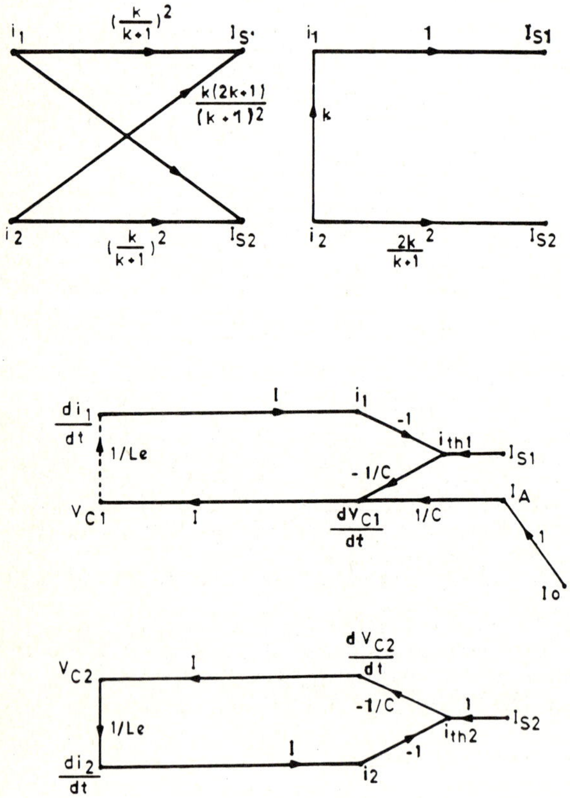

Fig. 7 : Signal flow graphs for the FBB configuration.

4. Simulation process.

The simulation programme needs the following data :
- the output frequency and current
- the input frequency and voltage
- the mutual coupling coefficient
- the inductance L and capacitance C
- the firing angle.

Initial conditions are calculated using these data. The line-to-line voltages can be gained through the values of the firing angle and the input frequency and voltage. The calculation of the capacitor's initial charge is more complex : at first it is necessary to calculate the conduction angle, what can be done thanks to the output frequency. Knowing the conduction angle, it is possible to calculate the initial capacitor voltage.

The next step to reach the solution is the determination of the rules to choose a configuration. Classic rules were used to know the state of the semiconductor.

For example, the state of a thyristor relies on several data. A thyristor is on during the next step of calculation only when it was on during the precedent step of calculation and when it carries a positive current, or when it is fired and carries a direct voltage. A thyristor turns off when its current is reduced to zero and a reverse voltage is applied during a sufficient period.

In the present case, the fact of firing thyristors is not tested, because during the whole simulation only Th 2 is fired. The diode bridges are tested as a whole, what means that each diode is not tested separately, but the bridge configuration is determined by the voltages applied to its terminals as shown in the table.

Condition	Configuration
$V_C > V_{out}$	B
$V_C = V_{out} \neq 0$	F
$V_C = -V_{out} \neq 0$	R
$V_C = V_{out} = 0$ and $i_{Th} > i_{out}$	CC

The configuration of a switch is determined by the combination of the test on the thyristor and on the diode bridge. Doing this for all switches gives the total configuration and calculations can be started. At each step of calculation tests are undergone to control the validity of the configuration. As soon as a test fails, another configuration is chosen. The final values of the variables of the first configuration are now the initial values of the variables of the next configuration. The calculations go on further in that way until the voltage of the incoming capacitor is decreased to zero. That means that the incoming switch is short circuited which terminates the commutation. That also means the end of calculation. If one likes to go on with further calculations at that moment, it is necessary to change the equivalent circuit.

In fact, some built assumptions are not allowed any more. For example, the assumption of considering the semi-conductors as ideal switches will lead to an indefinitely circulating excedentary current in the thyristor and the diode bridge. It is necessary to introduce dissipative elements in the equivalent circuit to model the losses of the semi-conductors and the mutual inductance.

Now, let's give a short outline of the structure of the programme. Earlier in this paper, it was told that a flow graph was made for each configuration and that the SIMULT programme generates a FORTRAN programme calculating all the values of the variables in the graph. Now, there are many configurations and the same number of graphs so that a programme that would use the generated programmes as subroutines would be very long. To solve this problem, all the flow graphs were assembled into one general flow graph with many independent sources. The union of the different flow graphs into a general one was reached as follows : Let us first point out that the variables are the same for all the different configurations. However, in some configurations not all the variables change so that they were not included in the graph of that particular configuration. In that respect, particular configurations had to be changed so that the most possible variables were present and were obtained in the most general way. For example, the voltage of a capacitor can be got in two different ways.

Firstly it is possible to integrate the derivative of the voltage, which is present while it is a main variable. Then, in configurations with two switches "on" it is possible to calculate directly the voltage using the line-to-line voltage. The first method was used because it was the most general. So far it was possible to superpose the different flow graphs and all the branches that were not the same in each graph were replaced by a multiplier. The multiplier allows to introduce the wanted transmittance through an independent source. Proceeding in that way, it is easy to get a particular configuration simply by changing the independent sources.

For example, if a particulary branch does not exist in a particular graph, it can be eliminated by setting the independent source to zero. Another particularity of the final programme is that the tests are done by special branches of the graph (RELAY). The state of a relay changes when the command variable reaches a given value. So all the tests necessary to detect the end of each particular configuration were included in the graph. While not all tests have to be done for each configuration, a multiplier was introduced to activate or disactivate the test.

Finally, in the generated programme, the switching of the relays are detected. Each particular relay and some other tests lead to a particular configuration, the values of the independent sources are altered and simulation can go on. At each change of configuration, a message is given on the listing.

5. Results of simulation.

On small output current, the simulation raised some problems. It seems that time constants became very small so that the calculation didn't progress in time any more.

To obtain an accurate experimental check, it is necessary to possess an equipment allowing to measure many parameters simultaneously (firing angle of the incoming thyristor, output current, capacitor voltage,...).

So the results are rather qualitative than quantitative.

Fig. 8 showns the results of simulation.

The values are not exactly the same because not all data were the same in the simulation and experimental check (Fig. 9).

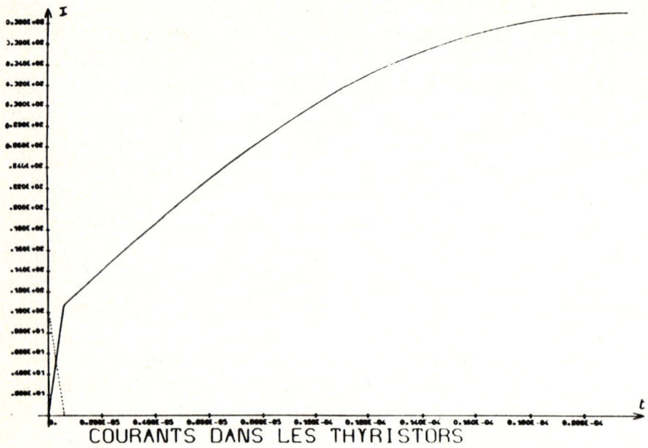

Fig. 8 : Currents carried by the incoming and the outgoing thyristor.

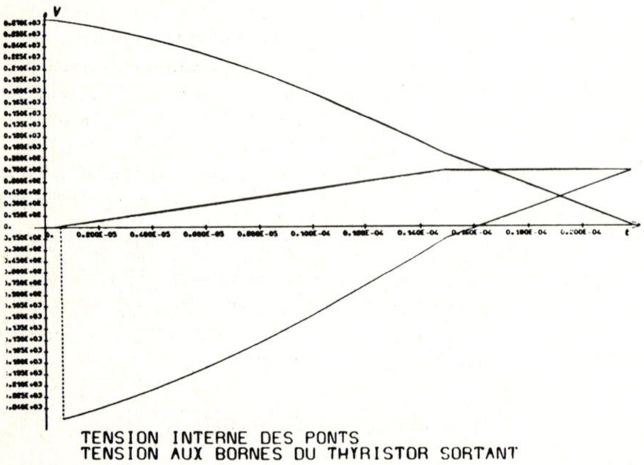

Fig. 8bis : Bridge's internal voltages and voltage applied to the outgoing thyristor.

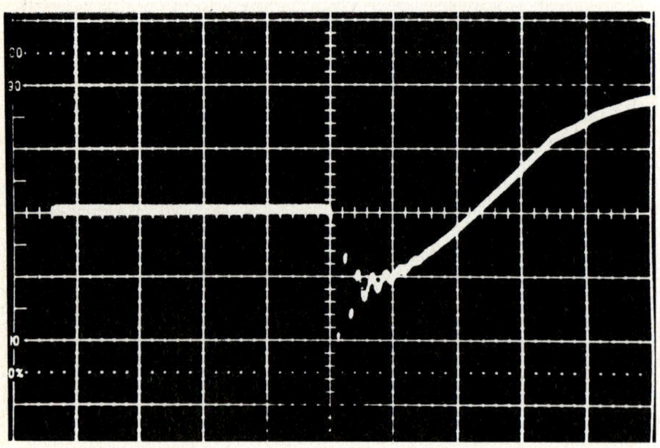

50 V/div. 10 µs/div.

Fig. 9 : Voltage applied to the outgoing thyristor.

6. Conclusions

The simulation gives satisfactory results. In spite of the assumptions, the reverse voltage time seems to cope with the experimental value, there is a little more divergence for the current carried through the incoming thyristor but it can be explained by the value of the mutual coupling inductance coefficient wich could be larger in the real case.

It's possible to extend simulation programmesto more general cases; for example, the calculation of the diminution of the internal circulating current. In that case, it is not possible to handle the semi-conductors as ideal switches any more. The optimal values of the components may be gained through the simulation programme.

Further it's possible to extend the programme to simulate a whole, command circuits, power circuits and the load.

References

[1] L. GYUGYI, B.R. PELLY, Static Power Frequency Changers, Wiley.

[2] P. CAPIEAUX, Etude et réalisation d'un changeur de fréquence statique pour moteur asynchrone, Travail de fin d'études 1980.

[3] J. DEPREZ, Les changeurs de fréquence statiques à conversion directe, Travail de fin d'études 1981.

[4] J. MAGAIN, Réalisation d'un variateur de vitesse avec moteur asynchrone et changeur de fréquence statique, Travail de fin d'études 1982.

[5] F. SPRONCK, Composants et circuits électroniques - Circuits électroniques de puissance, Notes de cours.

[6] J. ROBERT, Compléments d'Electrotechnique, Notes de cours.

[7] W. LEGROS, Applications spéciales de l'Electrotechnique, Notes de cours.

[8] F. SPRONCK, Les convertisseurs de fréquence en électronique de puissance, Revue HF, vol. XII n° 5.

[9] R. LE DOEUFF, S.A. MAHMOUD, Simulation numérique d'un cycloconvertisseur triphasé "Système Cyclope", Sept. 1982.

Session 5 A

FIELD COMPUTATION IN ELECTRICAL MACHINES (II) AND SIMULATION OF STATIC MAGNETIC DEVICES

Chairman: J.C. SABONNADIERE

ELECTRICAL MACHINES AND CONVERTERS - Modelling and Simulation
H. Buyse and J. Robert (editors)
Elsevier Science Publishers B.V. (North-Holland)/© IMACS, 1984

2D AND 3D FINITE ELEMENT MODELLING OF SMALL ELECTRICAL MACHINES

G. MEUNIER, J.L. COULOMB and J.C. SABONNADIERE

E.N.S.I.E.G. LABORATOIRE D'ELECTROTECHNIQUE (LA 355)
B.P. 46 - 38402 SAINT-MARTIN D'HERES, FRANCE

The determination of a magnetic field is essential for the design of electrical devices. Classical methods, analytic or analog are not satisfactory for the analysis of micromotors.
Two and three dimensionnal magnetic field computations using a finite element method are reported here.
Time integration is made to solve dynamic equation.
Some comparisons with experiment are report.

1. DESCRIPTION OF A MICROMOTOR

The figure 1 shows an example of synchronous micromotor, the analysis of wich is presented hereunder. A coil supplied with an alternating current creates an axial magnetic flux through the ferromagnetic teeth, the box cover being the yoke

The rotor is made of a ferrite permanent magnet wich is locked on the field created by the excitation coil.

Generally this kind of motor may start rotating in both directions and need a specific mean to create particular sense of rotation.

The main feature in the analysis of this kind of structure is the complexity of the magnetic circuit which is an actual three dimensionnal one. This paper describes various models of magnetic circuit we used in the overall mechanical analysis of the motor.

2. GENERAL ASSUMPTIONS

All electromagnetic structures are governed by Maxwell's equations. These equations are not easy to solve, even by numerical method in general cases. Thus, we shall examine the conditions in which it is possible to derive law from the local model of Maxwell's equations, the global Kirschoff's law model characterized by the concept of electric and its equations.

These last equations expresses the relation between the current I flowing in the circuit, the terminal voltage u and the induced electromotive force :

$$u = RI + N\frac{d\phi}{dt} \quad (1)$$

where R is the circuit resistance, ϕ the flux crossing the N turns of the circuit.

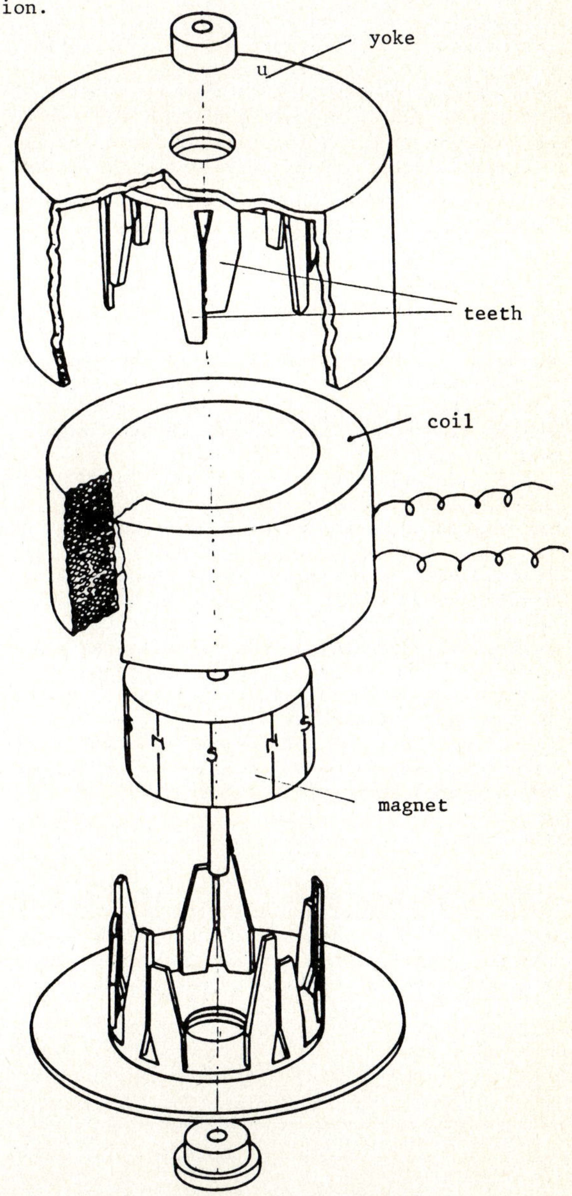

Figure 1 : A example of micromotor.

The electromagnetic system is then modelled by a set of electric circuit lenked by the various fluxes flowing through them. The analysis of a system can therefore be made according an analysis of the linkages wich exists between the various circuits in order to set up and solve equations (1).

In fact if the flux paths may be clearly identified the magnetic circuit can be modelled by a reluctance graph. However, in some problems this kind of model is not enough accurate ; this occurs particulary when the complexity of the geometric shape of the materials does not allows the identification of flux paths and when the flux distribution is affected by the saturation of the materials.

In the analysis of micromotors these special features lead us to the use of the finite element method for modelling the magnetic circuit and the permanent magnet by a local definition of the magnetization very useful for a multipolar magnetization along curvilinear field lines.

3. THE MODEL OF THE MAGNETIC CIRCUIT

3.1 Flux paths

If we look at the flux paths of the micromotor shown on figure 1 it appears that the major lines of flux created by the magnet and the coil flows through the yoke (figure 2). Conversely, looking at the interaction between the teeths, magnet and air gap, it is very difficult to make reasonable assumptions on the flux paths which morever depend on the relative position between the teeth and the poles of the magnet. This situation hint to reduce the finite element model to the restricted area in order to reduce the cost of the simulations. Two kind of finite element model may be built according to the kind of motor under analysis : a general 3 Dimensionnal analysis taking into account the actual tooth shape [2] or a bidimensionnal analysis in a plane orthogonal to the shaft of the motor.

Finite element computation allow to define magnet by its local magnetic behaviour :

$$B = \mu H + B_r$$

where B and B_r are respectively the total and residual flux densities and μ the magnetic permeability of the magnet. A first run which simulates the magnetization process where flux lines are set, gives the residual density at any point of the ferrite magnet [1].

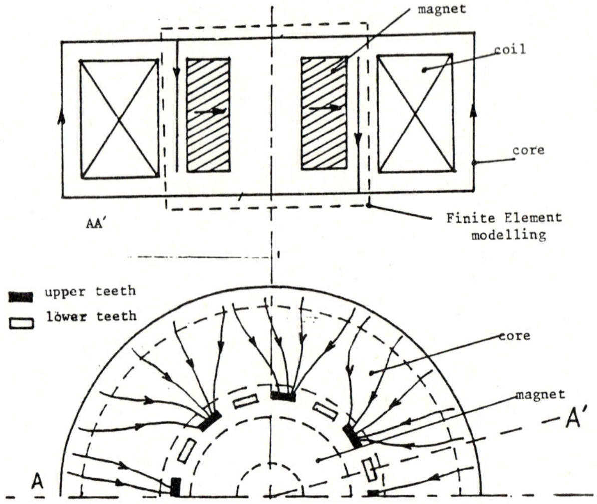

Figure 2 : Flux paths in a micromotor

3.2 Three dimensionnal finite element modelling

Figure 3 presents a model of magnetic circuit in wich the external part of the motor is merely represented by a reluctance R_C

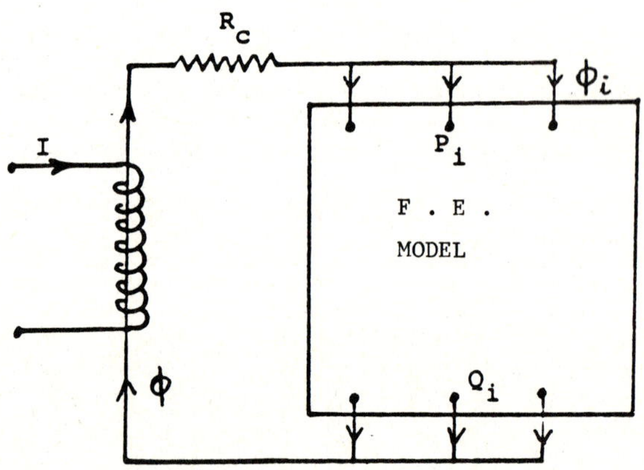

Figure 3 : Model of magnetic circuit (with 3D modelling)

The various ports P_i, Q_i represents the basis of upper and lower teeth. Each point P_i, Q_i are respectively at the magnetic potential V_1 and V_0. The relations between V_1, V_0, ϕ and ϕ_i are :

$$\phi = \sum_{\text{upper teeth}} \phi_i = \sum_{\text{lower teeth}} \phi_i \qquad (2)$$

$$V_1 - V_0 = N I - R_C \phi$$

The finite element model will be used to determine the relations

$$\Phi_i = \phi(V_1, V_0, \theta) \qquad (3)$$

where θ is the angular position of the magnet. The relations (2) and (3) define completely the magnetic circuit.

For the application of the finite element method it is recommended to use the proportion of symetries and periodicity to reduce as far as possible the domain of analysis. This reduction is made by the use of Dirichlet and Neuman conditions associated with periodicity boundary conditions. The figures 4 present an example of a domain modelling by three dimensional finite elements and figure 5 some magnetic equipotential plots.

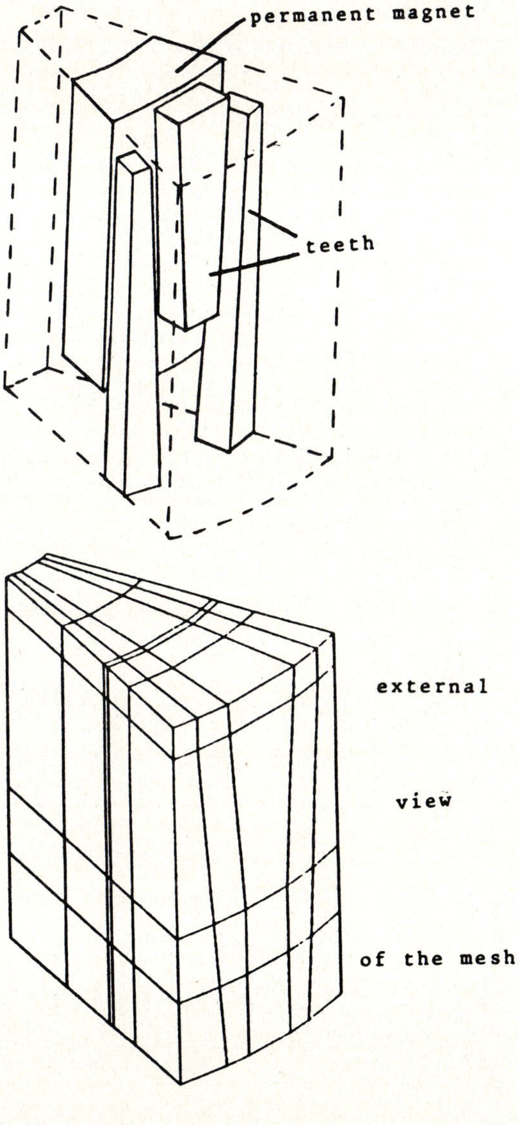

Figure 4 : Domain modelling with 3D finite element

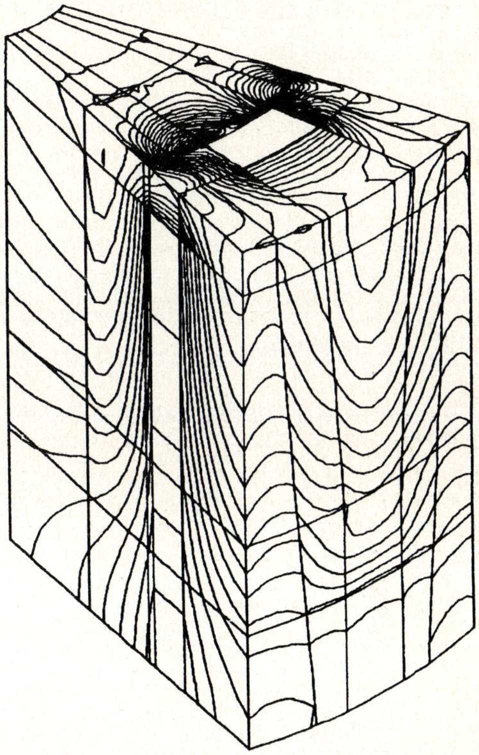

Figure 5 : Magnetic equipotential plots

Figure 6 shows a comparison between computed and measured curves of the flux generated by the rotor through the coil.

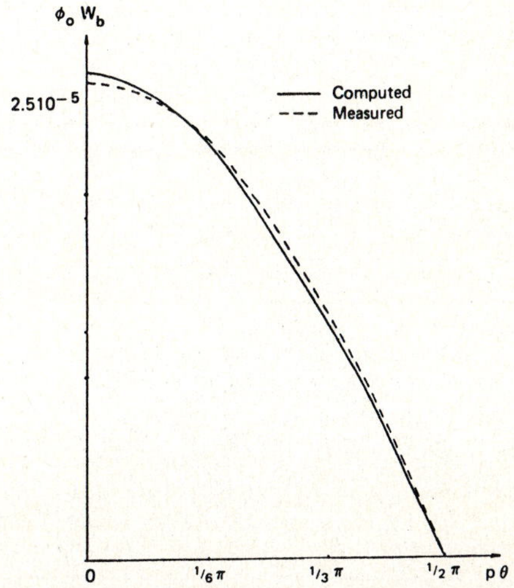

Figure 6 : Flux generated by the rotor through the coil.

3.3 Two-dimensionnal finite element modelling

The definition of this model implies the assumption wich states that all along the teeth fluxes are orthogonal to the axis of the motor. Similary the flux created by the coil flows from a tooth to an other in the same plane. The analysis is thus made in this plane in wich we consider the projection of teeth as scalar magnetic equipotential lines. This model of magnetic circuit is shown on figure 6 where the strenght of the flux at the basis of the teeth lead us to consider that the reluctance of the teeth (R_d) may be a function of the flux, and in which the leakage flux at the top of the teeth is represented by a reluctance (R_{extd})which is computed by application of the finite element method in the plane of the teeth.

In the plane orthogonal to the axis, the finite element resolution gives the relations :

$$\Phi_i = \phi(V_1, \ldots\ldots, V_n, \theta) \tag{4}$$

between the fluxes Φ_i and the scalar magnetic potential V_i (figure 7).

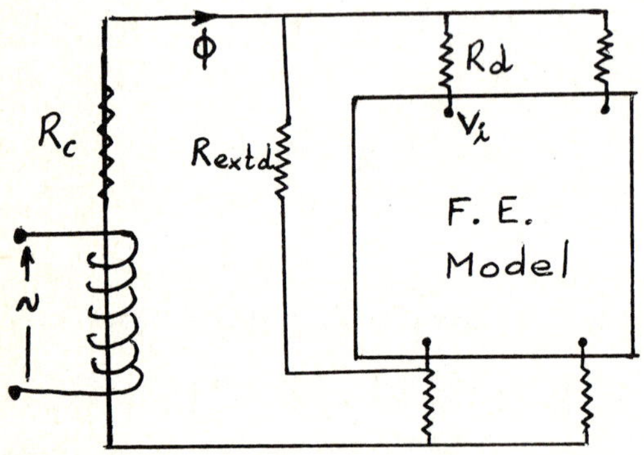

Figure 7 : Model of magnetic circuit (with 2D modelling)

Relations (4) coupled with the external circuit reluctances gives a complete model of the magnetic circuit. The figure 8 shows the scalar magnetic potential plots obtained from a finite element solution with an irregular distribution of teeth.

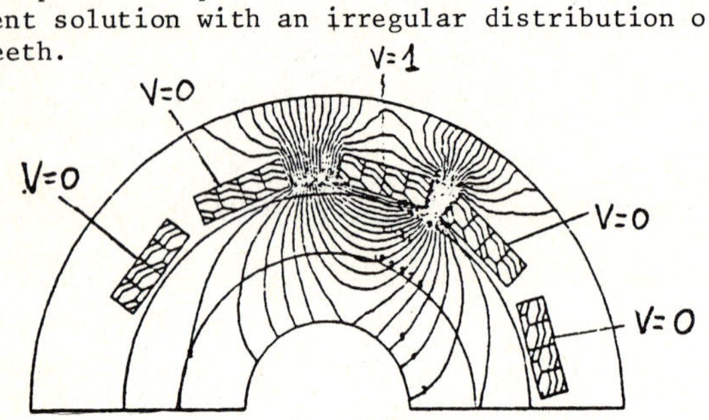

Interaction between teeth

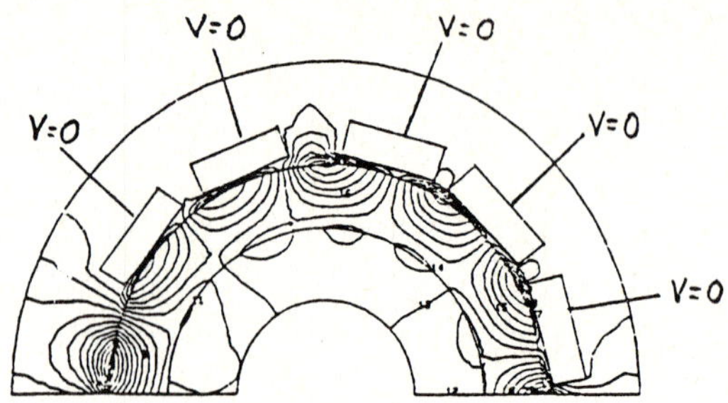

Interaction between permanent magnet and teeth.

Figure 8 : Magnetic equipotential plots

4. DYNAMIC OPERATION

The model of magnetic circuit described in the previous section is a magnetostatic one. A dynamic model must take into account the effect of the starting ring. This is done by assuming that the magnetic circuit experienced by the rings, is the model previously defined [1]. An additional ring may be introduced to model the effect of eddy current losses on the dynamic behaviour of the motor.

Then the equations of magnetic circuits associated to electrical and mechanical equations allows the simulation of the dynamic operation of the motor [6]. The results presented on figure 9 may be considered as satisfactory for this kind of problem.

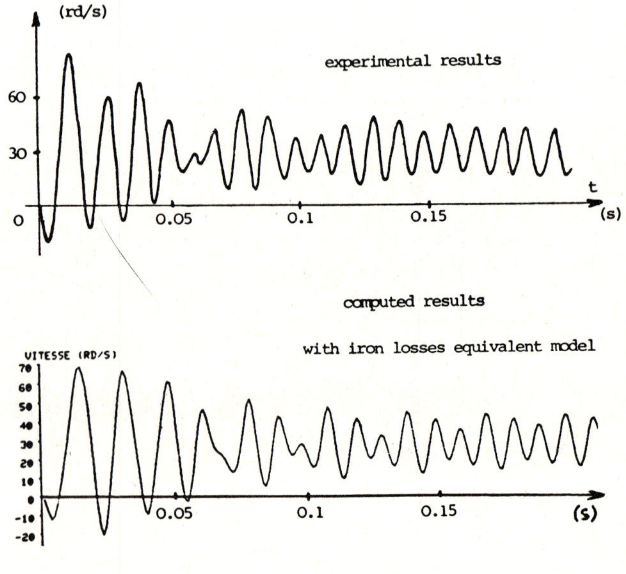

Figure 9 : Comparison of speed curves.

5. CONCLUSION

The methodology presented here is an approach of integration of the finite element method in a complete model of synchronous micromotor. The model defined software build on this model make possible the prediction of the behaviour of a micromotor before the construction of a prototype.

The geometrical structure of micromotors is so complicated that a 3D finite element model is generally necessary.

Nevertheless, according to such physically reasonable assumptions a twice 2D analysis may be use at better cost.

REFERENCES

[1] G. MEUNIER
"Application de la méthode des éléments finis au calcul des micromoteurs"
Thèse Docteur-Ingénieur - Grenoble, 1981

[2] J.L. COULOMB, G. MEUNIER
"Finite element analysis of a self starting synchronous micromotor"
IEEE Trans.Mag. Vol 17, n° 6, 1981

[3] J.L. COULOMB, G. MEUNIER
"Three dimensional analysis of a micromotor by the finite element method"
S.M.M.5 - Grenoble, 1981

[4] J.C. SABONNADIERE, G. MEUNIER, B. MOREL
"Flux : A general interactive Finite Element package for 2D electromagnetic field"
COMPUMAG, CHICAGO, 1981

[5] J.L. COULOMB
"Analyse tridimensionnelle des champs électriques et magnétiques par la méthode des éléments finis"
Thèse Docteur d'Etat, Grenoble, 1981

[6] B. ANCELLE, G. MEUNIER
"Dynamic simulation of synchronous micromotor using magnetic field modelling with finite elements"
Symposium on Electrical Machines for Special Purposes, Bologna, 1981.

ELECTRICAL MACHINES AND CONVERTERS - Modelling and Simulation
H. Buyse and J. Robert (editors)
Elsevier Science Publishers B.V. (North-Holland)/ ©IMACS, 1984

FINITE ELEMENT ANALYSIS OF A LINEAR TUBULAR MOTOR WITH COMPOSITE ROTOR

Azza A. Fahim*

Iowa State University
Ames, Iowa.

Mahmoud A. Saleh

National Research Centre
Cairo, Egypt.

In this paper a detailed prediction of magnetic fields and force for a linear tubular motor with composite rotor is obtained utilizing a first order finite element analysis. an appropriate energy functional was chosen taking into account the axisymmertic nature of the motor model and the motional emfs. The solution matrices are derived algebracilly with all associated integrals performed directly to reduce computational time. Results concerning the effects of the slotted structure of both stator and rotor on flux distribution, evaluation of the effect of the initial position of the rotor on the standstill thrust and the dynamic characteristics of the motor model are presented.

1.INTRODUCTION:

A tubular motor with a composite rotor consisting of alternate rings of copper and iron mounted on a solid iron core holds considerable promise as a linear actuator in control systems.

Analysis of such a motor with different degrees of approximation has been carried out by Saleh et al [1],[2] and design procedures have been proposed [3]. In the one dimensional solution [1], simplifying assumptions are necessary on both the stator and the rotor sides to allow an analytical solution. In the two-dimensional numerical solution, based on the finite difference technique [2], restrictions were imposed only on the stator side. The edge effect due to the finite lenth of excitation and the inhomogeneity in the rotor material were also taken into consideration. However, the analysis was carried out for standstill operation and neglected both the distributed current density of the excitation winding and the slotting of the stator surface.

To overcome some of the shortcomings of this earlier analysis, the finite element method is applied to the motor model in this paper. The machine is represented with full complexity in geometry and materials distribution, which allows improved accuracy in prediction of the electromechanical performance of the motor model.

2.PROBLEM FORMULATION:

A cutaway view outline of the tubular motor with composite rotor under investigation is shown in Fig.1. The travelling magnetic field in the motor is created by the stator polyphase winding and the rotor side, as shown in Fig.1, contains both magnetic and conducting materials to form the composite rotor structure. The rotor moves at speed V in the z direction.

Utilizing the symmetrical nature of the motor model in the circumferential direction (α), a two dimensional magnetic field solution in the z-r plane can be obtained based on the assumptions that all field components vary sinusoidally with time, the iron reluctivity and conductivity are neglected, the rotor extends infinitely in the z direction, the current density and the magnetic vector potential A are everywhere α directed, and the magnetic flux density B possesses two components B_r and B_z.

From Maxwell's equations and the definition of the magnetic vector potential, the general differential equation governing the magnetic field for the motor region will be :

$$\bar{\nabla} \times [\nu (\bar{\nabla} \times \bar{A})] = \bar{J} \qquad (1)$$

* Visiting Scholar from the National Research Centre, Cairo, Egypt.

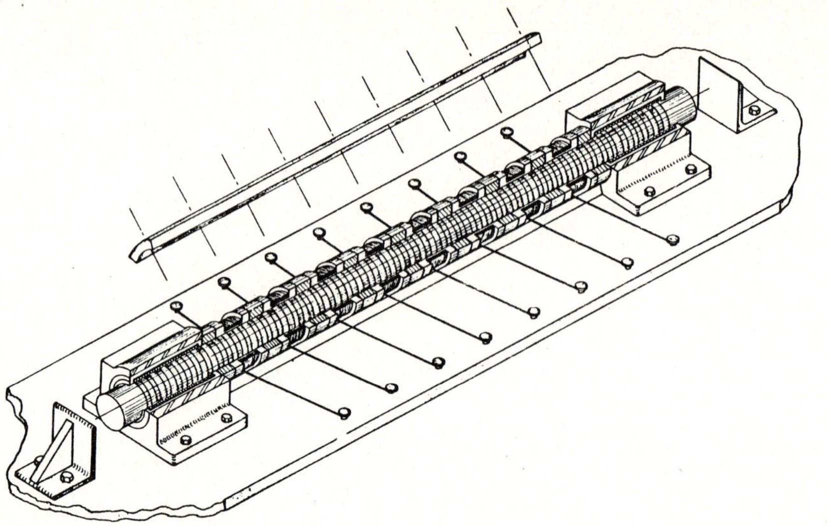

Fig 1. A cutaway view outline of the tubular motor with composite rotor.

where

$$\nu = \frac{1}{\mu}$$

Using the cylindrical coordinates (r,z,α) with consideration of model symmetry in the axial direction and with the assumption of linear magnetic material, equ (1) can be expanded in the form :

$$\nu\left[\frac{\partial^2 A}{\partial r^2} + \frac{\partial^2 A}{\partial z^2} + \frac{1}{r}\frac{\partial A}{\partial r} - \frac{A}{r^2}\right] = -J \qquad (2)$$

where A and J now denote the component of the magnetic vector potential and the current density respectively.
For the different regions of the motor,

$$J = J_s \qquad (3)$$

is the excitation current density within the stator slots,

$$J = J_r = \sigma\left(\frac{\partial A}{\partial t} + V\frac{\partial A}{\partial z}\right) \qquad (4)$$

is the induced current density within the rotor conducting rings,

and

$$J = 0 \qquad (5)$$

within the air gap and the iron parts.

3. TWO - DIMENSIONAL FINITE ELEMENT FORMULATION:

Following the finite element technique [4],the problem has to be formulated in variational terms. A functional expression which yields the field differential equation (2) as it's Euler equation will be (Appendix A) :

$$\mathcal{F} = 2\pi\iint_R \left\{\frac{\nu r}{2}\left[\left(\frac{\partial A}{\partial r}\right)^2+\left(\frac{\partial A}{\partial z}\right)^2\right]\right. \qquad (6)$$

$$\left. + \frac{\partial A}{\partial z}A + \frac{A^2}{r}\frac{\nu}{2} - rAJ_o\right\}drdz$$

where

$$J_o = J_s + J_r$$

and the region of integration R is the area of definition of the problem in the r - z plane.

The field region is discretized into triangular elements which has the advantage of fitting very well the geometrical boundaries of the motor model. Fig 2 shows cross sectional view of the motor illustrating the various material interfaces which must be modeled. Within each element the potential function is approximated as a linear interpolation of the nodal values and some weighting functions (area coordinates) [5], such that

$$A = \sum \xi_i A_i \qquad (7)$$

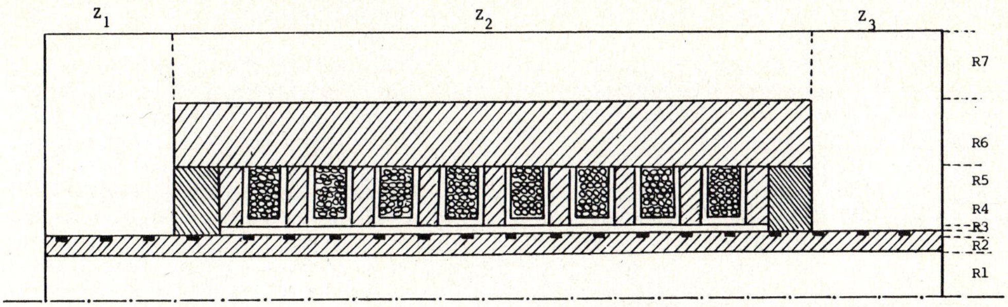

Fig 2. Motor regions discretization.

where i ranges over the vertices of the triangle and,

$$\xi_i = \frac{1}{2\Delta}(a_i + b_i z + c_i r) \qquad (8)$$

$$a_i = z_j r_m - z_m r_j$$

$$b_i = r_j - r_m \qquad (9)$$

$$c_i = z_m - z_j$$

with Δ as the triangular area with i,j,m as the node vertices.
Also, the original coordinate r can be expressed in terms of the area coordinates as [6],

$$r = \sum \xi_i \, r_i \qquad (10)$$

The functional defined in equ (6) is minimized by setting its first derivative with respect to each of the nodal potentials to zero. Evaluate the integral over one triangle at a time, one obtains:

$$\frac{\partial \mathcal{F}^e}{\partial A_i} = \iint_\Delta \Big[\nu r \frac{\partial A}{\partial r}\frac{\partial}{\partial A_i}\frac{\partial A}{\partial r} + \nu r \frac{\partial A}{\partial Z}\frac{\partial}{\partial A_i}\frac{\partial A}{\partial z}$$

$$+ \nu A \frac{\partial}{\partial A_i}\frac{\partial A}{\partial r} + \nu \frac{\partial A}{\partial r}\frac{\partial A}{\partial A_i} + \nu \frac{A}{r}\frac{\partial A}{\partial A_i}$$

$$+ 2j\omega\sigma A \frac{\partial A}{\partial A_i} + \sigma V r A \frac{\partial}{\partial A_i}\frac{\partial A}{\partial z} + \sigma V r \frac{\partial A}{\partial z}\frac{\partial A}{\partial A_i}$$

$$- J_s \, r \frac{\partial A}{\partial A_i}\Big] \, dr \, dz = 0 \qquad (11)$$

The above process is carried out for each of the nodal potentials Ai,Aj,...for every finite element. Substituting for A and r from equs (7) and (10) respectivelly and computing the integrals in equ (11)(see Appendix B),the resulting set of equations for every element in the entire region reduces to the form

$$[S]^e\{A\}^e + [G]^e\{A\}^e + [T]^e\{A\}^e = \{F\}^e \qquad (12)$$

where $[s]^e$,$[G]^e$ and $[T]^e$ are geometrical coefficient materices, materices $[G]^e$ and $[T]^e$ include the conductivity of the rotor conductor material and $[T]^e$ contains the speed effect. $\{A\}^e$ is the required solution vector of the nodal potential at the element vertices and $\{F\}^e$ is the forcing function or current source.

When an equation corresponding to (12) has been written for each of the triangles in the subdivision, the resulting assembly of algebraic equations may be combined into a single matrix equation.

$$[C]\,\{A\} = \{F\} \qquad (13)$$

where $\{A\}$ is the column vector scanning all the nodes in the domain. The final step in computing is to solve the set of linear algebraic equations defined by equ (13). The Gaussian triangular decomposition technique was chosen to be used because of its simplicity, numerical stability and economy of computation.

4.RESULTS:

The two-dimensional finite element analysis described above was applied to a linear tubular motor with composite rotor structure as shown in Fig 3. The motor was designed as an actuator according to the specific design procedure in referece [3]. The motor data is:

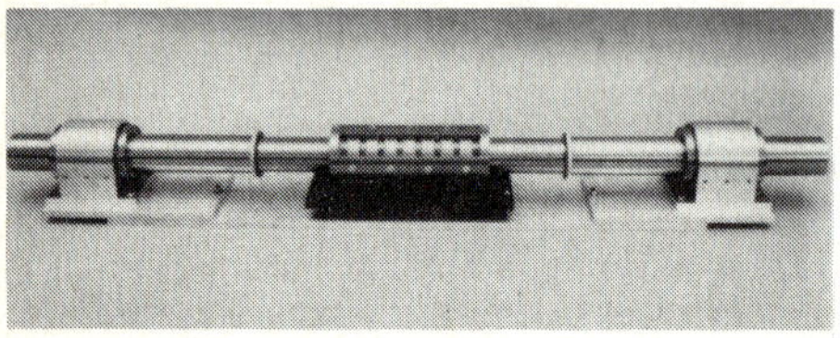

Fig 3 Experimented motor.

stator length	= 0.24	m
stator outer diameter	= 0.08	m
stator inner diameter	= 0.06	m
stator iron backing strip width	= 0.010	m
number of coils (slots)	= 8	
stator coil width	= 0.016	m
stator slot pitch	= 0.030	m
number of turns / coil	= 300	
air gap length	= 0.001	m
rotor outer diameter	= 0.058	m
rotor inner diameter	= 0.050	m
rotor copper ring width	= 0.001	m
rotor slot pitch	= 0.005	m
rotor copper ring depth	= 0.0005	m
number of poles	= 4	
number of phases	= 2	

For the above motor data, Fig 4 and Fig 5 show the flux lines at two different instants of time at zero speed. In Fig 6 the flux lines are plotted at V = 4.2 m/sec. All three figures show the effect of the slotted structure in both stator and rotor sides. Fig 6 clearly shows the effect of the speed on the flux distribution.

In Fig 7 the variation of the standstill force as a function of rotor position is shown. Variations up to 11 % of the maximum force were found. The variation of force with speed is shown in Fig 8.

5.CONCLUSIONS:

The method of finite elements proposed for the analysis of a tubular induction motor with composite rotor is a general one; the real structurre of the motor model with many different material interfaces is incorporated and the problem of induction in a moving media is taken into consideration. An appropriate energy functional for the tubular motor has been derived. The method can be utilized to evaluate different approximate approaches for representing the motor. The low computing costs allow repeated analysis for various machine dimensions and evaluation of their effects on performance.

References:

[1] Saleh, M. A., Ali, S. M., Sakr, M.F., Fahim, A. A., Electric tubular motor with composite rotor, Electric Machines and Electromechanics Jrnl. 4(1979) 47-57.

[2] Saleh, M. A., Ali, S. M., Fahim, A. A., Numerical analysis of tubular induction motor with composite rotor, Electric Machines and Electromechanics Jrnl. 4(1979) 101-111.

[3] Saleh, M. A., Day, A. L., Fahim, A. A., An approach to the design of tubular induction actuators with composite secondaries, IEEE PES Midwest power Symposium, Madison, Wisconsin (1982).

[4] Chari, M. V., Silvester, P. P., Finite elements in electrical and magnetic field problems (New York: John Wiley and Sons Inc. 1980).

[5] Armor, A. F., Chari, M. V. K., Heat flow in the stator core of large turbine generators by the method of three dimensional finite elements, IEEE Trans. PAS-95 (1979) 1648-1656.

[6] Dally, P., Finite elements for field problems in cylindrical co-ordinates, Int. Jrnl. for Numerical Methods in Eng. 6(1973) 168-178.

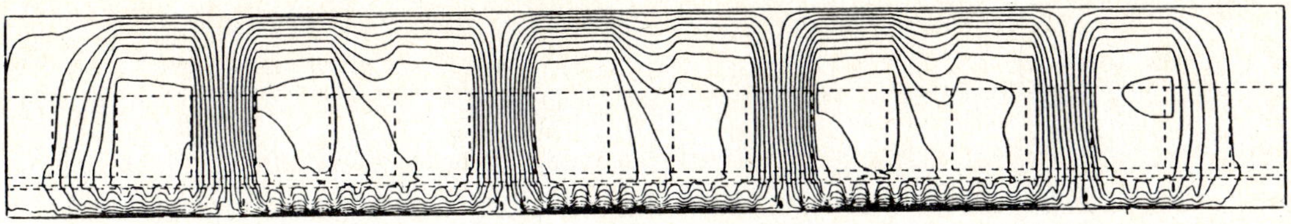

Fig 4. Flux plot at V = 0.0 ($\omega t = \pi/2$)

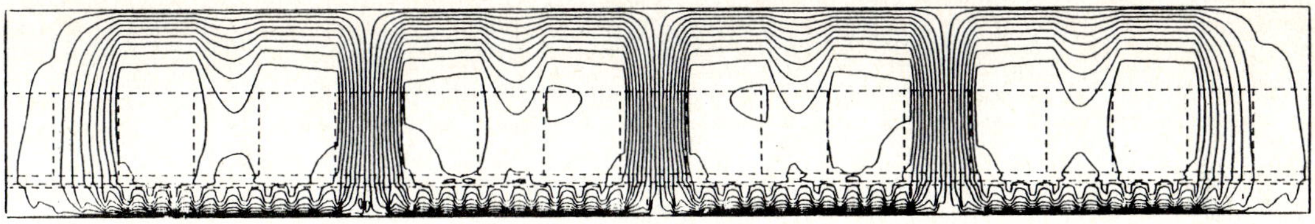

Fig 5. Flux plot at V = 0.0 ($\omega t = 3\pi/4$)

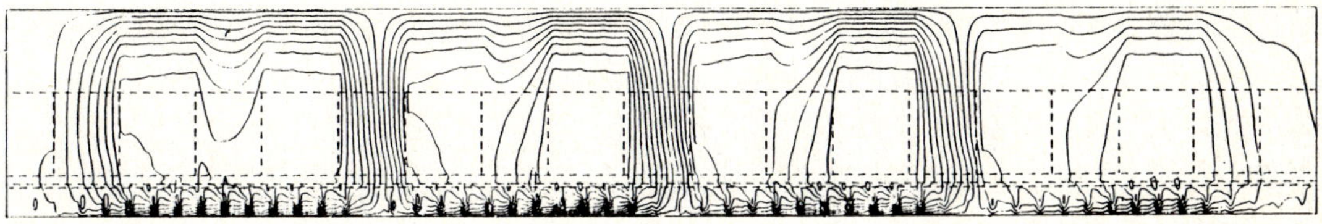

Fig 6. Flux plot at V = 4.2 m/sec ($\omega t = 0$)

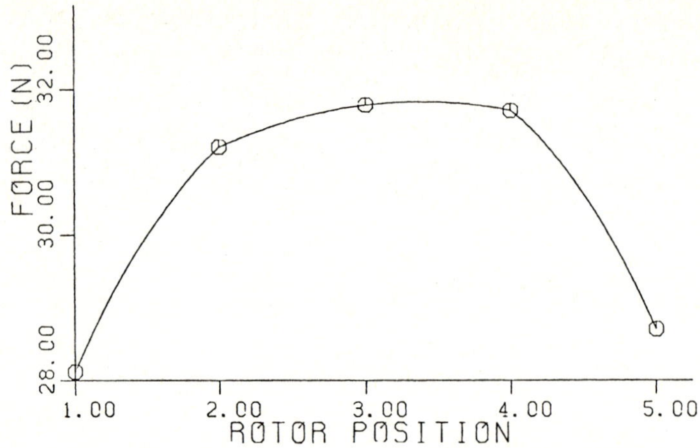

Fig 7. Variation of standstill force with rotor positions.

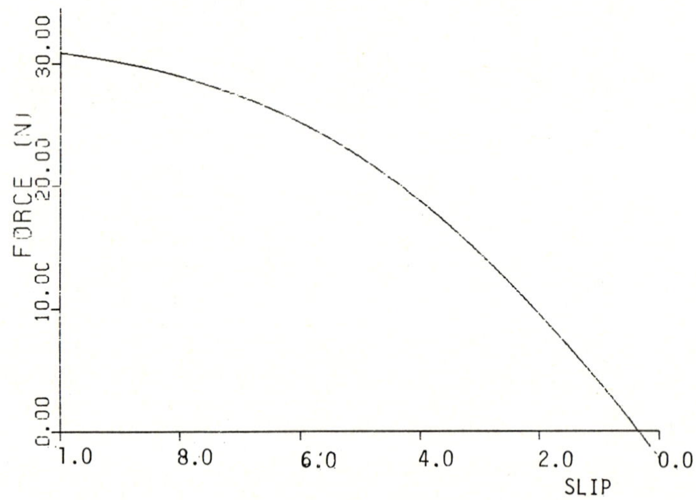

Fig 8. Force - Slip characteristic.

Appendix A:

The Functional expression.

For the electromagnetic fields, the functional expression corresponds to the energy stored in the system; therefore for linear conditions one obtains:

$$\mathscr{F} = \frac{1}{2} \iiint_{\Omega} \nu |B|^2 \, d\Omega - \iiint_{\Omega} A J_o \, d\Omega \qquad (A.1)$$

where $d\Omega$ is the unit volume of integration.
In cylindrical coordinates, the magnetic field B can be expressed as :

$$\overline{B} = -\left(\frac{\partial A}{\partial z}\right) \hat{a}_r + \left(\frac{\partial A}{\partial r} + \frac{A}{r}\right) \hat{a}_z \qquad (A.2)$$

where $\hat{a}_r$ and $\hat{a}_z$ are unit vectors in the r and z directions respectively.

Because of the symmetry in the α direction, the basic elemental volume $d\Omega$ becomes a torus of volume:

$$d\Omega = 2\pi r \, dr \, dz \qquad (A.3)$$

substituting for $|B|^2$ and $d\Omega$, in equ A.1 and using equ A.2 and A.3 provides equ 6 as the functional expression.

Appendix B:

Coefficients of the elemental matrix equation:

Integrating equ (11), collecting the terms and rearranging in matrix form, the coefficients of the elemental matrix equ (12) will be:

For matrix $[S]^e$:
Self coefficient S_{ii} is

$$S_{ii} = \frac{\nu}{12\Delta}(r_i + r_j + r_m)\, c_i^2 + \frac{\nu}{12\Delta}(r_i + r_j + r_m) b_i^2 + \frac{\nu}{3} c_i + 2\Delta\nu\, Q_{ii} \qquad (B.1)$$

where

$$Q_{ii} = \frac{r_i}{6(r_i - r_m)^2 (r_i - r_j)^2}\left[r_i^2 - 3r_i r_m - 3r_i r_j + 5r_j r_m\right] - \frac{r_m^3}{3(r_j - r_m)(r_i - r_m)^3}\log_e \frac{r_i}{r_m} + \frac{r_j^3}{3(r_j - r_m)(r_i - r_j)^3}\log_e \frac{r_i}{r_j}$$

The mutual coefficient S_{ij} is

$$S_{ij} = \frac{\nu}{12\Delta}(r_i + r_j + r_m)\, c_i c_j + \frac{\nu}{12\Delta}(r_i + r_j + r_m)\, b_i b_j + \frac{\nu}{6} c_i + \frac{\nu}{6} c_j + 2\Delta\, Q_{ij} \qquad (B.2)$$

where

$$Q_{ij} = \frac{r_i^2 (r_j - r_m) + r_j^2 (r_i - r_m)}{6(r_i - r_m)(r_j - r_m)(r_i - r_j)^2} - \frac{r_m^3}{6(r_i - r_m)^2 (r_j - r_m)^2}\log_e \frac{r_i}{r_m}$$
$$\frac{r_j^2 [r_m (r_i - r_j) - 2r_i (r_j - r_m)]}{6(r_j - r_m)^2 (r_i - r_j)^3}\log_e \frac{r_i}{r_j}$$

The mutual coefficient S_{im} will be similar to S_{ij} with node m replacing j.

For matrix $[G]^e$:
Self coefficient G_{ii} is

$$G_{ii} = \frac{\Delta}{30} j\omega\sigma\,(6r_i + 2r_j + 2r_m) \qquad (B.3)$$

Mutual coefficient G_{ij} is

$$G_{ij} = \frac{\Delta}{30} j\omega\sigma(2r_i + 2r_j + r_m) \qquad (B.4)$$

For matrix $[T]^e$:
Self coefficient T_{ii} is

$$T_{ii} = \frac{V\sigma}{24}[2b_i(2r_i + r_j + r_m)] \qquad (B.5)$$

Mutual coefficient T_{ij} is

$$T_{ij} = \frac{V\sigma}{24}[b_j(2r_i + r_j + r_m) + b_i(r_i + 2r_j + r_m)] \qquad (B.6)$$

The left hand side of equ(12) will be

$$F_i = \frac{\Delta}{12} J_S\,(2r_i + r_j + r_m) \qquad (B.7)$$

ELECTRICAL MACHINES AND CONVERTERS - Modelling and Simulation
H. Buyse and J. Robert (editors)
Elsevier Science Publishers B.V. (North-Holland)/© IMACS, 1984

AN EQUIVALENT CIRCUIT FOR COILS WITH MAGNETICALLY NON-LINEAR CORES

D S Thompson[1]

A M H Nasr[2]

1 University of Dundee, Scotland

2 Khartoum Polytechnic, Sudan

The analysis of circuits containing coils with ferromagnetic cores normally proceeds on the basis of numerical integration techniques which in contrast to the equivalent circuit approach of linear circuit theory masks the underlying theory of the circuits behaviour.
An equivalent circuit whose operation accounts for the basic observed phenomena which occur with iron-cored coils in electric circuits is developed. The circuit provides separate paths for increasing and decreasing currents to account for the effects of magnetic hysteresis. The application of the circuit to steady periodic excitation is found to predict current wave forms which show excellent agreement between measured and predicted values, as are some of the results for transient conditions.

1. INTRODUCTION

A recurring theme in the behaviour of electrical systems with iron-cored coils is, to the uninitiated, the unexpected occurrence of high-energy transient currents which frequently isolate the coil from its source. Investigation of the source of these unexpected isolations reveals the effect of remanent magnetism in the cores of such coils as the cause of the problem. In addition to these transient effects other sustained phenomena associated with the non-linear magnetising characteristics of the cores occur, e.g. (1) the non-sinusoidal current taken from a sinusoidal voltage source, (2) the excessive no-load current drawn by over excited transformers and (3) ferro-resonant oscillations. All of these are related to the complex magnetising characteristic of ferromagnetic materials.
For many purposes it is acceptable to overlook these non-linearities, as in the steady-state equivalent circuit of a transformer. For the quantitative analysis of the non-linear phenomena, their representation is vital to the success of the analysis.

2. SOME PHYSICAL CONCEPTIONS OF THE BASIC PHENOMENA

To develop a feeling for the representation of an iron-cored coil it is worth looking at the wave-forms of supply voltage and coil current, as shown in Fig. 1 for one particular coil.

Correspondingly Fig. 2 shows the wave-forms of core flux and current for the same conditions as Fig. 1. From Fig. 2 it can be concluded that the non-zero current at the instant of zero flux is due to the magnetic hysteresis of the core.

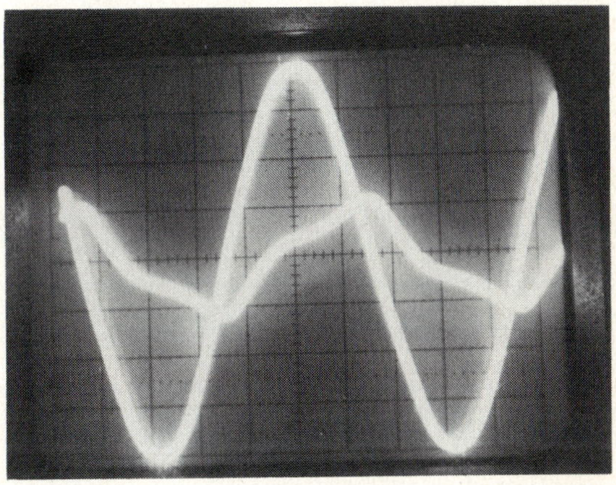

Figure 1 : Voltage and current wave-forms for an iron-cored coil

From the viewpoint of circuit behaviour the significant aspects of an iron-cored coil are associated with the representation of the non-linear hysteresis of the core and the core's losses. Fig. 1 clearly shows that the coil's inductive properties cannot be represented by a coil of constant inductance: the sudden transition in the values of the current suggests that it may be possible to represent it by two values of inductance one of which is periodically switched in to and out of circuit. It is worth noting that the cores cannot be represented by a constant resistance.

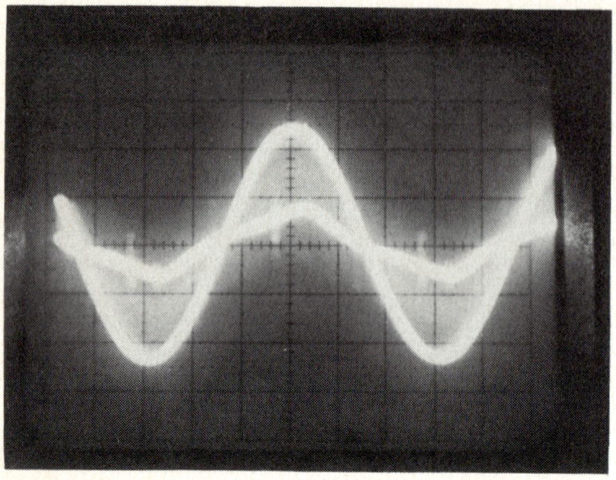

Figure 2 : Flux and current wave-forms for same coil and conditions as Figure 1

The representation of the effects of magnetic hysteresis forms the basis of an extensive literature, some of which is described by Nasr (1). Nasr (1) points out that ideally any model should reflect the physical processes under consideration. In this sense any model of magnetic hysteresis should represent the physical behaviour of the material. The accepted behaviour of magnetic materials is explained by the domain theory. According to this theory magnetic materials are thought of as assemblies of domains whose axes are randomly oriented until the application of an external magnetic field rotates them into the direction of the applied field by the growth of suitably directed domains at the expense of the other domains. Slemon (2) describes a model to explain this effect quantitatively. However its direct application to construct a simple equation to describe hysteresis is too difficult. Nasr (1) suggests that the re-interpretation of Slemon's model in terms of a suitably defined bulk permeability μ for any sample it may be related to a statistical model proposed by Thompson (3) who represents increasing and decreasing magnetisation processes by an appropriate Gaussian Distribution Function:-

$$\mu_i = \mu_m \exp\left[-\tfrac{1}{2}\left(\frac{H - \bar{H}}{\sigma}\right)^2\right]$$

where μ_m is the maximum value of the permeability, H the applied field, $\bar{H}$ the mean value of H for which $\mu_i = \mu_m$ and σ the variance of the distribution. This equation represents the increasing magnetisation and a corresponding one is required for decreasing magnetisation. By a mathematical analysis of this permeability function Nasr (1) derives a simple equation to represent each side of a hysteresis loop.

By a further averaging process a simple equation to represent the classical magnetisation curve is derived from these equations and is found to be of the form:

$$B = \alpha \tan^{-1}\beta H^3 + \gamma \tan^{-1}\delta H + \varepsilon H$$

For one sample of Silicon iron it is found that

$$B = 0.78 \tan^{-1}(7 \times 10^{-8} H^3) + 0.162 \tan^{-1}(10^{-5} H) + 2.4 \times 10^{-5} H$$

gives excellent agreement over the whole range of the magnetisation curve between measures and predicted values, at no part of the curve does the difference between the measured and predicted curves differ by more than 1%.

Effectively this analysis by Nasr is a demonstration from theoretical principles of the relationship between the curve-fitting descriptions of observed non-linear magnetising curves and domain behaviour under conditions of applied magnetic fields.

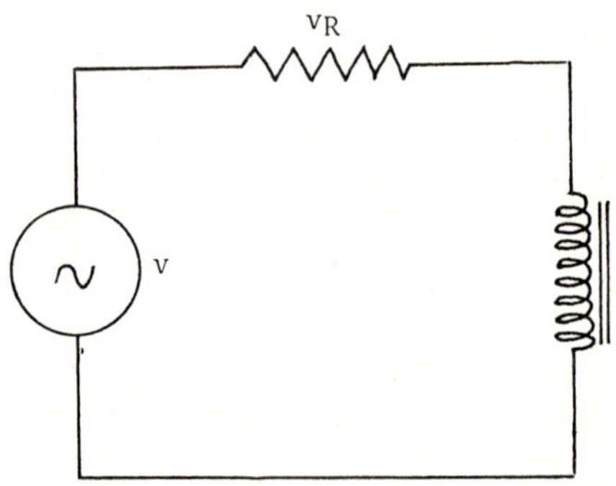

Figure 3 : Equivalent circuit of a coil

The analysis of a circuit comprising a voltage source and an iron-cored coil must proceed on the basis of accepted laws of electric and magnetic circuits. For the electric circuit of Fig. 3 Kirchoff's Second Law requires

$$v_R = v + e_L \tag{1}$$

and in terms of the rate of change of flux linkages equation 1 may be re-written as

$$v = Ri + N\frac{d\phi}{dt} \tag{2}$$

If it is now assumed that the flux ϕ is uniformly distributed across the iron, equation 2 may be expressed in terms of the conventional magnetic parameters of the iron-cored coil's magnetic circuit as

$$v = R\frac{H}{N} + NA\frac{dB}{dH}\cdot\frac{dH}{dt} \tag{3}$$

where B is the flux density and H the corresponding magnetising force, for which B and H are related by the appropriate mathematical function whose derivation may be found in reference (1).

If the magnetic properties of the iron-cored coil are linear the solution to equation 3 is the well-known one associated with linear circuit theory. For non-linear magnetic properties the direct solution to equation 3 involves the use of a step-by-step procedure of numerical integration. For many problems it is common practice to use Runge-Kutta or Predictor-Corrector methods for the solution to this equation which may be used to predetermine the wave-form of transformer magnetising currents. An alternative approach to the prediction of such wave-forms from the solution of equation 3 is to use a method due to Gautschi (4). This method is applicable only to situations for which a periodic solution is appropriate since it is based on the use of trigonometric interpolating polynomials in the integration process.

Although numerical integration techniques can predict wave-forms to any desired degree of accuracy they fail to provide the necessary physical insight into the behaviour of the system. More importantly, they don't always provide the correct solution on account of an instability inherent in the method: for this reason an alternative approach is desirable. Now the approximation of the current wave-form of Figs. 1 or 2 by the linear segments

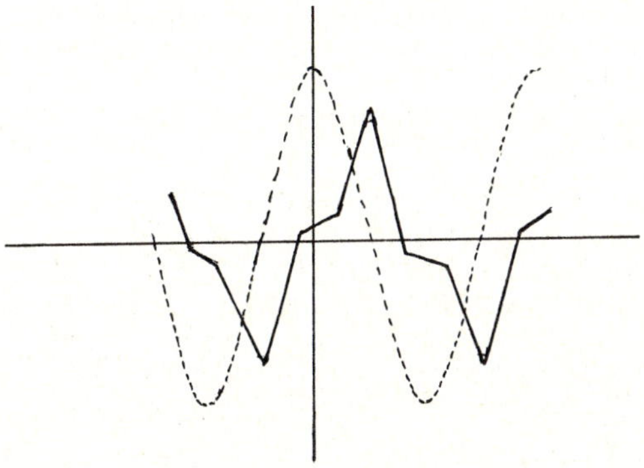

Figure 4 : Segmented representation of the current wave-form of Fig. 1

of Fig. 4 suggests that a combination of inductances, some of which are switched in and out of circuit may provide a basis for the analysis of these devices. It is the development of this idea that forms the basis of this contribution.

3. DEVELOPMENT OF AN EQUIVALENT CIRCUIT

The commentary in the previous section makes reference to the behaviour of a non-linear core in terms of resistance and inductance. The non-linear properties of the core require recourse to the instantaneous values of resistance and inductance and thus a mathematical model of the hysteresis loop. If the boundaries of a hysteresis loop are approximated by linear segments, the computational effort required to solve equation 3 is so reduced that it becomes helpful to develop the analysis in equivalent circuit terms. The approach to the development of the equivalent circuit may be thought of as the reverse of the well-known text-book demonstration of the non-sinusoidal current drawn from a sinusoidal voltage source by an iron-cored coil.

3.1 Requirements of an equivalent circuit

An important requirement of any equivalent circuit is its ability to represent the features of the device modelled. For iron-cored coils with magnetic non-linearities it is desirable that any equivalent circuit should model the non-linear inductive effects, the hysteresis and the losses.

With these requirements noted it is desirable to develop a model which will reproduce the features of the device, in particular a current wave from which suffers no mathematical discontinuities, i.e. the circuit parameters of an actual coil change continuously with current time.

3.2 The representation of hysteresis

The hysteresis loop for the coil to which Figs. 1 and 2 relate is shown in Fig. 5. Let this loop be approximated by the sequence of

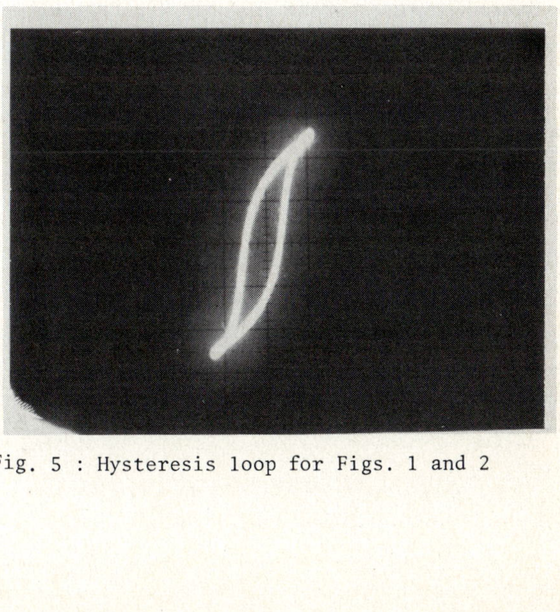

Fig. 5 : Hysteresis loop for Figs. 1 and 2

of linear segments lm, mn, np, pq, qa and al as shown in Fig. 6A i.e. for this approximation the values of dB/dH are constant for each segment.

For ease of development of the equivalent circuit attention is focussed initially on the inductance effects, i.e. equation 3 reduces to

$$v = \frac{dB}{dH} \cdot \frac{dH}{dt} \quad \text{or} \quad v = \frac{dB}{di} \cdot \frac{di}{dt}$$

since N/ℓ is the same constant in the differentials dH.

With reference to Fig. 6 it is apparent that dB/dH is small for saturated conditions and large for permeable conditions.

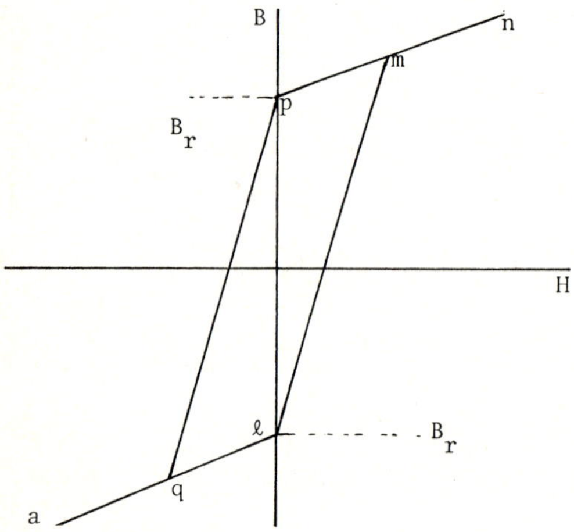

Figure 6A: Segmented form of the hysteresis look of Fig. 5

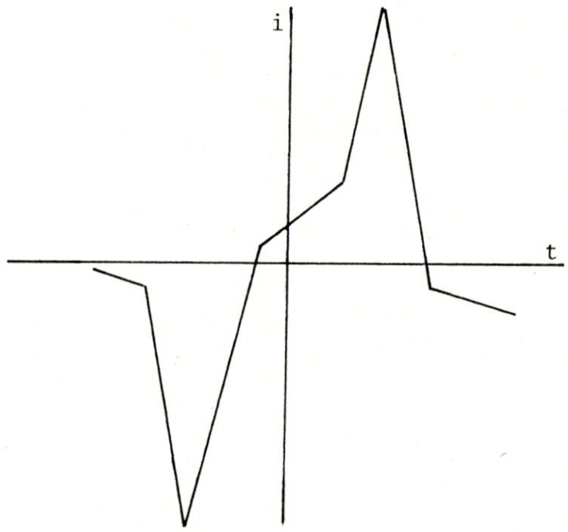

Figure 6B : Segmented form of the current derived from 6A assuming a lossless coil.

In terms of circuit considerations. a coil whose hysteresis loop is modelled by the linearised loop of Fig. 6A draws a large current from the supply during the instants of saturated core conditions, as represented by al and np to balance its self-induced emf with the supply voltage. With the onset of permeable core conditions represented by lm and pq, when dB/dH is large, the necessary current to balance the supply voltage is small. Thus the changes in level of the coils currents shown in Fig. 6B are accounted for. At the points m, p, q and l where the transitions from permeable core conditions to saturated-core conditions, or vice versa, occur it is clear that the electrical equivalent of the mechanical idea of conservation of momentum requires a discontinuous change of current; this is inconsistent with the stated requirements. However at these points when core losses are included it turns out that the effect of the discontinuity is less marked.

It is worth noting here that an important consequence of the linear segmentation of the hysteresis loop neglects completely the points of inflexion of the actual loop. A consequence of this and the instantaneous change from saturated to unsaturated core conditions leads to an imperfect prediction of the waveform of a transformer's magnetising current as shown in Fig. 7. However in studies of ferro-resonant oscillations these imperfections are found as in Figs. 10 and 11 to be less pronounced.

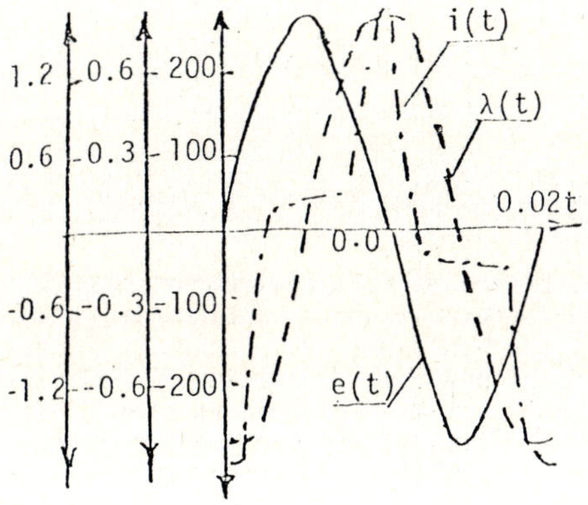

Figure 7 : Predicted wave-form of magnetising current using complete equivalent circuit

With this linear segmented hysteresis loop to model the core's hysteresis it is found that only two values of inductance are required. However to ensure that the instantaneously zero values of flux and current are not co-incident then different paths are required to represent the rising and falling portions of hysteresis loop.

In terms of the emf developed by the coil over a complete cycle of hysteresis a basic form of the equivalent circuit is shown in Fig. 8. The emf sources have the property that their emfs also determine the required instantaneous current. The operation of the circuit is related to the position on the approximated hysteresis loop of Fig. 6. In the following explanation attention is focussed separately on the coil's behaviour for the effects of increasing and decreasing current.

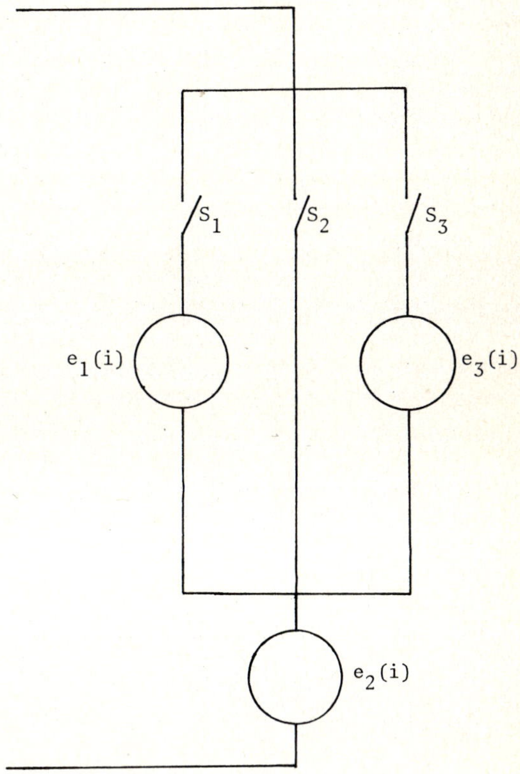

Figure 8 : Basic form of the equivalent circuit

(a) Consider increasing current conditions

(i) For saturated core conditions only the emf e_2 is in circuit 1 thus switch S_2 is closed and switches S_1 and S_3 are open.

(ii) With the onset of permeable core conditions dB/di increases and so the necessary emf to be developed requires S_3 closed and S_2 open.

(iii) At the onset of saturated conditions again S_2 is closed and S_3 opened.

(b) For decreasing current conditions the reverse process continues except that e_1 is brought into circuit for permeable core conditions.

In practice the realisation of the emf generators, together with the problems arising from the interruption of the currents by opening the switches S_1 and S_3 of Fig. 8 makes this circuit difficult to handle. The direct replacement of the emf sources by coils with suitable L-i characteristics is accompanied by equally many problems of a similar nature. At the instant of current reversal the need to isolate one coil and introduce another without the introduction of extraneous transcient effects into the supply current wave-form can be met by the insertion of current generators. Fig. 9 shows a complete equivalent circuit to meet this requirement.

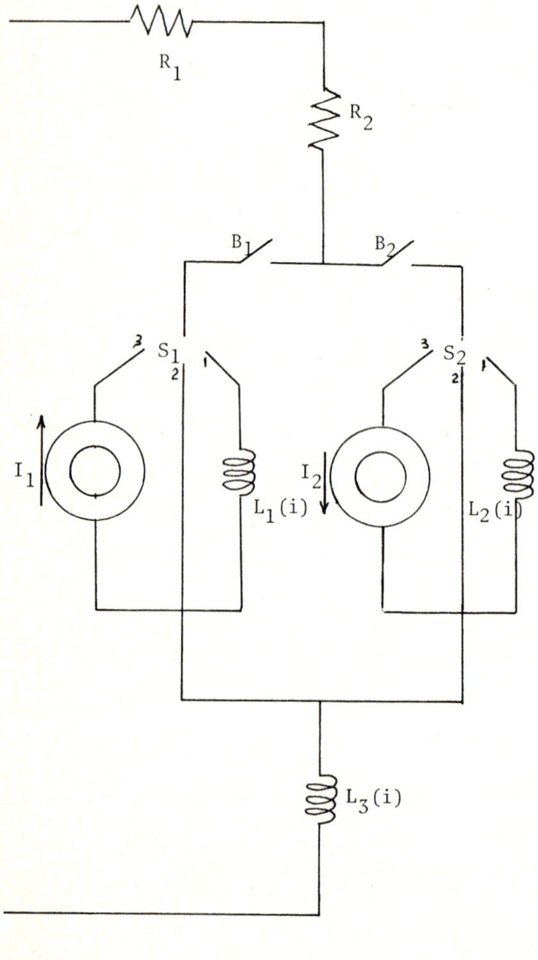

Figure 9 : Complete equivalent circuit

The switches S_1 and S_2 have three positions, viz. (i) to connect the coil into the circuit, (ii) to short circuit the coil, (iii) to link to current generator I_2 (or I_1) to L_2 (L_1).

The operation of the circuit for steady ac conditions is as follows. For current increasing from an instantaneous zero L_2 and L_3 are on circuit with S_2 in position (i) with B_2 closed and B_1 open. At the onset of saturation only L_3 is in circuit and S_2 moves to position (ii) to short-circuit L_2. At the instant of current reversal it is required to bring the combination of L_1 and S_1 and I_1 into operation. Unwanted transients are avoided if I_1 at this instant is circulating through L_1 a current equal to the current in the rest of the circuit. At instants of current reversal S_1 switches to position (ii), S_2 to position (iii) B_1 closed and B_2 open. This continues until S_1 switches to position (i) and cycle repeats the previous half cycle.

The resistance of the circuit includes the direct winding resistance R_1 together with a variable resistance R_2 to account for the variation of the core losses with magnetising current.

3.4 Analytical procedures with the equivalent circuit

The application of this equivalent circuit to the analysis of various situations forms the basis of the next section. However the details of the analytical procedures need consideration; clearly they involve step-by-step procedures which are performed conveniently by computer. Since the approximation to the hysteresis loop is by a sequence of linear segments the values of dB/dH or dB/di for any segment are constant and so the model effectively may be considered to have linear properties. Consequently for any particular time interval provided remains unchanged, the solution for that time interval is the solution to the well-known linear transient with the appropriate initial condition viz. the current at the start of the interval under consideration is the current at the end of the previous interval. The details of a scheme of calculation are indicated in the Appendix.

3.5 Asymmetrical operating conditions

To extend the application of the equivalent circuit to abnormal and transient conditions of operation some further elaboration is required.

Let the magnetising current be slowly reduced from point n of Fig. 6A to zero. During this period L_2 is connected to I_2 whose value is the value of the supply current and B_2 is open and L_1 is short-circuited. If the supply is interrupted at the current zero the coil's magnetic circuit is left with a remanent flux of density B_r. If on reconnection to the supply the current is slowly increased from zero the path traced is along and for this situation L_1 remains short circuited and L_2 connected to I_2 with B_2 open. For all such conditions, including zero remanent flux the initial path in the B-H plane is taken to be parallel to the ones shown in Fig. 6A for saturated core conditions until it meets the appropriate permeable core line.

If now interruption is considered to take place at a non-zero current, say at point m, on the B-H plane the path returns to p with the same remanent flux. In this case switch B_2 is left closed and S_1 moves to connect I_1 to L_1 to circulate the value of current corresponding to the value at the instant of interruption and so account for the energy of the remanent flux (note this does not include the situation just discussed). For calculations it is essential to check on the value of the flux at the end of each interval: if the flux direction remains unchanged the path for current flow is unchanged until the flux direction reverses. Some further details are found in the Appendix.

4. RESULTS AND DISCUSSION

It is clearly desirable to demonstrate the validity of the equivalent circuit in terms of results obtained by its application. This section considers the results for steady-state ac conditions and then for some transient conditions. As already shown the results of the application of this equivalent circuit to the prediction of a transformer's magnetising current is, for the reasons already given not as good as the stated criteria demand. Its application to ferro-resonant circuits is found to give more promising results.

4.1 The fundamentals of ferro-resonance reviewed

When a capacitor is connected in circuit with an iron-cored coil and the excitation is sufficient to force the coil to operate partly on the linear and the non-linear regions of its magnetising characteristic then the wave form of current experiences even more pronounced peaks than those already observed for the coil alone. Such a condition is a ferro-resonant oscillation rather than the simple resonant oscillation of elementary theory.

The equation describing the behaviour of a series ferro-resonant circuit involves a simple modification to equation 1, viz.

$$
\begin{aligned}
v &= Ri + N\frac{d\phi}{dt} + \frac{1}{C}\int i\,dt \\
&= R\frac{H}{N} + NA\frac{dB}{di}\cdot\frac{di}{dt} + \frac{1}{CN}\int H\,dt
\end{aligned}
$$

For this case it is to be noted that the pds across resistance, inductance and capacitance are essentially non-sinusoidal but must sum for a sinusoidal supply to a sinusoidal form; consideration of ferro-resonance in 3-ph circuits is deferred to the next paragraph.

Results for one situation for a single-phase ferro-resonant circuit are shown in Fig. 10. The agreement between the measured and predicted wave-forms is typical of the results obtained for a wide variety of 3-ph circuits.

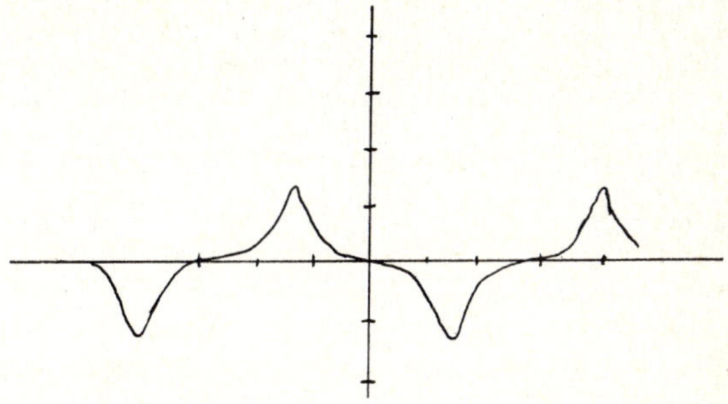

(a) Experimental

Each division is 7.5 mm.
Each vertical division = 4a.
Each horizontal division = 5 ms.

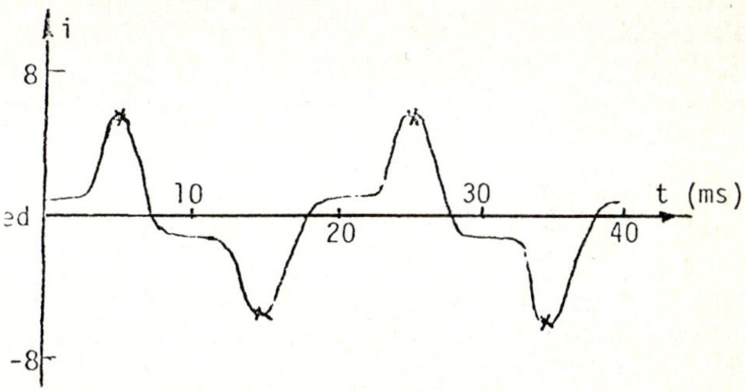

Figure 10 : Showing measured and predicted wave-form for a 1-ph resonant circuit

4.2 Ferro-resonant oscillations in three-phase circuits

One of the more complex and fascinating applications of the equivalent circuit is to the analysis of ferro-resonant phenomena in 3-ph circuits. In practice one of the more startling occurrences of ferro-resonance is at the terminals of a transformer located at the end of a de-energised transmission line which runs parallel to an excited line. The mutual capacity between the lines providing the necessary capacitive link between coil and source. For this situation the constraints imposed on the fluxes in the core limbs of the core itself together with the effect of the electrical circuit connection form a severe test of the equivalent circuit model.

For the 3-ph circuit the number of equations to solve increases very significantly. However by inclusion of the electric and magnetic circuit constraints the number of simultaneous equations reduces to 5.

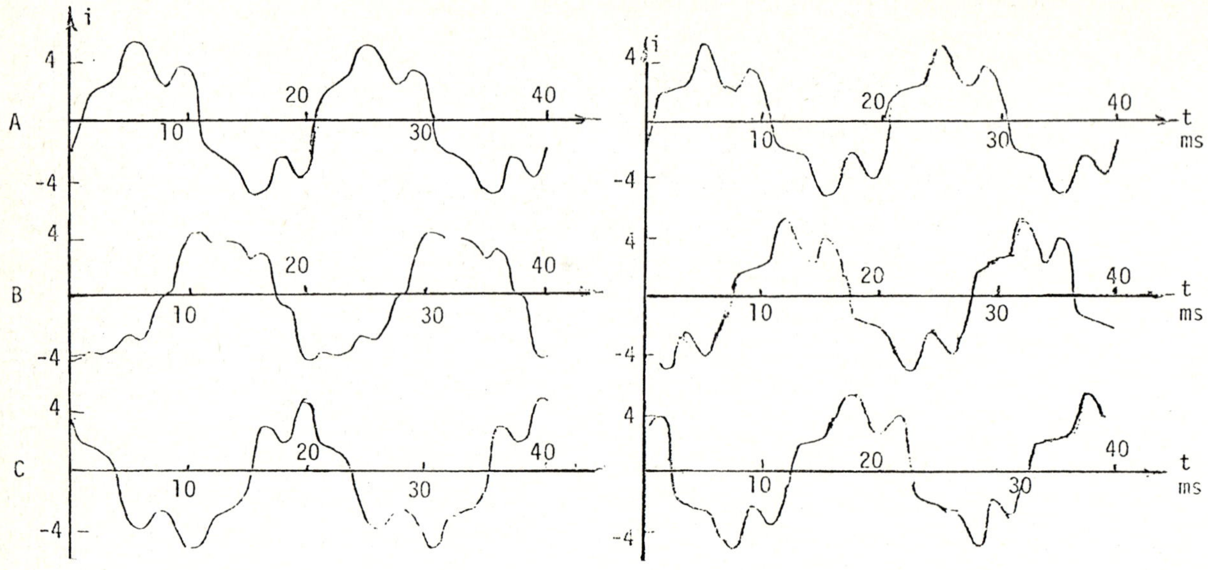

Figure 11

For a 3-ph system the analysis proceeds on the basis of the application of the result for a 1-ph system to each of the three phases simultaneously and including the effects of the constraints imposed by the electric and magnetic circuits, viz:

(a) the sum of the fluxes is zero at all instants and for a connected electric circuit with an isolated neutral

(b) the sum of currents is zero at all instants

Ideally each phase is considered in isolation with the system constraints inserted into each phase as parts of the boundary conditions of the integration for each phase.

Measured and predicted results for both star and delta connection of the transformer's windings are shown in Figs. 11 and 12. An important feature of the predicted results is the neglect of the effect of inter-phase magnetic coupling between the phases of the transformer's windings. The level of agreement between these sets of results is remarkable: especially when set against the results obtained by Teape (5) and his co-workers who include this phenomenon in their analysis of ferro-resonance in 3-ph circuits with core-type 3-ph coils. At this point it must be stated that none of our experiments took the supply voltage up to the value appropriate to the transformer's rating, thus the effect of this phenomenon in our experiments in all probability was of very minor importance only. Equally we are convinced that the accuracy of our modelling of the hysteresis loop itself is of at least equal significance to the results we predict, this despite the neglect of the points of inflexion in the hysteresis.

Despite the limitations already noted of this model, the results obtained by it for the steady state analysis of circuits excited by constant sinusoidal sources are sufficiently encouraging to make it a useful addition to the repertoire of available equivalent circuits. The application of the equivalent circuit to the analysis of steady state ac conditions provides results which are both acceptable as to wave-form and amplitude for a limited number of situations. The extension of the theory to include transient phenomena involves no change of principle for the situation of zero remanent flux, and results of acceptable error. When however the more common situation of connecting an iron-cored coil with remanent flux in its magnetic circuit to a supply occurs it turns out that the results obtained are less adequate on account of the restricted representation of the coil's hysteresis loops by linear segments.

5. CONCLUSIONS

An equivalent circuit which gives adequate results for steady state conditions is found to be incomplete in relation to its performance for the complete range of transient conditions which occur in practice.

6. REFERENCES

(1) Nasr, A.M.H., A general analysis of ferroresonance in single-phase and three-phase circuits including the ferro-resonant behaviour of an induction motor. Ph.D. Thesis, University of Dundee 1980.

(2) Slemon, G.R., Magnetoelectric Devices, Transducers, Transformers and Machines (John Wiley & Sons, New York 1966).

(3) Thompson, W.I., Mathematical Model for Non-Linear Cores at Low Frequencies. IEEE Trans Magnetics. 9. 1974 p 322.

(4) Gautschi, W. Numerical integration of differential equations based on trigonometric polynomials. Numerische Mathematische 3 (1961) 381-397.

(5) Teape, J.W., R.D. Slater, R.R.S. Simpson, W.S. Wood. Hysteresis Effects in Transformers, including ferroresonance. Proc IEE 123, 1976 p 153.

7. APPENDIX: THE CALCULATION PROCEDURE

In this appendix the procedure is given for the calculation of the current in a 1-ph ferro-resonant circuit. For magnetisation curve calculations only the effect of the capacitor is omitted.

For zero initial charge and remenant flux the circuit works with S_1 and S_2 in position 3. If $e = E\cos(\omega t + \psi)$, the steady component of current i' that the circuit would tend to demand is

$$i' = \frac{E}{Z}\cos(\omega t + \alpha) \text{ where } Z = (R^2 + (\omega L - \frac{1}{\omega C})^2)^{\frac{1}{2}}$$

and $\alpha = \tan^{-1}\dfrac{1/\omega C - \omega L}{R}$. Now at t = 0,

$i' = E/Z \cos\alpha$ and for which the corresponding capacitor voltage is $e_{c_o} = E/\omega CZ \sin\alpha$.

The corresponding transient component is determined from this and the associated boundary conditions which must determine for each step the inductance of the coil and the nature of the forced response. Now the instantaneous total current i_o at t = 0 less i' is not zero but some value i" such that

$$\frac{d^2i''}{dt^2} + \frac{R}{L}\frac{di''}{dt} + \frac{i''}{LC} = 0 \qquad (4)$$

whose solution depends on that value of $((R/2L)^2 - 1/LC)^{\frac{1}{2}}$ which in turn provides the necessary simultaneous conditions to determine the value of i" at t = 0, viz:

$$I_2'' = \frac{e_{c_o}'' + i_o''(L\alpha_1 + R)}{L(\alpha_1 + \alpha_2)}$$

$$I_1'' = i_o'' - I_2''$$

where (i) α_1 and α_2 are the roots of $S^2 + R/L\, S + 1/LC = 0$, (ii) I_1'' and I_2'' are the constants of integration for equation and

$$e_{c_o}'' = -I_1''(L\alpha_1 + R) - I_2''(L\alpha_2 + R).$$

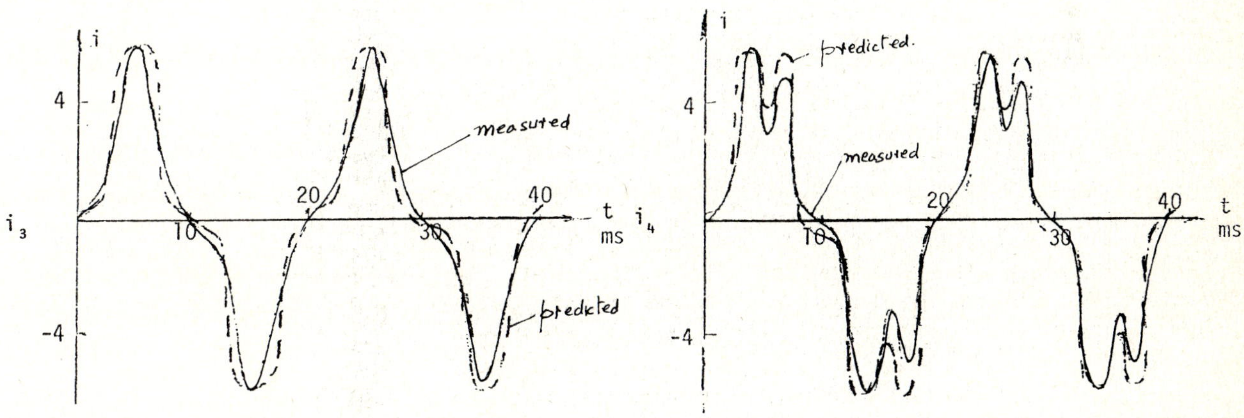

Figure 12

ELECTRICAL MACHINES AND CONVERTERS - Modelling and Simulation
H. Buyse and J. Robert (editors)
Elsevier Science Publishers B.V. (North-Holland)/ © IMACS, 1984

ANALYTICAL MODEL OF HV CASCADE CONNECTED TEST TRANSFORMERS

G. Olivier G. Roy R.-P. Bouchard Y. Gervais

Ecole Polytechnique de Montréal, P.O. Box 6079, Station A, Montréal, Canada, H3C-3A7

Cascade connected transformers are widely used in HV laboratories to produce very high ac voltages. However this connection is plagued with three serious problems: voltage distribution, frequency response and short-circuit currents. This paper proposes an analytical method to evaluate the voltage and current distribution in any order cascade submitted to various load or short-circuit conditions. Experimental results confirm the validity of the model.

INTRODUCTION

Cascade connected transformers, Fig. 1, are widely used in high voltage laboratories all around the world to produce the high ac test voltages. Because of their relatively small nominal power rating and their extremely large turn ratio (100/1), they cannot be considered as conventional transformers. As example, the stray capacitance of the high voltage winding often represents 20 to 30% of the nominal load of the transformer. Furthermore, this arrangement presents its own specific problems: voltage distribution, frequency response and short-circuit impedance.

The insulation of each individual transformer of any cascade is not submitted to the same level of electrical stress. The unit at the upper end exhibits a voltage often much more higher than the first unit. It is extremely important to be able to accurately predict the exact voltage distribution in order not to submit the most stressed unit to voltages larger than its nominal value. The second most severe problem encountered with this type of connection is the overall frequency response of the cascade. Due to the excessively high value of parasitic capacitances, the cascade may behave as a tuned filter with the peak frequency any where between 60 Hz and 2 kHz. If a resonant frequency happens to be close to a predominant line harmonic frequency, the HV waveforms becomes heavily distorted. This resonance phenomenom may cause serious unexpected problems if no countermeasures are taken. Finally, in order to meet several test specifications such as those performed under contaminated conditions on insulators and bushings, the HV ac source must be able to provide a specified level of short-circuit current. Since direct measurements are often unpractical, the short-circuit impedance must be evaluated from the impedance of the individual units.

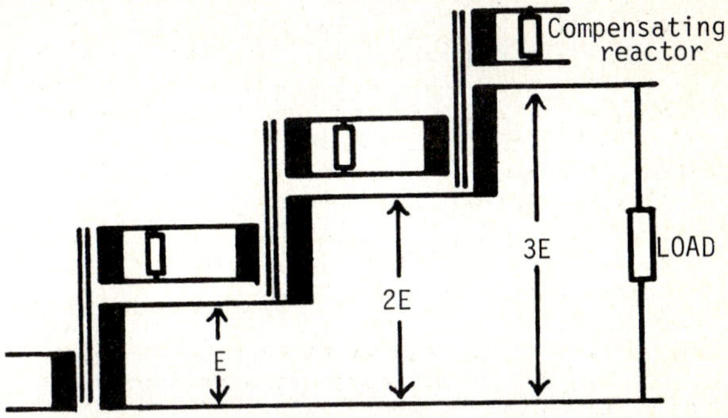

Fig. 1: Cascade connection of three transformers.

This paper presents an analytical model of cascade connected transformers constituted of an arbitrary number of basic transformers. The knowledge of the equivalent circuit of the basic units enables the calculation of the voltage distribution along the cascade, its frequency response and the short-circuit impedance. Furthermore, the model includes the various parasitic capacitances and the compensating reactors which are often added across the tertiary winding to compensate the effect of the HV winding capacitance.

This paper summarizes the work undertaken several years ago in the HV laboratory of the Electrical Engineering Department of Ecole Polytechnique de Montréal. The analytical method presented here constitutes the extension of analytic equations proposed in 1979 for a two transformer cascade [1].

EQUIVALENT CIRCUIT OF A CASCADE

Fig. 1 shows the interconnections between the various units of a three stage cascade. The first unit is directly fed from the ac supply while the others are fed from the previous unit through a tertiary (coupling) winding connected to the high voltage end of the secondary winding. Each transformer develops approximately one third of the total voltage applied to the load which normally exhibits a capacitive nature. This type of arrangement is preferred to a single, full size, transformer for two fundamental reasons. First, remembering that nominal power of the individual units is ridiculously low in comparison of the voltage level, the construction of single full voltage unit with the high turn-ratio needed is extremely difficult to achieve at reasonable cost mainly due to the insulation problem of the HV winding. The use of smaller units designed for lower voltage results in a more realistic construction. Secondly, the use of several small units is much more versatile: they may be connected in various combinations to obtain the voltage and power capacity required [3].

The various parisitic capacitances are describded in Fig. 2. The most predominant one is the stray capacitance of the HV winding followed by the capacitance of the field smoothing electrode and by the capacitance of the interstage electrodes. These capacitances often represent an intrinsic load of 0,20 to 0,30 p.u. for the transformer cascade.

Fig. 3 shows the equivalent circuit of a N-stage cascade, where:

Z_{Pi}, Z_{Ti}, and Z_{Si} represent the series impedance of the three windings;
Z_{Hi}, the magnetization branch;
Z_{Ci}, the impedance of the compensating reactor.

Furthermore, ideal coupling transformers are needed to take account of the insulating properties of the coupling windings. These ideal transformers have to be eliminated from the equivalent circuit in order to apply conventional circuit analysis methods.

Fig. 4 illustrates the equivalent circuit of a two-stage cascade without compensating reactors. This circuit will be used to demonstrate the analysis yielding the

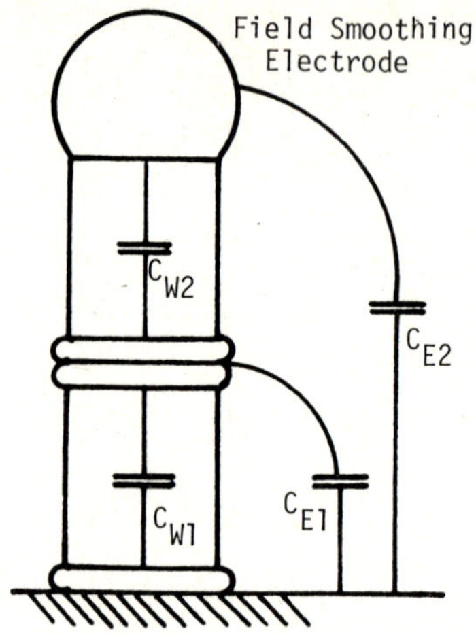

Fig. 2: Parasitic capacitances.

elimination of the coupling transformer. The first step consists in the replacement of load and the stray capacitance currents by two current sources I_1 and I_2 (Fig 4b). In order to preserve the amplitude of the various currents, these sources must provide the following currents:

$$I_1 = I_2 + I_{E2} \quad (1)$$

and $$I_2 = I_L + I_{E1} + I_{E2} \quad (2)$$

The currents flowing through the various impedances are not modified. However, the load current does not flow anymore directly to the ground but through the secondary of the coupling transformer and through the source I_1. Moreover, there is no more current flowing into the "wire" connecting the primary and the secondary of the coupling transformer (dotted line in Fig. 4b). This connection can, therefore, be eliminated. Since the secondary is now floating with respect to the primary, the impedances connected to the secondary side may be referred to the primary without affecting any current or voltage.

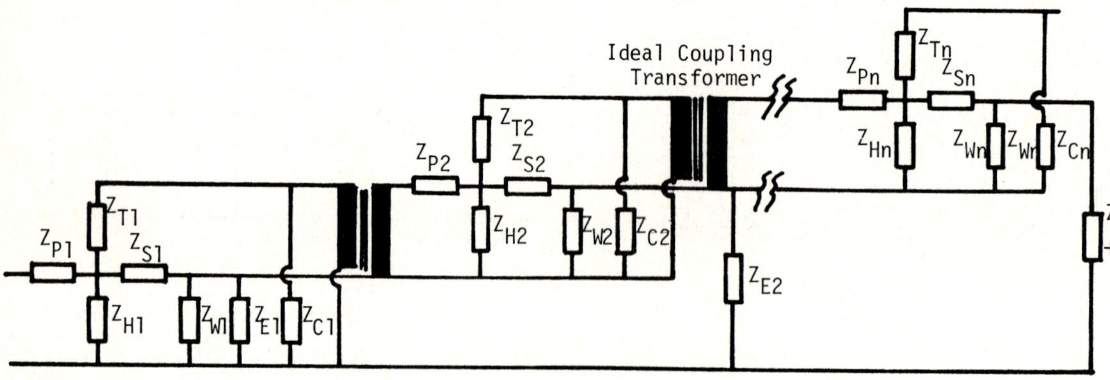

Fig. 3: Equivalent circuit for a cascade of "N" transformers.

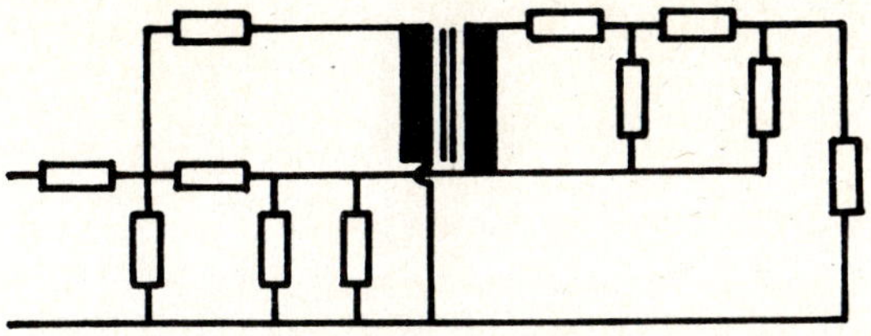

a) Equivalent circuit of a two transformer cascade.

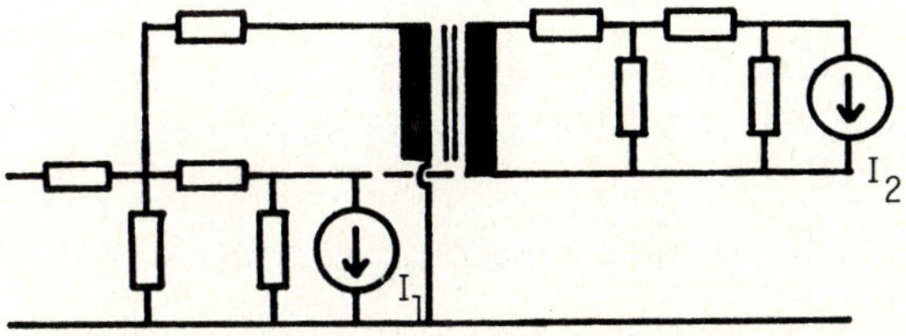

b) Addition of the current sources.

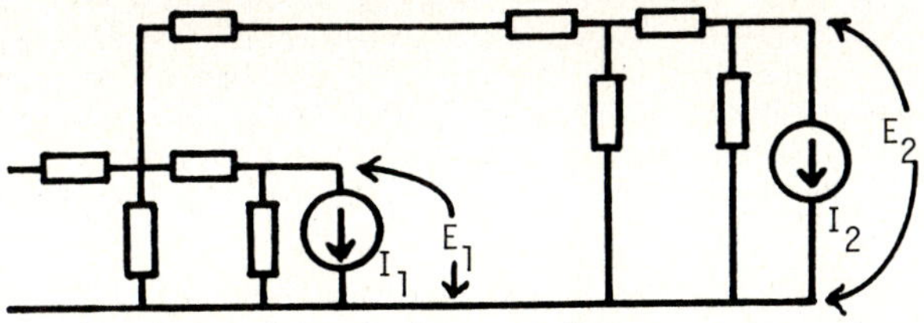

c) Elimination of the transformer.

Fig. 4: Elimination process of the ideal coupling transformers.

This yields to the final equivalent circuit of Fig. 4c. This reasoning may be generalized to any order cascade resulting, as example, in the equivalent circuit of Fig. 5 for a 5th order cascade.

Since the ideal coupling transformers have been eliminated, conventional circuit analysis methods may be applied to evaluate the voltage and current distribution, the frequency response and the short-circuit currents. The only major difference is that the voltages E_1 to E_N do not represent any more the voltages across the first N units but rather the individual stage voltages. The output voltage becomes the summation of the individual voltages.

$$E_L = \sum_{i=1}^{N} E_i \tag{3}$$

FREQUENCY RESPONSE AND VOLTAGE DISTRIBUTION

The set of equations corresponding to the circuit of Fig. 5 is given in matrix form in (4). The various node voltages, E_1 to E_{15}, are defined in Fig. 5. The admittance matrix has been obtained by applying the usual rules for node method. The size of this matrix is 3N x 3N if compensating reactors are used or 2N x 2N otherwise. The solution of this set of equations gives the voltage across each secondary winding (E_1 to E_5) thus the voltage and current distributions and the load voltage.

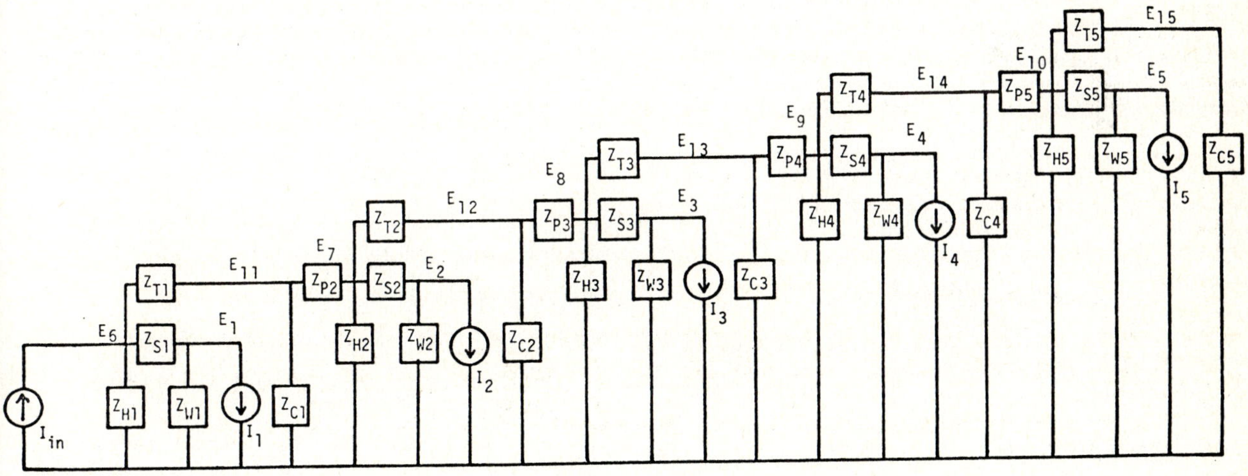

Fig. 5: Complete Equivalent Circuit for a 5th order Cascade

where: $I_{in} = E_{in}/Z_{P1}$

$$I_i = I_L + \sum_{j=i}^{N} E_j/Z_{E_j}$$

$$I_L = \sum_{i=1}^{N} E_i/Z_L$$

$$
\left|\begin{array}{ccccccccccccccc}
A_1+B_1 & A_2 & A_3 & A_4 & A_5 & -Y_{S1} & \text{--} & \text{--} & \text{--} & \text{--} & \text{--} & \text{--} & \text{--} & \text{--} & \text{--} \\
A_2 & A_2+B_2 & A_3 & A_4 & A_5 & \text{--} & -Y_{S2} & \text{--} & \text{--} & \text{--} & \text{--} & \text{--} & \text{--} & \text{--} & \text{--} \\
A_3 & A_3 & A_3+B_3 & A_4 & A_5 & \text{--} & \text{--} & -Y_{S3} & \text{--} & \text{--} & \text{--} & \text{--} & \text{--} & \text{--} & \text{--} \\
A_4 & A_4 & A_4 & A_4+B_4 & A_5 & \text{--} & \text{--} & \text{--} & -Y_{S4} & \text{--} & \text{--} & \text{--} & \text{--} & \text{--} & \text{--} \\
A_5 & A_5 & A_5 & A_5 & A_5+B_5 & \text{--} & \text{--} & \text{--} & \text{--} & -Y_{S5} & \text{--} & \text{--} & \text{--} & \text{--} & \text{--} \\
-Y_{S1} & \text{--} & \text{--} & \text{--} & \text{--} & C_1 & \text{--} & \text{--} & \text{--} & \text{--} & -Y_{T1} & \text{--} & \text{--} & \text{--} & \text{--} \\
\text{--} & -Y_{S2} & \text{--} & \text{--} & \text{--} & \text{--} & C_2 & \text{--} & \text{--} & \text{--} & -Y_{P2} & -Y_{T2} & \text{--} & \text{--} & \text{--} \\
\text{--} & \text{--} & -Y_{S3} & \text{--} & \text{--} & \text{--} & \text{--} & C_3 & \text{--} & \text{--} & \text{--} & -Y_{P3} & -Y_{T3} & \text{--} & \text{--} \\
\text{--} & \text{--} & \text{--} & -Y_{S4} & \text{--} & \text{--} & \text{--} & \text{--} & C_4 & \text{--} & \text{--} & \text{--} & -Y_{P4} & -Y_{T4} & \text{--} \\
\text{--} & \text{--} & \text{--} & \text{--} & -Y_{S5} & \text{--} & \text{--} & \text{--} & \text{--} & C_5 & \text{--} & \text{--} & \text{--} & -Y_{P5} & -Y_{T5} \\
\text{--} & \text{--} & \text{--} & \text{--} & \text{--} & -Y_{T1} & -Y_{P2} & \text{--} & \text{--} & \text{--} & D_1 & \text{--} & \text{--} & \text{--} & \text{--} \\
\text{--} & \text{--} & \text{--} & \text{--} & \text{--} & \text{--} & -Y_{T2} & -Y_{P3} & \text{--} & \text{--} & \text{--} & D_2 & \text{--} & \text{--} & \text{--} \\
\text{--} & \text{--} & \text{--} & \text{--} & \text{--} & \text{--} & \text{--} & -Y_{T3} & -Y_{P4} & \text{--} & \text{--} & \text{--} & D_3 & \text{--} & \text{--} \\
\text{--} & \text{--} & \text{--} & \text{--} & \text{--} & \text{--} & \text{--} & \text{--} & -Y_{T4} & -Y_{P5} & \text{--} & \text{--} & \text{--} & D_4 & \text{--} \\
\text{--} & \text{--} & \text{--} & \text{--} & \text{--} & \text{--} & \text{--} & \text{--} & \text{--} & -Y_{T5} & \text{--} & \text{--} & \text{--} & \text{--} & D_5
\end{array}\right|
\times
\left|\begin{array}{c} E_1 \\ E_2 \\ E_3 \\ E_4 \\ E_5 \\ E_6 \\ E_7 \\ E_8 \\ E_9 \\ E_{10} \\ E_{11} \\ E_{12} \\ E_{13} \\ E_{14} \\ E_{15} \end{array}\right|
=
\left|\begin{array}{c} \text{---} \\ \text{---} \\ \text{---} \\ \text{---} \\ \text{---} \\ I_{IN} \\ \text{---} \\ \text{---} \\ \text{---} \\ \text{---} \\ \text{---} \\ \text{---} \\ \text{---} \\ \text{---} \\ \text{---} \end{array}\right|
$$

(4) Set of equations for normal conditions,

where:
$$A_K = Y_L + Y_{E(k)} + Y_{E(k+1)} + \dots + Y_{E(N)}$$
$$B_K = Y_{S(K)} + Y_{W(K)}$$
$$C_K = Y_{H(K)} + Y_{P(K)} + Y_{S(K)} + Y_{T(K)}$$
$$D_K = Y_{C(K)} + Y_{T(K)} + Y_{P(K+1)}$$

The frequency response is obtained by resolving (4) for various frequencies. Each computation involves the inversion of a 3N x 3N complex matrix and normally at least 25 points are necessary to draw one complete frequency response curve. However the computation time required is affordable even for a 5th order cascade.

SHORT-CIRCUIT IMPEDANCE

The short-circuit currents are determined in a very similar way. The only modifications are as follows:

1: I_L becomes I_{sc}

2: The short-circuit current is equal to the voltage appearing at the magnetization branch of the last unit divided by the impedance of the secondary winding of this unit minus the current flowing in its stray capacitance. For a 5th order cascade:

$$I_{sc} = (E_1+E_2+E_3+E_4+E_{10})/Z_{S5} - E_5/Z_{W5} \qquad (5)$$

3: The amplitude of the various current sources becomes:

$$I_i = I_{sc} + \sum_{j=i}^{N-1} E_j / Z_{Ej} \qquad (6)$$

The set of equations corresponding to the short-circuit case is given in (7) for a 5th order cascade. The matrices have been reorganized to give a symmetrical admittance matrix.

EXPERIMENTAL WORK

Various experimental tests have been performed to confirm the validity and the usefulness of the model presented in this paper. First, the analytical results computed for our two-stage, 200 kV, 5 kVA cascade were checked against experimental data obtained several years ago. The agreement between both sets of data is similar to the results already published in [1].

$$
\begin{vmatrix}
C_1 & -Y_{S1} & \text{--} & \text{--} & \text{--} & \text{--} & \text{--} & \text{--} & \text{--} & \text{--} & -Y_{T1} & \text{--} & \text{--} & \text{--} & \text{--} \\
-Y_{S1} & F_1 B_1 & \text{--} & F_2 & \text{--} & F_3 & \text{--} & F_4 & Y_{S5} & -Y_{S5} & \text{--} & \text{--} & \text{--} & \text{--} & \text{--} \\
\text{--} & \text{--} & C_2 & -Y_{S2} & \text{--} & \text{--} & \text{--} & \text{--} & \text{--} & \text{--} & -Y_{P2} & -Y_{T2} & \text{--} & \text{--} & \text{--} \\
\text{--} & F_2 & -Y_{S2} & F_2 B_2 & \text{--} & F_3 & \text{--} & F_4 & Y_{S5} & -Y_{S5} & \text{--} & \text{--} & \text{--} & \text{--} & \text{--} \\
\text{--} & \text{--} & \text{--} & \text{--} & C_3 & -Y_{S3} & \text{--} & \text{--} & \text{--} & \text{--} & \text{--} & -Y_{P3} & -Y_{T3} & \text{--} & \text{--} \\
\text{--} & F_3 & \text{--} & F_3 & -Y_{S3} & F_3 B_3 & \text{--} & F_4 & Y_{S5} & -Y_{S5} & \text{--} & \text{--} & \text{--} & \text{--} & \text{--} \\
\text{--} & \text{--} & \text{--} & \text{--} & \text{--} & \text{--} & C_4 & -Y_{S4} & \text{--} & \text{--} & \text{--} & \text{--} & -Y_{P4} & -Y_{T4} & \text{--} \\
\text{--} & F_4 & \text{--} & F_4 & \text{--} & F_4 & -Y_{S4} & F_4 B_4 & Y_{S5} & -Y_{S5} & \text{--} & \text{--} & \text{--} & \text{--} & \text{--} \\
\text{--} & Y_{S5} & \text{--} & Y_{S5} & \text{--} & Y_{S5} & \text{--} & Y_{S5} & C_5 & \text{--} & \text{--} & \text{--} & \text{--} & -Y_{P5} & -Y_{T5} \\
\text{--} & -Y_{S5} & \text{--} & -Y_{S5} & \text{--} & -Y_{S5} & \text{--} & -Y_{S5} & \text{--} & -Y_{S5} & \text{--} & \text{--} & \text{--} & \text{--} & \text{--} \\
-Y_{T1} & \text{--} & -Y_{P2} & \text{--} & \text{--} & \text{--} & \text{--} & \text{--} & \text{--} & \text{--} & D_1 & \text{--} & \text{--} & \text{--} & \text{--} \\
\text{--} & \text{--} & -Y_{T2} & \text{--} & -Y_{P3} & \text{--} & \text{--} & \text{--} & \text{--} & \text{--} & \text{--} & D_2 & \text{--} & \text{--} & \text{--} \\
\text{--} & \text{--} & \text{--} & \text{--} & -Y_{T3} & \text{--} & -Y_{P4} & \text{--} & \text{--} & \text{--} & \text{--} & \text{--} & D_3 & \text{--} & \text{--} \\
\text{--} & \text{--} & \text{--} & \text{--} & \text{--} & \text{--} & -Y_{T4} & \text{--} & -Y_{P5} & \text{--} & \text{--} & \text{--} & \text{--} & D_4 & \text{--} \\
\text{--} & \text{--} & \text{--} & \text{--} & \text{--} & \text{--} & \text{--} & \text{--} & -Y_{T5} & \text{--} & \text{--} & \text{--} & \text{--} & \text{--} & D_5
\end{vmatrix}
\times
\begin{vmatrix} E_6 \\ E_1 \\ E_7 \\ E_2 \\ E_8 \\ E_3 \\ E_9 \\ E_4 \\ E_{10} \\ E_5 \\ E_{11} \\ E_{12} \\ E_{13} \\ E_{14} \\ E_{15} \end{vmatrix}
=
\begin{vmatrix} I_{IN} \\ \text{---} \\ \text{---} \\ \text{---} \\ \text{---} \\ \text{---} \\ \text{---} \\ \text{---} \\ \text{---} \\ \text{---} \\ \text{---} \\ \text{---} \\ \text{---} \\ \text{---} \\ \text{---} \end{vmatrix}
$$

(7) Set of equations for short-circuit conditions,
where B_K, C_K, D_K as in (4) and $F_K = Y_{C(N)} + Y_{E(K)} + Y_{E(K\ 1)} + \ldots + Y_{E(N-1)}$

In order to verify the method in the case of higher order cascades, small reduced-scale three winding transformers have been designed and built to our specifications. These 400 VA models have primary and tertiary voltages of 12 V and a secondary of 600 V. Furthermore, two 72 µH inductors and a 90 nF capacitor were added to each individual unit to increase the value of the leakage inductances of the primary and tertiary windings and to simulate the stray capacitance of the HV winding. The impedances of the individual units were measured by standard short-circuit tests [1,5]. However, exceptional cares were taken to obtain extremely accurate values. Appendix 1 gives the average values of the various impedances.

Table 1 shows the experimental and theoritical voltages and currents in the case of a five transformer cascade with a load of 48 kΩ and 118 nF. The analytical values are extremely close to the measured ones. However, in order to obtain such results it was necessary to include in the computer program an iterative calculation of the magnetization currents. Magnetization current of each individual transformer was approximated by a fifth order polynomial. Two to five iterations are needed to reach a satisfactory value corresponding to the operating point of the transformers.

E_{in}		6 V	10 V	12 V
E_1	C	309 V	504 V	596 V
	M	309 V	505 V	598 V
E_2	C	335 V	524 V	602 V
	M	333 V	523 V	603 V
E_3	C	357 V	544 V	612 V
	M	355 V	540 V	610 V
E_4	C	374 V	558 V	622 V
	M	370 V	553 V	617 V
E_5	C	382 V	567 V	630 V
	M	378 V	561 V	623 V
I_{P1}	C	21,5 A	26,7 A	26,8 A
	M	21,0 A	25,8 A	25,4 A
I_{P2}	C	16,7 A	20,2 A	20,3 A
	M	16,4 A	19,6 A	19,3 A
I_{P3}	C	12,4 A	14,8 A	15,2 A
	M	12,1 A	14,3 A	14,0 A
I_{P4}	C	8,3 A	9,9 A	10,1 A
	M	8,1 A	9,5 A	9,3 A
I_{P5}	C	4,2 A	5,1 A	5,2 A
	M	4,1 A	5,1 A	4,8 A

Table 1: Voltage and current distribution for a five transformer cascade . Load: 48 kΩ // 118.4 nF. (C: Computed ; M: Measured)

The saturation of the cores is responsible for the fact that the voltage distribution in the cascade changes with the input voltage. In the example shown, for primary voltages of 6 and 12 volts, the voltages across the HV winding of the upper unit are respectively 1,24 and 1,06 time the voltage across the HV winding of the lower unit. At high input voltage, the magnetic circuits are more saturated and the increase of the reactive power level cancels out the effect of the stray and the load capacitances. The exact knowledge of the magnetization curve is, therefore, essential to a correct evaluation of the voltage distribution.

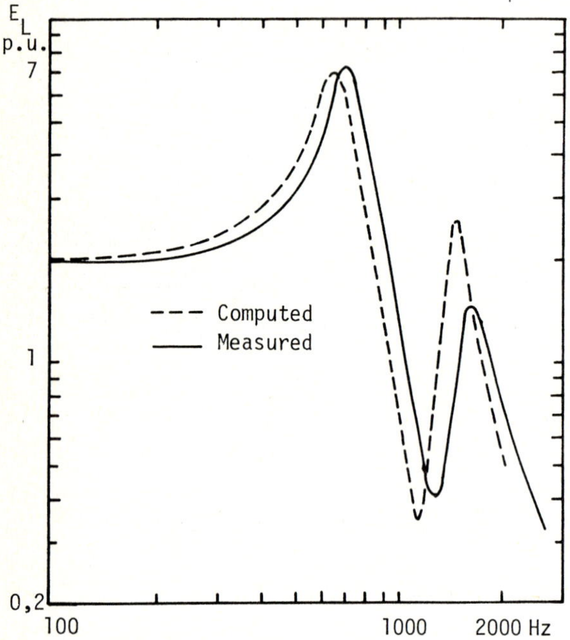

Fig. 6: Frequency response of a two transformer cascade, unloaded.

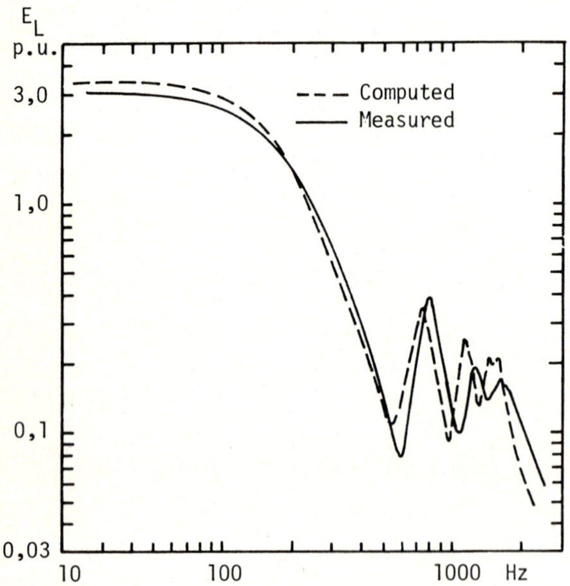

Fig. 7: Frequency response of a three transformer cascade for a load of 8 kΩ // 103 nF.

Figures 6 and 7 are typical frequency responses of cascade connected transformers. The first one was obtained for a second order unloaded cascade and the second one for a cascade of three transformers with a load consisting of a 8 kΩ resistor in parallel with a 103 nF capacitor. In the first case, resonance occurs at a frequency near 600 Hz with a peak amplitude of 7 p.u. In consequence the residual 9th and 11th harmonics, always present in the 60 Hz excitation voltage, are amplified and distincly appear on the output voltage causing unexpected measurement errors. In the second case, no amplifying effect occurs . Even more, due to the limited bandwith, the amplitude of the harmonics are filtered out. In general, the frequency response depends largely on the load, the unloaded case being the worst.

Finally, a last verification was obtained by using our model to predict the voltage distribution of the 550 kV transformers of the Hydro-Quebec Research Institute (IREQ) laboratories.[3,4,2]. The parameters of these transformers are summarized in Appendix 2. Data published by IREQ [3] were used to compare the results obtained from our computations with the exact voltage distribution. The results for a 4th order cascade with 300 kVA compensating reactors are shown in Figures 8 and 9. The first one presents the overall voltage ratio of the cascade while the second shows the voltage appearing across the upper unit for capacitive loads varying between 0 and 2 nF. These results and many more, for different other configurations, confirmed the validity of our analytical model.

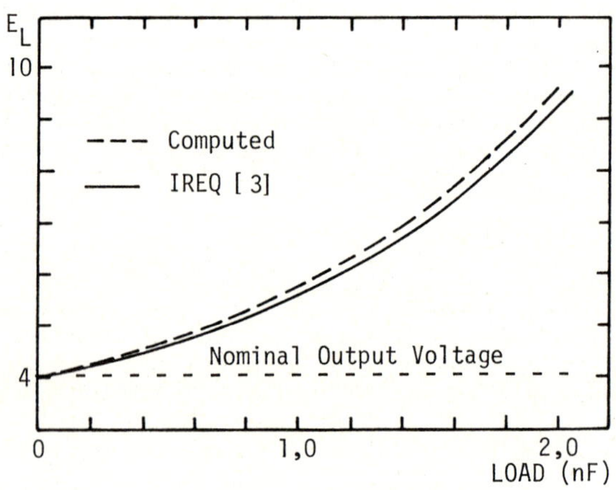

Fig. 8: Output voltage as function of the load capacitance for a cascade of four transformers with 300 kVA compensating reactors, in p.u.

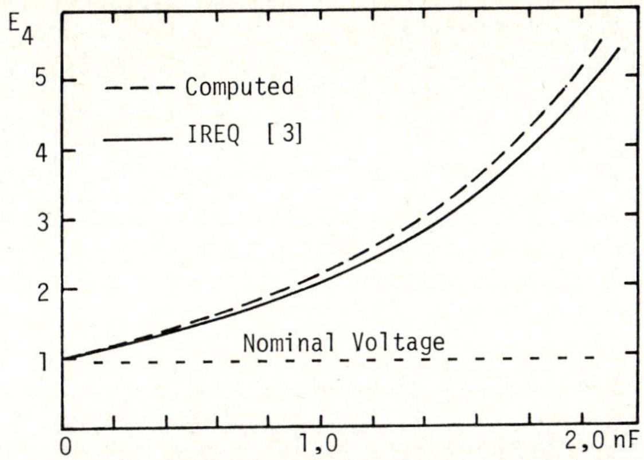

Fig. 9: Voltage in p.u. across the upper unit for a four transformer cascade with 300 kVA compensating reactors.

CONCLUSION

The analytical model proposed in this paper enables an accurate prediction of the voltage distribution, the frequency response and the short-circuit current of any order cascade connected HV test transformers. The voltage distribution and the frequency response largely depends on the cascade order, on the impedance of the object under test and, unexpectedly, on the level of saturation of the cores. At high input voltages, the increase of the magnetization currents caused by the saturation of the magnetic circuits partially cancels out the effects of the stray capacitances and drastically modify the voltage distribution. In further developments, the method will be extended to include the transients produced by a flashover of the test object.

ACKNOWLEDGEMENTS

This work was supported by the National Sciences and Engineering Research Council of Canada and the Ministry of Education of the Province of Québec. The authors wish to express their appreciation to the following students who worked on the project: S. Caron, C. Lafond and Y. Nazon.

NOMENCLATURE

E_{IN}: Input voltage
E_L : Load voltage
I_{IN}: Input current
I_L : Load current
I_{sc}: Short-circuit current
N : Cascade order
Z_C (Y_E): Compensating reactor impedance (admittance)
Z_E (Y_E): Electrode capacitance
Z_H (Y_H): Magnetization impedance
Z_P (Y_P): Primary impedance
Z_S (Y_S): Secondary impedance (HV)
Z_T (Y_T): Tertiary impedance (Coupling)
Z_W (Y_Z): HV winding capacitance

REFERENCES

[1] Olivier G. , Bouchard R.-P., Gervais Y. and Mukhedkar D.; Frequency Response of HV Test Transformers and the Associated Measurement Problems", IEEE Trans. App. Syst., Vol. PAS-99, No. 1, Jan. 1980, pp. 141-146.

[2] Hertlein H.-P. and Landers E.-U. ; "Harmonic Resonance Phenomena in Multi-Stage High Voltage Testing Transformers", IEEE 1972 Winter Meeting, New York, Jan 30 - Feb. 4 1972, Paper No. C 72 126-6.

[3] Train D. and Vohl P.-E.; "Determination of Characteristics of Cascade Connected Transformers," IEEE Trans. App. Syst., Vol. PAS-95, No. 6, Nov.-Dec. 1975, pp. 1911-1918.

[4] Train D. and Trinh G.; "Prevention of Recovery Overvoltages on High-Voltage Transformers", IEEE Trans. App. Syst., Vol. PAS-92, No. 5, Sept.-Oct. 1973, pp. 1631-1641.

[5] Kind D. ; "An Introduction to High Voltage Experimental Technique", Vieweg, Germany, 1978.

APPENDIX 1

Parameters of the model transformers (referred to the primary side).

C_W : 90,1 nF
Z_P : 9,1 mΩ + 75 μH
Z_S : 19,8 mΩ + 8.3 μH
Z_T : 8,3 mΩ + 83 μH
Turn-ratios: secondary/primary: 50
tertiary/primary: 1,0
Nominal primary voltage: 12 V
Nominal power: primary: 400 VA
secondary: 100 VA
tertiary: 300 VA

APPENDIX 2

Parameters of the test transformers of IREQ [3] (referred to the HV side).

Nominal output voltage: 550 kV
C_W: 1720 pF
Z_P: 481 Ω + 42,5 H
Z_S: 1224 Ω + 61,9 H
Z_T: 461 Ω + -12,3 H

Session 5 B

SIMULATION AND MODELLING OF CONVERTERS (II)

Chairman: R. LE DOEUF

ELECTRICAL MACHINES AND CONVERTERS - Modelling and Simulation
H. Buyse and J. Robert (editors)
Elsevier Science Publishers B.V. (North-Holland)/© IMACS, 1984

SIMULATION OF SWITCH-FUNCTIONING SEMICONDUCTOR ELECTRIC CIRCUITS
STUDY OF WORKING FAULTS

F. Bordry, H. Foch, J.C. Alarcon

Laboratoire d'Electrotechnique et d'Electronique Industrielle
ERA au CNRS n° 536
ENSEEIHT
2 rue Camichel
31071 TOULOUSE Cédex (FRANCE)

This paper presents a method for simulating switch-functioning semiconductor electric circuits. The method only requires knowledge of the circuit structure, its parameters and the control order of the semiconductors being controlled.
The automatic linking of the functioning sequences is obtained through the functional properties of the semiconductor models.
The originality of the method lies in the definition of a generalized switch. This notion rests on the representation of a semiconductor by a logical sequential system which leads to the definition of a state of the semiconductor, and to the inclusion in this state of a characteristic variable of nature of the semiconductor (e.g. diode, thyristor, transistor...). Using the notion, working faults and especially those leading to a degradation of the semiconductor (i.e. state changes occasioned by passing the absolute limits of the semiconductor) can be analyzed.
Details of the modelling by Petri nets of this generalized switch, as well as the numerical resolution allowing an accurate study of the propagation of working faults are provided.
An application of the method simulating an electric energy static convertor which demonstrates its interest in the course of the appearance of a fault leading to destructive failure is presented.

I. INTRODUCTION

In the last few years the study and simulation of linear electric circuits have become highly developed. The most frequently employed representation is that by state variable system as it lends itself perfectly to numerical calculation (i.e. matrix calculation) (1), (2), (3).

Switch-functioning semiconductors are non-linear elements. The main difficulty lies in determining the existence and the elementary linking sequences, defined by the conduction or non-conduction of these components (non-linear systems with piecewise linear structures). The simulation methods of static convertors can be divided into two groups :

(a) the method based on the splitting up of the functioning role into a succession of known sequences ("sequence simulation") (4). Although these methods include a number of advantages (especially in calculation time) during the study of classical set-ups, they are difficult to use in the design and analysis of new systems.
(b) the method which requires only knowledge of circuit structure, its parameters and the control orders of the controlled switches ("global simulation") (5), (6), (7), (8). In this method the semiconductors are individually modelled, and linking of the functioning sequences are carried out by the functional properties of the models.

In this article a global simulation method which pays particular attention to the modelling of switches, is presented. Through the modelling account can be taken of their deterioration. An example demonstrates the possible applications of the simulation programme.

2. SIMULATION METHOD

2.1. Modelling of semiconductors

2.1.1. Introduction

The choice of the semiconductor electric model is fundamental to the design of a simulation method. The present authors consider switch-functioning semiconductors (the notion of the switch in opposition with a linear component). This hypothesis involves the representation of a semiconductor by a mono-output logical system (switch On or Off).

The binary representation is conducive to a number of models (5), (6), (7), (8).

The present authors chose semiconductors modelled in the form of a binary resistance varying on a large scale according whether they are Off (high resistance : R off) or conductors (low resistance : R on). This modelling is interesting because of its very physical aspect which lends itself especially to considerable simplicity of implementation.

A semiconductor has to be considered as a sequential system. The state of the semiconductor obviously depends on the sensitivity of the chosen model (inclusion of reverse voltage bias time, recovered charges, etc.). The switch output (On or Off) only identifies with the state of the switch for simplified models ; it is generally a state vector component.

It should be noted that in the modelling of a semiconductor in binary resistance form, the input vector is constituted of the voltage at the bounds of the power electrodes (Vak) (this vector also represents current) and control signal (G).

With the aim of having a single model for all the semiconductors the present study defined the notion of the generalized switch. It rests on the introduction of a characteristic variable of the switch in its state vector (9).

Using this definition a study can be made of functioning faults especially those leading to the deterioration of the semiconductors (state changes due to exceeding the absolute limits of the semiconductors).

The present authors chose the Petri net representation, which offers an excellent compromise between generality and simplicity of implementation, as a functional representation tool for the generalized switches.

2.1.2. Representation of switches by Petri nets

Defining the Petri net representing a generalized switch depends on the nature of the semiconductors under consideration (i.e. diode, thyristor, transistor, etc) and the sensitivity of the chosen model. The places characterize the state of the switch and the transitions have the associated receptivities which are functions of the switching conditions. A receptivity is a logical variable define by a function of the characteristic variables are either real variables (Vak, Iak, tinv...) or logical variables (G, ...).

A safe Petri net (one token maximum per place) and a pure Petri net (a place cannot be both an input and an output place of transition) are considered to represent a switch.

By defining logical states 1 and 0 for an On and an Off switch (switch output) an output matrix So is associated to the Petri net. Output S of a switch will be determined from marking M at that instant and of So : S = So.M. This logical signal S determines the binary resistance value (Rsw) modelling the semiconductor (Rsw = S.Ron + S.Roff).

A transition is called enabled by a given marking if each of the preceding places includes a token. An enabled transition is called firable if its receptivity is true (or validated). Receptivity validations by characteristic real variables of a switch arise from inequations of the form : (1) $L > L\,max$; (2) $L \geqslant L\,max$; (3) $L < L\,max$ or (4) $L \leqslant L\,max$ (L max being a characteristic limit of the switch : 0, Vd max, Ir max, tq ...).

It will be stated that a receptivity of this type is "just-verified' (and by extension that the transition is "just-verified") in the following case : (1) $L \in]Lmax, Lmax{+}s]$; (2) $L \in [Lmax, Lmax{+}s]$; (3) $L \in [Lmax{-}s, Lmax[$ and (4) $L \in [Lmax{-}s, Lmax]$ (s being a prefixed threshold attached to the transition receptivity).

The functional representation of a set of switches will be made by the definition of a couloured Petri net ; the number of colours in the net corresponds to the number of switches. Figure 1 Shows an example of a Petri net graph.

Places	Transitions
P1. Diode ON	T1. $Vak > 0$
P2. Diode OFF	T2. $Iak \leqslant 0$
P3. Thyristor ON	T3. $(Vak > 0$ and $G = 1)$ or $(Tinv < Tq$ and $Vak > 0)$ or $(Vak \geqslant Vak\,max)$
P4. Thyristor OFF	T4. $Iak \leqslant 0$
P5. Transistor ON	T5. G = 1
P6. Transistor OFF	T6. G = 0
P7. Dual thyristor ON	T7. $Vak \leqslant 0$
P8. Dual thyristor OFF	T8. $Iak > 0$ and G = 1
P9. Thyristor on overloading of Id max	T9. Iak Iak max
P10. Thyristor out of order by overloading of I^2t	T10. $W \geqslant (I^2.t)max$
P11. Thyristor out of order by excessive reverse voltage	T11. $Vak \geqslant Vak$ reverse max

Sub-grap 1 (places P1, P2, ... P6 and transitions T1, ... T6) are used to take into account the three classical semiconductors.

Sub-graph 2 (P7, P8 and T7, T8) represents a dual thyristor (11).

Sub-graph 3 (P3, P4, P9, P10, P11 , T3, T4, T9, T10, T11, T12) shows an example of a thyristor representation which is used to study faults (destruction by overloading of I^2t and by exceeding the reverse voltage maximum).

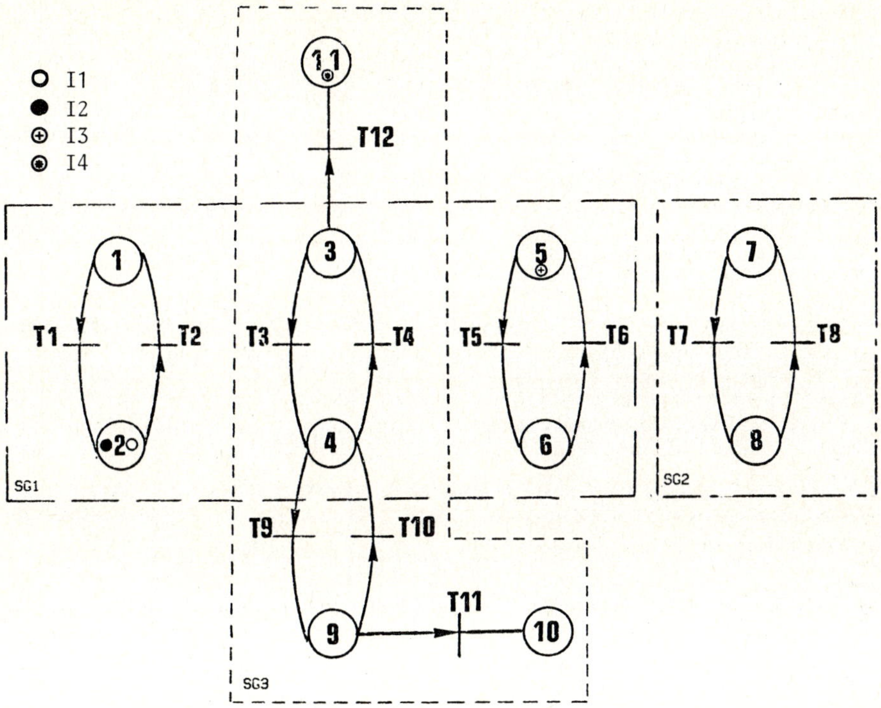

Fig. 1

The net thus defined (graph and marking) therefore represents a system comprising four switches which at the initial instant are :

I1 : diode On
I2 : diode On
I3 : transistor Off
I4 : thyristor out of order by excessive reverse voltage, that is to say an open circuit (R off)

This representation, which is linked to the notion of the generalized switch, gives a high evolutionary capacity to the simulation software ; a user could define his own Petri net (e.g. introduction of new switches : dual thyristor, self-turn on transistor ; inclusion of composite semiconductors as switches : anti-parallel thyristor diode, series transistors with a diode ; study of semiconductor deterioration).

2.2. Simulation method

2.2.1. Expressing the device in equations

The automatic expression in equations of a circuit is a determining aspect in the use (and consequently the diffusion) of a simulation programme. Its aim is to define the equations of a closed-circuit in state form (1) :

$$dX/dt = A.X + B.U + E.\ dU/dt$$
$$Y = C.X + D.U$$

from a simple topological description (incidence matrix) and parameter values.

State vector X is formed of the voltages at the bounds of the capacitors belonging to the tree and of the currents in the inductances not belonging to it. Input vector U contains the voltage and current sources. Output vector U represents the voltage or the current of any component.

State matrixes A, B, C, D and E are calculated from the topology and the value of the device elements (including representative resistances of the switches).

The choice of a resistance-form model for the semiconductors allows a fixed topology of the circuit, i.e. a single state vector (thus continuity of the state vector during configuration changes ; only the state matrixes need to be recalculated).

2.2.2. Resolution of state equations

For a given configuration, and knowing the state equations of the system, the following solution in the form of a recurrent equation is obtained :

$$X(t+T) = \exp(A.T).X(t)+A^{-1}.(\exp(A.T)-I).BU(t) + A^{-1}.(\exp(A.T)-I).E.dU/dt(t)$$

Calculation of matrix exp(A.T) has to be accurate and rapid (5).

2.2.3. Determination of instants of state changes in the semiconductors

A change of state in a semiconductor takes place if a enabled transition is fired, i.e. if the event corresponding to this transition occurs (firable transition). It is important to know the instant when the transition becomes firable. Two types of case are possible :

(a) the instant is known (control signal) ; a repositioning is then made to the instant using a reduced step ;

(b) the instant is not known (a passage by zero of an anode-cathode voltage, the anode-cathode voltage becoming higher than the avalanche voltage, etc.) ; as soon as an event of this type is detected, a search is made for the instant where the transition is just-verified, proceeding by refinement of the step. The threshold attached to the transition is chosen as a function of the value of the On and Off resistances.

2.2.4. Configuration changes - Evolution of the marking

The voltages at the bounds of the semiconductors are calculated from the matrix relationship :

$$Y = CX + DU$$

At the instant of the configuration change state vector (X) and sources (U) cannot undergo discontinuities (instantaneous commutations). Conversely, the coefficients of matrixes C and D, as functions of the resistances of the resistances of the semiconductors, can vary. (Y) can thus undergo discontinuities.

At that instant it is therefore necessary to cause the marking to evolve (compatibility tests) to the point where no transition is firable (state marking).

The example of a transistor chopper shows that during a compatibility test originating in the turn-off of the transistor, the voltage at the bounds of the diode becomes positive. It is therefore necessary to carry out a new compatility test involving the turning on of the diode. The marking position obtained is then stable.

An overvoltage occurs at the bounds of the transistor during the first test. It could lead, during a fault study, to an erroneous final configuration where the diode is turned on and where the transistor cracks up because of excessive forward voltage.

This example shows that it is necessary to determine the occurence sequence of the transition firings during a commutation.

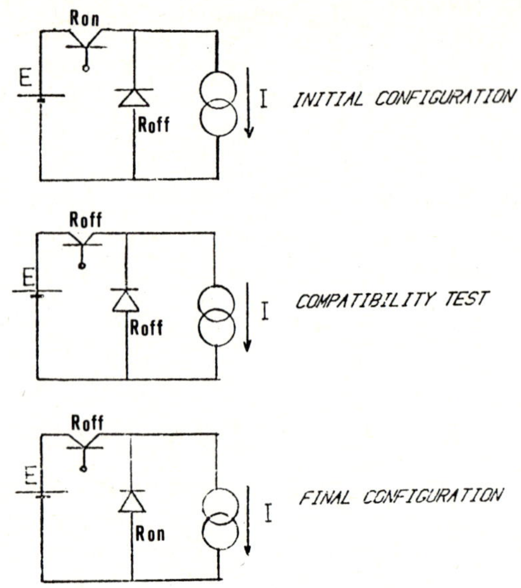

Fig. 2 Chopper

Notion of a gradual commutation

(a) definitions

When in a compatibility test several transitions are firable while their receptivities are not all "just-verified", they are considered to be in "conflict".

The transitions whose firing led to a compatibility test involving a conflict are called "conflict transitions".

(b) Methodology

A gradual commutation consists, when a conflict occurs to look for the resistance values, of switches associated to the conflict transitions which are going to suppress the conflict (all the firable transitions having at that point their receptivities just verified).

This search is carried out by sampling the resistance values, between R On and R Off, of the initial value towards the final value.

The firing of validated transitions are carried out. In the case where a conflict appears in the subsequent compability test, they are in their turn considered as conflict transitions but of level 1.

A new commutation is then effected from resistance values which validated the receptivities of level 1 conflict transitions.

This procedure is repeated up to the arrival of

a compability test when the notion of conflict will have disappeared.

In the example to the chopper, the turn off of the transistor involves a conflict between the transition turning on the diode and the one which causes the transistor to crack up. A gradual commutation is carried out on the transistor, from R on to R off. A follow-up of the evolution of the commutation is given in figure 3.

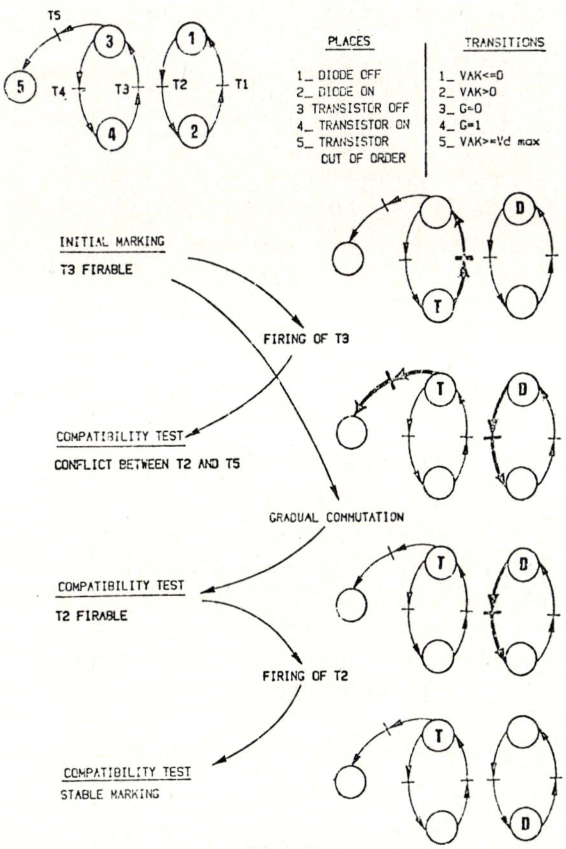

Fig. 3

3. SIMULATION PROGRAMM

It was decided to elaborate the following :

(a) a structured programm so as to include a high adaptation flexibility (i.e. maintenance and diffusion). Figures 4, 4a, 4b and 4c describe, in form of Nassi's and Schneidermann's graphs (12), the general architecture of the programm.
(b) a conversational programm in the form of hierarchical menus. Figure 5 illustrates this conversational aspect.

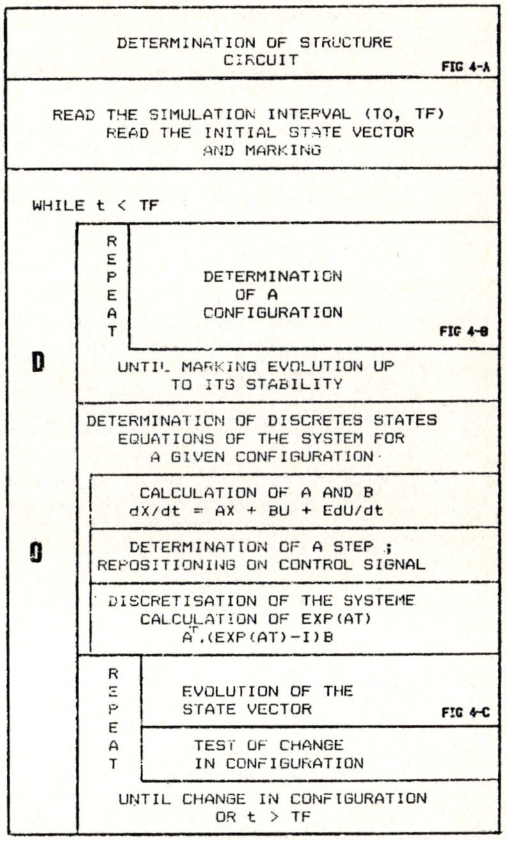

Fig. 4

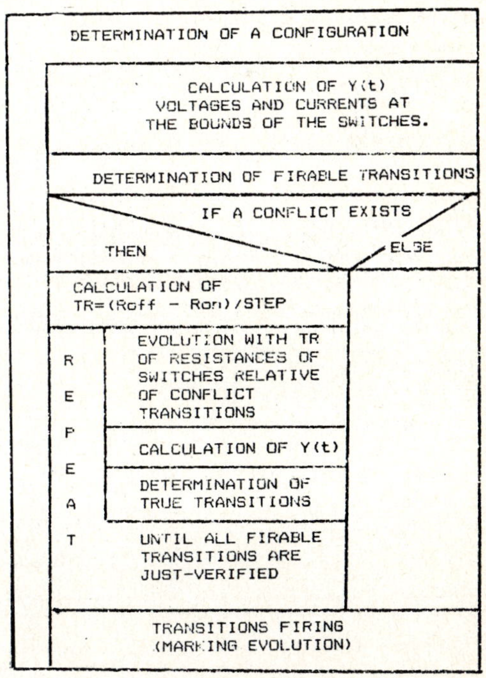

Fig. 4b

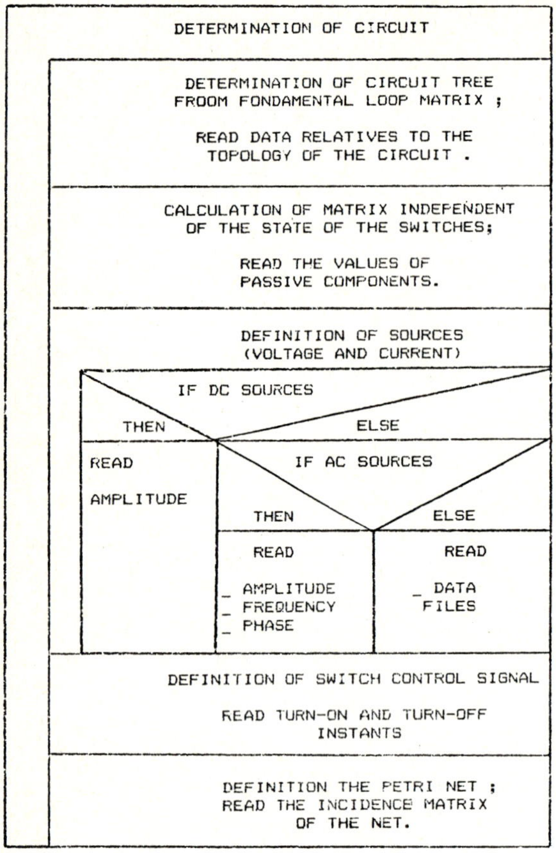

Fig. 4 a

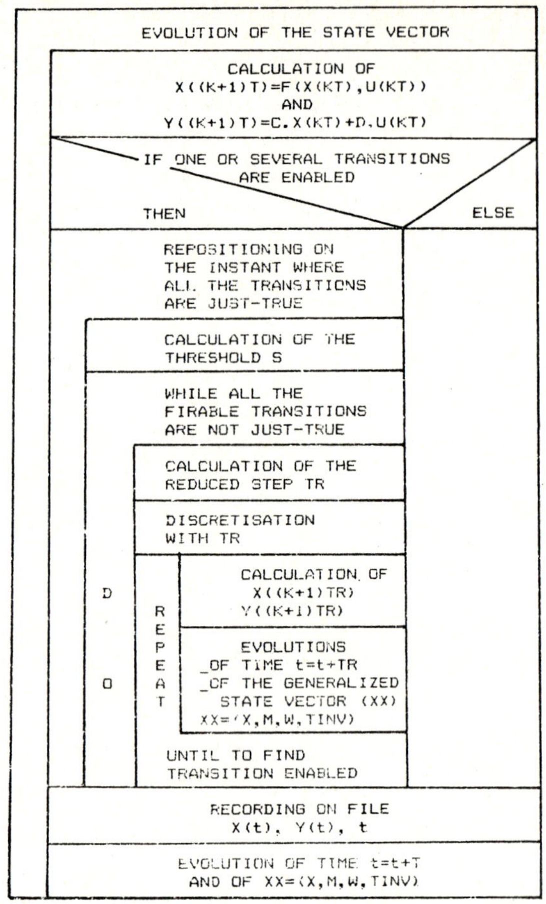

Fig. 4 c

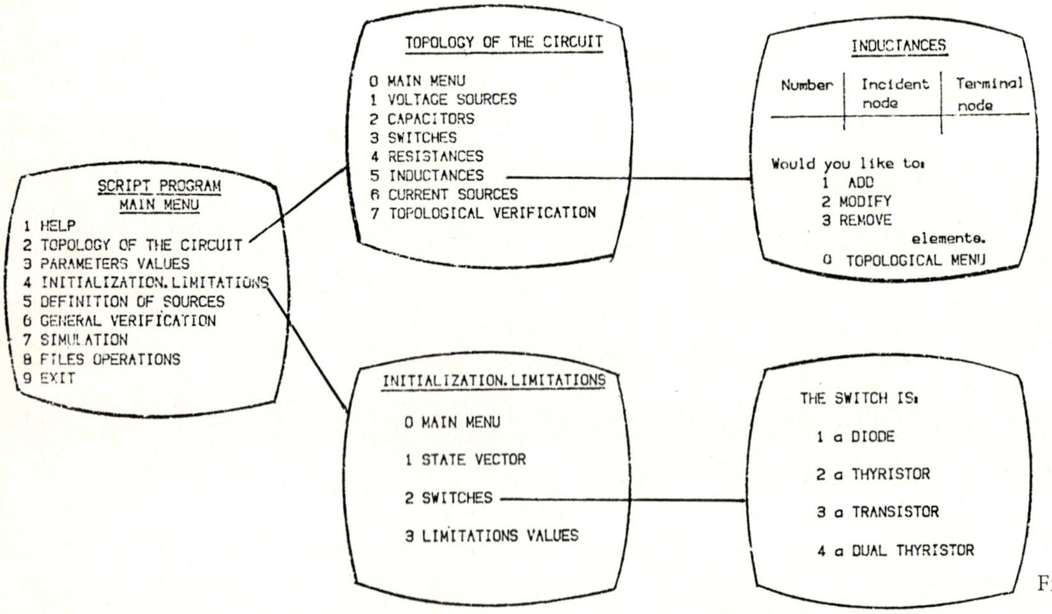

Fig. 5

4. APPLICATION EXAMPLE

A natural self-adaptative thyristor chopper is presented as an application (Fig. 6). Fig. 7 illustrates a transitory voltage regime showing a loss of control after the third functioning cycle (the reverse voltage bias time at the bounds of Tp becomes lower than its recovery time) and the destruction of the main thyristor by overloading of (I^2t). The fault is propagated and leads to the destruction of the inversion thyristor (Ta 2).

The Petri net graph used for this simulation is described in Fig. 1.

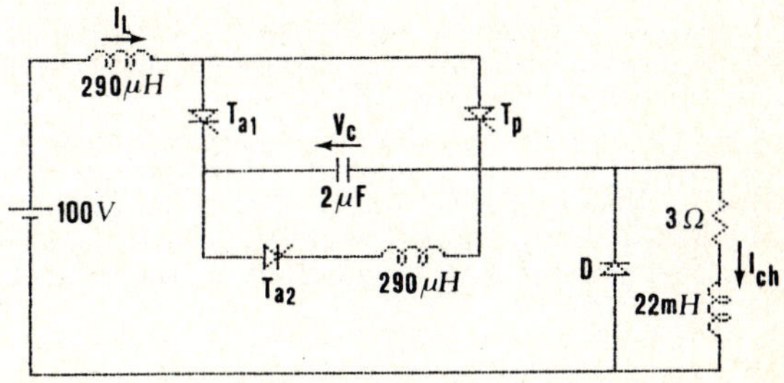

Figure 6

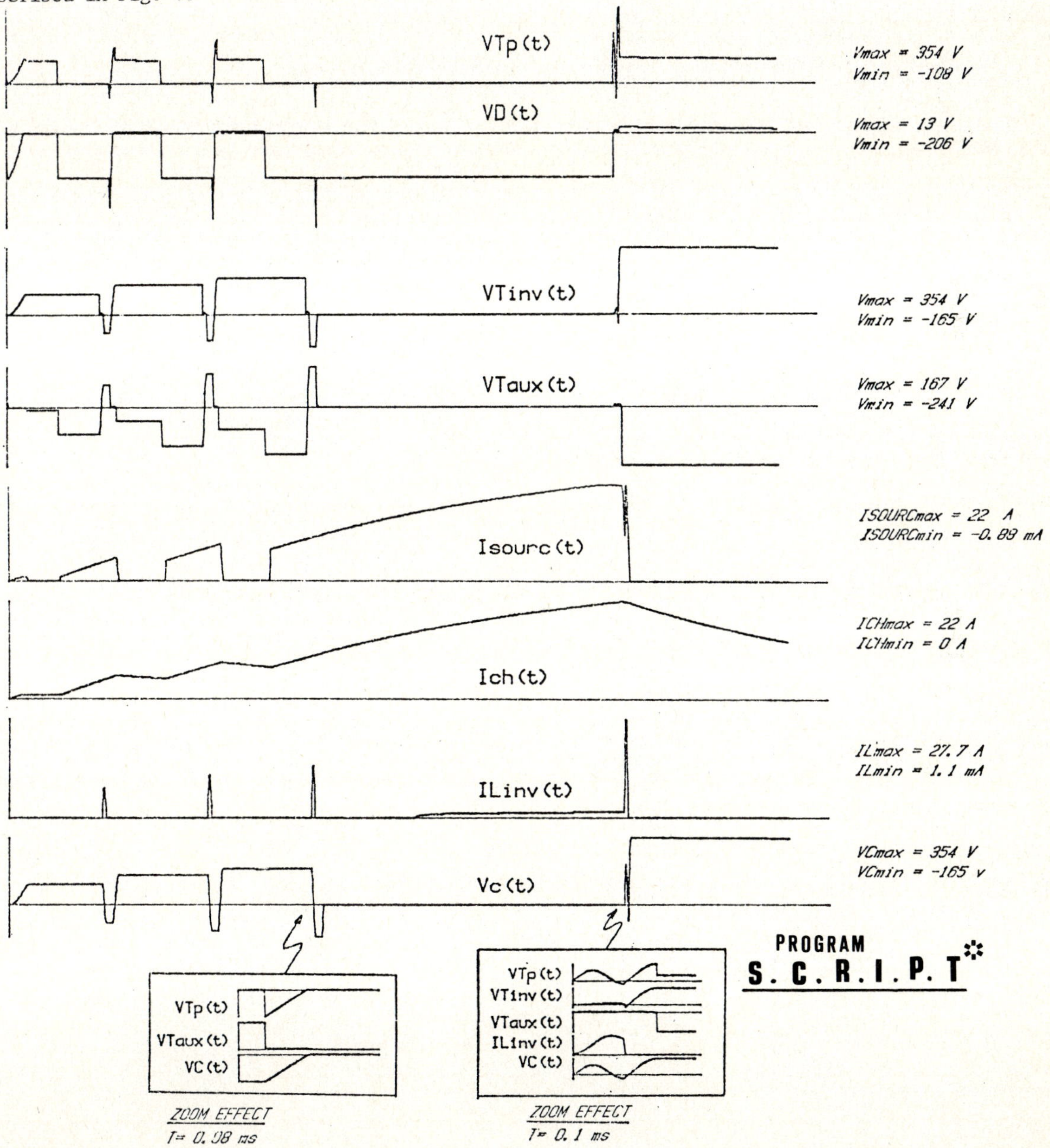

Figure 7

5. CONCLUSION

An example has demonstrated the aid which a global program simulation can bring to the study of a semiconductor electric circuit. In addition, the use of a "generalized switch", represented by Petri nets, within the framework of global simulations allows the study of all types of transistories and especially those conducive to a functioning fault (cataleptic and deterioration faults).

This simulation method therefore seems to the present authors to be a very powerful means of investigation of the functioning of even very perturbed static convertors. It can thus be used in the study of the functioning security of such systems, for an impoved conception of their protection, and for a more rational dimensioning of their components.

6. REFERENCES

(1) BALABANIAN N., BICKART T.
Electrical network theory, Wiley, New York, 1979

(2) FELDMANN M.
Théorie des réseaux et systèmes linéaires, Eyrolles, Paris, 1981

(3) ADBY P.R.
Applied circuit theory, Wiley, 1980

(4) DE FORNEL B., CRUZ G., HAPIOT J.C.
Programme de simulation des ensembles convertisseurs statiques machines alternatives Journées SEE, 5 novembre 1981, Paris

⁂ S.C.R.I.P.T.

Simulation des Convertisseurs statiques

Représentation des Interrupteurs par

réseau de PéTri.

(5) SCHONEK J.
Simulation numérique des convertisseurs statiques. Application à la conception et l'optimisation des convertisseurs, Thèse de Docteur Ingénieur, INP Toulouse, 1977

(6) REVANKAR G.N.
Topological approach to thyristor circuit analysis, Proceedings IEE, Vol. 120, n° 11, nov. 1973

(7) RAJAGOPALAN V., SANKARA K.
User's manual on topological analysis of power electronic converter systems, ATOSEC, Université de Québec à Trois Rivières, Canada

(8) LAKATOS L.
A new method for simulating power semiconductor circuits, IEEE Trans. Ind. Electron. Contr. Instrum. Vol. IECI-26, n° 1, Fev. 1979

(9) BORDRY F., FOCH H., METZ M.
Simulation des circuits électriques à structure linéaire par morceaux. Représentation par réseau de Pétri des semiconducteurs fonctionnant en commutation. Symposium IMACS Simulation in Engineering Sciences, Nantes, mai 1983

(10) DIAZ M.
Modeling and analysis of communication and cooperation protocolo using Petri net based models. Computer Networks, 6 (1982) pp. 419-441

(11) FOCH H., MARTY P., ROUX J.
Use of duality rules in the conception of transitorised converters, Power conversion 80, Munich, 3-5 sept. 1980 4B3, 1-11

(12) NASSI I., SCHNEIDERMANN
Iteration graphs, ACM Sigplan, Vol. 8, n° 8, pp. 12-26, Aug. 1973

ELECTRICAL MACHINES AND CONVERTERS - Modelling and Simulation
H. Buyse and J. Robert (editors)
Elsevier Science Publishers B.V. (North-Holland)/ ©IMACS, 1984

A PETRI NET COMPUTATION STRUCTURE FOR TRANSISTOR INVERTER AND INDUCTION MOTOR

J.P. HAUTIER et G. MANESSE

Université des Sciences et Techniques de Lille
Laboratoire de Systèmes Electromécaniques
59655 VILLENEUVE D'ASCQ CEDEX - FRANCE

ABSTRACT

In industrial logic, Petri nets are easily implemented on numerical processors. In this paper the concepts and methodology of the Petri net theory are used in order to describe a particular static converter. Implementation rules are adapted to the formal language selected and computer constraints. A computation graph or simulation graph is defined by the initial Petri net and a program is directly obtained for simulating the systems behaviour

1. INTRODUCTION

For the purpose of writing a simulation program with respect to the functional description of the static inverter [1], it is necessary to define implementation rules adapted to the Petri net language [2] and numerical computer constraints.
We propose a method based upon initial fracturing into two nets severally accounting for behaviour of the control-system on the one hand and the several stages in the working of the convertor on the other hand.

Next a computation graph is built in accordance with the computer chosen.
Correct coding of results gives the simulation program.

2. FUNCTIONAL DESCRIPTION OF INVERTER

The twelve controlled stages of the inverter (figure 1) are figured out on table 1 which moreover shows those transistors which have become saturated.

G \ X	1	2	3	4
1	T11-T02-T03	T01-T12-T13	T11-T02	T01-T12
2	T12-T03-T01	T02-T13-T11	T12-T03	T02-T13
3	T13-T02-T02	T03-T11-T12	T13-T01	T03-T11

Table 1

A control sequence of n length is figured out by the Petri net on figure 2, which correspond to the control graph of inverter.

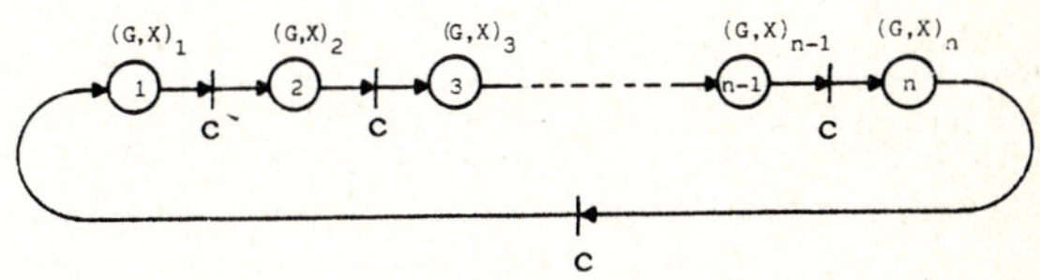

Figure 2

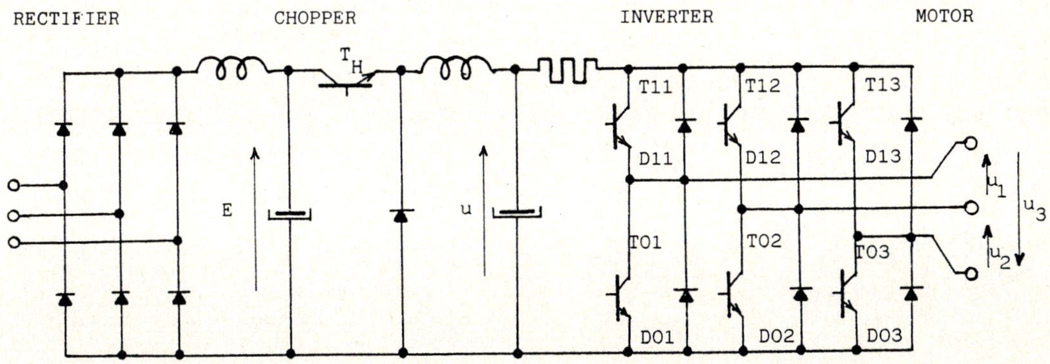

Figure 1

Token evolution is fonction of receptivity as shown by a signal clock.

The working graph of the inverter shows four macroplaces according to the number of transistors saturated at a given moment (figure 3). Definition of control brings out three modes corresponding to all possible firing sequences (figure 4).

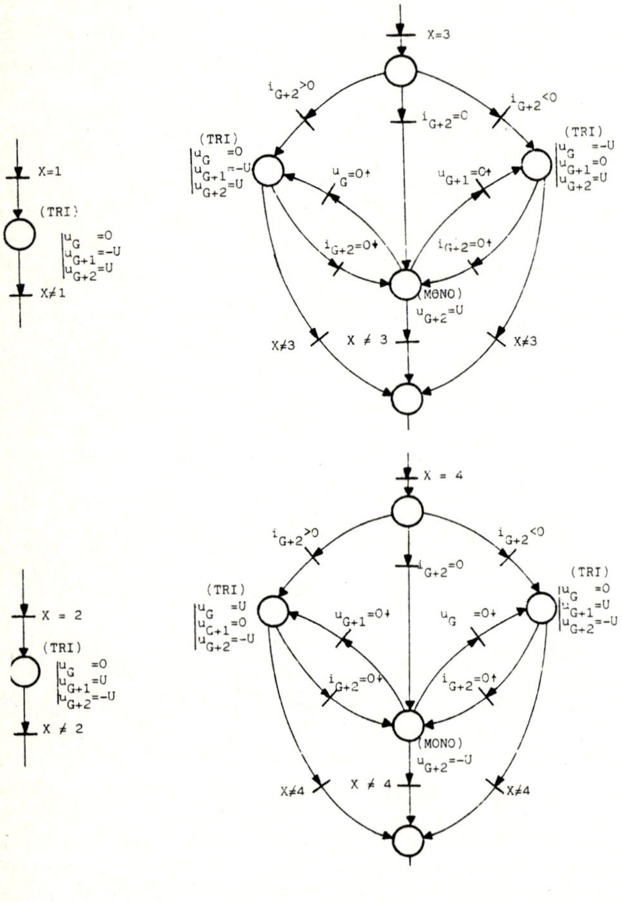

Figure 3

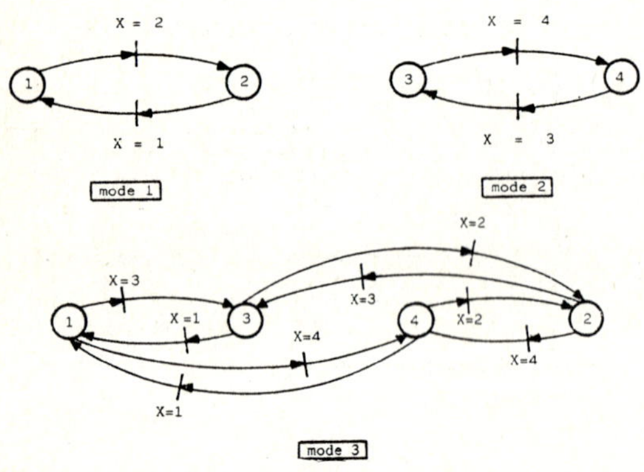

Figure 4

The lumped voltage vector across the machine is given by reference to the vector Y (figure 5).
Accordingly the permutation table of indexes allows us to deduce statoric data from parameter G (Table 2)
For each value of X, the referential vector is stated on table 3.

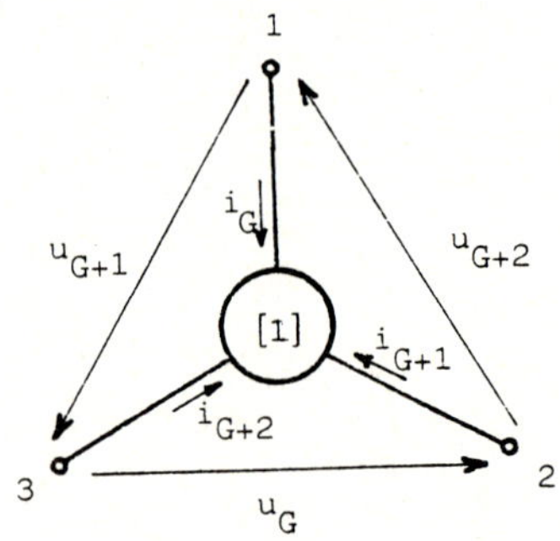

Figure 5

G	G+1	G+2
1	2	3
2	3	1
3	1	2

Table 2

X	1	2	3			4		
y_1	0	0	C	-U	*	U	0	*
y_2	-U	U	-U	0	*	0	U	*
y_3	U	-U	U	U	U	-U	-U	-U
i_{G+2}	–	–	>0	<0	=0	>0	<0	=0

Table 3

The three lumped voltages are always determined by load with X=1 and X=2. With X=3 and X=4 the conduction of the two transistors fixes only one of those voltages. The other two depend upon the state of the diodes in the inverter arm corresponding to the transistors locked.

The functional description postulates knowledge of the statoric equations ruling the working patterns, either three-phase or single-phase, of the usual couplings of windings.

3. THE PRINCIPLE OF NUMERICAL TREATMENT

The natural asynchronism of the process is to be transposed on the computer which is a synchronous system.
So, it is imperative at each step of the calculator

- to produce the control signal by working out the corresponding graph
- to test the condition of evolution of the working graph with lastest control and state pa-

rameters
- to process the pattern defined by lastest token position.

Those operations lead to the organizing of the data flow (figure 6).

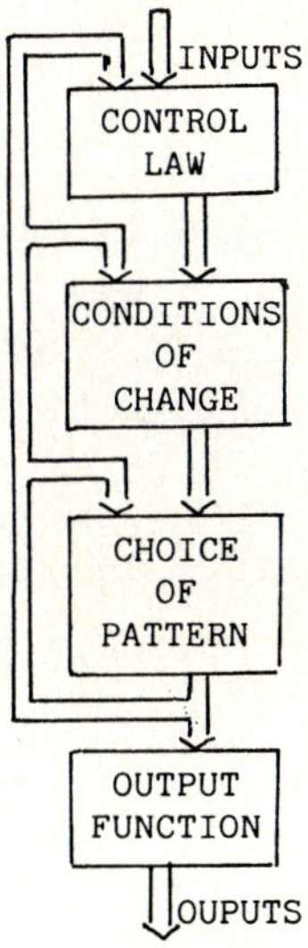

Figure 6

On the initial graph, two types of transition come out, the first being determined by transistor control (controlled commutations), the second bound up with the system's generalized state (natural commutations).
Wherever a number of transition fire simultaneously, conflict is ruled out by control right-of-way.

4. DESCRIPTION OF TREATMENT

A t variant incremented at each step defines time on a given ladder. Time delays (instants defined by clock C) are gained by means of comparative tests.
The t variant is brought back to zero wherever its numerical value reaches defining limit.
When the distribution of tokens on the graph brings about a change of pattern one must make sure there is no jump in numerical value of state variables. To that end the state variables have to be actualized in the new marking. An extra place is introduced to bring about the new pattern. If the change of pattern is determined by the control graph the new place will be positioned on the aforesaid graph for simplification sake, the working graph being preserved.
A binary flag (IH) assigned to the new places holds the time in the unaffected graph (figure 7).

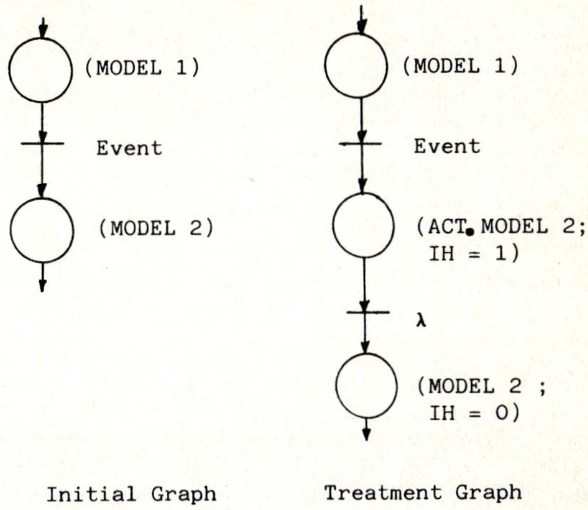

Figure 7

The initial graph thus modified develops into a treatment -or- simulation multigraph.
The foregoing deductions result in an overall treatment flowchart separating time control from both simulation graphes (figure 8).

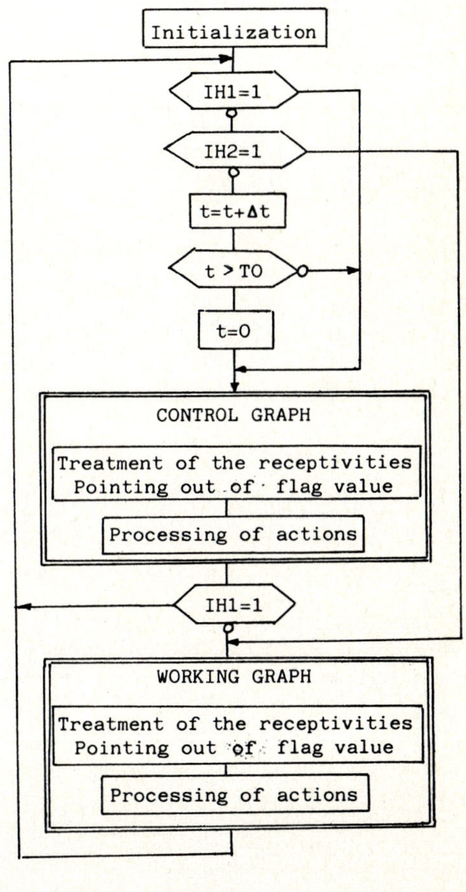

Figure 8

The test effected on T_0 determines the period of treatment. IH1 and IH2 stand for the actualization flags positioned respectively for treatment of control and working.
Graph implementation first requires processing the sum total of its receptivities in order to trigger off the tasks defined by the marking.
To each place two subprograms correspond
- the first covers the firing conditions of all its output transitions.
- the second states the aggregate of its tasks.

For each graph, a selector points to the treatment to be undergone in each list obtained according to graph marking.
For instance, in basic language, the instruction ON < Selector > GOSUB < SP, > determines access to each subprogram by means of a selector positioned by relation to the marked place.
The instruction IF < True receptivity > THEN < selector = numerical value > insures the coding transitions and assigns the new marking.
Figure 9 shows the main treatment of a graph made up of i places (P1,...,Pi) and I transition groups (T1,...Ti).

```
ON S GOSUB T1,T2,...,Ti

ON S GOSUB P1,P2,...,Pi
```

Figure 9

5. SIMULATION MULTIGRAPH

Regarding the inverter two mathematical models are defined (THREE PHASE) and (SINGLE PHASE) according to the number of phases connected.
The actualization tasks dependent upon natural commutations are included in the macroplaces X=3 and X=4. In the case of controlled commutations, the actualization tasks are marked down on the control graph.

5.1.Control simulation graph

The cyclical character of the control graph (figure 2) along with the repetitive character of the tasks to be performed leads to the adoption of a definition time interval equal to T/n.
The n (G,X) doublets are arranged in a table and designated by the n_0 value of a selector.
A binary index (M) positioned in the places affected by the single phase working mode conditions the actualization of the three phase variables whenever X changes value.
Two macroplace selectors PM1 and PM2 for X=3 and X=4 are initialed with each controlled commutation.

Under the circonstances any control system will be accounter for by figure 10

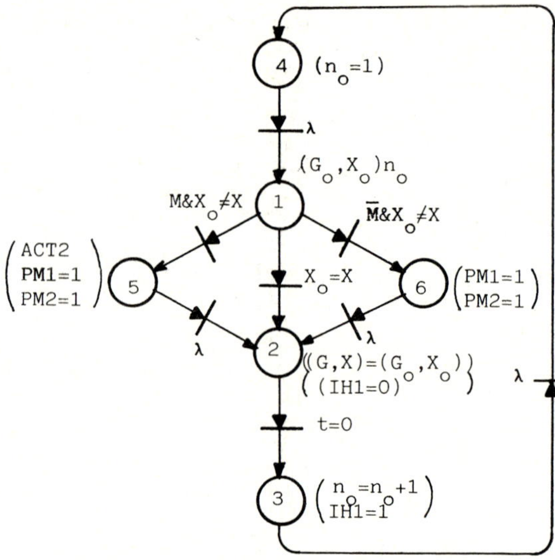

Figure 10

Places	Remarks
1	Decyphering of table
2	Assigning G_1X
3	Index incrementation
4	Index initialization
5	Actualization fo 3 phase model Initialing PM_1 and PM_2
6	Initialing PM_1 and PM_2

5.2. Working simulation graph

The Petri net working graph naturally evolues into working simulation graph.
- Initial description of X=1 and X=2 is inaltered because actualization of 3 phase model is carried along to graph control.
- For X=3 and X=4, the actualization tasks bound up with natural commutations (ACT.1 and ACT.2) produce the graphes on figure 11.

6. SIMULATION EQUATIONS

Motor equations come out in matricial form [3] (chart 4).

Actualization tasks ACT.1 and ACT.2 are implemented from the equations in chart 5 by means of change over matrix below

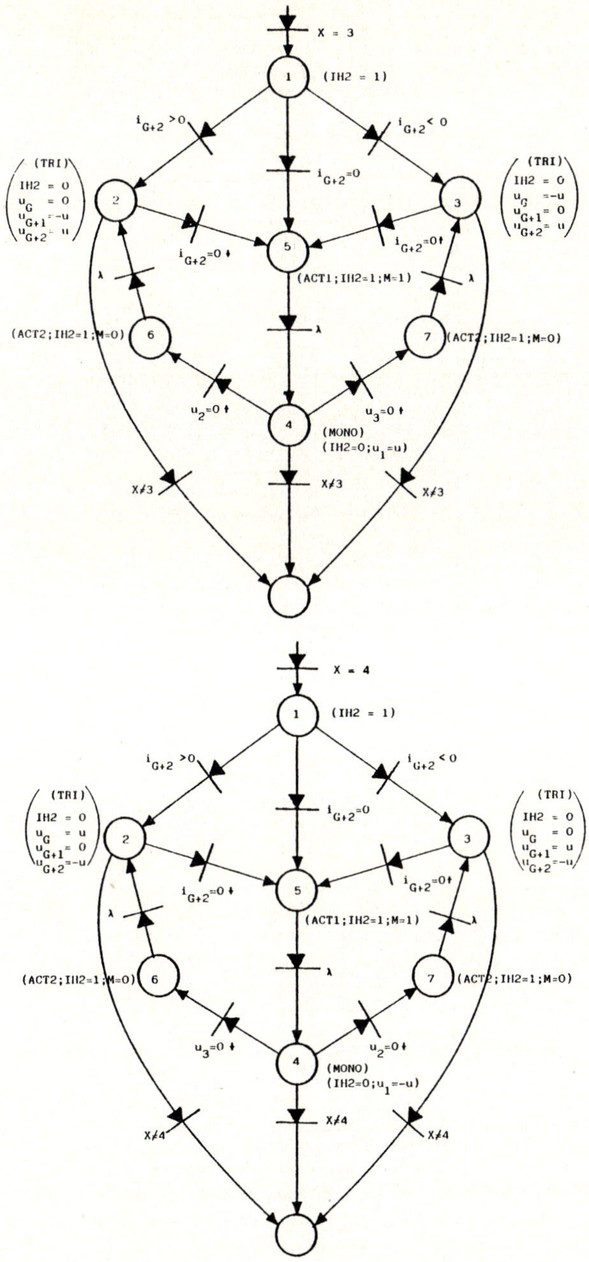

Figure 11

Coupling	Model	Equations
TRIANGLE	Three-Phase	$[V_T]=\{[R_T]+p\frac{d\theta}{dt}[\alpha_T][L_T]\}[I_T]+[L_T]\frac{d}{dt}[I_T]$
	Single-Phase	$[V_{T1}]=\{[R_{TM}]+p\frac{d\theta}{dt}[\alpha_{TM}][L_{T1}]\}[I_{T1}]+[L_{T1}]\frac{d}{dt}[I_{T1}]$
STAR	Three-Phase	$[V_E]=\{[R_E]+p\frac{d\theta}{dt}[\alpha_E][L_E]\}[I_E]+[L_E]\frac{d}{dt}[I_E]$
	Single-Phase	$[V_{E1}]=\{[R_{E1}]+p\frac{d\theta}{dt}[\alpha_{EM}][L_{E1}]\}[I_{E1}]+[L_{E1}]\frac{d}{dt}[I_{E1}]$

Chart 4

Coupling	Passage	Equations of actualization
TRIANGLE	3 PHASE SINGLE PHASE (ACT.1)	$[I_{T1}]=[P(\theta_r)]\,[I_{T(G+2)}]$
	SINGLE PHASE 3 PHASE (ACT.2)	$[I_{T(G+2)}]=[P(\theta_r)]^{-1}\,[I_{T1}]$
STAR	3 PHASE SINGLE PHASE (ACT.1)	$[I_{E1}]=[P(\theta_r)]\,[I_{E(G+2)}]$
	SINGLE PHASE 3 PHASE (ACT.2)	$[I_{E(G+2)}]=[P(\theta_r)]^{-1}\,[I_{E1}]$

Chart 5

A single single-phase model defined by matrixes [ZT1] or [ZE1] results in a univocal relation between angle θ_r and permutation index G (Chart 6).

G	θ_r
1	$-\frac{2\Pi}{3}$
2	0
3	$+\frac{2\Pi}{3}$

Chart 6

7. OVERALL NUMERICAL PATTERN

A cascade chopper is introduced at this point (figure 1) in order to produce an input voltage for the filter cell.

This input voltage has a steady frequency $f_1=\frac{1}{T1}$ and a variable cyclical ratio.

The chopper and the inverter act as two coupled simultaneous-evolution systems (figure 12).

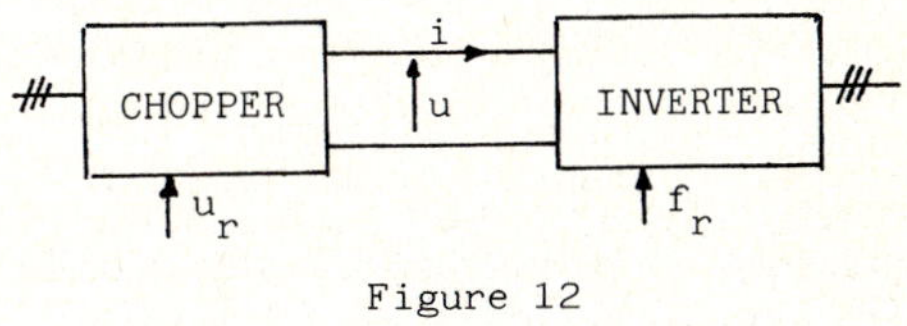

Figure 12

Input variables are respectively ($f_r=\frac{n}{T}$) frequency for the inverter and reference voltage (u_r) for the chopper. The coupling variable here are voltage u and current i.

In order to implement such parallelism, three time variables are required.

- inverter time t,
  increment $\Delta t = \frac{T}{xn}$(x:integer),
  definition interval $\frac{T}{n}$.

- Chopper time t1,
  increment $\Delta t1 = \frac{T1}{xn}$ (x1 : integer),
  definition interval T1,
- Chopper inspection time t2,
  increment $\Delta t1$,
  definition interval Δt.

The treatment cycle unrolls as follows :

At the end of each Δt interval, knowledge of the inverter input voltage is requisite time (T2) allows for chopper treatment inside interval (Δt) and thus insures synchronisation of exchanges between the two systems.

Petri net in figure 13 shows time and treatment implementation.

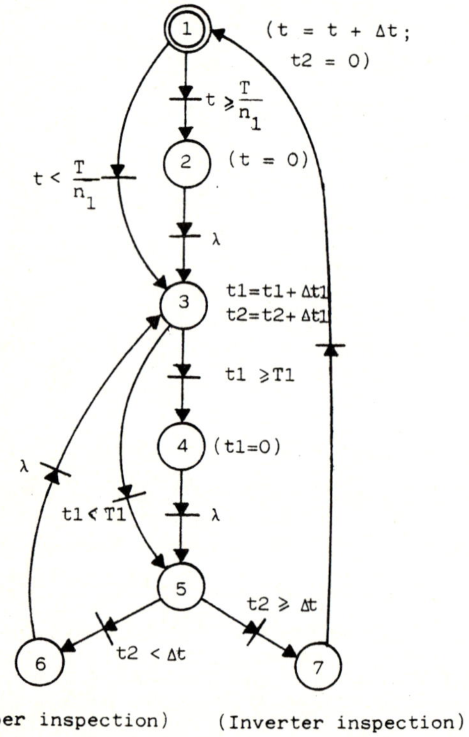

Figure 13

The programming flowchart of the overall system is built in accordance with the principles enunciated above with the proviso that chopper treatment predetermines inverter treatment (figure 14).

8. CONCLUSION

The breakdown and coding suggested above result, whatever the computer used, in a straightforward and quasi systematic formulation of the simulation of program.
Its implementation is made easy by continuous aprreciation of the position of all the graph selectors, which allows observation of the concatenated phases of the process.

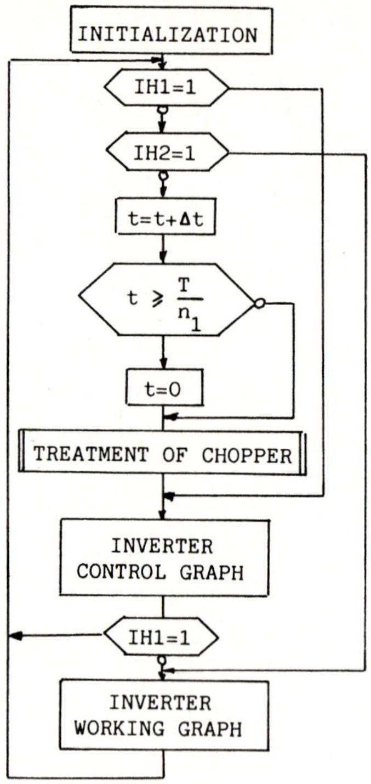

Figure 14.a

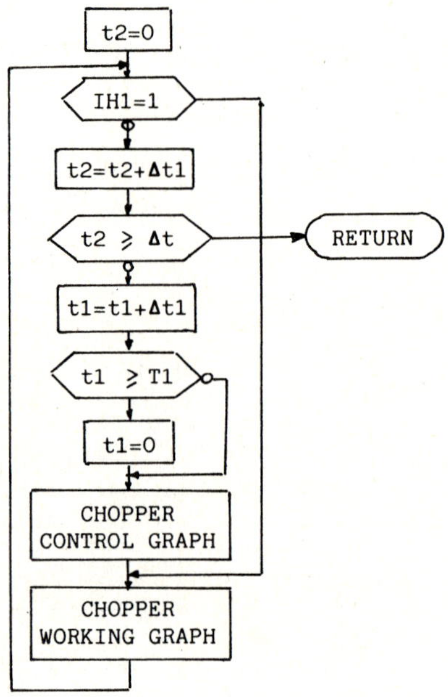

Figure 14.b

The numerical model thus elaborated intimately imitates the actual dynamics of the system, whatever the initial state and control inputs. Thus can the global simulation of complex systems including among other things their control regulators be anticipated thanks to an easy implementation of working parallelisms.

BIBLIOGRAPHIE

-o-o-o-o-o-

[1] : G. MANESSE et C. MAIZIERES
Sur la description fonctionnelle des convertisseurs statiques. Applications aux montages redresseurs.
C.R. Acad. Sc. Paris, t.295, Série II, p963-966 (1982).

[2] : J.M. TOULOTTE et S. THELLIEZ
GRAFCET et logique industrielle programmée
EYROLLES

[3] : J.P. HAUTIER, G. MANESSE et C. MAIZIERES
Sur la description fonctionnelle, par réseau de Pétri, d'un onduleur en tension à transistors
C.R. Acad. Sc., Paris (sous presse)

[4] : J.P. HAUTIER, G. MANESSE et C. MAIZIERES
Description fonctionnelle et modélisation numérique d'un onduleur à transistor alimentant un moteur asynchrone
Congrès IASTED - Lugano - Juin 1983

[5] : J.P. HAUTIER et G. MANESSE
Sur le contrôle par microprocesseur d'un moteur asynchrone alimenté par un onduleur à transistor
Congrès A.I.M. - Liège - Octobre 1983.

ELECTRICAL MACHINES AND CONVERTERS - Modelling and Simulation
H. Buyse and J. Robert (editors)
Elsevier Science Publishers B.V. (North-Holland)/ ©IMACS, 1984

ON PARK'S MODEL OF A THREE-PHASE DIODE BRIDGE RECTIFIER

M.S. Garrido

INIC, 2 Av. Gama Pinto, 1699 Lisboa Codex, Portugal
and UCL, 3 Place du Levant, 1348 Louvain-la-Neuve, Belgium

Park's model has been successfully used to describe the association of three-phase sinusoïdal power sources with diode bridge rectifiers.
However, the model is not suitable, in general, for describing the behaviour of many-phase energy sources associated with power rectifiers and other methods are necessary.
In this paper Park's model is first situated in the context of more general approach, giving a very complete and precise description of rectifier circuits.
The simplification implicit in Park's model is then analysed.

1. Introduction

A complete detailed simulation of rectifier circuits is necessary if one wants to know the instantaneous constraints on the different components of the circuits.

But for system analysis, it is often better to adopt global models, which are simpler and well adapted to the classical techniques of system theory.

The Park's model of a rectifier belongs to these global models and it is useful for studying three-phase diode rectifiers when supplied by sinusoïdal sources |1|.

However, the model presents strong limitations if one tries to study polyphase non-sinusoïdal systems and other global models are then necessary |2|.

The reasons why this happens and the relationships between the different global models that can be used is not immediate.

Therefore, in this paper, we will start from a very detailed model of the rectifier |3| giving the instantaneous AC behaviour of the system, and show how to derive to the DC global models used in the system theory approach.

The Park's global model will then be analysed in this general context.

In the last part, the paper presents some experimental data and compares the results obtained by simulation, using the detailed model and the Park's global model.

2. The AC and the DC models of a three-phase diode bridge rectifier

Let us consider a three-phase diode bridge rectifier.

In order to simplify the presentation, we will adopt a purely resistive system, as represented in figure 1, but our results can be easily extended to inductive systems.

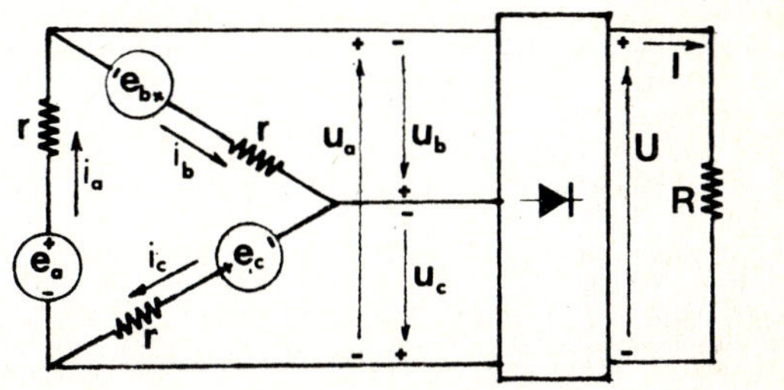

Fig. 1 : Three-phase diode bridge rectifier.

If the real characteristics of the diodes were considered, there would be 6 independent currents in the system.

But, adopting an ideal model with perfect diodes, the number of independent currents will be reduced to 3 : the currents i_a, i_b and i_c (figure 1).

It has been shown in reference |3| that, in such conditions, we have

$$(1) \quad I = \sum_k \gamma_k i_k \qquad k = a,b,c$$

where

$$(2) \quad \gamma_k = \lim_{\alpha\to\infty}\left(\frac{1}{1+e^{\alpha(i_{k-1}-i_k)}} - \frac{1}{1+e^{\alpha(i_k-i_{k+1})}}\right)$$

It has also been established that

$$(3) \quad u_k = \gamma_k U \qquad k = a,b,c$$

which relates the AC phase voltages to the rectified output voltage.

Let us consider the DC load equation

$$(4) \quad U = RI$$

If we introduce it in (3) and use the expressions (1) for the rectified current, we obtain

$$(5) \quad u_k = \sum_j \gamma_k \gamma_j R i_j \qquad j,k = a,b,c$$

These are the AC load equations introduced by the rectifier.

This result shows that the system in figure 1 can be studied by means of the equivalent AC model represented in figure 2.

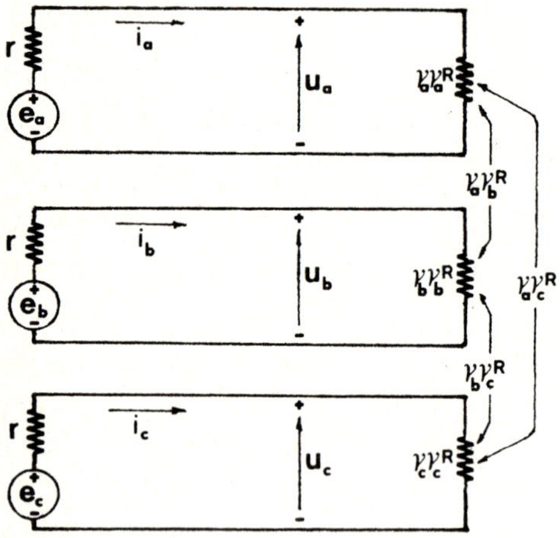

Fig. 2 : Equivalent AC model of the rectifier.

The AC equations of the system are, in matrix form :

$$(6) \quad \begin{vmatrix} e_a \\ e_b \\ e_c \end{vmatrix} = \begin{vmatrix} r & 0 & 0 \\ 0 & r & 0 \\ 0 & 0 & r \end{vmatrix} \begin{vmatrix} i_a \\ i_b \\ i_c \end{vmatrix} + \begin{vmatrix} u_a \\ u_b \\ u_c \end{vmatrix}$$

with the load equations

$$(7) \quad \begin{vmatrix} u_a \\ u_b \\ u_c \end{vmatrix} = R \begin{vmatrix} \gamma_a\gamma_a & \gamma_a\gamma_b & \gamma_a\gamma_c \\ \gamma_b\gamma_a & \gamma_b\gamma_b & \gamma_b\gamma_c \\ \gamma_c\gamma_a & \gamma_c\gamma_b & \gamma_c\gamma_c \end{vmatrix} \begin{vmatrix} i_a \\ i_b \\ i_c \end{vmatrix}$$

These equations permit a direct numerical calculation of the AC variables if a finite value is given to the constant α in the γ functions : this finite value is to be chosen by precision requirements and for a three-phase system a precision of the order of 1% is obtained with α such that $\alpha I \sim 100$.

The rectified current I is related to the alternating currents i_a, i_b, i_c by the transformation (1) and this suggests that it should also be possible to obtain, at least formally, an equivalent DC model of the rectifier system.

In order to introduce this model, let us consider the load equations (7) and calculate the eigenvalues of the impedance matrix.

We obtain

$$(8) \quad \lambda_1 = \Gamma^2 R \qquad \lambda_2, \lambda_3 = 0$$

where Γ stays for :

$$(9) \quad \Gamma^2 = \gamma_a^2 + \gamma_b^2 + \gamma_c^2$$

The eigenvector associated with λ_1 is easily calculated and we have :

$$(10) \quad \vec{\Gamma} = (\gamma_a, \gamma_b, \gamma_c)/\Gamma$$

As the other eigenvalues are zero, we can form a new basis by associating with $\vec{\Gamma}$ any other two orthonormed vectors. Let us call them

$$(11) \quad \vec{\Delta} = (\delta_a, \delta_b, \delta_c)/\Gamma \qquad \vec{O} = (o_a, o_b, o_c)/\Gamma$$

Then, we can introduce a new set of electrical currents by the unitary transformation

$$(12) \quad \begin{vmatrix} I_\gamma \\ I_\delta \\ I_o \end{vmatrix} = \frac{1}{\Gamma} \begin{vmatrix} \gamma_a & \gamma_b & \gamma_c \\ \delta_a & \delta_b & \delta_c \\ o_a & o_b & o_c \end{vmatrix} \begin{vmatrix} i_a \\ i_b \\ i_c \end{vmatrix}$$

Comparing with (1), we see that the new current I_γ is associated with the rectified current I by

$$(13) \quad I_\gamma = I/\Gamma$$

The new voltages are defined by a similar transformation

$$(14) \quad \begin{vmatrix} U_\gamma \\ U_\delta \\ U_o \end{vmatrix} = \frac{1}{\Gamma} \begin{vmatrix} \gamma_a & \gamma_b & \gamma_c \\ \delta_a & \delta_b & \delta_c \\ o_a & o_b & o_c \end{vmatrix} \begin{vmatrix} u_a \\ u_b \\ u_c \end{vmatrix}$$

Taking (3) and the orthogonality of the vectors $\vec{\Gamma}$ and $\vec{\Delta}$, $\vec{O}$ into account, the voltages U_δ, U_o are zero :

$$(15) \quad U_\delta = U_o = 0$$

All the rectified power is associated to the γ component.

In fact, we obtain by (14) :

$$(16) \quad U_\gamma = \frac{1}{\Gamma} \sum_k \gamma_k u_k \qquad k = a,b,c$$

Using (3) and (9), we immediately deduce

$$(17) \quad U_\gamma = \Gamma U$$

and then, according to (13) ,

$$U_\gamma I_\gamma = UI \quad (18)$$

After transformation, the AC load equation (7) gives :

$$\begin{vmatrix} U_\gamma \\ 0 \\ 0 \end{vmatrix} = \Gamma^2 \begin{vmatrix} R & 0 & 0 \\ 0 & 0 & 0 \\ 0 & 0 & 0 \end{vmatrix} \begin{vmatrix} I_\gamma \\ I_\delta \\ I_o \end{vmatrix} \quad (19)$$

which is the very simpler DC load equation.

The source equations (6) become, after transformation :

$$\begin{vmatrix} E_\gamma \\ E_\delta \\ E_o \end{vmatrix} = \begin{vmatrix} r & 0 & 0 \\ 0 & r & 0 \\ 0 & 0 & r \end{vmatrix} \begin{vmatrix} I_\gamma \\ I_\delta \\ I_o \end{vmatrix} + \begin{vmatrix} U_\gamma \\ 0 \\ 0 \end{vmatrix} \quad (20)$$

Equations (19) and (20) are the DC equations of the rectifier system and they can be represented by the equivalent circuits in figure 3.

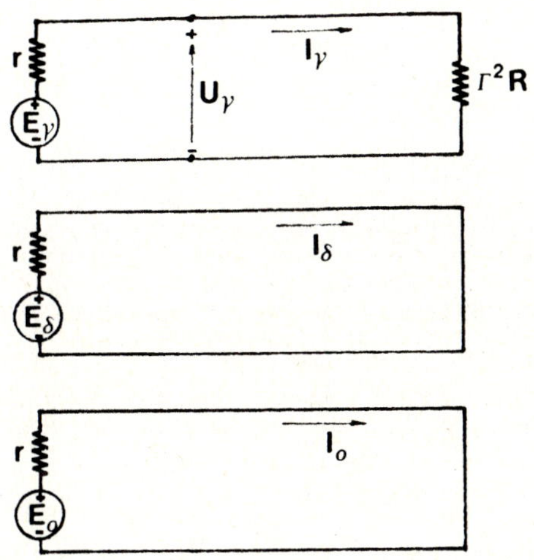

Fig. 3 : Equivalent DC model of the rectifier.

For the moment we have not fixed the δ,o variables.

That can be done according to the nature of the source and we are going to choose the eigenvector $\vec{0}$ such that I_o be the normal homopolar current, which has the advantage of being zero for balanced sources :

$$\vec{0} = \left(\frac{\Gamma}{\sqrt{3}}, \frac{\Gamma}{\sqrt{3}}, \frac{\Gamma}{\sqrt{3}}\right)/\Gamma \quad (21)$$

The eigenvector $\vec{\Delta}$ is, then

$$\vec{\Delta} = \left(\frac{\gamma_b - \gamma_c}{\sqrt{3}}, \frac{\gamma_c - \gamma_a}{\sqrt{3}}, \frac{\gamma_a - \gamma_b}{\sqrt{3}}\right)/\Gamma \quad (22)$$

With this choice of the eigenvectors, the equivalent DC model reduces to the γ, δ circuits for the normal, balanced, sources.

The DC model of the rectifier system appears to be advantageous, because it concentrates the main power flow in a single circuit; the other circuits only account for some losses in the source and will have, normally, a less important role.

This kind of separation is almost always interesting in practice, because it makes the measurement of the parameters easier and allows approximate calculations.

One can even expect that, in some cases, it would be possible to keep only the main circuit, as an approximate global model of the system.

Unfortunately, the DC model is not directly usable as we introduced it, because the transformation (12) depends on the primitive AC variables : so, we cannot calculate the transformed parameters and source voltages included in the DC model without first solving, completely, the AC problem.

In order to use the DC model we must introduce some a priori conjectures and this can be done in two different directions.

The former is to start directly from the DC model and by physical considerations define the corresponding variables and parameters. This method is explored in the theory of sliding contact circuits and has given interesting results, namely in the study of polyphase rectifiers with non-sinusoïdal source |2|.

The second direction is to replace the non-linear transformation (12) by a time dependent transformation, giving the γ_k an a priori form in function of the time. This method is explored in the Park's theory and was successfully applied to the study of three-phase rectifiers, with a sinusoïdal source |1|.

In this paper we are only concerned by the second method.

3. Park's model of a three-phase diode bridge rectifier with sinusoïdal balanced source

The Park's model of a three-phase diode bridge rectifier with a sinusoïdal balanced source is obtained by introducing two assumptions in the general transformation (12).

The first is that the γ_k are sinusoïdal functions of time, with the same frequence ω as the sources, with amplitude 1 as the exact γ_k and with a certain phase ρ :

$$\begin{aligned} \gamma_a &= \cos(\omega t - \rho) \\ \gamma_b &= \cos(\omega t - \rho - \tfrac{2\pi}{3}) \\ \gamma_c &= \cos(\omega t - \rho - \tfrac{4\pi}{3}) \end{aligned} \quad (23)$$

With this assumption and using the eigenvectors (21), (22), the transformation matrix becomes

$$\sqrt{\frac{2}{3}}\begin{vmatrix} \cos(\omega t-\rho) & \cos(\omega t-\rho-\frac{2\pi}{3}) & \cos(\omega t-\rho-\frac{4\pi}{3}) \\ -\sin(\omega t-\rho) & -\sin(\omega t-\rho-\frac{2\pi}{3}) & -\sin(\omega t-\rho-\frac{4\pi}{3}) \\ 1/\sqrt{2} & 1/\sqrt{2} & 1/\sqrt{2} \end{vmatrix} \quad (24)$$

which is the classical Park's transformation matrix. One immediately sees that Γ becomes a constant equal to $\sqrt{3}/2$.

The second hypothesis to be introduced is that the phase ρ is such that there is no power associated to the δ circuit and this implies

$$(25) \quad I_\delta = 0$$

This last condition fixes the phase ρ of Park's transformation.

In the very simple case of a resistive source we obtain the single equivalent circuit represented in figure 4, where E is the source voltage amplitude.

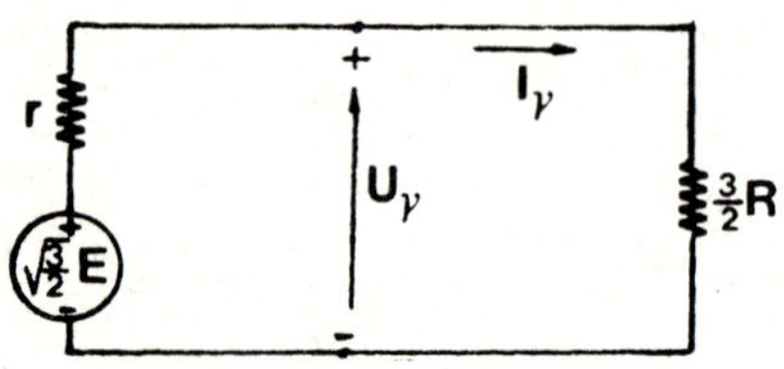

Fig. 4 : Park's model of a resistive rectifier.

Using the value of Γ, we can now change the scaling in order to use directly the rectified variables U, I and this gives the equivalent DC circuit in figure 5.

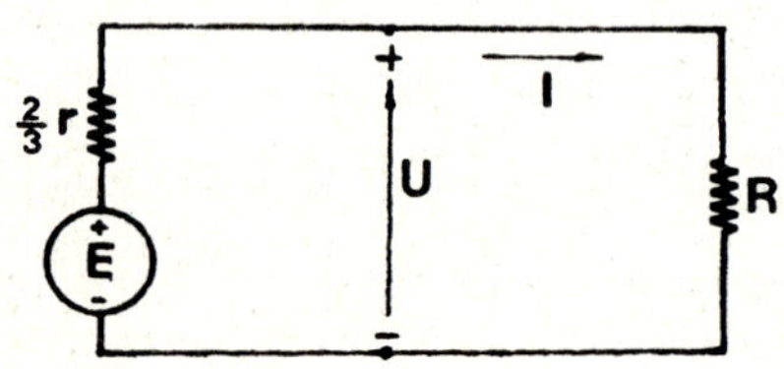

Fig. 5 : Park's model of a resistive rectifier.

It is interesting to see what the cost is in precision, of this simple DC model.

In figure 6 two γ_a functions are represented, for α such that $\alpha I \sim 100$ and $\alpha I \sim 2.5$.

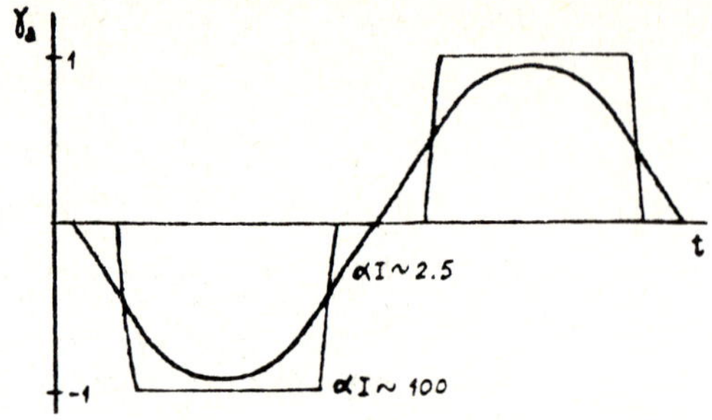

Fig. 6 : Function γ_a for different values of αI

This shows that, by reducing α, γ_a tends to a sinusoïdal function of time.

It is easy to derive the order of magnitude of α which gives a sinusoïdal approximation.

For sinusoïdal γ_j, the currents are also sinusoïdal and we can write

$$(26) \quad \frac{1}{1+e^{\alpha I \sin\omega t}} - \frac{1}{1+e^{\alpha I \sin(\omega t+\frac{2\pi}{3})}} \sim \sin(\omega t+\beta)$$

A series development limited to the second order term gives

$$(27) \quad \frac{1}{4}\,\alpha I\,[\sin\omega t - \sin(\omega t+\frac{2\pi}{3})] \sim \sin(\omega t+\beta)$$

and finally

$$(28) \quad \alpha I \sim 2.5$$

So, the Park's model of a three-phase diode bridge rectifier corresponds to a quite poor approximation and one can expect errors in the mean rectified voltage and current between 5 and 10%.

But the model gives a fairly good description of the dynamics of the system, as we will see and this is important in the transient analysis of the rectifier systems.

4. Application to the transient analysis of a rectified output alternator

Let us consider, now, an application of the previous models to the electrical transient analysis of a cylindrical rotor alternator with rectified output, as used in brushless excitation systems.

Let us represent the armature by the equivalent Δ circuit, in order to make direct use of the asymptotic model developed in reference |3|.

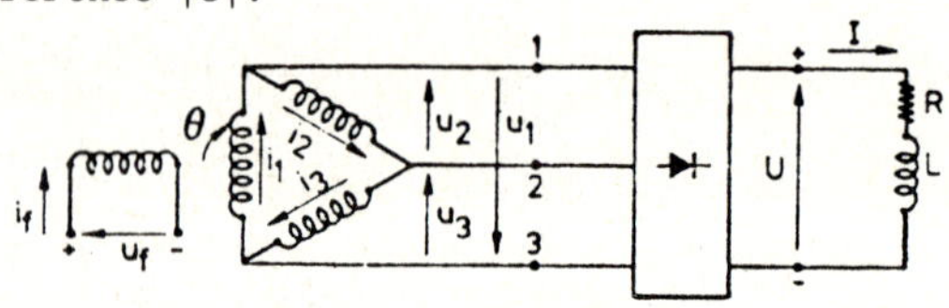

Fig. 7 : Rectified output alternator.

The electrical AC equations of the alternator are :

$$\begin{vmatrix} u_1 \\ u_2 \\ u_3 \\ u_f \end{vmatrix} = \begin{vmatrix} R^{ss}+L^{ss}p & -M^{ss}p & -M^{ss}p & M^{sr}p\cos\theta \\ -M^{ss}p & R^{ss}+L^{ss}p & -M^{ss}p & M^{sr}p\cos(\theta+\frac{2\pi}{3}) \\ -M^{ss}p & -M^{ss}p & R^{ss}+L^{ss}p & M^{sr}p\cos(\theta+\frac{4\pi}{3}) \\ M^{sr}p\cos\theta & M^{sr}p\cos(\theta+\frac{2\pi}{3}) & M^{sr}p\cos(\theta+\frac{4\pi}{3}) & R^{rr}+L^{rr}p \end{vmatrix} \begin{vmatrix} i_1 \\ i_2 \\ i_3 \\ i_f \end{vmatrix} \quad (29)$$

where $p = d/dt$.

The AC load equations, according to the asymptotic model of the rectifier |3|, become

$$\begin{vmatrix} -u_1 \\ -u_2 \\ -u_3 \end{vmatrix} = \begin{vmatrix} \gamma_1\gamma_1 & \gamma_1\gamma_2 & \gamma_1\gamma_3 \\ \gamma_2\gamma_1 & \gamma_2\gamma_2 & \gamma_2\gamma_3 \\ \gamma_3\gamma_1 & \gamma_3\gamma_2 & \gamma_3\gamma_3 \end{vmatrix} (R+Lp) \begin{vmatrix} i_1 \\ i_2 \\ i_3 \end{vmatrix} \quad (30)$$

The rectified current I can be obtained from the AC currents by equation (1).

Using this model, we simulated a loading transient.

The result presented in figure 8 has been obtained with $\alpha I \sim 100$ and the comparison with experimental data shows a very good precision.

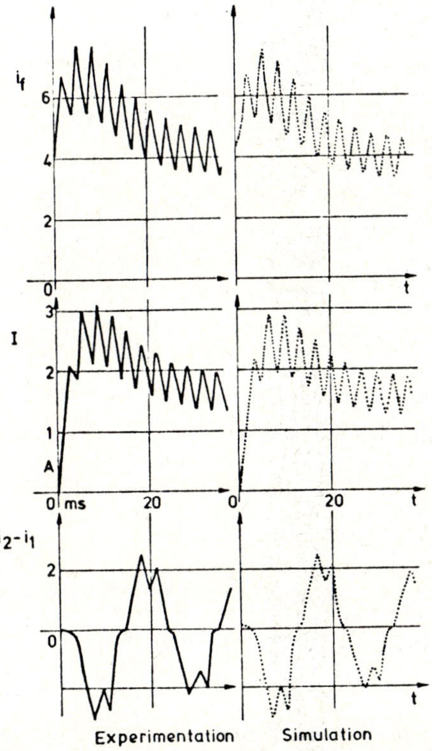

Fig. 8 : Loading of a rectified output alternator.

A numerical step of 10^{-4} sec was used with a simple Euler's method and the simulation took about 15 min/period on a 8 bits microcomputer.

In order to observe the transition to Park's approximation, the same simulation has been repeated for a smaller value of α , such that $\alpha I \sim 10$. The results are represented in figure 9 and compared with the previous ones.

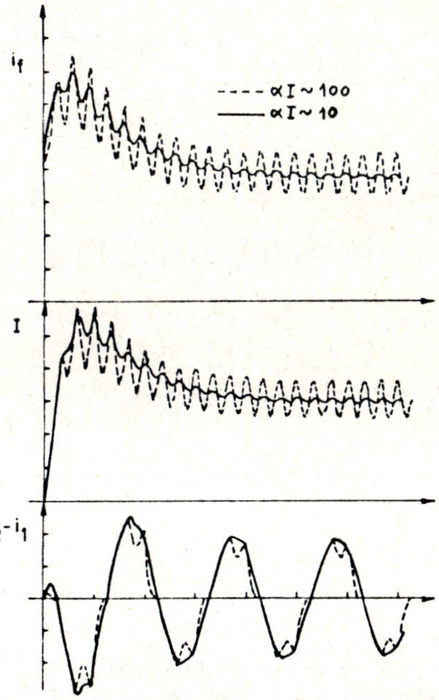

Fig. 9 : Loading of a rectified output alternator.

We observe that the commutation details are partialy lost and the AC currents tend to become sinusoïdal.

If a smaller precision is accepted, we can adopt the sinusoïdal approximation and use Park's transformation in order to work with the DC model.

The Park's transformation of the machine's equations (29) is classical. Introducing the conditions (15) and (25), $U_\delta, I_\delta = 0$ and using the rectified variables U,I instead of U_γ, I_γ, one obtains :

$$\begin{vmatrix} U \\ 0 \\ u_f \end{vmatrix} = \begin{vmatrix} R_s+L_sp & (\omega-\dot\rho)L_s & M\cos\rho p - \omega M\sin\rho \\ -(\omega-\dot\rho)L_s & R_s+L_sp & -M\sin\rho p - \omega M\cos\rho \\ M\cos\rho p - \dot\rho M\sin\rho & -M\sin\rho p - \dot\rho M\cos\rho & R_f+L_fp \end{vmatrix} \begin{vmatrix} I \\ 0 \\ i_f \end{vmatrix} \quad (31)$$

where $R_f = R^{rr}$, $L_f = L^{rr}$, $R_s = \frac{2}{3} R^{ss}$,

$L_s = \frac{2}{3} (L^{ss} + M^{ss})$ and $M = M^{sr}$.

The transformation of the load equation (30) evidently gives :

(32) $-U = RI + LpI$

As a curiosity, we can observe that the AC load matrix (30) contains, for sinusoïdal γ_k, the second harmonic terms similar to those existing in a salient pole machine, but with $L_d = \frac{3}{2} L$ and $L_q = 0$.

The results of simulation with Park's model are represented in figure 10.

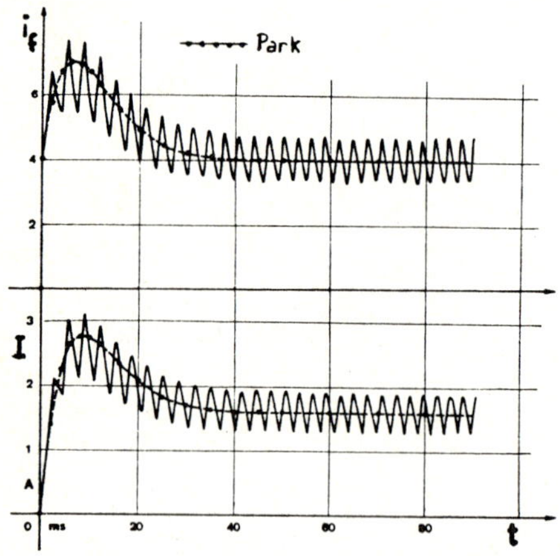

Fig. 10 : Loading of a rectified output alternator.

The simulation by Park's model ignores all the commutation details, but it gives a good dynamical description of the system.

The computing time with this simplified model is about 1/10 of the time required by the detailed asymptotic model.

The same results as those represented in figure 10 can also be obtained by the theory of sliding contact circuits |4|.

Conclusion

The use of Park's transformation for studying three-phase, sinusoïdal, power rectifiers has given some good results, but serious difficulties appear if one intends to apply the same method to more complicated systems.

In this paper a detailed analysis of a rectifier system has been presented which explains why Park's model is suitable for the three-phase sinusoïdal rectifiers.

The general method, to be used for less trivial cases, shows that Park's model is, in such a context, a quite particular simplification.

References

|1| M.S. Garrido : "Simplified transfer function of the brushless exciter of synchronous machines", in proc. International Conference on Electrical Machines, Part 2, pp. 954-961 (University of Athens, Greece, 1980).

|2| L. Pierrat et al. : "Dynamic models of polyphase rectified output alternators" in W.F. Ames & R. Vichnevetsky (Eds), Modelling and Simulation in Engineering, vol. III, pp. 221-227 (North-Holland, Amsterdam, 1983).

|3| M.S. Garrido : "Nouvelle modélisation d'un redresseur idéal polyphasé", to appear in G. Mesnard (Ed.), Modelling and Simulation, vol. 2 (AMSE Press, Tassin-la-Demi-Lune, 1983); see also the appendix in the present paper.

|4| H. Buyse & M. Kemler : "Etude de la génératrice à redresseur en régime transitoire", Revue E, vol. VII, n°3, pp. 71-78, 1972.

Appendix

The asymptotic model of a polyphase diode rectifier is based on a certain representation of the diode voltages in function of the source currents and the output rectified voltage.

Let us consider one phase of the source and the associated diodes.

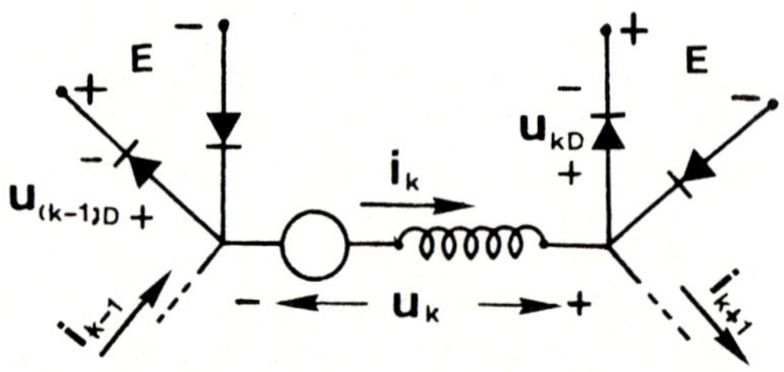

The total voltage u_k in the phase is

$$u_k = u_{kD} - u_{(k-1)D}$$

where u_{kD}, $u_{(k-1)D}$ are the forward voltages in the up diodes.

For perfect diodes, these voltages can be expressed, asymptotically, as functions of the phase currents i_j and of the output rectified voltage E.

In fact, for a perfect diode model, u_{kD} is such that

If $i_k - i_{k+1} > 0$ then $u_{kD} = 0$

If $i_k - i_{k+1} < 0$ then $u_{kD} = -E$

If $i_k - i_{k+1} = 0$ then $0 \geqslant u_{kD} \geqslant -E$

These conditions are equivalent to the Heaviside function represented in the next figure.

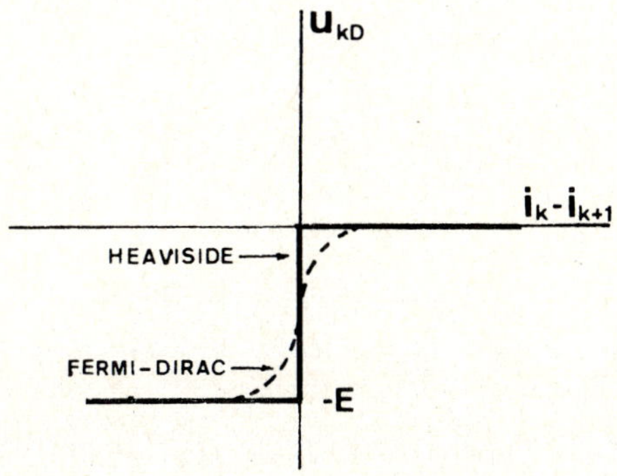

We can now replace the Heaviside function by an asymptotic Fermi-Dirac function and write

$$u_{kD} = \lim_{\alpha\to\infty} \frac{-E}{1 + e^{\alpha(i_k - i_{k+1})}}$$

We have thus

$$u_k = \lim_{\alpha\to\infty} \left(\frac{1}{1 + e^{\alpha(i_k - i_{k+1})}} - \frac{1}{1 + e^{\alpha(i_k - i_{k+1})}}\right)E$$

which is the basic expression used before.

For the rectified current we have

$$I = \sum_k \lim_{\alpha\to\infty} \frac{-1}{1 + e^{\alpha(i_k - i_{k+1})}} (i_k - i_{k+1})$$

This can also be written,

$$I = \sum_k \lim_{\alpha\to\infty} \left(\frac{1}{1 + e^{\alpha(i_{k-1} - i_k)}} - \frac{1}{1 + e^{\alpha(i_k - i_{k+1})}}\right).i_k$$

which is the expression adopted in this paper.

ELECTRICAL MACHINES AND CONVERTERS - Modelling and Simulation
H. Buyse and J. Robert (editors)
Elsevier Science Publishers B.V. (North-Holland)/© IMACS, 1984

DIGITAL SIMULATION OF THE STEADY STATE BEHAVIOUR OF A 3Φ FULL CONTROLLED D.C.DRIVE

George Kontos, Athanassious Dratsas

Electrical Engineering Department National Technical University of Athens 42, 28th October Str. - Athens - Greece

A digital computer simulation method for the study of the steady state behaviour of a full wave, full controlled 3-phase bridge. In the proposed solution no idealizations for the power supply are made. The solution is based on Kron's method of analysis and tensor theory is used. It must be pointed out that only continuous conduction in both modes of operation- rectifier and inverter - is considered. The main phenomena appearing during both modes of operation are examined and relevant results are obtained, approaching in a satisfactory way the problem. The validity of the method is proved also in the case of the study of such a system.

LIST OF SYMBOLS

V_A, V_B, V_C : 3-Φ supply voltages

i_A, i_B, i_C : 3-Φ currents

i_L, V_L : 3- Φ load current and voltage

i_a, i_b, i_c, i_d, i_e : independent mesh current of the general network.

$i_1, i_2, i_3, i_4, i_5, i_6$: thyristor currents.

All the above quantities are instantaneous.

Z_A, Z_B, Z_C : operational impedances of 3-phase supply.

$Z_1, Z_2, Z_3, Z_4, Z_5, Z_6$: operational impedances of the load.

E = K.W: counter EMF of the D.C.motor.

K : armature voltage constant (V.S)

W: angular velocity of the D.C. motor (rad/s.)

1. INTRODUCTION

A lot of problems appear during theoretical analysis of a 3-phase full wave thyristor bridge when it drives dynamic load as a D.C. motor. This is due to the fact that the above mentioned bridge presents different modes of operation depending on the thyristor firing angle, the nature of the load and the impedances of system feeding the bridge.

Digital computer offers the advantage of quick solution of all differential equations appearing during the successive modes of operation of the bridge. It takes also into account the important problems posed by the big number of differential equations involved, the commutation and the conduction time of the thyristors of the bridge as well.

In this paper we develop a computer program elaborated in the purpose to completely describe the steady-state behaviour of the D.C. drive in the case of continuous conduction of the thyristor bridge.

In that case we analyse the operation of the bridge in both quadrants of its V-i characteristic, that is when it operates as a rectifier or as an inverter as well.

The study is based on tensor analysis and the Kron's method of solution is applied (5).

2. TENSOR ANALYSIS OF A 3-PHASE D.C. DRIVE BY KRON'S METHOD

The basic idea of that method of analysis is the establishment and solution of the different sets of equations, corresponding to the successive modes of operation of the system (2), (4).

Kron's method consists in the use of mesh analysis for the establishment in the successive phases of operation of the bridge the corresponding matrices of system's impedances, currents and voltages (5)

Considering a general network of the bridge we form via a transformation tensor the appropriate matrices describing the successive modes of system's operation, according to the firing and extinction time of every thyristor of the bridge.

In Fig. 1 the generalized network is considered where all thyristors are taken as conducting. It is evident that this is an impossible operation (2).

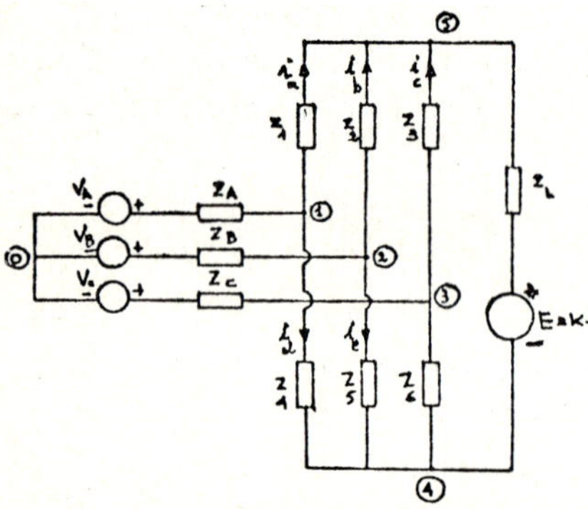

Fig.1: 3-phase full controlled bridge.

In order to use the mesh analysis method for the network of Fig.1, we have:

Number of nodes, $n_o = 6$

Number of branches, $b = 10$

Number of independent equations and currents, $b - n - 1 = 5$

We choose as independent currents, the thyristor currents i_1, i_2, i_3, i_4, i_5.

The independent meshes and their corresponding currents are then: (015430): i_a, (025430): i_b, (3543): i_c, (01430): i_d, (02430): i_e.

The general matrix equation of the network is then:

$$[V_{iE}] = [Z_i] \cdot [i_i] \qquad (1)$$

where:

$$[V_{iE}] = \begin{bmatrix} V_A - V_B - E \\ V_B - V_C - E \\ -E \\ V_A - V_B \\ V_B - V_C \end{bmatrix}, \quad [i_i] = \begin{bmatrix} i_a \\ i_b \\ i_c \\ i_d \\ i_e \end{bmatrix}$$

$$[Z_i] = \begin{bmatrix} Z_{11} & Z_{12} & Z_{13} & Z_{14} & Z_{15} \\ Z_{21} & Z_{22} & Z_{23} & Z_{24} & Z_{25} \\ Z_{31} & Z_{32} & Z_{33} & Z_{34} & Z_{35} \\ Z_{41} & Z_{42} & Z_{43} & Z_{44} & Z_{45} \\ Z_{51} & Z_{52} & Z_{53} & Z_{54} & Z_{55} \end{bmatrix}$$

The elements Z_{ij} of the matrix $[Z_i]$ are defined as follows:

$$Z_{11} = Z_A + Z_C + Z_1 + Z_6 + Z_L$$

$$Z_{12} = Z_{21} = Z_C + Z_6 + Z_L$$

$$Z_{13} = Z_{31} = Z_6 + Z_L$$

$$Z_{14} = Z_{41} = Z_A + Z_C + Z_6$$

$$Z_{15} = Z_{51} = Z_C + Z_6$$

$$Z_{22} = Z_B + Z_C + Z_2 + Z_6 + Z_L$$

$$Z_{23} = Z_{32} = Z_6 + Z_L$$

$Z_{34} = Z_{43} = Z_6$

$Z_{35} = Z_{53} = Z_6$

$Z_{45} = Z_{54} = Z_C + Z_6$

$Z_{33} = Z_3 + Z_6 + Z_L$

$Z_{24} = Z_{42} = Z_C + Z_6$

$Z_{25} = Z_{52} = Z_B + Z_C + Z_6$

$Z_{44} = Z_A + Z_C + Z_4 + Z_6$

$Z_{55} = Z_B + Z_C + Z_5 + Z_6$

From eq.(1) and if we consider the steady-state operation of the D.C. drive (E=K.W.) we have that:

$$\left[V_{iE}\right] = \left[V_i\right] - \begin{bmatrix} E \\ E \\ E \\ 0 \\ 0 \end{bmatrix} = \left[V_i\right] - W \begin{bmatrix} K \\ K \\ K \\ 0 \\ 0 \end{bmatrix} \quad (2)$$

where we define the matrix $\left[V_i\right]$ as

$$\left[V_i\right] = \begin{bmatrix} V_A - V_C \\ V_B - V_C \\ 0 \\ V_A - V_C \\ V_B - V_C \end{bmatrix}$$

The so defined independent mesh currents $\left[i_i\right]$ are related to the currents flowing in every branch of the system by the matrix $\left[C_p\right]$, according the equation:

$$\left[i_p\right] = \left[C_p\right] \left[i_i\right] \quad - (3)$$

where we define:

$$\left[i_p\right] = \begin{bmatrix} i_1 \\ i_2 \\ i_3 \\ i_4 \\ i_5 \\ i_6 \\ i_A \\ i_B \\ i_C \\ i_L \end{bmatrix}, \left[C_p\right] = \begin{bmatrix} 1 & 0 & 0 & 0 & 0 \\ 0 & 1 & 0 & 0 & 0 \\ 0 & 0 & 1 & 0 & 0 \\ 0 & 0 & 0 & 1 & 0 \\ 0 & 0 & 0 & 0 & 1 \\ -1 & -1 & -1 & -1 & -1 \\ 1 & 0 & 0 & 1 & 0 \\ 0 & 1 & 0 & 0 & 1 \\ -1 & -1 & 0 & -1 & -1 \\ 1 & 1 & 1 & 0 & 0 \end{bmatrix}$$

As it has been mentioned above, Eqs (1) and (2) are describing the operation of the system in the hypothetical case where all thyristors are simultaneously conducting. Because in every successive phase of operation of the bridge maximum number of three thyristors can conduct simultaneously (2) we must establish the appropriate sets of equations describing the operation of the D.C. drive.

It seems to us that the best method used in that purpose is the so-called Kron's tensor approach, because of the stability of the obtained numerical solutions, their accuracy and a relative short computing time.

Refering to Fig. 1 we point out that in any change of state of the thyristors of the bridge, the topology of the network alters and consequently the number of meshes change. If we take into account that every non-conducting thyristor has an infinite impedance, in order to overcome the mathematical problem appearing in that case, we have to develop a systematic method of analysis in the purpose to reduce the independent set of variables $\left[V_i\right]$, $\left[i_i\right]$ and $\left[Z_i\right]$ to a new set which is valuable after the change of topology (5).

If we consider the example of Fig. 2 where thyristors 2, 3 and 1 are conducting we finish by establishing the basic transformation tensor $\left[C_n\right]$ relating the independent currents of the basic configuration of Fig. 1 to that of Fig. 2.

We have that:

$$\left[C_n\right] = \begin{bmatrix} 0 & 0 \\ 1 & 0 \\ 0 & 1 \\ -1 & -1 \\ 0 & 0 \end{bmatrix}$$

and $\left[i_i\right] = \left[C_n\right] \left[i_n\right]$ (4)

where $\left[i_n\right] = \begin{bmatrix} i_b \\ i_c \end{bmatrix}$

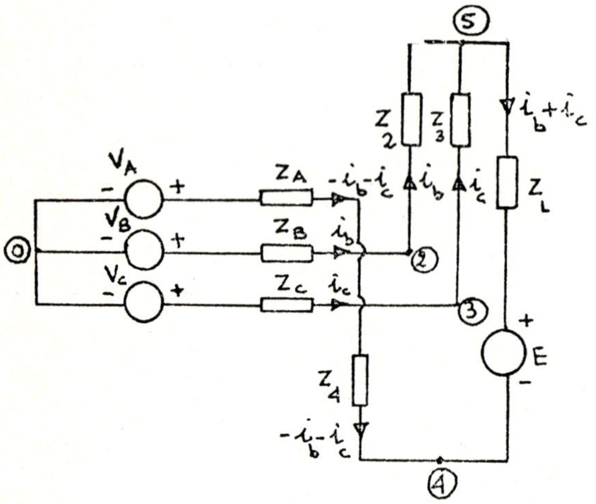

Fig.2: 3-phase, full controlled bridge. Only thyristors 2,3 and 4 are conducting.

We define the tensors corresponding to the configuration of Fig.2.

$\left[i_n\right]$: independent current tensor of the network.

$\left[V_{nE}\right]$: voltage tensor of the network.

$\left[Z_n\right]$: impedance tensor of the network

Using the basic transformation tensor $\left[C_n\right]$ and applying the usual transformation techniques we get the expressions:

$$\left[V_{nE}\right] = \left[C_n\right]^T \cdot \left[V_{iE}\right] \quad (5)$$

$$\left[Z_n\right] = \left[C_n\right]^T \cdot \left[Z_i\right] \cdot \left[C_n\right] \quad (6)$$

For the network of Fig. 2, mesh equations are given by:

$$\left[V_{nE}\right] = \left[Z_n\right] \cdot \left[i_n\right] \quad (7)$$

Where:

$$\left[V_{nE}\right] = \begin{bmatrix} V_B - V_A - E \\ V_C - V_A - E \end{bmatrix}$$

$$Z_n = \begin{bmatrix} Z_A + Z_B + Z_2 + Z_4 + Z_L & Z_A + Z_4 + Z_L \\ Z_A + Z_4 + Z_L & Z_A + Z_C + Z_3 + Z_4 + Z_L \end{bmatrix}$$

We observe that matrices $\left[V_{nE}\right]$ and $\left[Z_n\right]$ of Eq. (7) are analogous to the matrices $\left[V_{iE}\right]$ and $\left[Z_i\right]$ of Eq. (1).

It is then obvious that the use of the above stated method of analysis, based on the developed transformation method leads to the solution of any configuration resulting from the changing topology of the D.C. drive system.

3. DEFINITION OF ELECTRICAL AND MECHANICAL MAGNITUDES OF THE SYSTEM

3.1. Definition of currents

Combining Eqs. (2) and (5) we get the equation

$$\left[V_n\right] = \left[V_{nE}\right] + \left[C_n\right]^T \begin{bmatrix} K \\ K \\ K \\ 0 \\ 0 \end{bmatrix} \omega \quad (8)$$

where $\left[V_n\right] = \left[C_n\right]^T \left[V_i\right]$ (9)

From Eq. (6) we have:

$$\left[V_n\right] = \left[Z_n\right]\left[i_n\right] + \left[C_n\right]^T \cdot \begin{bmatrix} K \\ K \\ K \\ 0 \\ 0 \end{bmatrix} \omega \quad (10)$$

If we put $\left[Z_n\right] = \left[R_n\right] + \left[L_n D\right]$

where $D = \frac{d}{dt}$, and if we make the appropriate mathematical operations we get:

$$\left[D i_n\right] = \left[L_n\right]^{-1} \left[V_n\right] - \left[R_n\right] \left[i_n\right] - \left[C_n\right]^T \begin{bmatrix} K \\ K \\ K \\ 0 \\ 0 \end{bmatrix} \omega \quad (11)$$

Eq. (11) is the electromechanical equation of the D.C. drive. It is of the form $\dot{y} = f(x,y)$ required for numerical integration. If we know the magnitude of angular velocity ω from the solution of the mechanical equation of the D.C. motor, as it will be developed furthermore, the new set of currents is determined from the solution of Eq. (11). It follows that the independent set of currents i_i is found from Eq. (4) and the currents of every branch of the network are given from Eq. (3).

3.2. Definition of thyristor voltages

In order to find thyristor voltages a new vector $[V_{th}]$ is defined as follows:

$$[V_{th}] = \begin{bmatrix} V_1 \\ V_2 \\ V_3 \\ V_4 \\ V_5 \\ V_6 \end{bmatrix}$$

As the 6-th thyristor carries all the independent mesh currents the network's topology can be considered to be as shown in Fig. 3.

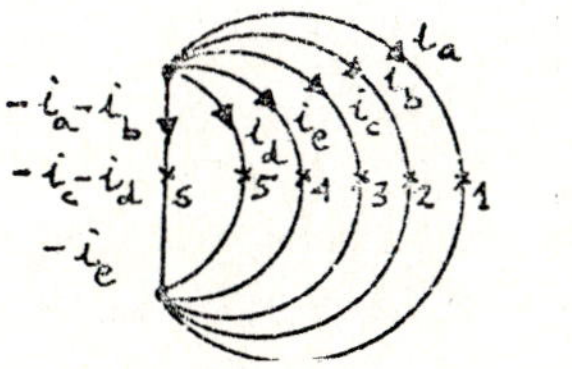

Fig.3: 3-phase full controlled bridge. Representation of bridge meshes the numbers 1 to 6 indicate the six thyristors

By applying of Kirchoff's voltage law in the general network, an array of mesh-sum thyristor voltages V_x can be found :

$$[V_x] = [V_i] - \omega \begin{bmatrix} K \\ K \\ K \\ 0 \\ 0 \end{bmatrix} - [Z_i]\ [i_i] \quad (12)$$

where $[V_x]$ is a column vector 5x1 with elements $V_i - V_6$ $(i = 1,\ldots,5)$

If the 6-th thyristor is conducting, then $V_6 = 0$ and the remaining thyristor voltages may be calculated from Eq. 11. In the case that the 6-th thyristor is non conducting then for calculating the thyristor voltages we proceed as follows:

We check which of the thyristors is conducting from the elements of the so-called system - state vector $[S]$. That vector comprises six row elements having the values 1 or 0, and corresponding to the conduction state of the six thyristors of the bridge (5). Then from the mesh containing the conducting j-th thyristor and the 6-th thyristor. We calculate the voltage V_6 of the 6-th thyristor, taking into account that $V_j = 0$.

In that way by knowing V_6 and by using Eq. 11. and Eq. 12 we calculate all thyristor voltages.

3.3. Definition of the angular velocity of the D.C.machine

It is known that the equations describing the operation of the separately excited D.C. machine are:

$$V_L = R_L i_L + L\frac{di_L}{dt} + K\omega \quad (13)$$

$$Ki_L = \frac{d\omega}{dt} + f\omega + T_L \quad (14)$$

It must be noted that the load on the shaft of the D.C. machine is considered as constant, T_L.

The electrical equation (13) of the D.C. machine is included in Eq.(1).

By rearranging the mechanical equation (14) we get:

$$D.\omega = \frac{K}{F}\, i_L - \frac{f}{F}\cdot \omega - \frac{T_L}{F} \quad (15)$$

which is comparable to the standard form $\dot{y} = f\,(x,y)$.

In the analysis made in this paper a continual communication is established between Eqs. (11) and (15). In that way for an initial value of angular velocity of the motor ω_o, we find the corresponding value for the load current i_L from Eq. (11) which applying to Eq. (15) gives the value of ω for the next step of operation of the drive. It will be noted from Fig. 1 that the current in the lower limbs of the bridge circuit has been defined in directions opposite to the normal flow. When testing in numerical solution for turn-on or turn-off of the 4-th, 5-th and 6-th thyristor the sign of the values of the associated currents and voltages must be inversed.

4. DIGITAL COMPUTER STUDY OF THE SYSTEM

4.1. Discussion of the flow -chart of the program.

Fig. 4 shows the flow chart of the computer program developed for the numerical solution of equations of section 3.

As the program enters in operation, it reads such data as voltage source characteristics (maximum value, frequency, impedances) circuit resistances and inductances, D.C. motor parameters and the thyristors firing angle as well. The last is converted to time. The execution of the program is continuously controlled by the system-state vector $[S]$.

The program start running at time to when we can find which of the thyristors are able to conduct. In that way an initial system-state vector is constructed. Then the transformation tensor $[C_n]$ is formed and after that tensors $[R_n]$, $[L_n]$, $[L_n]^{-1}$. When a topology change accurs during the last step of calculation a new system-state vector is formed and then the new tensors $[C_n]$, $[R_n]$, $[L_n]$ e.t.c. In every change of topology the values of the independent currents are forming the new vector $[i_n]$ in order that the correct initial conditions be available for the next step of numerical integration. That is realized as follows: The new $[S]$ produced, is scanned and if its j-th element is equal to 1, the j-th element of tensor $[i_i]$ will be the i-th element of the new tensor $[i_n]$, where index i is the i-th 1 that it has been found to scan $[S]$. Then, by using the new values of the input phase voltages, the new voltage tensor is obtained. After that we can proceed to the numerical integration of system's equations, taking into account that the angular velocity of the D.C. machine is known from the solution in the previous loop of the program of the mechanical equation of the drive. In the next step of the program, currents of all branches and the thyristor voltages are calculated, and tensors $[i_p]$ and $[V_x]$ are formed from tensor $[i_n]$.

Then, each non-conducting thyristor is tested for turn-on by investigating the branch voltage and the firing angle.

A time counter begins counting when V_{th} becomes positive and the thyristor changes state when $t_{counter} \geq \frac{\alpha}{\omega}$ where α is the firing angle.

From the other hand each conducting thyristor is tested for turn-off by investigating the sign of its current. That is if $i_{th} \leq 0$ the thyristor changes state.

Finally the duration of the program is increased by the step length and it is tested if it stops or not. It depends on whether a change of the system state has been occured for reentering the program loop, as it has been shown.

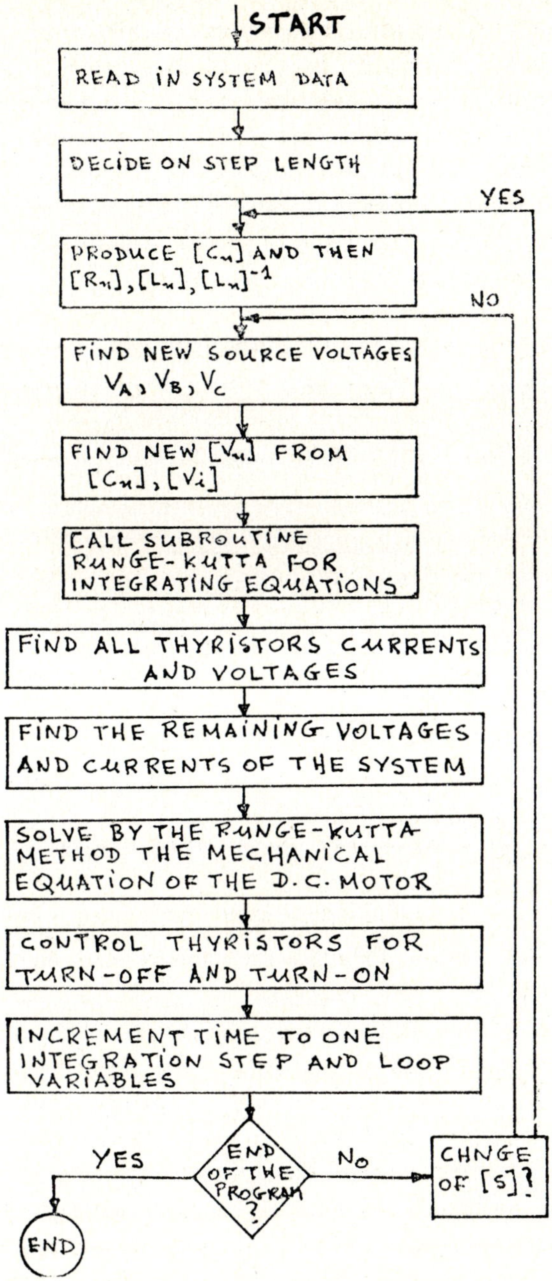

Fig.4: Flow-chart of the program.

4.2. Numerical integration and step length.

As discussed in (5) the most appropriate method of solution is the 4-th order. Runge-Kutta method, because as a one-step method is preferable from multistep or iterative methods. Between other advantages permits to the user readily available estimation of the truncation error. It has been prooved that the limit of stability for the numerical solution is reached with a step length of 2.8 times the smallest time constant of the system. The time constants of the system are found by arranging the system equations in the standard state-variable form:

$$[\dot{x}] = [A]\,[x] + [B]\,[u]$$

and finding the eigen values of $[A]$, λ_n. These appear in the solution as $\epsilon^{-\lambda_n t}$ terms, where λ_n is the eigen-value associated with n -th order local solution.

The largest eigen-value is the reciprocal of the smallest time constant.

It must be noted that in the solution of the mechanical equation the integration step length is the same as for Eq. (5).

4.3. Initial start-up conditions

To ensure correct initial conditions, we made the assumption that calculation starts at time t = 0.0119s from the begining of the time counter operation. We consider that at that time $i_L = 0$ and $\omega_o = 0$.

Under the above mentioned conditions the thyristor voltages are equal to:

$$V_1 = -V_4 = V_A$$

$$V_2 = -V_5 = V_B$$

$$V_3 = -V_6 = V_C$$

5. EXPERIMENTAL RESULTS AND PROGRAM COMPARISON

Fig. 5 shows two typical cases obtained from the operation of the bridge as a rectifier, with firing angle equal to 0° and 50°, or as an inverter with firing angle equal to 120° and 150°.

The D.C. motor used has the following characteristics:

R_M = 1.426 Ω

L_M = 0°0104 H

F = 0.0678 $N.m/s^2$

f = 0.007 Nm/rad/s

K = 1.437 N.m/A or V/rad/s

The D.C. motor operates in its nominal angular velocity equal to 1150 rpm or 120.4 rad/s.

In series with the motor there is a smoothing inductor having R_p = 10 Ω and L_p = 0.4 H.

The high accuracy of prediction was also obtained for many other combinations of circuit parameters and other modes of operation as well. This gives a high degree of confidence in the proposed method of analysis.

6. CONCLUSIONS

The method described in this paper is an extension of the method described in (5) for a 3-phase, full controlled bridge driving a D.C. motor. It is evident that the advantages of the method mentioned in the above section as in (5) make possible the extension of the method for the solution of more complex thyristor circuit configuration, different modes of operation and also different situations - normal or abnormal - of operation of such systems.

In a next stage of research a complete study on the overlap phenomenon will be attempted, trying different operation curves for the overlap angle u for different values of the inductunces of the motor and of the line feeding the converter. The study can be realized for both modes of operation of the converter and the complete four quadrant operation of the drive with two back-to-back converters can be encountered.

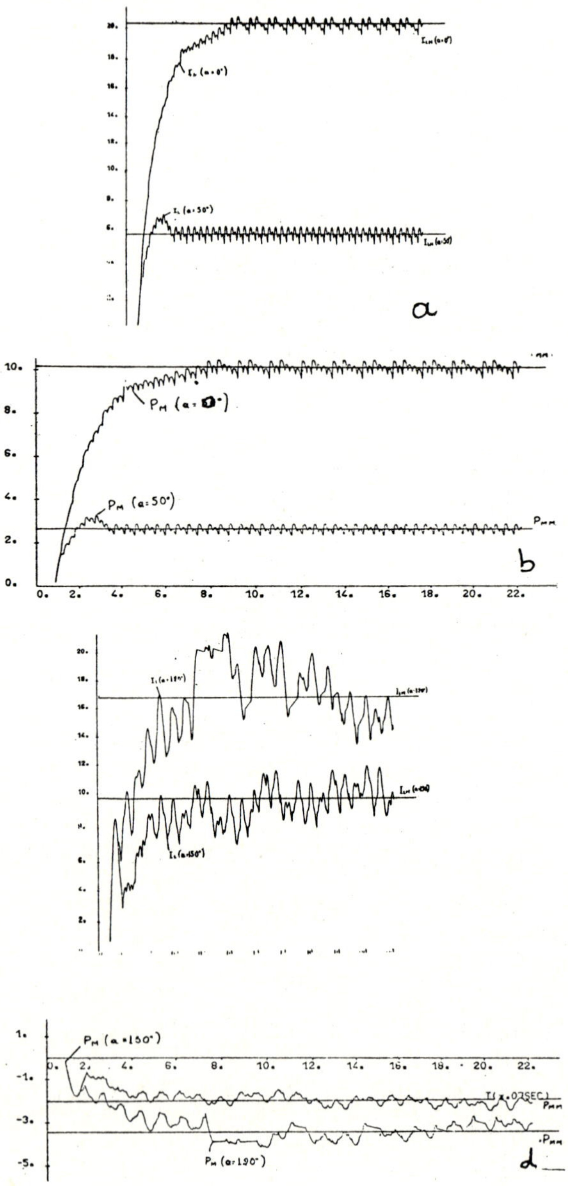

Fig. 5 : Experimental and computed results for 3-phase full controlled D. C. drive.
380 V, 50 Hz supply.
a. Rectifier operation of the bridge. Firing angles 0° and 50° - Motor armature current. Computation time compile 1.8m, 19 s. Mean values : $\alpha=0^o$, I_L =20,29A $\alpha=50^o$, =5,80A

b. Rectifier operation of the bridge. Firing angles 0° and 50° - Motor shaft power. Mean values:

$\alpha=0^o$, P_M = 5,06 KW
$\alpha=50^o$, P_M = 1,33 KW.

c. Inverter operation of the bridge. Firing angles 120° and 150°.
Generator armature current. Computation time compile 2,10 m, 2qs.
Mean values.
$\alpha = 120^o$ I_M= 17 A
$\alpha = 150^o$ I_M= 9.8 A

d. Inverter operation of the bridge. Firing angles 120° and 150°. Power delivered from the D.C. machine.
Mean values

$\alpha = 120^o$ P_M = - 3,37 KW
$\alpha = 150^o$ P_M = - 2,04 KW

REFERENCES

(1) DASKALOPOULOS G.D., Applied Linear Algebra - Part 1 (Athens 1975)

(2) DEWAN S.B - STRAUCHEN A., Power Semiconductor circuits (Wiley-interscience 1975)

(3) SKORDALAKIS E., ANSI FORTRAN Manual (Athens 1972)

(4) KUSKO A., Solid state D.C. motor drives (Mc Graw Hill 1969)

(5) WILLIAMS S. - SMITH I.R., Fast digital computation of a 3-phase thyristor bridge, Pro I.E.E. Vol. 120, No. 7, July 1973, p.p. 791 - 796.

Session 6

MODELLING OF SYNCHRONOUS MACHINES AND CONVERTERS (I)

Chairman: B. TRANNOY

ELECTRICAL MACHINES AND CONVERTERS - Modelling and Simulation
H. Buyse and J. Robert (editors)
Elsevier Science Publishers B.V. (North-Holland)/© IMACS, 1984

GLOBAL SIMULATION METHOD OF SOLID ROTOR MACHINES FED BY STATIC CONVERTERS

B. Davat, O. Hounkonnou, M. Lajoie-Mazenc

Laboratoire d'Electrotechnique
et d'Electronique Industrielle
2 rue Camichel
31071 TOULOUSE CEDEX
FRANCE

The authors present a global simulation method of converter machine set with the resolution of both the space time differential equations which define the field in the machine and the electrical equations representing the behaviour of the static converter. The simplification of the stator structure presented leads to an important decrease of the state variables issued from the machine equations and gives the possibility of modelling the whole system. This method is illustrated by the simulation of the functionning of a permanent magnet synchronous machine fed by a current inverter.

1. INTRODUCTION

Many papers have shown the advantages of global simulation method for the modelling of converter fed electric machines (1), (2), (3). Generally based on a semi-conductor binary model, these methods do not need the knowledge of the different functionning sequences of the static converter before the simulation. This method is then well adapted to the study of converter machine set when the converter contains many semi-conductors or to the study of unusual functionning sequences (4).

The authors have already shown in (5) that these methods can be used for the modelling of electric machines with solid parts in the rotor where the harmonics delivered by the static converter induce eddy currents. The modelling is then based on the resolution of both the space time differential equations which define the electromagnetic field in the machine and the electrical equations representing the behaviour of the static converter.

The description, of the machine by its geometrical dimensions and of the converter by its electrical circuits, leads to a differential equation in which the state variables come from both machine and converter. These state variables are the vector potentials at the nodes of the meshes where eddy currents flow, the feeding currents in the stator windings, the voltage accross the capacitors and the currents through the inductances of the converter.

If the integration of the differential equation which represents the whole system leads to the knowledge of any variable concerning the machine or the converter, the number of state variables issued from the machine equations is very important.

For this reason the authors present in this paper a method to modelize th machine converter set where a simplification of the machine structure allows an important decrease of the number of state variables. This method can be used for electric machines where the rotor is built of different parts of constant thickness. The proposed method is illustrated by the simulation of the functionning of a permanent magnet synchronous machine fed by a current inverter.

2. MACHINE MODELLING

2.1. Principle

The modelling of the machine is based on a simplification of the structure of the machine. This method is used for smooth rotor electrical machines where the rotor is built of different parts of constant thickness. The simplified stator structure is obtained by replacing the conductors of each phase by the fundamental of a conductor density placed on the surface of an infinite permeability stator (figure 1). The machine is supposed to be enough long to allow a 2 D field modelling.

The current density at the stator surface is then in a rotating frame :

$$k(\theta,t) = \sum_{n=1}^{\infty} i_{sn}(t)\; k_o \sin\left(p\theta + \omega t - (n-1)\frac{2\pi}{m}\right) \qquad (1)$$

In any point of the structure, the vector potential a is defined by :

$$\Delta a = \sigma\mu \frac{\partial a}{\partial t} \qquad (2)$$

Let :

$$\bar{k}(\theta,t) = \sum_{n=1}^{\infty} i_{sn}(t)\cdot k_o \cdot e^{j\left(p\theta + \omega t - (n-1)\frac{2\pi}{m}\right)}$$
$$= \bar{K}\cdot k_o \cdot e^{j(p\theta+\omega t)} \qquad (3)$$

with :

$$K = M\; I$$
$$M = \left| 1 \;\; \ldots \;\; e^{-j(n-1)\frac{2\pi}{m}} \right| \qquad (4)$$

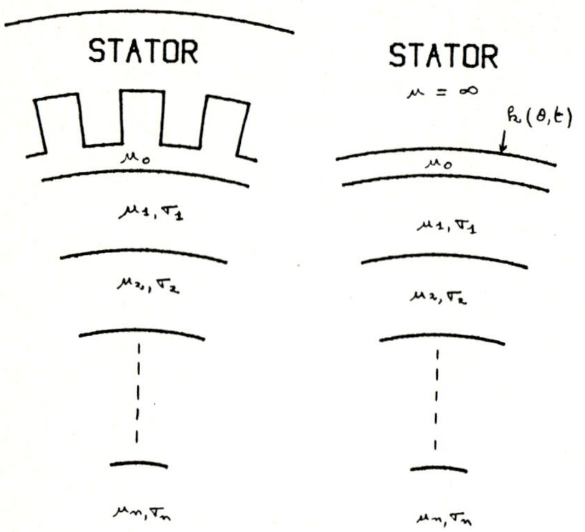

Figure 1 : Simplified stator structure

The vector potential $\bar{a}$ which verifies (2) and the boundary conditions :

$$\left(\frac{\partial \bar{a}}{\partial r}\right)_{r=R_0} = \mu_0 \bar{k} \qquad (\bar{a})_{r=0} = 0 \tag{5}$$

is linked to the real vector potential a by :

$$a(\theta,t) = \Im m(\bar{a}(\theta,t)) \tag{6}$$

For the considered structure and the current density $k(\theta,t)$ which is sinusoïdal in θ , the vector potential $\bar{a}$ can be expressed as :

$$\bar{a}(\theta,t) = \bar{\mathcal{A}}(r,t) \cdot k_0 \cdot e^{j(p\theta+\omega t)} \tag{7}$$

The knowledge of $\bar{\mathcal{A}}(r,t)$ gives the value of the vector potential in any point of the structure. The discretisation of the integral equation which defines the vector potential :

$$\int_c \left(\frac{\nu}{r}\frac{\partial \bar{a}}{\partial \theta} dr - \nu r \frac{\partial \bar{a}}{\partial r} d\theta\right) + \iint_S \sigma \frac{\partial \bar{a}}{\partial t} r\, dr\, d\theta = 0 \tag{8}$$

is performed by the finite difference method (figure 2).

The matrix equation obtained links the value of $\bar{\mathcal{A}}(r,t)$ at the nodes. If we split these discrete values in two vectors according to wether the node belongs to a part where eddy currents are induced (vector A_1) or not (vector A_2), (8) leads to :

$$B_0 \dot{A}_1 + (B_1 - j\omega B_0) A_1 + B_2 A_2 = 0$$
$$C_1 A_1 + C_2 A_2 = N_2 \bar{K} \tag{9}$$

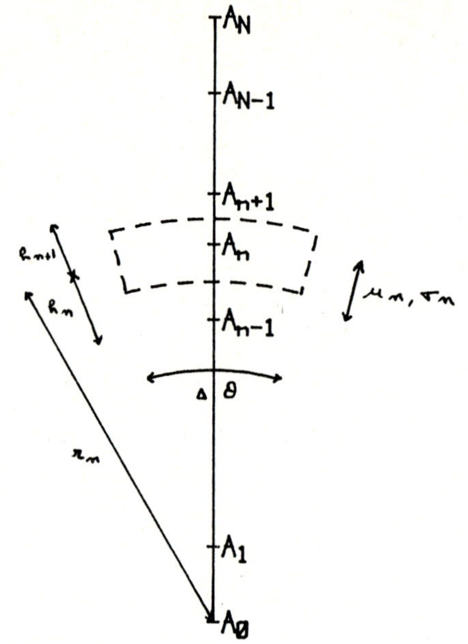

Figure 2 : Discretisation principle

The voltage V accross the windings of the machine is :

$$V = RI + L\dot{I} + \dot{\Phi} + E \tag{10}$$

The flux Φ can be obtained from the value of the vector potential at the surface of the stator. In the phase n :

$$\varphi_n = \int_0^{2\pi} \ell\, a(\theta, R_0, t) \cdot k_0 \cdot \sin\left(p\theta + \omega t - (n-1)\frac{2\pi}{m}\right) r\, d\theta$$
$$= \frac{\ell k_0^2 \pi R_0}{2}\left(\bar{A}_N e^{j(n-1)\frac{2\pi}{m}} + \bar{A}_N^* e^{-j(n-1)\frac{2\pi}{m}}\right) \tag{11}$$

Hence :

$$\Phi = \beta\,(M^{*T} D A_2 + M^T D A_2^*) \tag{12}$$

with :

$$\beta = \frac{\ell k_0^2 \pi R_0}{2} \qquad D = |\,0\ 0 \cdots 0\ 1\,|$$

The different equations (9), (10) and (12) lead to the equation which defines the functionning of the machine :

$$\dot{I} = Z_1 I + Z_2 A_1 + Z_2^* A_1^* + Z_3 (V - E)$$
$$\dot{A}_1 = Z_4 A_1 + Z_5 I \tag{13}$$

with :

$$Z_1 = -Z_3^{-1}(R - 2\,\Re(M^{*T} D C_2^{-1} C_1 Z_5))$$
$$Z_2 = Z_3^{-1}(\beta M^{*T} D C_2^{-1} C_1 Z_4) = V_2 + j W_2$$
$$Z_3 = L + 2\beta\,\Re(M^{*T} D C_2^{-1} N_2 M)$$
$$Z_4 = -B_0^{-1}(B_1 - B_2 C_2^{-1} C_1) - j\omega \mathbb{1} = V_4 - j\omega \mathbb{1}$$
$$Z_5 = -B_0^{-1} B_2 C_2^{-1} N_2 M = V_5 + j W_5$$

2.2. Validity of the model

To show the validity of the model we have compared the simulation results with the analytical ones for a voltage feeding by sinusoïdal waves of the structure representated in figure 3. For this structure and for a sinusoïdal feeding, the steady state can be obtained analytically (6).

We have calculated the feeding current, the vector potential on the stator surface, an eddy current density near the airgap (figure 4) and the joule losses in all the conductive part (figure 5) for different feeding frequencies (the feeding voltage was choosen to obtain the same current in the stator coils).

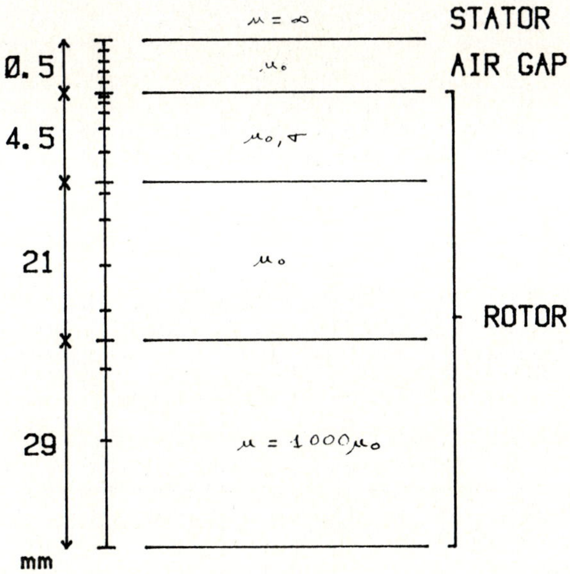

Figure 3 : Testing structure and discrete points

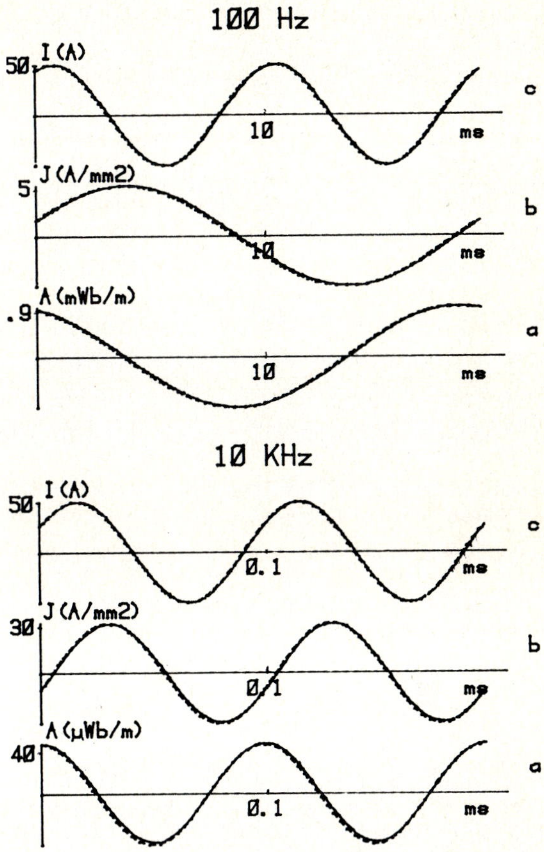

Figure 4 : Simulation results and calculated results (in dashed lines)
Ω = 1 500 rpm, p = 2
a : vector potential (y = 0)
b : eddy current (y = - 0.0006)
c : feeding current in phase 1

3. CONVERTER MACHINE SET EQUATION

By using a binary model to represent the semiconductors we can describe the behaviour of the converter by :

$$\dot{X}_c = A_c X_c + B_{c1} U_c + B_{c2} I \tag{14}$$

in which X_c , U_c and I are the state variables of the convertor circuits (current through the inductances and voltage accross the capacitors), the supply current or voltage of the converter and the current through the stator windings of the machine.

The voltage accross the windings of the machine can be obtained from the converter equations as

$$V = Q_0 X_c + Q_1 U_c + Q_2 I \tag{15}$$

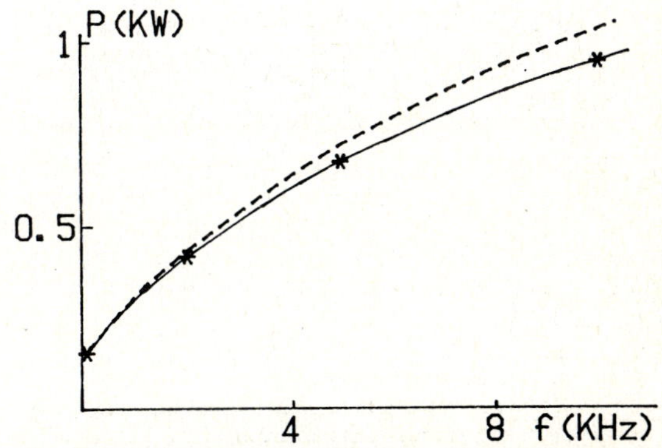

Figure 5 : Eddy current losses as a function of the feeding frequency : Imax = 50 A (analytical results in dashed line)

The different equations (13), (14) and (15) lead to :

$$\dot{X} = F X + G U \qquad (16)$$

with :

$$X = \begin{vmatrix} X_c \\ I \\ R_1 \\ Y_1 \end{vmatrix} \qquad A_1 = R_1 + j Y_1$$

$$U = \begin{vmatrix} U_c \\ E \end{vmatrix}$$

$$F = \begin{vmatrix} A_c & B_{c2} & \mathbb{0} & \mathbb{0} \\ Z_3 Q_0 & Z_1 + Z_3 Q_2 & 2.V_2 & -2W_2 \\ \mathbb{0} & V_5 & V_4 & \omega \mathbb{1} \\ \mathbb{0} & W_5 & -\omega \mathbb{1} & V_4 \end{vmatrix}$$

$$G = \begin{vmatrix} B_{c1} & \mathbb{0} \\ Z_3 Q_1 & -Z_3 \\ \mathbb{0} & \mathbb{0} \\ \mathbb{0} & \mathbb{0} \end{vmatrix}$$

This matrix equation represents the whole system and allow to determine any variable concerning the machine of the converter. The knowledge of the converter state variables gives the value of any current or voltage in the converter. The real and imaginary parts of the discrete value of $\mathcal{A}(x,t)$ lead to the vector potential in any point of the machine and then to the eddy current densities, the losses and the electromagnetic torque

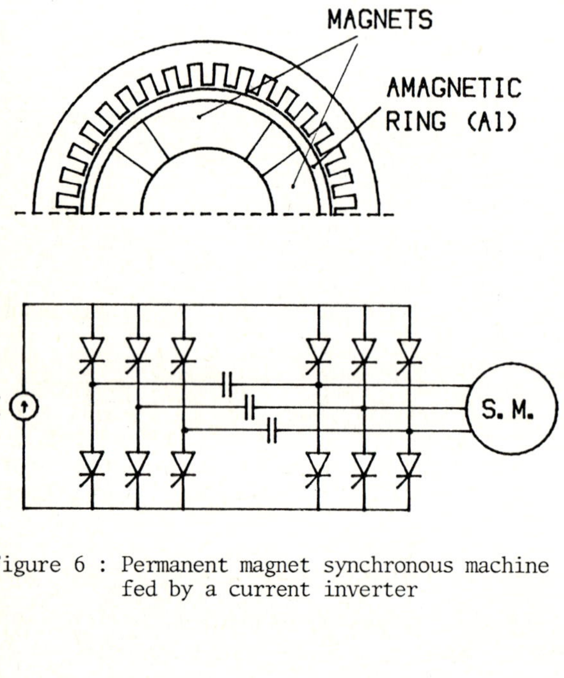

Figure 6 : Permanent magnet synchronous machine fed by a current inverter

4. SIMULATION EXAMPLE

The converter set which will illustrate this study is presented in figure 6. It is composed of a permanent magnet synchronous motor fed by a current inverter. The order of the differential system which represents the whole system is 20. The state variables are :

- the voltage accross the capacitors (3),
- the current in the stator coils (3),
- the real and imaginary parts of $\mathcal{A}(x,t)$ (7x2).

The studied system beeing linear between two charges of state of semi-conductors, the integration of the differential equation (16) is performed with a matrix experimental :

$$X(t+T) = e^{FT}.X(t) + F^{-1}.\left(e^{FT} - \mathbb{1}\right).G.U \qquad (17)$$

Different results concerning both converter and machine are presented in figure 7.

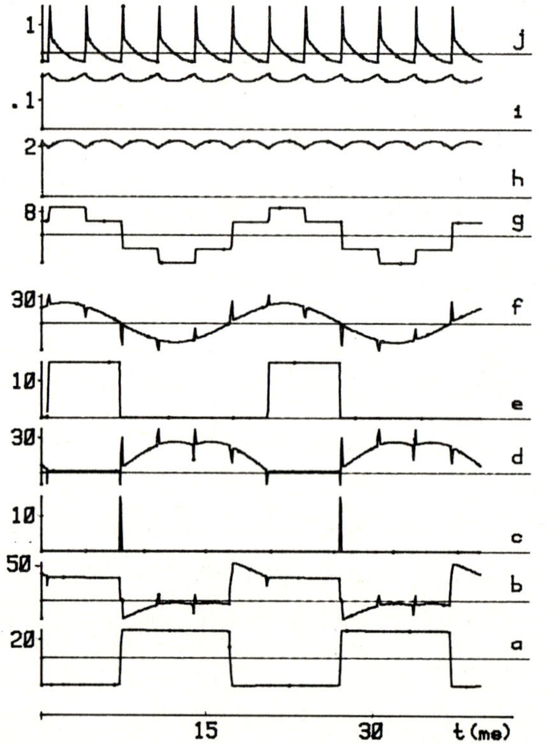

Figure 7 : Simulation example :
a : capacitor voltage (V)
b : auxilary thyristor voltage (V)
c : auxilary thyristor current (A)
d : main thyristor voltage (V)
e : main thyristor current (A)
f : machine voltage (V)
g : machine current (A)
h : electromagnetic torque (N.m)
i : potential vector on the stator surface (mWb/m)
j : eddy current density near the surface of the ring (A/mm2)

5. CONCLUSION

For the modelling of converter machine set with a field characterisation of the machine we have shown the possibility to simplifie the structure of the machine in the case of a smooth rotor. This simplification leads to reduce the order of the differential matrix equation which represent the behaviour of the whole system.

The state variables for the machine are no longer the vector potential in the conductive parts but only a one dimension space variable. Then, the simulation duration for the study of the converter machine set decrease but the machine is stil representated by its geometrical dimensions.

6. NOMENCLATURE

a : vector potential
$\bar{a}$: complex vector potential
$\bar{\mathcal{A}}$: one dimensionnal vector potential variable
E : e.m.f. vector
Φ : flux vector
I : stator current vector
i_{s_n} : stator current in phase n
k : current density on the stator surface
$\bar{k}$: complex current density
ℓ : machine lengh
L : leakage inductance vector
m : number of phases
Ω : rotor speed
$2p$: number of poles
R : stator resistance vector
R_0 : stator radius
σ : conductivity
τ : integrating step
μ_0 : permeability of air
μ : permeability
ν : reluctivity
U_c : convector input vector
X_c : convector state vector
ω : electrical speed
θ : mechanical angle

7. REFERENCES

(1) Demerdash, N.A., Nehl, T.W., "Discrete time dynamic modeling of brushless DC motor transistorized power conditioner units for actuation and drive application", CONUMEL 80, Lyon, 28-30 avril 80

(2) Davat, B., Foch, H., Trannoy, B., "Analysis of a converter fed asynchronous machine by a global simulation method", Electric machines and Electromechanics, Vol. 5, pp 201-210, 1980

(3) Johns, T.M., "Torque production in permanent-magnet synchronous motor drives with rectangular current excitation", IEEE-IAS, Mexico, Annual Meeting 1983, pp 476-487

(4) Davat, B., Hapiot, J.C., Foch, H., "Application of a global simulation method to the study of faults in the function of a static converter", IEE Proc., Vol 128, Pt. B, n° 6, nov. 1981

(5) Davat, B., Lajoie-Mazenc, M., Hector, J., "Magnetic structure and feeding circuit modelling", COMPUMAG, 30 may-2 june, 1983

(6) Davat, B., Rezine, H., Lajoie-Mazenc, M., "Eddy currents in solid rotor permanent magnet synchronous motors fed by voltage inverter", Electric machines and Electromechanics, Vol. 7, pp 115-124, 1982

ELECTRICAL MACHINES AND CONVERTERS - Modelling and Simulation
H. Buyse and J. Robert (editors)
Elsevier Science Publishers B.V. (North-Holland)/© IMACS, 1984

NUMERICAL SIMULATION OF PERMANENT MAGNET SYNCHRONOUS MACHINES ASSOCIATED WITH VOLTAGE OR CURRENT INVERTERS

Francis Piriou and Adel Razek

LABORATOIRE DE GENIE ELECTRIQUE DE PARIS
L.A. 127 C.N.R.S., E.S.E.,
Universités de Paris VI et XI
Plateau du Moulon - 91190 GIF-SUR-YVETTE -
(FRANCE)

In the proposed paper, the different inductances in a permanent magnet synchronous machine are calculated from the field distribution obtained from finite element calculations. These inductances are used in the machine differential equations which are solved by numerical integration. The complete simulation of the variable speed system can be done if the machine differential equations are solved through the operating equations of the static convertor. In the present work such simulation is given for both cases of current and voltage source convertors associated with the permanent magnet machine. Application on a 6.2 kVA machine is done and numerical results are compared with experience.

1. INTRODUCTION

The development of high energy permanent magnets of the rare earth cobalt type has made the application of permanent magnets as sources of excitation in synchronous machine very attractive without penalty in weight and size. The elimination of rotor winding and excitation auxiliaries, has resulted in a machine package which exhibits significantly improved reability and rotor life.

For variable frequency applications the self controlled synchronous machines associated with static convertors are currently used nowadays and have particular advanteges over other forms of combinations in many applications. The predetermination of machine parameters is necessary to study the characteristics of the variable speed system. Such predetermination can be done using the well known classical methods supposing the magnetic induction sinusoidaly distributed in the machine air-gap. Such method use constant values of inductances following the two machine magnetic axes |1| , |2| , |3| , and consider the machine as an idealised one.

The use of such methods in the predetermination of the machine performance under non sinusoidal conditions |4| and in the presence of magnetic saturation |5| may lead to serious errors both in the form and magnitude of the different calculated variables. Under these conditions, the idealised machine theory fails and only numerical methods can be used to solve directly the variable coefficient differential equations of the machine |6| . The different inductances have to be used in these equations are functions of the geometry and saturation of the magnetic circuit ; in other words, the rotor position, the no load saturation level and the armature reaction |7||8| .

The purpose of the work reported here, was to develop a numerical model for a variable speed drive consists of a self controlled voltage or current fed machine.

2. SYSTEM EQUATIONS

The studied system is given in figure 1. The inverter in such a system is a voltage fed one or a natural commutation current fed one.

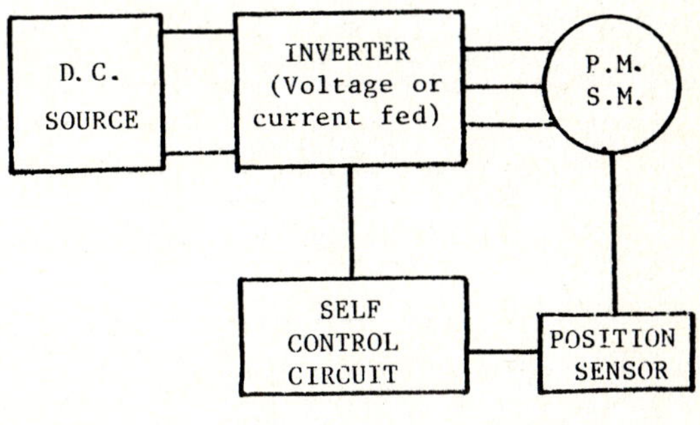

Figure 1. : Studied system

2.1 Machine differential equations

The variable coefficient differential equations of a permanent magnet synchronous machine without dampers for a given speed N take the form:

$$v = ri + \frac{d\emptyset}{dt} + e \qquad (1)$$

where vectors v, i, $\emptyset$, e and matrix r have widely accepted meanings.

The saturation and dissymetry of the magnetic circuit make the flux dependant of the local

permeability and the rotor angle position.
For a given specific problem, the solution of equation (1) can be done by numerical integration associated with a local flux calculation accounting for saturation.

2.2 Inverter operating equations

2.2.a Voltage source

Figure 2a shows schemas of the machine with a voltage source inverter and its operating sequence for a $2\pi/3$ full wave rectangular voltage.
Figure 2b shows the equivalent circuit of the machine during a conduction period ($\pi/3$, the first conduction in figure 2a). The commutation period is supposed negligible. From figure 2b the next equations can be written ;

$$\begin{aligned} 0 &= v_a - v_c \\ 0 &= i_a + i_b + i_c \\ U &= v_a - v_b \end{aligned} \qquad (2)$$

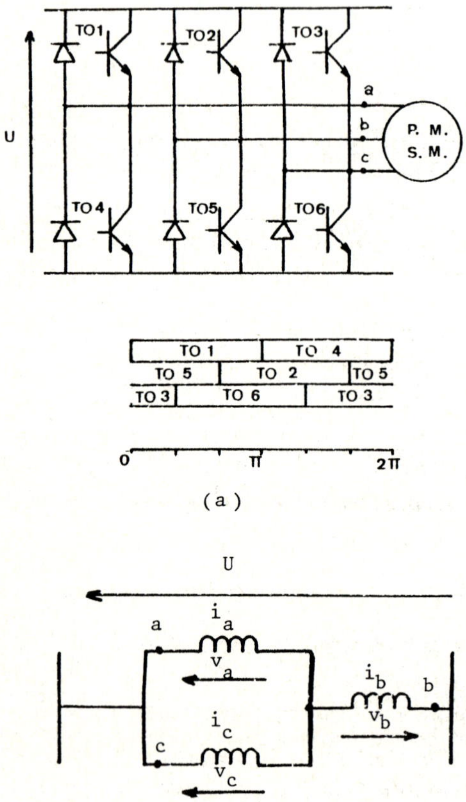

Figure 2. Schemas and equivalent circuit during conduction period of a voltage fed machine.

2.2.b Current source

Figure 3.a shows the schemas of the machine with a current source natural commutation inverter and its operating sequence. Figure 3.b shows the machine equivalent circuits during $\pi/3$ for succesive commutation and conduction periods (first one in figure 3.a). During commutation (phases c and a) one can write ;

$$\begin{aligned} 0 &= v_a - v_c \\ 0 &= i_a + i_b + i_c \\ U &= v + v_a + v_b \end{aligned} \qquad (3)$$

and during conduction (phases a and b)

$$\begin{aligned} 0 &= i_a + i_b \\ U &= v + v_a - v_b \end{aligned} \qquad (4)$$

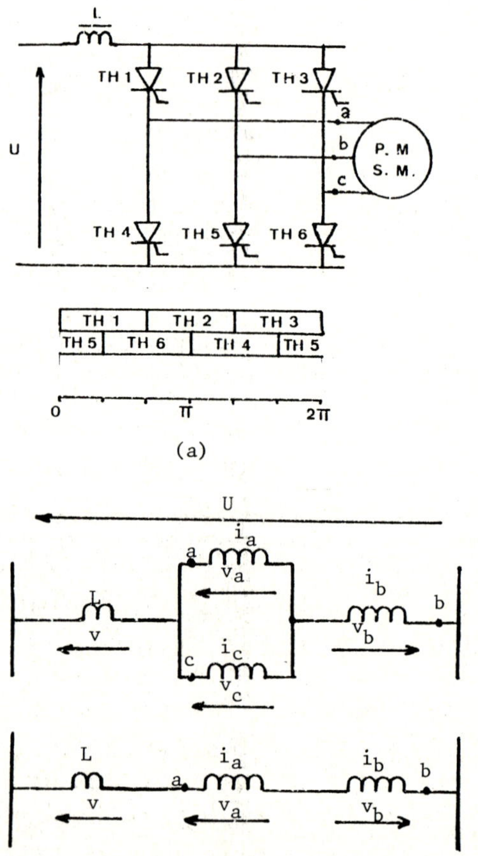

Figure 3. Schemas and equivalent circuits during commutation and conduction periods of a current fed machine.

2.3 Machine inductances

The flux linkage in the winding "a" in equation (1) can be written under the form :

$$\phi_a = \sum_x \varphi_{ax} \qquad (5)$$

where x represents the different windings in the armature ;the permanent magnets are represented by equivalent windings. The flux linkages φ_{ax} in (5) can be written as :

$$\varphi_{ax} = d_{ax}\, i_x \qquad (6)$$

d_{ax} in (6) are functions of local saturation and rotor position.We call such functions equivalent saturated inductances.

2.4 Finite element calculation

From Maxwell equations and the characteristics of magnetic materials, relative to the case of magnetostatic problem, in electric machines we can write the following Poisson's equation :

$$\vec{\nabla}_\wedge (\nu \vec{\nabla}_\wedge \vec{A}) = \vec{J} \qquad (7)$$

where $\vec{J}$: the current density

$\vec{A}$: the vector potential

$\nu = 1/\mu$: the magnetic reluctivity

The finite element method is used to calculate the local flux distribution in bidimensional cross section of the machine |9| , |10| . For a given operating point, the static solution of Poisson's equation |7| gives values of local permeabilities. These values are then used to determine linearely the local vector potentials due to each circuit current for different rotor angles over one pole pitch |7| . It may be noted that the later calculations are not time consuming due to absence of iteration. The values of the equivalent saturated inductances can be calculated from vector potential values, |7| , |8|

In this calculation each magnet is replaced by a piece of material with the same magnet reluctivity in addition to a fictitious winding (for which the ampere turns are equal to the product of the magnet coercitivy times its height in the direction of flux).

Equivalent inductance can be expressed under a general expanded series, of the form : (8)

$$d(\theta,\nu) = \sum_{n=0}^{\infty} f_n(\nu)\cos 2n\theta + g_n(\nu)\sin 2n\theta$$

$$0 < \theta < 2\pi$$

The functions f_n and g_n in (8) depend on the operating point.

3. NUMERICAL SIMULATION

In the last section the inverter operating equations have been developed. It can be noticed that the equations (2) simulating conduction in voltage source case are the same as those simulating commutation (3) in current source case if the smoothing inductance is omitted.

Under these conditions the system simulation for voltage fed or current fed cases can be done by combining equations (1) and (3), or (1), (3) and (4) respectively. The inductances in these equations are considered by expressions as (8).

The solution of the system equations can be done using step-by-step numerical integration. The flow chart describing such solution is shown in figure 4.

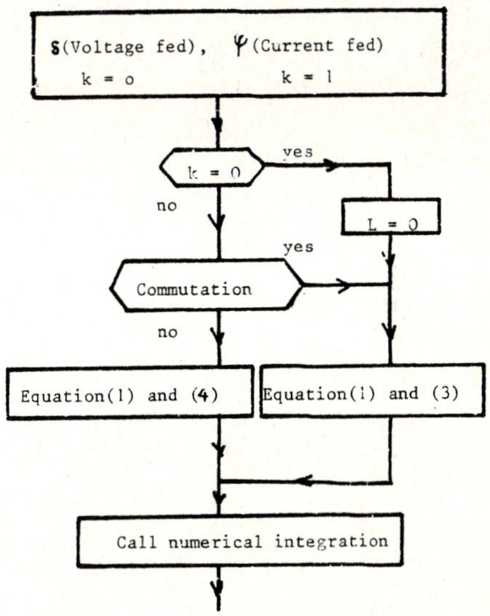

Figure 4. Flow chart ; δ (internal load angle) ψ(phase angle of current and e.m.f.)

4. RESULTS

The proposed model has been applied to a 6.2 kVA permanent magnet machine: 500 V, 1000 Hz, 4 poles. The machine is fed by a transistor inverter or a natural commutation thyristor inverter.

4.1 Armature inductance variations with the rotor position

As explained above the inductances are calculated from vector potentials in the machine windings. In these calculations, the different rotor positions have been considered, for commodity, in the next way : the rotor is considered stationary, the slots occupied by each of the different phases are reconsidered for each new corresponding rotor position.

The time variations of d_{aa} and d_{bc} for the studied machine operating under nominal conditions are shown in figure 5. The armature inductances in this figure are given in per unit value based on the maximum value of the armature self inductance. It may be noted in this figure that the value of d_{aa} is greater for θ = 0 (pole axis) than for θ = $\pi/2$ (quadrature axis).

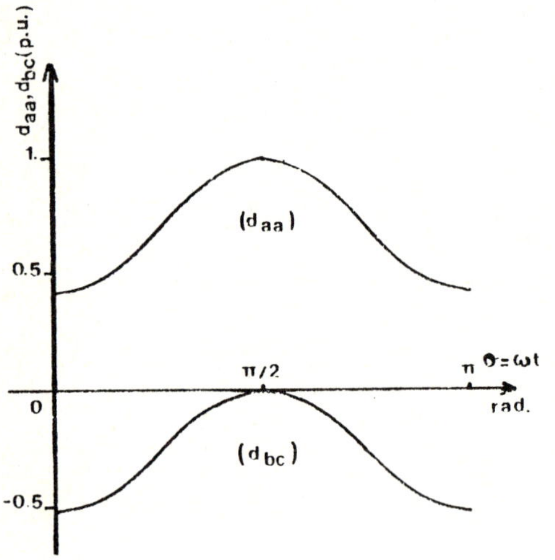

Figure.5 Values of d_{aa} and d_{bc} (p.u.) as a function of the rotor position, for the nominal operating point.

4.2 Voltage fed system

Figure 6.a shows the experimental current wave form for a voltage control angle (internal angle) δ = 20° elec. and a speed of 3100 r.p.m.. Figure 6.b shows the corresponding calculated wave form. Figure 6.c shows the calculated torque wave form.

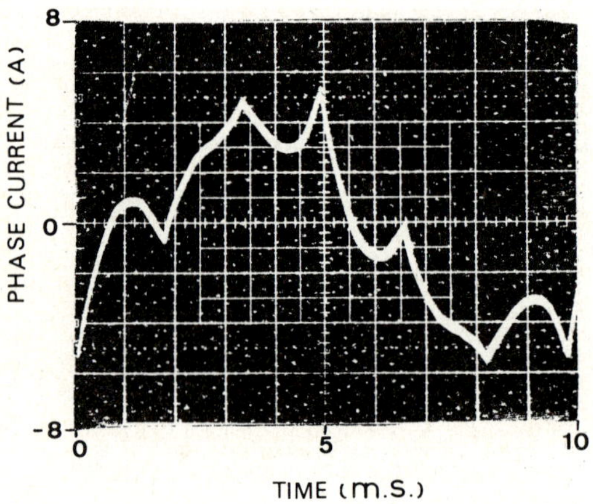

$I_{r.m.s.}$ = 3.32 (A)

Figure 6.a

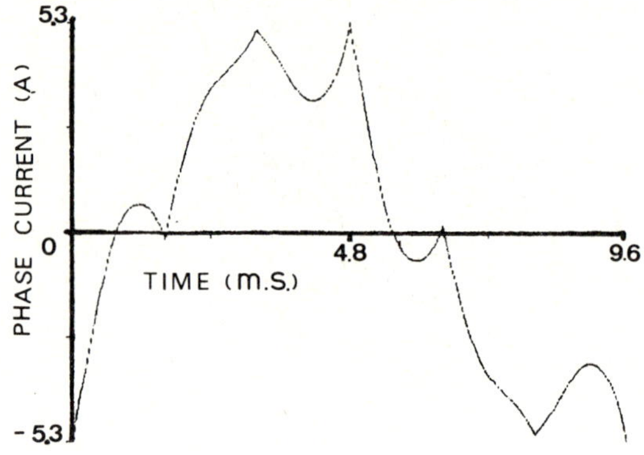

$I_{r.m.s.}$ = 3.30 (A)

Figure 6.b

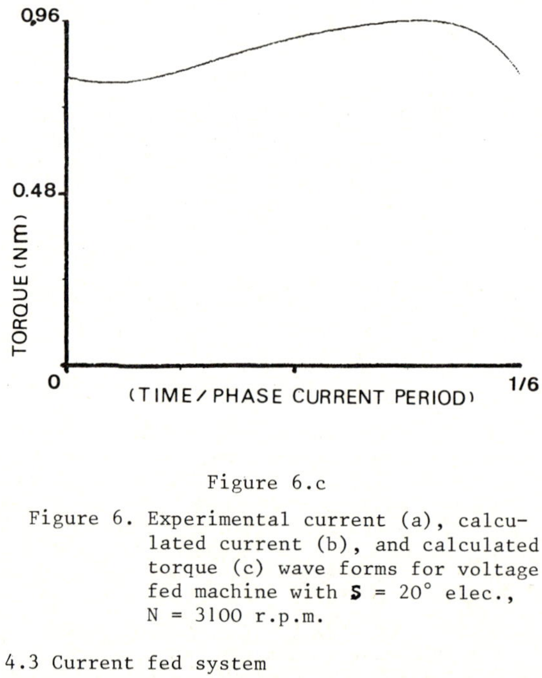

Figure 6.c

Figure 6. Experimental current (a), calculated current (b), and calculated torque (c) wave forms for voltage fed machine with δ = 20° elec., N = 3100 r.p.m.

4.3 Current fed system

Figures 7a and 7b show the mesured current and voltage wave forms for a current control angle (phase angle of current with respect to e.m.f.) ψ = 55° elec. and a speed of 3100 r.p.m.. Figures 8a and 8b show the corresponding calculated wave forms and figure 9 shows the calculated torque wave form.

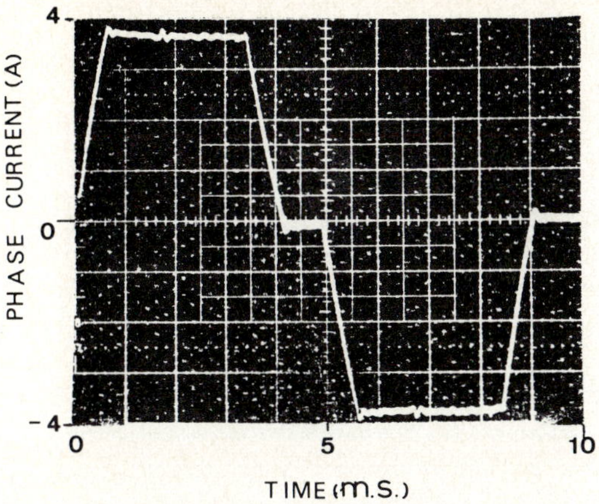

$I_{r.m.s.}$ = 3.19 (A)

(a)

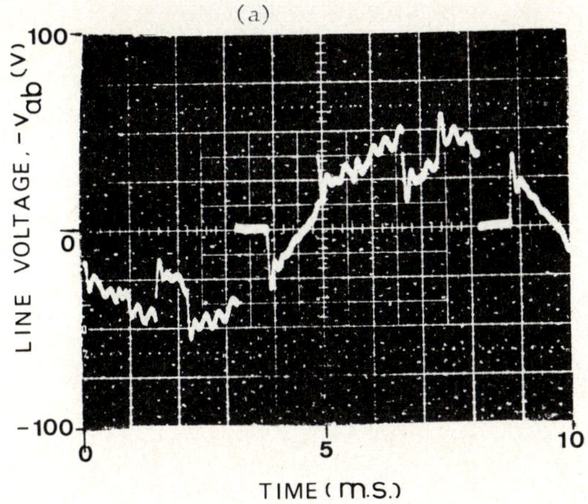

(b)

Figure 7. Experimental current (a) and voltage (b) wave forms for current fed machine with φ = 55° elec. N = 3100 r.p.m.

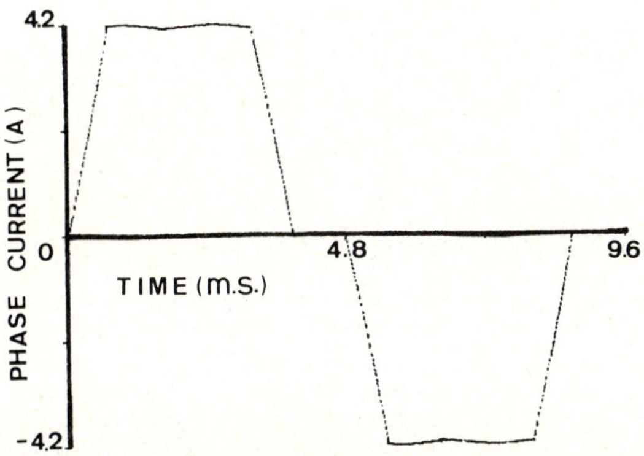

$I_{r.m.s.}$ = 3.45 (A)

Figure 8.a

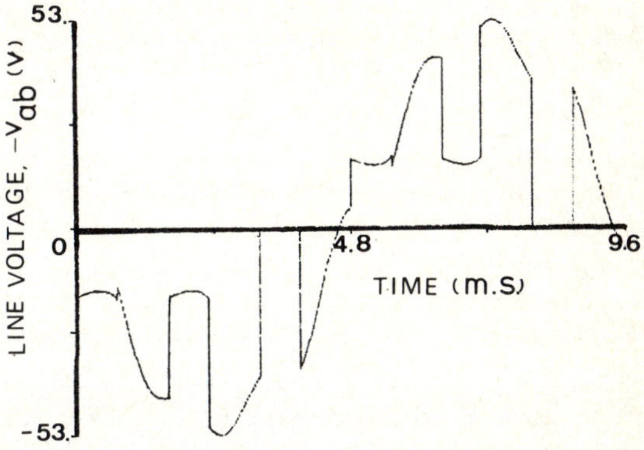

Figure 8.b

Figure 8. calculated results corresponding to Figure 7.

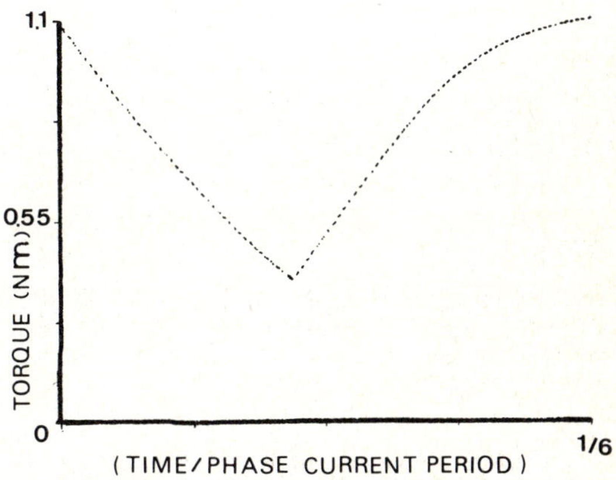

Figure 9. Calculated torque wave form corresponding to figure 8.

5. CONCLUSION

In the paper a numerical model of a variable speed permanent magnet synchronous machine drive is discussed. This model is valid for both cases of voltage and current fed inverters. The used numerical procedure associated with the finite element method is simple and efficient. Results obtained using the proposed model and those given by experience are in agreement.

6. REFERENCES

1 Park, R.H., Definition of an ideal synchronous machine and formula for the armature flux linkages , G.E.R. vol. 48 (1928) 332-334.

2 Concordia, C., synchronous machines (Wiley New-York 1951).

3 Lewis, W.A., A basic analysis of synchronous machines. I., trans. A.I.E.E. vol. 77 (1958) 436-456.

4 Razek, A. and Poloujadoff, M., Analytical approach to the operation of a synchronous machine associated with a thyristor bridge, E.M.E. vol. 2 (1978) 167-184.

5 Jokl, A.L., Amstuitz, L.J. and Erdelyi, E.A., Nonsymetrical nonlinear aerospace alternator on nonlinear loads, trans I.E.E.E. vol. AES 11 (1975) 298-315

6 Chassande, J.P., Razek, A. and Poloujadoff, M., Various pratical results concerning the operation of inverter fed self controlled synchronous machines, trans I.E.E.E. vol. PAS 101 (1982) 4645-4655.

7 Piriou, F. and Razek, A., A numerical model for saturated inductances in synchronous machines, E.M.P.S. vol. 8 (1983) 215-224.

8 Piriou, F. and Razek, A., Calculation of saturated inductances for numerical simulation of synchronous machines, trans. I.E.E.E. MAG. 19 (1983)

9 Chari, K. and silvester, P., Analysis of turboalternator magnetic fields by finite elements, trans. I.E.E.E. vol. PAS 90 (1971) 454-464.

10 Razek, A., Coulomb, J.L., Feliachi, M. and Sabonnadiaire, J.C., Conception of an air-gap element for the dynamic analysis of the electromagnetic field in electric machines, Trans. I.E.E.E., vol. MAG 18 (1982) 655-659.

ELECTRICAL MACHINES AND CONVERTERS - Modelling and Simulation
H. Buyse and J. Robert (editors)
Elsevier Science Publishers B.V. (North-Holland)/© IMACS, 1984

SIMULATION PROGRAM FOR AN INVERTER LOADED SYNCHRONOUS MACHINE

Manfred Manowarda

Technical University Graz
Institute for Electromagnetic Energy Conversion
Graz, Austria

A digital simulation program for a thyristor-bridge loaded synchronous machine is presented. An advantageous method of inverter modelling allows simulation of any possible connection state in a simple manner. The R-L-elements representing the thyristors in the equation system change between very high and very low values according to the state of thyristors (Method of rigid topology). Equation system modification in the case of connection state change-over is confined to alteration of only a few coefficients.

1. INTRODUCTION

The usage of systems consisting of electric machines and inverters becomes more and more common. There is still an interest in finding instruments for simulation and prediction of stresses in the machine as well as in the inverter components (torques, harmonics, losses, currents in the case of a defect). For the system investigated in this paper (synchronous machine connected with 3-phase thyristor bridge) some simulation programs have been developed, already [1,2,3,4,5]. Some of them use far-reaching simplifications, as the machine-inverter system in them is only a component of a greater system, others are confined to the connection state sequence of normal operation (2 or 3 valves conducting), while programs which are able to consider all possible connection states mostly are complicated. A lot of different differential equation sets must be considered, and the logical structure for change-over between them makes these programs hardly comprehensible (method of varied topology). The program presented here operates with the same equation system for all connection states; inverter connection state consideration is effected by proper choice of R-L-elements in the valve branches.

2. THE INVESTIGATED SYSTEM

A three-phase synchronous machine with a damper winding is connected with a naturally commutated thyristor bridge. Firing commands for the thyristor bridge are obtained by rotor position sensing, adding proper shift angles for each thyristor, and an arbitrary delay angle. An optional torque-speed characteristic of the driven resp. driving aggregate can be used in the simulation program. The inverter is fed by resp. feeds a constant voltage source over a smoothing reactance and a resistor:

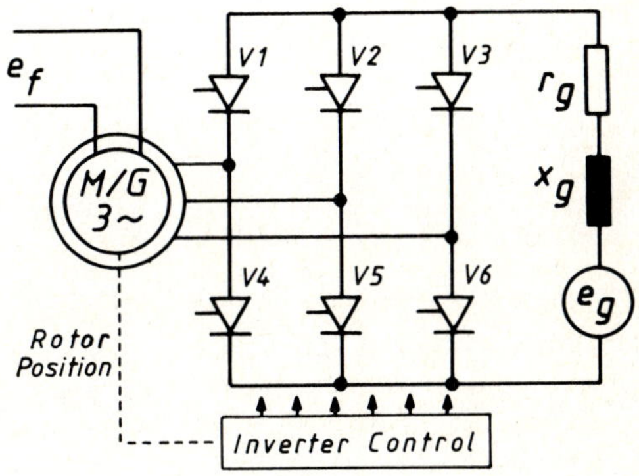

Fig. 1: Investigated system

3. MATHEMATICAL MODEL OF THE MACHINE

The synchronous machine is represented by the d-q-equations, with proper normalization as follows (normalized variables are written in italics):

a) Differential equations:

$$e_{sd} = - r_s \cdot i_{sd} - \frac{d\psi_{sd}}{d\tau} + \omega \cdot \psi_{sq}$$

$$e_{sq} = - r_s \cdot i_{sq} - \frac{d\psi_{sq}}{d\tau} - \omega \cdot \psi_{sd}$$

$$e_f = - i_f - \frac{x_{fd}}{r_f} \cdot \frac{d\psi_f}{d\tau}$$

$$0 = r_{rd} \cdot i_{rd} + \frac{d\psi_{rd}}{d\tau}$$

$$0 = r_{rq} \cdot i_{rq} + \frac{d\psi_{rq}}{d\tau}$$

Motion equations:

$$\frac{d\vartheta}{d\tau} = \omega$$

$$\frac{d\omega}{d\tau} = \frac{1}{T_m}[m - (i_{sd} \cdot \psi_{sq} - i_{sq} \cdot \psi_{sd})]$$

b) Algebraic equations:

$$\psi_{sd} = x_{sd} \cdot i_{sd} + x_{hd} \cdot i_{rd} - i_f$$

$$\psi_{sq} = x_{sq} \cdot i_{sq} + x_{hq} \cdot i_{rq}$$

$$\psi_f = -\frac{x_{sfd}^2}{x_{fd}} \cdot i_{sd} - \frac{x_{sfd} \cdot x_{rfd}}{x_{fd}} \cdot i_{rd} + i_f$$

$$\psi_{rd} = x_{hd} \cdot i_{sd} + x_{rd} \cdot i_{rd} - \frac{x_{rfd}}{x_{sfd}} \cdot i_f$$

$$\psi_{rq} = x_{hq} \cdot i_{sq} + x_{rq} \cdot i_{rq} ,$$

where:

e_s ... normalized stator voltage (EMF)

$\mathrm{e_s} = e_s \cdot \mathrm{E_N} \cdot \sqrt{2}$

$\mathrm{E_N}$ ... rated stator voltage

e_f ... normalized field voltage (EMF)

$$\mathrm{e_f} = e_f \cdot \mathrm{E_N} \cdot \sqrt{2} \cdot \frac{r_f}{x_{sfd}}$$

i .... normalized currents

$\mathrm{i} = i \cdot \mathrm{I_N} \cdot \sqrt{2}$ (except i_f)

$\mathrm{I_N}$ ... rated stator current

$$\mathrm{i_f} = i_f \cdot \frac{\mathrm{I_N} \cdot \sqrt{2}}{x_{sfd}}$$

r, x .. resistances and reactances in p.u.

ψ ... normalized flux linkages

$$\Psi = \psi \cdot \frac{\mathrm{E_N} \cdot \sqrt{2}}{\omega_N} \quad \text{(except } \psi_f\text{)}$$

ω_N ... rated angular frequency

$$\Psi_f = \psi_f \cdot \frac{\mathrm{E_N} \cdot \sqrt{2}}{\omega_N} \cdot \frac{x_{fd}}{x_{sfd}}$$

τ .... normalized time, $t = \frac{\tau}{\omega_N}$

ω ... normalized angular frequency

$\omega = \omega \cdot \omega_N$

ϑ ... electric rotor position angle

T_m .. mechanical time constant

$$T_m = \frac{J \cdot \omega_N^3}{S_N \cdot p^2}$$

J ... moment of inertia

S_N .. rated power

p ... number of pole pairs

m ... normalized driving torque

$$\mathrm{m} = m \cdot \frac{S_N \cdot p}{\omega_N}$$

Subscripts

s ... stator

r ... rotor (damper winding)

f ... field

d ... direct axis

q ... quadrature axis

However, one must bear in mind the simplifying assumptions of the d-q-model:

- sinusoidal distribution of all windings
- infinite number of slots.

There are some other simplifications in the mathematical model, used here, too:

- saturation is not considered,
- zero flux is not considered.

4. MODEL OF THE INVERTER

There are two possibilities of modelling the discontinuous behaviour of an inverter in a digital simulation program [6,7]:

- method of varied topology:

  The system is analized separately for each considered connection state of the inverter. Branches with nonconducting valves simply do not exist. Changing the connection state at beginning or end of a commutation entails exchanging the equation system. Calculation of "dummy equations" is avoided.

- method of rigid topology:

  All valves are represented by

R-L-elements, the system is analized as a whole. Values of the R-L-elements are changed in the case of connection state change-over. The equation system remains the same for all possible connection states and contains some "dummy equations" most of the time.

For the simulation program presented here the second method is used. Although the first method would operate with an equation system of lower order most of the time, the change-over between connection states and the detection of its necessity effects high complexity of the program, especially if all possible connection states must be considered. The cases "no valve conducting" and "all valves conducting" must be treated in different ways by the first method, while the second method comprises them without additional effort.
Arbitrary connection states as well as simulation of valve faults do not affect clearness of the program or its course. However, computation time is shorter in the case of varied topology; so the main advantage of the program presented here lies in its clearness, which is very desirable for education and teaching, and facilitates usage of the program by students intending further studies and development.
With the R-L-valve-representations the equivalent network of the inverter becomes:

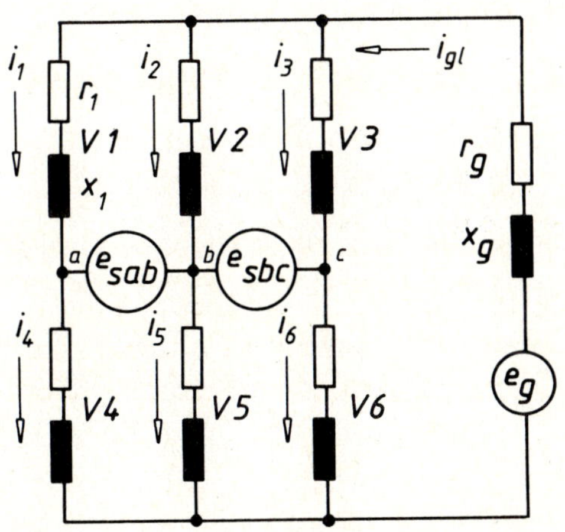

Fig. 2: Equivalent network of inverter

V1 ÷ V6 ... valves

e_g .... D.C. voltage source

The network analysis gives ($\frac{di}{d\tau} = $ "$\dot{i}$")

$$e_{sab} + x_1 \cdot \dot{i}_1 + r_1 \cdot i_1 - r_2 \cdot i_2 - x_2 \cdot \dot{i}_2 = 0$$

$$e_{sbc} + x_2 \cdot \dot{i}_2 + r_2 \cdot i_2 - r_3 \cdot i_3 - x_3 \cdot \dot{i}_3 = 0$$

$$e_{sab} - r_4 \cdot i_4 - x_4 \cdot \dot{i}_4 + x_5 \cdot \dot{i}_5 + r_5 \cdot i_5 = 0$$

$$e_{sbc} - r_5 \cdot i_5 - x_5 \cdot \dot{i}_5 + x_6 \cdot \dot{i}_6 + r_6 \cdot i_6 = 0$$

$$e_g + x_6 \cdot \dot{i}_6 + r_6 \cdot i_6 + x_3 \cdot \dot{i}_3 + r_3 \cdot i_3 + r_g \cdot i_{gl} + x_g \cdot \dot{i}_{gl} = 0$$

e_{sab}, e_{sbc} ... normalized voltage between terminals of the machine

x_g .... smoothing reactance in p.u.

5. SIMULATION PROGRAM

By retransformation of the e_d and e_q equations, the machine model can be "embedded" in the inverter network equations.
According to the number of equations, only three of the currents in the inverter network equations can be chosen as new state variables.
So i_4 ... i_6 and i_{gl} are expressed in termes of the valve currents i_1 ... i_3 (the new state variables) and the stator currents i_{sd} and i_{sq}, which must be retransformed, too.
In a comprehensive form the system of ten differential equations now may be rewritten as:

$$\underline{\underline{H}} \cdot \frac{d}{d\tau} \underline{\underline{Y}} = \underline{\underline{K}} \cdot \underline{\underline{Y}} + \underline{\underline{S}} \ ,$$

where:

$\underline{\underline{H}}$, $\underline{\underline{K}}$ .. coefficient matrices (10x10)

$\underline{\underline{Y}}$ ..... vector of state variables (10)

$\underline{\underline{S}}$ ..... vector of disturbance variables (10)

For application of the Fourth Order Runge-Kutta integration method, explicit evaluation of $\frac{d}{d\tau} \underline{\underline{Y}}$ is required. For each integration step this must be executed 5 times, as some of the elements of $\underline{\underline{H}}$ and $\underline{\underline{K}}$ contain functions of $\underline{\underline{Y}}$ (nonlinear terms of the system).

If the representation of the valves does not contain an inductance, the inversion of the $\underline{\underline{H}}$-matrix becomes impossible.
To keep clearness of the program there are inductances in all valve branches, although mathematically two inductances would be sufficient, assuming a smoothing reactance not equal to zero.

In the case of a conducting valve the inductance should be very low to avoid influence on simulation result. The resistance value can be zero or simulate the forward valve resistance. Yet loops with a time constant shorter than time step of integration must be avoided to keep numerical stability. In the case of nonconducting valves this becomes very important. A high resistance value in a valve representation requires a high inductance, too. So valve simulation by altering resistance values only is impossible [8].
Change from the non-conducting state to the conducting state of a valve depends on three conditions:

- voltage sign permits a forward current,
- a firing command exists,
- the valve is defined as functioning at current moment.

The end of a conducting period depends on two conditions:

- current sign has become negative,
- the valve is defined as functioning at current moment.

The end of a conducting period of a valve mostly occurs after a commutation. During commutation currents change fast in the valves concerned by it. At the end of the last integration step of commutation the current in the valve to be turned off may have a considerable negative value. In this case course of program is interrupted and the last integration step is repeated with a computed step length. For step length computation the last positive current value and its derivative are evaluated:

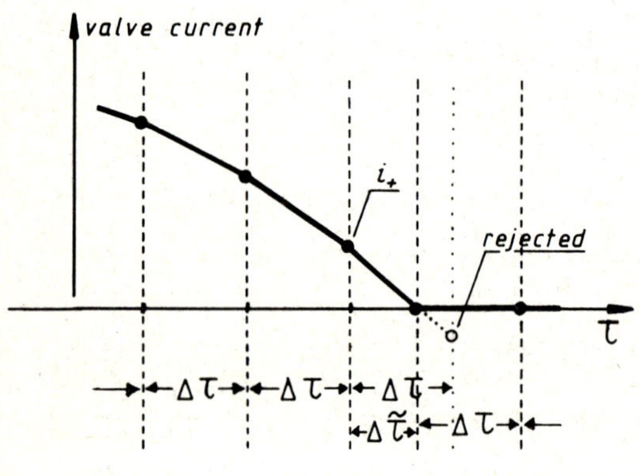

Fig. 3: Integration step length computation

i_+ ... last positive valve current

$\Delta\tau$ ... normal integration step length

$\Delta\tilde{\tau}$ ... computed integration step length, $\Delta\tilde{\tau} = i_+ \Big/ \frac{di_+}{d\tau}$

As an additional facility of the program each valve can be defined as functioning or defect; if it is defined as defect it will keep its initial state. Two points of time can be chosen for simulation of "special events", such as valve short circuit or interruption. One of the following simulation examples makes use of these facilities.

6. SIMULATION EXAMPLES

In the following diagrams all variables are given in normalized units; time is given in ms. The machine parameters are drawn from a laboratory synchronous machine with rated data as follows:

E_N = 220 V f = 50 Hz

I_N = 28,9 A n = 1500 $\min^{-1}$

List of machine parameters:

x_{sd} = 1.1 p.u.	x_{sq} = 0.8 p.u.
x_{rd} = 1 p.u.	x_{rq} = 0.7 p.u.
$x_{hd} = x_{sfd}$ = 0.9 p.u.	x_{hq} = 0.6 p.u.
r_s = 0.06 p.u.	$r_{rd} = r_{rq}$ = 0.1 p.u.
x_{fd} = 1.1 p.u.	$T_f = \frac{x_{fd}}{r_f}$ = 85 radians
T_m = 600 radians	

For explanation see sect. 3.

Other parameters common to all examples:

Resistance of smoothing reactor

r_g = 0.11 p.u.

Integration step length

$\Delta\tau$ = 0.01 radians

Valve representation, on-state

$x_V = 10^{-4}$ p.u.

$r_V = 0$

Valve representation, off-state

x_V = 100 p.u.

$r_V = 10^4$ p.u.

The torque-speed characteristic of the coupled aggregate is shown in Fig. 4; it approximates the characteristic of the driving D.C. machine in the setup for the experimental verification of simulation example 3.

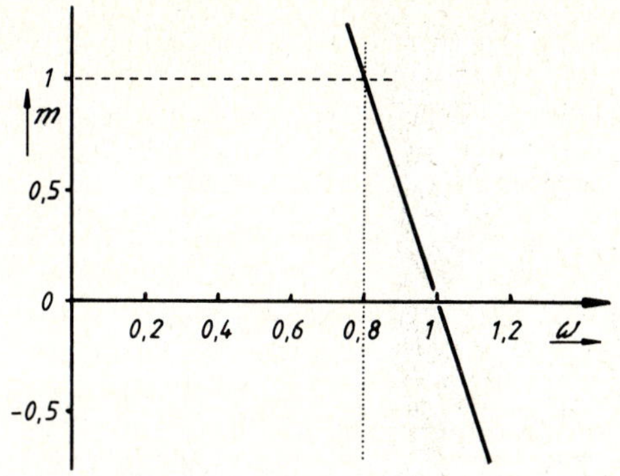

Fig. 4: Torque-speed characteristic of load/drive.

6.1 Motor operation; commutation failure caused by excitation drop

Supplementary parameters

x_g = 1.5 p.u.

e_f = 0, $i_f(t=0)$ = 1.75 p.u.

e_g = -1.87 p.u.

α = 120°

α ... firing impulse delay angle, referring to rotor position.

This example shows simulation of an unpredictable connection state sequence.

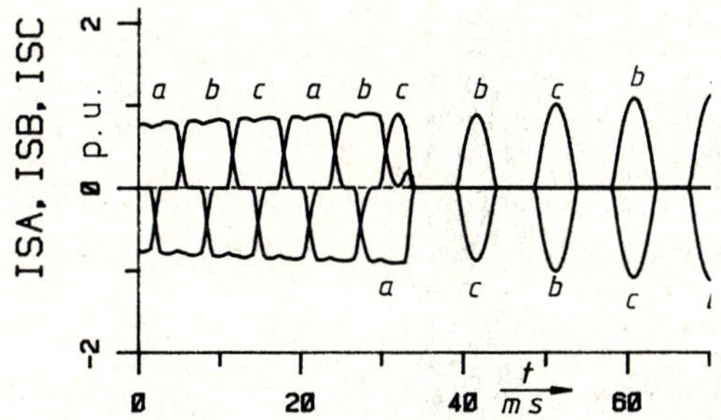

Fig. 5: Stator currents of the machine

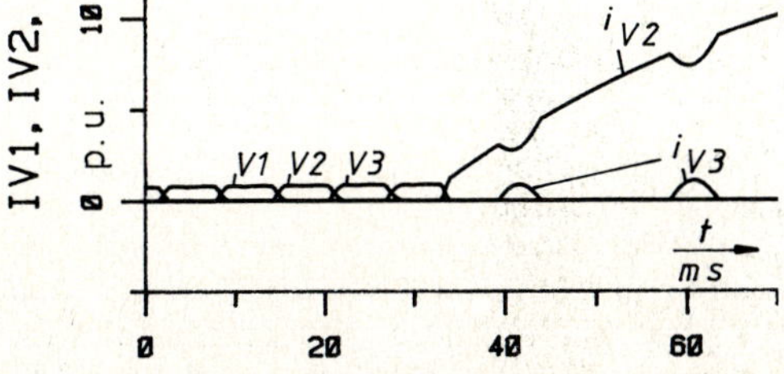

Fig. 6: Valve current of V1 ... V3

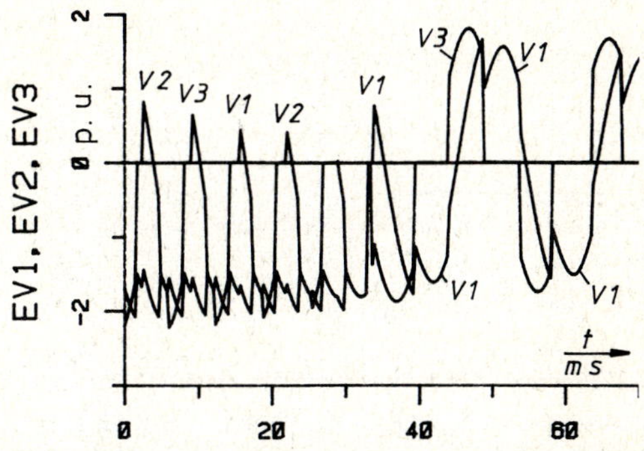

Fig. 7: Valve voltage of V1 ... V3

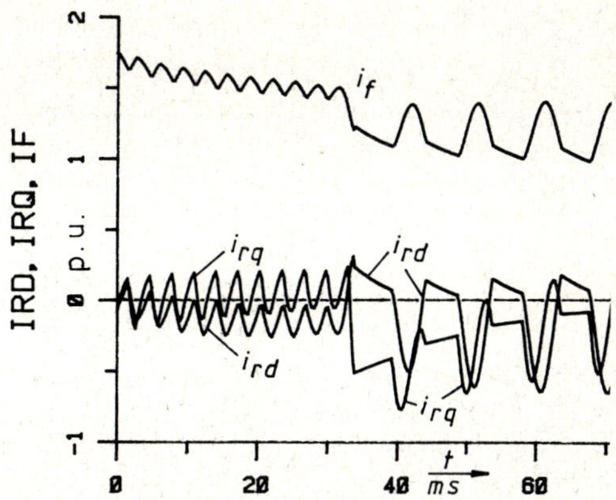

Fig. 8: Damper currents and field current

The machine is not endangered by the stator currents, but the damper currents will cause additional heating if fault is only cleared by disconnection of the D.C. source.

6.2 Generator operation, valve short-circuit of valve 2, cleared by fuse after 10 ms

Supplementary parameters

x_g = 1.0 p.u.

e_g = 0.94 p.u.

e_f = 1.73 p.u.

α = 40°

This example makes use of the additional facilities of the program. Fuse clearing is simulated by raising the resistance in the affected valve branch with a time constant of 0.5 ms up to the off-state value of 10^4 p.u., starting from a value of 0.1 p.u. .

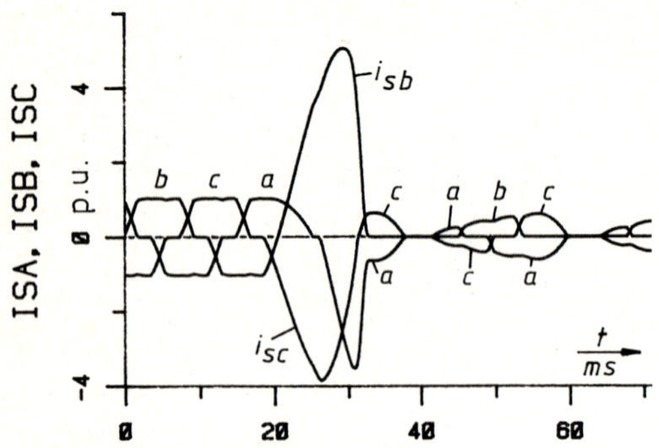

Fig. 9: Stator currents

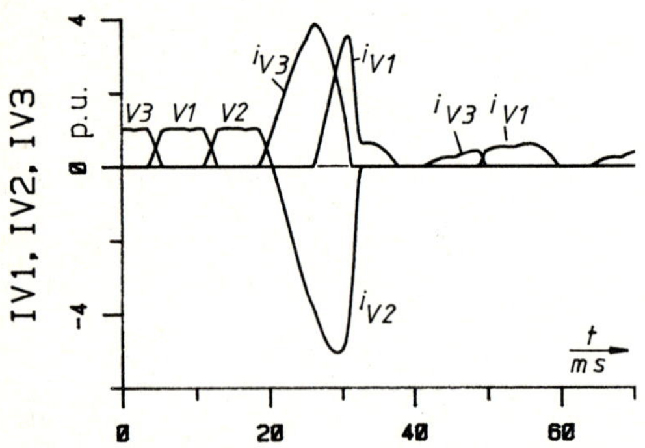

Fig. 10: Valve current of V1 ... V3

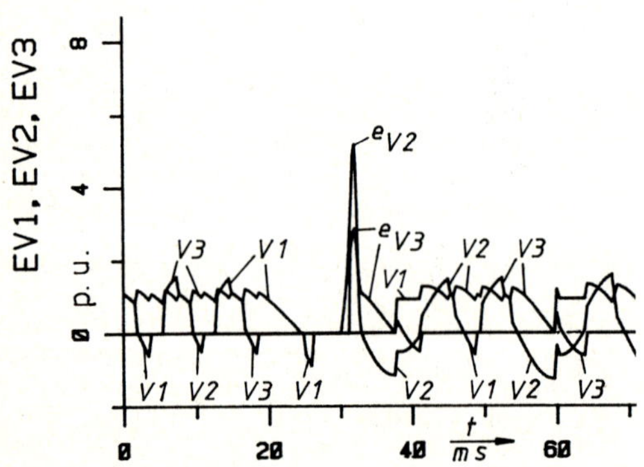

Fig. 11: Valve voltage of V1 ... V3

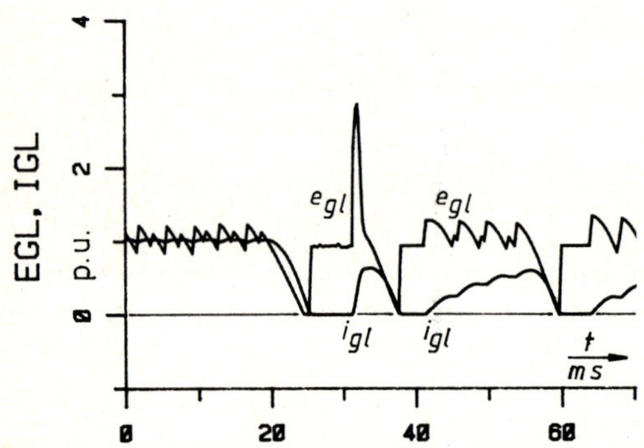

Fig. 12: D.C. voltage and D.C. current

The dangerous voltage peaks naturally will depend on the fuse characteristic.

6.3 Rectifier operation after turning on of excitation of non-loaded machine, covering a period of discontinuous current; comparison with measurement

Supplementary parameters

x_g = 0.07 p.u.
e_g = 1.406 p.u.
e_f = 1.733 p.u., $i_f(t=0) = 0$
α = 0° (rectifier op.)

To emphasize discontinuous current, a low smoothing reactance was used. In the calculated diagrams time counts from turning on of excitation. For the first 800 ms of simulation the unsaturated values of the reactances and time constants were used. This makes possible a rough consideration of saturation.

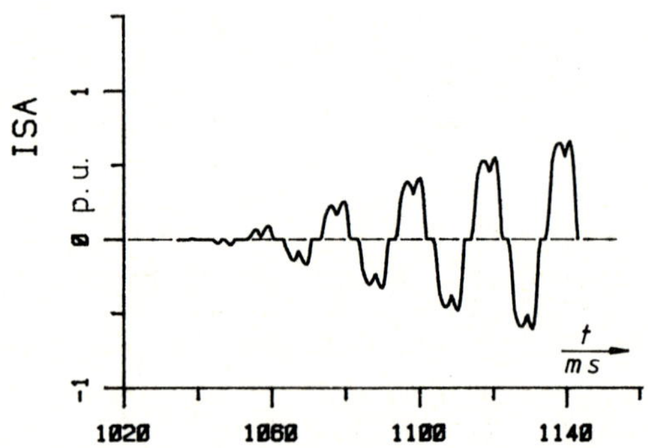

Fig. 13: Stator current, calculated

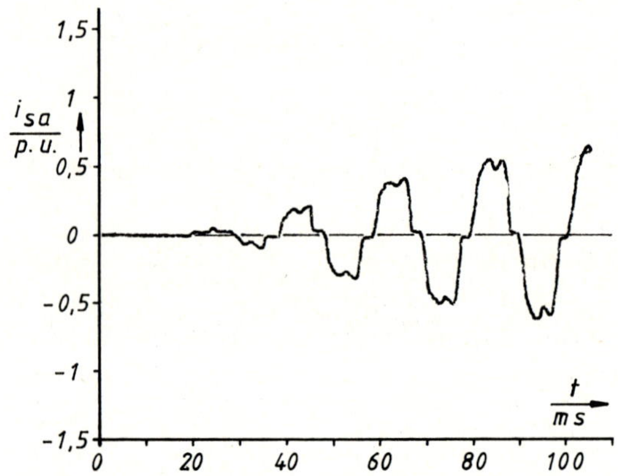

Fig. 14: Stator current, measured

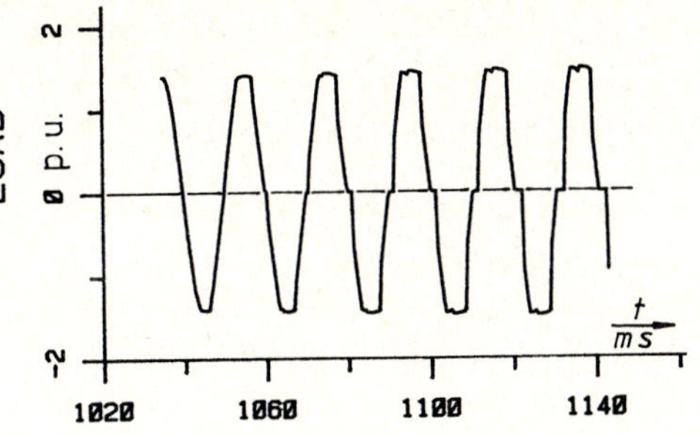

Fig. 15: Stator voltage, calculated

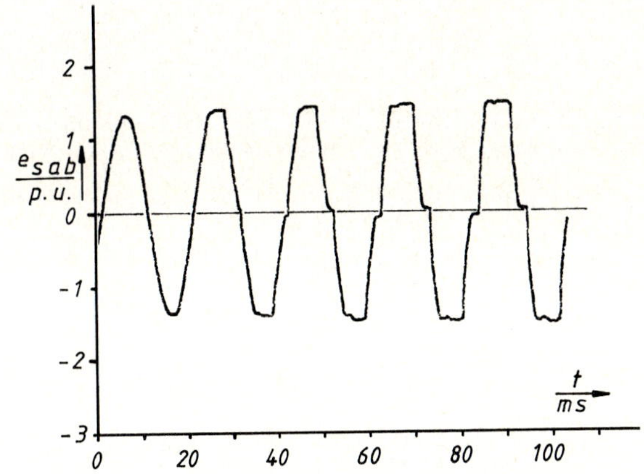

Fig. 16: Stator voltage, measured

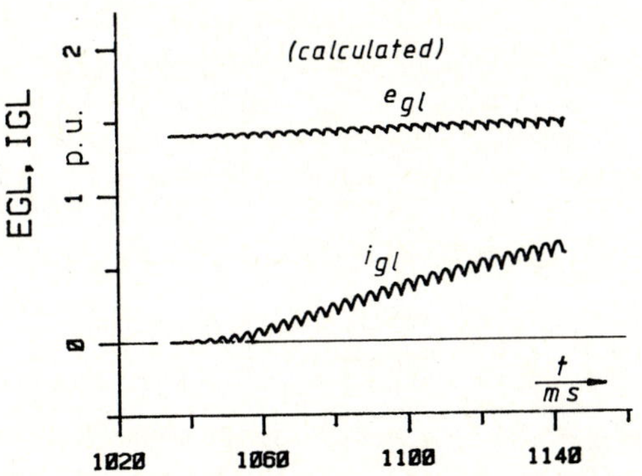

Fig. 17: D.C. voltage and D.C. current

The measurement results are slightly affected by hysteresis effects of the measuring device.

7. CONCLUSION

Employing rigid topology method in a simulation program for an electric drive system containing thyristors facilitates program development, as no considerations are necessary, whether a connection state may occur or not and how to describe it by equations. The program can be surveyed easily by the user intending a modification. Simulation of faults can be implimented in a simple manner.

But the rigid topology method does not support fast simulation, as the equation system contains redundancy most of the time, due to the inclusion of branches with nonconducting valves.

If a simulation program must serve as a part of a control or realtime system or if it is used in parameter optimization, the varied topology method may be more effective.

REFERENCES

[1] Aichholzer,G., Huber,G.N., Moskon,G.: Synchronmaschine mit Gleichrichterbelastung; ein Sonderfall transienten Verhaltens. Elektrotechnik und Maschinenbau 94 (1977), no. 6, pp. 239 - 243.

[2] Mutschler,P.: Untersuchung zur direkten Generatoreinspeisung auf einen 12-pulsigen HGÜ-Stromrichter. ETZ-A 97 (1976), no. 6, pp. 427 - 431.

[3] Mahmoud,S.A., El-Hefnawy,A.A., El-Sattar,A.A.: A digital simulation of synchronous generator feeding dynamic D.C. load. Proc. ICEM 1980 (Athens), part 2, pp. 1003 - 1011.

[4] Subrahmanyam,V., Gopalakrishnan,N.: Analysis of inverterfed self-controlled synchronous motor. Proc. ICEM 1980 (Athens), part 2, pp. 1020 - 1027.

[5] Safacas,A., Giavassoglou,Kl.: Behaviour of a synchronous motor fed by nonsinusoidal currents. Proc. ICEM 1980 (Athens),part 2, pp. 1043 - 1051.

[6] Vogt,F.: Die Simulation von Stromrichtern. ETZ-A 94 (1973), no. 8, pp. 479 - 482.

[7] Mutschler, P.: Verfahren zur digitalen Simulation beliebiger Stromrichterschaltungen. ETZ-A 95 (1974), no. 11, pp. 610 - 614.

[8] Eisenack,H., Cordes,D.: Digitale Nachbildung der Vorgänge in Stromrichterschaltungen. ETZ-A 92 (1971), no. 2, pp. 120 - 122.

ELECTRICAL MACHINES AND CONVERTERS - Modelling and Simulation
H. Buyse and J. Robert (editors)
Elsevier Science Publishers B.V. (North-Holland)/© IMACS, 1984

A GLOBAL MODEL OF THE POLYPHASE CONVERTER FED SYNCHRONOUS MACHINE

B. Detemmerman J. Van Hulse

ACEC
Charleroi

H. Buyse, M. Compère, J.M. Thiry

Université Catholique de Louvain
Louvain-la-Neuve

The communication presents a global model of a round rotor, sinusoïdally coupled polyphase self controlled synchronous machine.
The combination machine-inverter can be considered as a conventional DC machine with shifted brushes.
For any number of phases, this model allows to take into account the influence of the commutation process and can be used for the study of the steady-state and transient behaviours.

1. Introduction

The association of static converters and and AC machines is more and more used for variable speed drives, particularly in applications where the collector of a DC machine should introduce significant limitations (high speed, high power, adverse environment).

A synchronous motor associated with a load commutated inverter (fig. 1) is well adapted to high power levels, since the inverter consists in a conventional thyristor bridge triggered synchronously with the revolution of the rotor of the machine. The motor which is fed by a current inverter, is affected by torque ripple and damper losses. It is possible to overcome those limitations by increasing the number of phases of the synchronous machine or by fitting it with several shifted armatures |1|.

The detailed analysis of such an association is generally possible only by numerical simulation, although the analysis of the behaviour of the drive in a more complex system necessitates simplified global models.

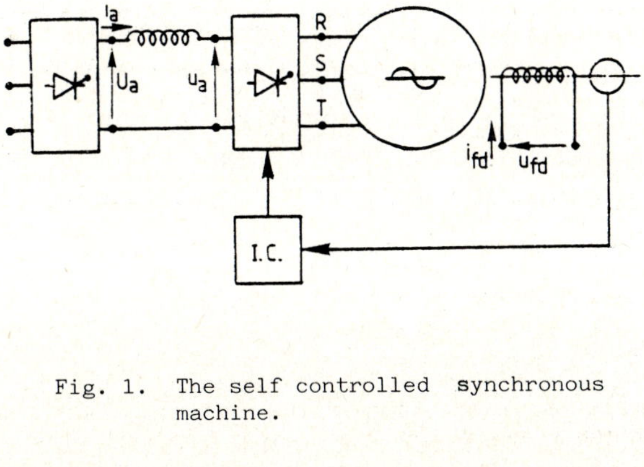

Fig. 1. The self controlled synchronous machine.

A classical approximation consists in considering only the fundamental sinusoïdal component of the quasi rectangular armature current waveform; this simplification leads to a satisfactory global model for a three-phase inverter fed synchronous machine |2|. But this method does not provide some interesting details such as torque ripple or damper losses; in addition, the extension of this model to machine having a number of phases higher then three or several armatures is not obvious; such machines are used for large power applications to reduce torque ripple and damper losses.

In former communications, another global model has been proposed |3|. It consists in approximating the current waveform by a rectangular waveform, considering an instantaneous commutation of the armature current between two phases.

The precision of this model is improved in this paper by taking into account the influence of overlap. The first part of this paper introduces the study of the commutation process and presents the analytical relations of the model in steady state.

The second part describes the application of this model for the study of the transient behaviour.

For any number of phases, the proposed method gives a good approximation for the average and instantaneous values of field and armature currents and voltages, for the mechanical torque of the machine, for the overlap and the commutation failure locus.

2. Statement of the model

The energy conversion in the motor involves at a given time only the part of the armature which is conducting a current. For this reason, we shall consider particularly the set of phases connected simultaneously to the DC terminals through two thyristors; it is called a branch of the armature.

Figure 2 shows the phase and branch currents for a three phase motor.

Each branch consists of the serie connection of two phases (star topology) or the parallel connection of two ways comprising together all the phases (polygonal topology).

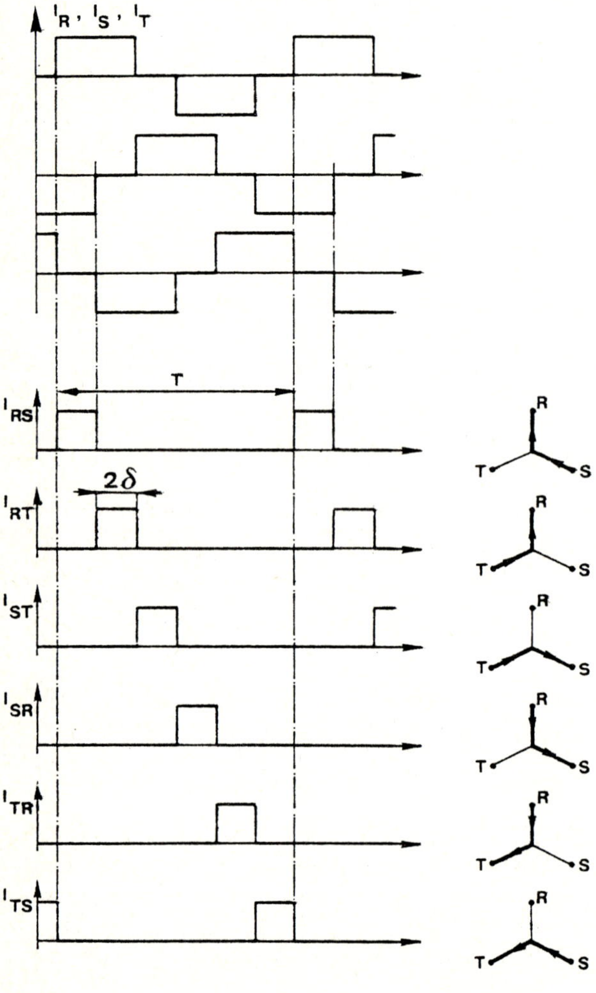

Fig. 2. Phase currents and branch currents for a three-phase machine.

If we choose an inverted representation of the synchronous machine, the analogy with a classical DC machine is clear; the operation of the inverter can be compared with the operation of the commutator of a DC machine (fig. 3).

The rectification angular interval is the period between two successive commutations and it is called 2δ with $\delta=\pi/n'$.

where n' equals the number of phases if it is even or twice this number if it is odd. For a three-phase machine, $\delta=\pi/6$.

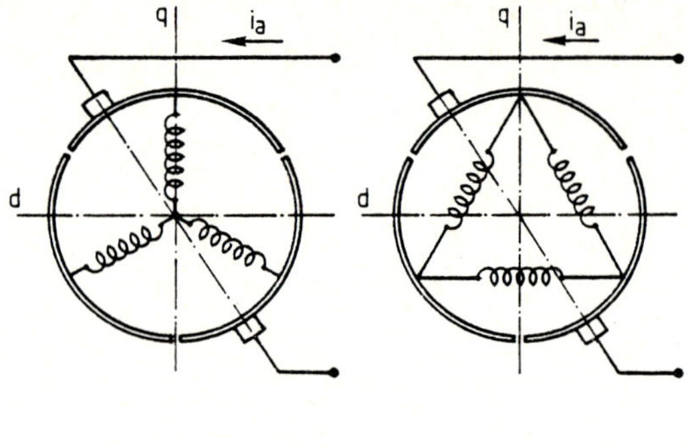

Fig. 3. Equivalent DC machine.

2.1. Basic flux equations

The detailed electrical behaviour of the machine is considered during one rectification angular interval 2δ; this period is a succession of a "commutation interval" or "overlap interval", with 3 thyristors simultaneously conducting and a "conduction interval", with only 2 conducting thyristors. In term of branches, the commutation interval corresponds to the conduction of two branches in parallel; during the conduction interval, the whole current flows through only one branch.

The equivalent circuit of the machine, with an inverted representation, is shown at figure 4.

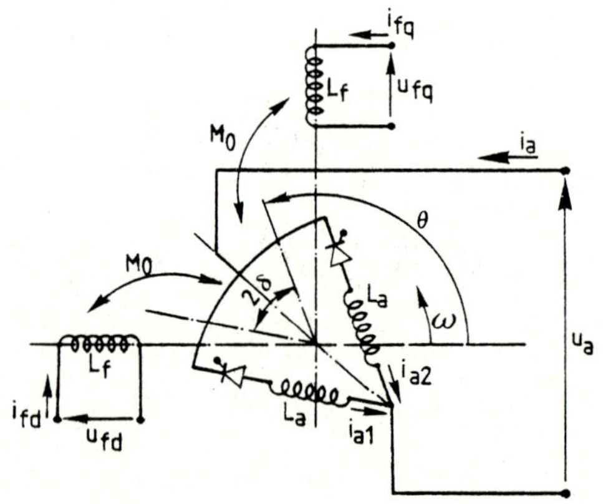

Fig. 4. Equivalent circuit and parameters.

Considering the simultaneous conduction of two branches, we can write the relation (1), between magnetic fluxes and currents :

$$\begin{bmatrix}\psi_d\\ \psi_q\\ \psi_{a1}\\ \psi_{a2}\end{bmatrix}=\begin{bmatrix}L_f & 0 & -M_o\cos(\theta+2\delta) & -M_o\cos\theta\\ 0 & L_f & M_o\sin(\theta+2\delta) & M_o\sin\theta\\ -M_o\cos(\theta+2\delta) & M_o\sin(\theta+2\delta) & L_a & L_a\cos2\delta\\ -M_o\cos\theta & M_o\sin\theta & L_a\cos2\delta & L_a\end{bmatrix}\begin{bmatrix}i_{fd}\\ i_{fq}\\ i_{a1}\\ i_{a2}\end{bmatrix} \quad (1)$$

ψ_d and ψ_q are the direct and quadrature axis field fluxes.

ψ_{a1} and i_{a1} are the flux and the current of the branch which completes its conduction period, and ψ_{a2} and i_{a2} are the flux and the current of the other branch, beginning to conduct; i_{fd} and i_{fq} are the direct and quadrature axis field currents.

L_a is the armature inductance; neglecting the magnetic leakage of the armature, we consider that the coupling between two branches depends only on their angular distance ($M_c = L_a\cos2\delta$).

M_o is the maximum value of the field-armature mutual inductance. θ is the angular coordinate of the conducting branch; the angular speed $\dot{\theta} = \omega$ is assumed to be constant.

From (1), we obtain (2)

$$\begin{bmatrix}\psi_{a1}\\ \psi_{a2}\end{bmatrix}=L'_a\begin{bmatrix}1 & \cos2\delta\\ \cos2\delta & 1\end{bmatrix}\begin{bmatrix}i_{a1}\\ i_{a2}\end{bmatrix}-\frac{M_o}{L_f}\begin{bmatrix}\cos(\theta+2\delta) & -\sin(\theta+2\delta)\\ \cos\theta & -\sin\theta\end{bmatrix}\begin{bmatrix}\psi_d\\ \psi_q\end{bmatrix} \quad (2)$$

with $$L'_a = L_a - \frac{M_o^2}{L_f} \quad (3)$$

2.2. Average voltage on the DC side of the inverter

Let ϕ be the average angular coordinate of a branch during the rectification angular interval; thus, its limits are $\theta = \phi-\delta$ and $\theta = \phi+\delta$.

The average voltage during this period is given by (4)

$$\bar{u}_a = \frac{1}{2\delta}\int_{\phi-\delta}^{\phi+\delta}\frac{d\psi_{a2}}{dt}\,d\theta = \frac{\omega}{2\delta}[\psi_{a2}(\phi+\delta) - \psi_{a2}(\phi-\delta)] \quad (4)$$

(ψ_{a2} is the flux of the branch beginning its conduction for $\theta = \phi-\delta$)

$$\begin{aligned}&\text{for} \quad \theta = \phi-\delta,\ i_{a2} = 0 \ \text{ and } \ i_{a1} = i_a\\ &\text{and for} \quad \theta = \phi+\delta,\ i_{a2} = i_a \ \text{ and } \ i_{a1} = 0\end{aligned} \quad (5)$$

Where i_a is the current flowing through the DC link circuit.

Since the field circuits have a small resistance and are connected to a voltage source or short circuited, their flux linkages are assumed to be constant during a rectification interval.

This consideration with (2) and (5) allows us to write (4) as follows : (6)

$$\bar{u}_a = \omega L'_a\frac{\sin^2\delta}{\delta}i_a + \frac{\omega\bar{M}_o}{L_f}(\psi_d\sin\phi+\psi_q\cos\phi) \quad (6)$$

with $$\bar{M}_o = M_o\frac{\sin\delta}{\delta} \quad (7)$$

The expression (6) can be compared with the corresponding for a machine having a perfect coupling between armature and field circuits ($M_o^2 = L_aL_f$) |1|, with two corrections : the addition of a commutation voltage proportional to i_a and a modification in the expression of the fluxes ψ_d and ψ_q, calculated further.

2.3. Instantaneous values

The instantaneous values of the armature current, the rectified voltage and the field currents can be calculated as a function of ψ_d and ψ_q.

2.3.1. Armature current

During the commutation interval, we have :

$$\frac{d\psi_{a1}}{dt} = \frac{d\psi_{a2}}{dt} \quad \text{or} \quad \frac{d\Delta\psi}{dt} = \frac{d(\psi_{a2}-\psi_{a1})}{dt} = 0 \quad (8)$$

(8) involves that $\Delta\psi$ is constant during the overlap interval.

Taking into account that $i_{a1}+i_{a2}=i_a$, and the expressions (2), $\Delta\psi$ can be written :

$$\frac{\Delta\psi}{2} = L'_a \sin^2\delta(i_{a2}-i_{a1})$$

$$- \frac{M_o \sin\delta}{L_f} (\psi_d \sin(\theta+\delta) + \psi_q \cos(\theta+\delta)) \quad (9)$$

As $\Delta\psi$ is constant during the commutation interval, it can be calculated for $\theta=\phi-\delta$; at this moment, $i_{a2}=0$ and $i_{a1}=i_a$.

Hence,

$$\frac{\Delta\psi}{2} = -L'_a \sin^2\delta\ i_a - \frac{M_o \sin\delta}{L_f}(\psi_d \sin\phi+\psi_q\cos\phi) \quad (10)$$

From (6), (10) can also be written :

$$\bar{u}_a = \frac{-\omega\Delta\psi}{2\delta} \quad (11)$$

Expressions (9) and (10) allows us to obtain the evolution of $i_{a2}-i_{a1}$ during the commutation interval :

$$i_{a2}-i_{a1} = -i_a+ \frac{M_o}{L_f L'_a \sin^2\delta}[\psi_d(\sin(\theta+\delta)-\sin\phi)$$

$$+\psi_q(\cos(\theta+\delta)-\cos\phi)] \quad (12)$$

or, with (11)

$$i_{a2}-i_{a1} = \frac{1}{L'_a\sin^2\delta}[\frac{-\bar{u}_a.\delta}{\omega} + \frac{M_o}{L_f}\sin\delta(\psi_d\sin(\theta+\delta)$$

$$+ \psi_q\cos(\theta+\delta))] \quad (13)$$

With the consideration that $i_{a2}+i_{a1}=i_a$, it is now possible to obtain the evolution of i_{a1} and i_{a2} during the overlap interval; during the conduction interval, we have :

$$i_{a2} = i_a \quad \text{and} \quad i_{a1} = 0 \quad (14)$$

2.3.2. Rectified voltage and field currents

The introduction of (13) or (14) in (1) gives easely the values of i_{fd}, i_{fq} and u_a during the whole rectification interval 2δ.

For the evolution of the field currents, we use the relations :

$$i_{fd} = \frac{1}{L_f}[\psi_d+M_o\cos\theta\cos\delta i_a+M_o\sin\theta\sin\delta(i_{a2}-i_{a1})] \quad (15.a)$$

$$i_{fq} = \frac{1}{L_f}[\psi_q-M_o\sin\theta\cos\delta i_a+M_o\cos\theta\sin\delta(i_{a2}-i_{a1})]$$

The instantaneous voltage on the DC side of the inverter is given by :

$$u_a=L'_a (\cos2\delta \frac{di_{a1}}{dt} + \frac{di_{a2}}{dt})+ \frac{M_o\omega}{L_f}(\psi_d\sin\theta+\psi_q\cos\theta) \quad (15.b)$$

The use of the equations (13) or (14) for the determination of i_{a1} and i_{a2}, introduces the distinction between the overlap interval and the conduction interval.

2.4. Determination of the field fluxes

The evolutions of all voltages and currents are now expressed as functions of the field fluxes ψ_d and ψ_q assumed to be constant. The evaluation of ψ_d and ψ_q can be made by a numerical method as in |4|; this method, suitable for a steady-state analysis, is less appropriate for a dynamic global model; an approximation made in other papers consists in evaluating ψ_d and ψ_q as if the commutations were instantaneous; we obtain then : (16)

$$\psi_d = L_f \bar{i}_{fd} - \bar{M}_o i_a \cos\phi$$

$$\psi_q = L_f \bar{i}_{fq} + \bar{M}_o i_a \sin\phi \quad (16)$$

In this communication, the precision is improved by introducing the influence of overlap in an instantaneous commutation model.

The detailed commutation process is replaced by an instantaneous commutation taking place when :

$$i_{a2} = i_{a1} \quad (17) \quad \text{(cf fig.5)}$$

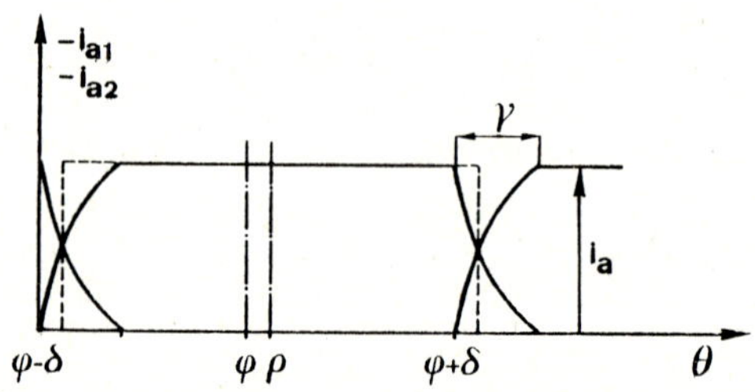

Fig. 5. Definition of the equivalent instantaneous commutation.

Let $\theta=\rho-\delta$ the angular position of this equivalent instantaneous commutation; if we introduce (17) in (12), we obtain :

$$i_a = \frac{M_o}{L_f L'_a \sin^2\delta}[\psi_d(\sin\rho-\sin\phi) + \psi_q(\cos\rho-\cos\phi)] \quad (18)$$

ψ_d and ψ_q are now given by

$$\psi_d = L_f \bar{i}_{fd} - \bar{M}_o i_a \cos\rho$$

$$\psi_q = L_f \bar{i}_{fq} + \bar{M}_o i_a \sin\rho \quad (19)$$

The system (18) and (19) allow us to obtain the analytical expressions of ρ, ψ_d and ψ_q.

From (13), we can also write (20)

$$\bar{u}_a = \frac{\omega \bar{M}_o}{L_f} (\psi_d \sin\rho + \psi_q \cos\rho) \qquad (20)$$

This relation gives another physical interpretation of the angle ρ.

2.5. Duration of the overlap and commutation failure locus

The overlap is caracterized by

$$|i_{a2}| < |i_a| \qquad (21)$$

If we neglect the turn-off time of the thyristors, we can say that we have a commutation failure if we have simultaneously (21) and

$$\frac{di_{a2}}{dt} = 0 \qquad (22)$$

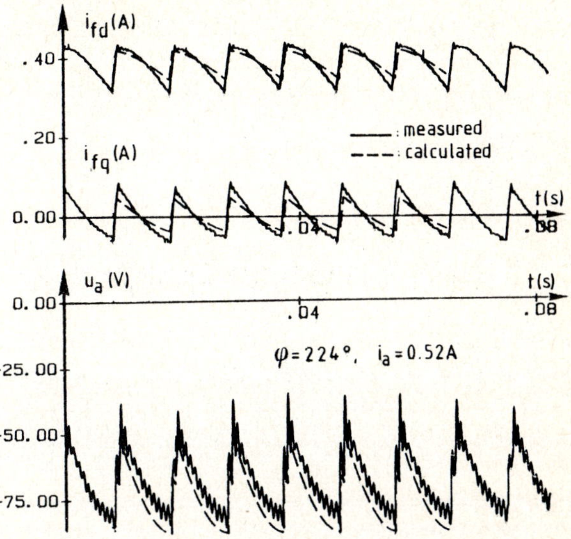

Fig. 6. Three-phase machine steady-state behaviour.

3. Experimental results

This model has been applied to a small laminated rotor synchronous machine connected as three-phase machine and nine-phase machine.

Machine ratings :

- 3000 rpm - 3 kVA
- armature : 220 V - 4.5 A
- field : 75 V - 2.5 A
- Three-phase machine (star connection) :

For the three phase machine, the parameters of the equivalent DC machine are related to the classical parameters of the synchronous machine by the relations :

$$L_a = 2\, L_d \quad ; \quad M_o = \sqrt{2}\, M_d \quad ; \quad R_a = 2\, R_d$$

Equivalent DC machine parameters :

$M_o = 2.32$ H , $L_a = 0.58$ H , $R_a = 2.2\ \Omega$

$L_f = 10.2$ H , $R_f = 28.9\ \Omega$

Classical parameters :

$M_d = 1.64$ H , $L_d = 0.29$ H , $R_d = 1.1\ \Omega$

$L_f = 10.2$ H , $R_f = 28.9\ \Omega$

- Nine-phase machine (polygonal connection) :

Equivalent DC machine parameters :

$M_o = 1.35$ H , $L_a = 0.21$ H , $R_a = 0.88\ \Omega$

$L_f = 10.2$ H , $R_f = 28.9\ \Omega$

Figures 6 and 7 represent the instantaneous values of the field and phase currents, the armature voltage in steady-state conditions for both modes of connection.

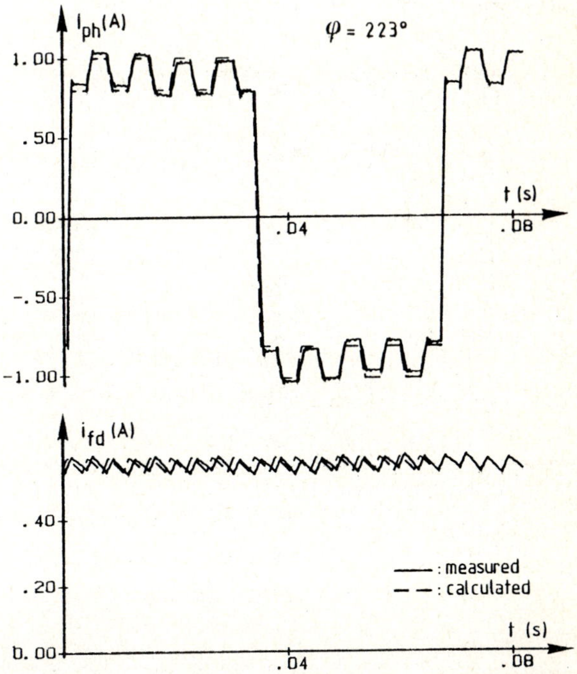

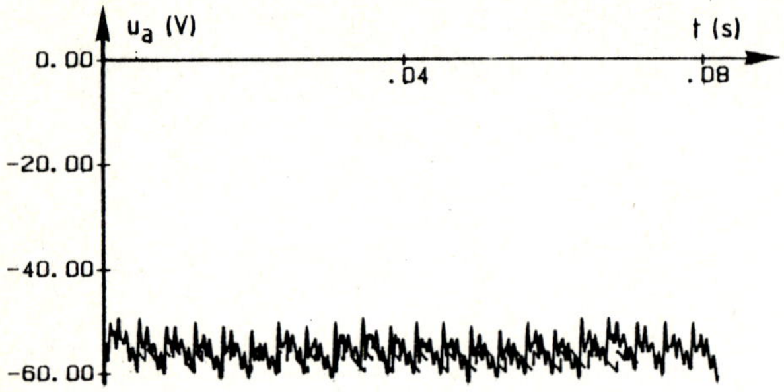

Fig. 7. Nine-phase machine steady-state behaviour.

In figure 8, we have measured the evolution of the overlap angle as a function of the current in the DC link circuit.

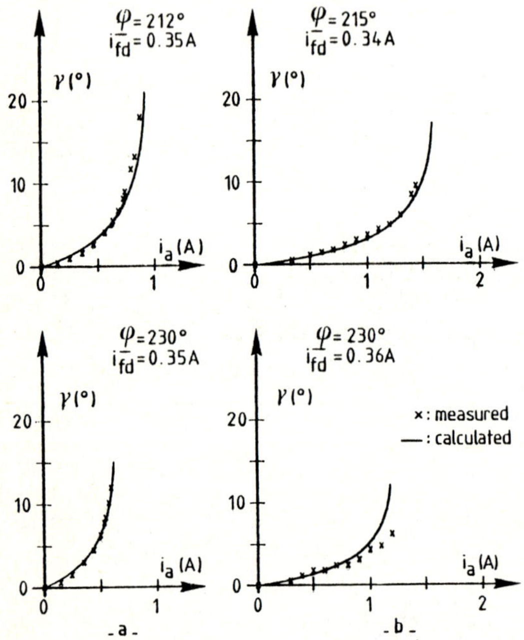

Fig. 8. Evolution of the overlap angle

a. three-phase machine

b. nine-phase machine.

5. Dynamic model

5.1. Introduction

The model obtained here is formally the same as in |5|, where the transient behaviour of the self controlled synchronous machine has been studied without the influence of the overlap.

The same assumptions are made :

a) $\frac{d\bar{I}_a}{dt}$ is constant during a rectification interval

b) ψ_d, ψ_q, ω are also constant during the same interval.

Expressions (19) and (20) characterize the quasi-static state of the machine.

5.2. Transient equations

In fact, all those variables are not constant but slowly time dependent. The magnetic flux linkages are given by the equations

$$U_f = R_f i_{fd} + \frac{d\psi_d}{dt} \simeq R_f \bar{i}_{fd} + \frac{d\psi_d}{dt}$$
$$0 = R_f i_{fq} + \frac{d\psi_q}{dt} \simeq R_f \bar{i}_{fq} + \frac{d\psi_q}{dt} \qquad (23)$$

The third member of these equations is obtained by neglecting the ohmic voltage drop due to the current ripples.

Neglecting the influence of the armature voltage ripple on the external armature circuit, we have :

$$U_a = R\bar{i}_a + L\frac{d\bar{i}_a}{dt} + \bar{u}_a \qquad (24)$$

The torque equation corresponds to the expression

$$T_r = \bar{T} - J\frac{d\omega}{dt} \qquad (25)$$

$$\text{with } \bar{T} = \frac{\bar{u}_a \bar{i}_a}{\omega} \qquad (26)$$

The quasi-static relations (19) and (20) are introduced in the expressions (23), (24), (25) and we consider that the angle ρ changes slowly :

$$\frac{d\cos\rho}{dt} = 0 \quad ; \quad \frac{d\sin\rho}{dt} = 0 \qquad (27)$$

We obtain a set of differential equations of the system :

$$\begin{bmatrix} U_f - R_f \bar{i}_{fd} \\ - R_f \bar{i}_{fq} \\ U_a - R\, i_a - \omega M_o (\bar{i}_{fd} \sin\rho + \bar{i}_{fq} \cos\rho) \\ T_r - M_o\, i_a (\bar{i}_{fd} \sin\rho + \bar{i}_{fq} \cos\rho) \end{bmatrix} = \begin{bmatrix} L_f & 0 & -M_o \cos\rho & 0 \\ 0 & L_f & M_o \sin\rho & 0 \\ 0 & 0 & L & 0 \\ 0 & 0 & 0 & -J \end{bmatrix} \begin{bmatrix} \frac{d\bar{i}_{fd}}{dt} \\ \frac{d\bar{i}_{fq}}{dt} \\ \frac{d\bar{i}_a}{dt} \\ \frac{d\omega}{dt} \end{bmatrix} \qquad (28)$$

The set of equations (28) must be resolved numerically. At each step of computation, the angle ρ is changed by a new value calculated by relation (18).

5.3. Instantaneous value of the electrical variables

In transient conditions, it is possible to derive the instantaneous value of the electrical variables i_{fd}, i_{fq}, u_a from the values of the state variables $\bar{i}_{fd}$, $\bar{i}_{fq}$, i_a and ω and from the position θ of the conducting branch.

We use the relations (15) where ψ_d and ψ_q are calculated by (19) for each new value of $\bar{i}_{fd}$, $\bar{i}_{fq}$ and ρ.

6. Verification

The dynamic model has been applied to the machine presented in 3.

Figure 9 represents the evolution of the field and armature currents and armature voltage for an electric transient of the three-phase machine.

Figures 10 and 11 show for a three-phase and a nine-phase machine the evolution of the armature and field currents and the speed after an armature supply voltage pulse.

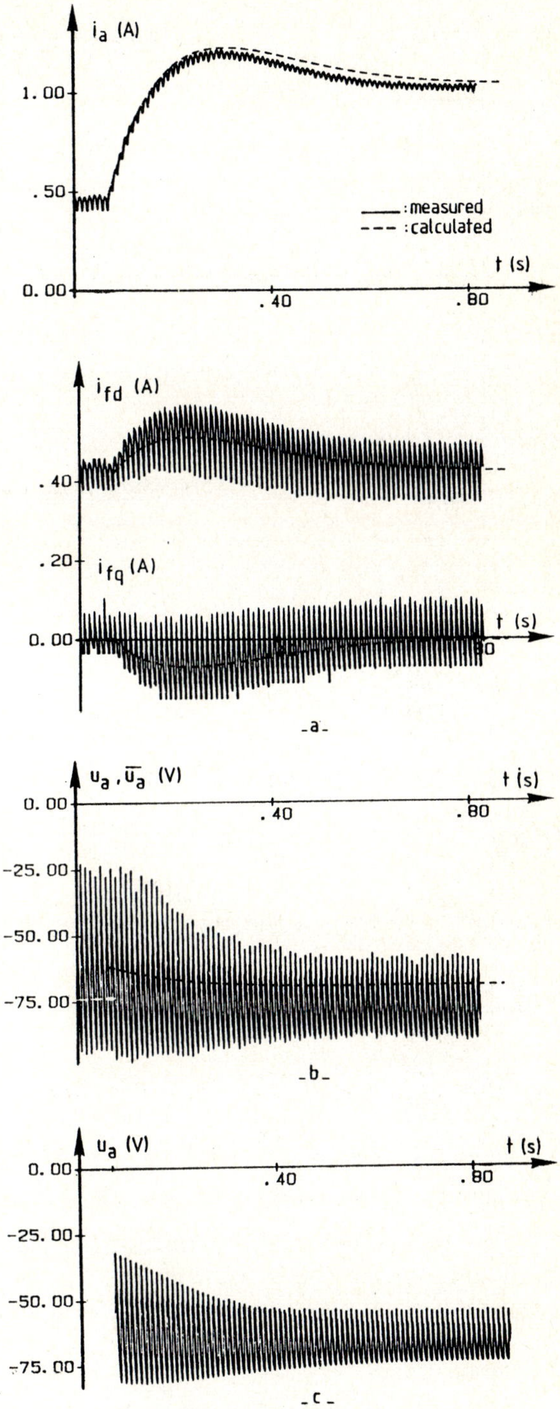

Fig. 9. Three-phase machine
Electrical transient behaviour
U_a : -59V → -69V ; U_f = 12.1V
R = 6.5Ω; L = 1.1H; Φ = 215°; N=968t/mn

a. Experimental and theoretical field and armature currents.
b. Experimental value and theoretical mean value of armature voltage.
c. Theoretical armature voltage.

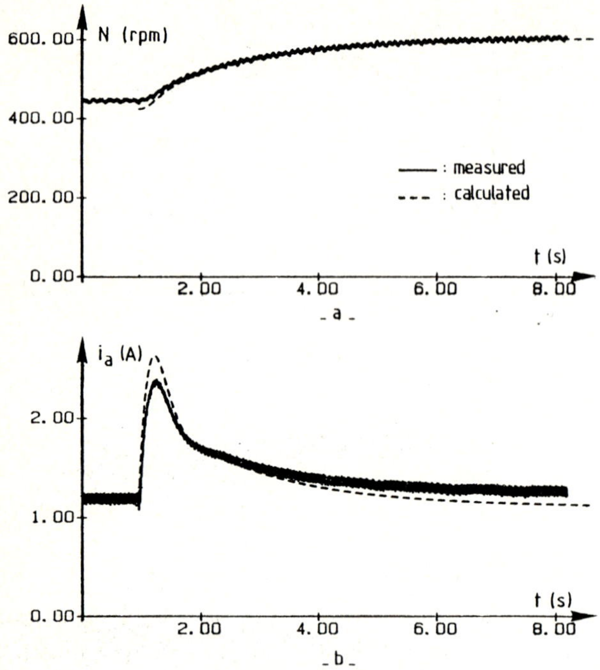

Fig. 10. Three-phase machine
Electromechanical transient behaviour
U_a: $-52V \rightarrow -72V$; $U_f = 26V$; $R = 6.5\Omega$;
$L = 1.4H$; $\phi = 213°$; $J = .15\ Nms^2$;
$T_r = 1.009+0.002\omega$ Nm

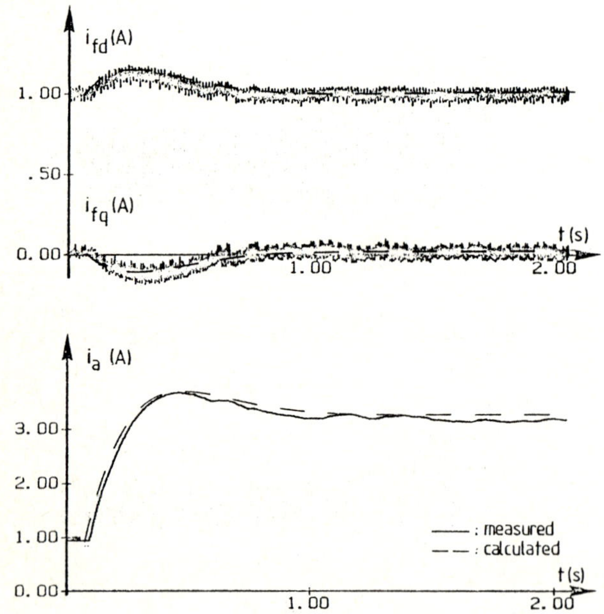

Fig. 11. Nine-phase machine
Electromechanical transient behaviour
U_a : $-47V \rightarrow -73V$; $U_f = 28.9V$
$L = 1.4H$; $R = 6.5\Omega$; $\phi = 217°$
$J = .15\ Nms^2$; $T_r = -40.4+0.53\omega$ Nm

7. Conclusions

The communication proposes a method for the analysis of the self-controlled synchronous machine in steady state and transient conditions. A DC machine with shifted brushes is considered as model.

This global model introduces in the modelling the influence of the commutation process and can be used for any number of phases.

The self-controlled synchronous machine can be described by a set of non-linear differential equations involving the average value of the electrical and mechanical variables. That method can be useful for system applications.

Comparisons with experimental measurements are made with a good precision, for a three-phase and a nine-phase constant airgap synchronous machine.

8. Acknowledgements

The authors wish to acknowledge financial assistance from the Belgian Research Institute for Industry and Agronomy (IRSIA) and from ACEC (Ateliers de Constructions Electriques de Charleroi).

9. References

|1| "A simplified model of the multiple armature, polyphase self-controlled synchronous machine"
J.M. Thiry, M. Compère, H. Buyse
3rd IFAC Symposium on Control in Power Electronics and Electrical Drives, Lausanne, September 1983.

|2| "Equivalent circuit representation of current inverter driven synchronous machines"
D.W. Novotny, T.L. King
IEEE Transactions on P.A.S., vol. PAS-100, June 1981, pp. 2920-2926.

|3| "A simplified model of the rectifier-loaded synchronous machine with damper windings"
H. Buyse, S. Jadin
International Conference on Electrical Machines, Athens, September 1980.

|4| "Enhancement of commutation of current-source inverter-fed synchronous machines by pole-face compensating windings"
F.C. Brockhurst
IEEE Transactions on P.A.S., vol. PAS-100, June 1981, pp. 2840-2853.

|5| "Contribution to the transient analysis of the self-controlled synchronous machine"
H. Buyse, J.M. Thiry, M.S. Garrido
International Conference on Electrical Machines, Budapest, September 1982.

Session 7

MODELLING OF SYNCHRONOUS MACHINES AND CONVERTERS (II), SIMULATION OF POWER GENERATION SYSTEMS

Chairman: M. CRAPPE

ELECTRICAL MACHINES AND CONVERTERS - Modelling and Simulation
H. Buyse and J. Robert (editors)
Elsevier Science Publishers B.V. (North-Holland)/ ©IMACS, 1984

STATIC AND DYNAMIC BEHAVIOUR OF SALIENT POLES RECTIFIED OUTPUT GENERATORS WITH A LARGE NUMBER OF PHASES

L. Pierrat F. Labrique B. Mauhin

Electricité de France,
Direction des Etudes et Recherches
1, Av. Gén. de Gaulle, F-92140 Clamart, France

Université Catholique de Louvain
B-1348 Louvain-la-Neuve, Belgium

This paper deals with the modelisation of salient poles rectified output generators with a large number of phases. It shows that the equivalent DC machine model previously developed by neglecting saliency effects can be adapted to take them into account. The resulting improvement in precision is made clearly apparent by experiments made on two test machines with very different power ratings.

1. INTRODUCTION

Rectified output generators with a large number of phases have been recently introduced for the brushless excitation of large turbo alternators |1,2|. Such generators are normally designed with salient poles.

In previous papers, a basic model neglecting saliency has been developped for this type of generators. According to this model, rectified output alternators with a large number of phases have been found to operate like a DC machine having two sets of moving brushes which position themselves according to the working conditions |3,4,5|.

The present paper introduces an improvement of this model by taking saliency into account.

In the first part, devoted to steady state behaviour, one demonstrates that saliency does not affect the fundamental aspects of the commutation processes. However by considering the output characteristic, it is shown that taking saliency effects into account substantially improves the precision of the model.

In the second part, the improved model is extended to transients. All the theoretical results are confirmed by experiments made on a laboratory test machine and on a power exciter.

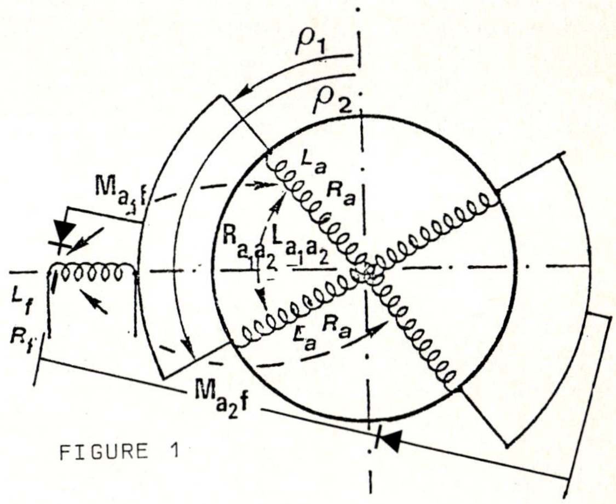

FIGURE 1

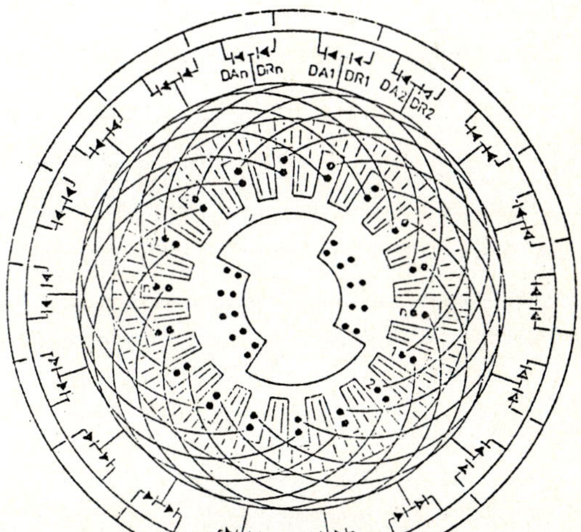

FIGURE 2

2. STEADY STATE BEHAVIOUR

In this section, saliency is introduced in the equivalent DC machine model (figure 1), and it is shown that the resulting equations are in agreement with the switching logic of the rectifier.

2.1. Parameters of the equivalent DC machine

The parameters of the equivalent DC machine (see figure 1) are derived under the following assumptions (figures 2 and 3) :

- infinite iron in permeance
- neither ends nor slotting effects
- airgap permeance as shown in figure 3 (the permeance between the poles is refered to the permeance under the poles by a coefficient s less than unity which characterises the importance of saliency.

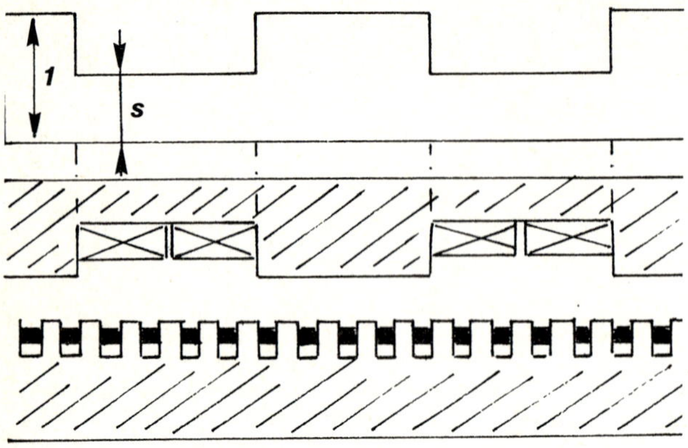

FIGURE 3

Figures 4 and 5 show as functions of the rotor position (ρ), the computed and measured parameters of an 18 phases laboratory test machine (an armature circuit of the equivalent DC machine corresponds to the circuit defined by two diametrically opposed armature terminals of the generator).

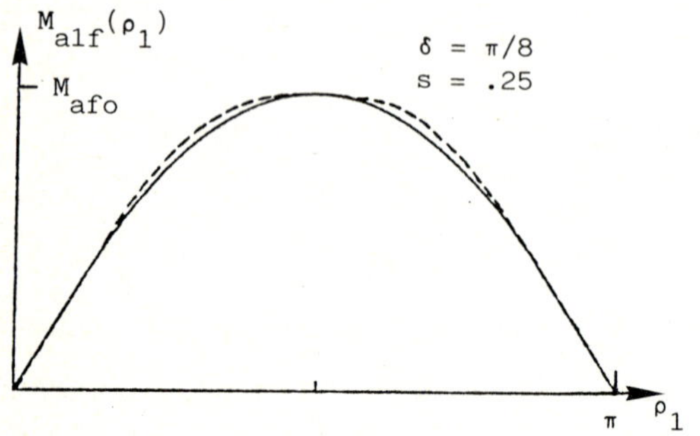

FIGURE 4

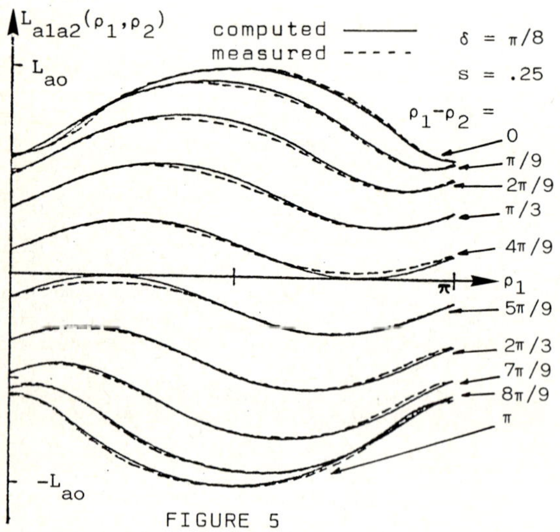

FIGURE 5

The good agreement between theoretical and experimental results confirm the validity of the adopted computation method.

2.2. Steady-state equations

Two mode of operation must be considered |3,4,5|.

During the first one, the two sets of brushes occupy, with respect to the field neutral axis, the same position ρ, which varies from 0 to δ, δ being the angular distance between the field neutral axis and the pole edge (figure 6). This position is fixed by expressing that the voltage across a fictitious set of brushes at position ρ' and with zero current has a derivative with respect to position equal to zero when $\rho'=\rho$.

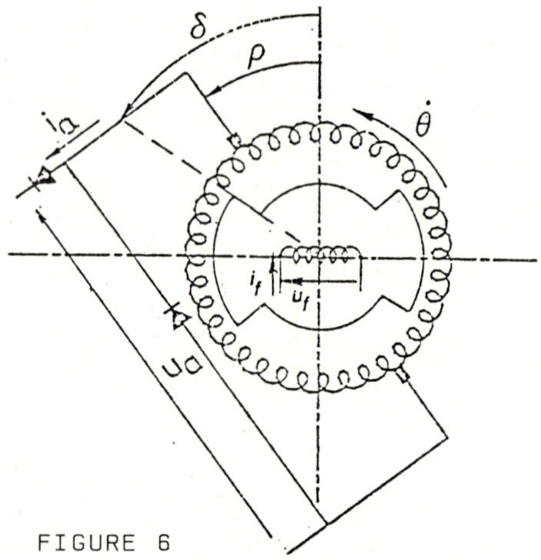

FIGURE 6

By using the general equation of sliding contacts circuits the following equations are obtained for the electrical behaviour :

(1a) $u_f = R_f\, i_f$

(1b) $$u_a = \dot\theta \frac{\partial M_{af}}{\partial \rho} i_f - \frac{1}{2}\dot\theta \frac{\partial L_a}{\partial \rho} i_a - R_a i_a$$

(1c) $$0 = \frac{\partial}{\partial \rho'}\left[\dot\theta \frac{\partial M_{a'f}}{\partial \rho'} i_f - \dot\theta \frac{\partial L_{aa'}}{\partial \rho'} i_a - R_{aa'} i_a\right]_{\rho'=\rho}$$

with (for $0 < \rho$, $\rho' < \delta$) :

(2a) $$L_a = L_{ao} \frac{1}{1-(1-s)\left(\frac{2\delta}{\pi}\right)^3} \times \left[12\left(\frac{\pi-2\delta}{\pi}\right)(1-s)\left(\frac{\rho}{\pi}\right)^2 + s + (1-s)\left(\frac{\pi-2\delta}{\pi}\right)\right]$$

(2b) $$L_{aa'} = L_{ao} \frac{1}{1-(1-s)\left(\frac{2\delta}{\pi}\right)^3} \times\left[4s\left(\frac{\rho'-\rho}{\pi}\right)^3 - 6s\left(\frac{\rho'-\rho}{\pi}\right)^2 + 12\left(\frac{\pi-2\delta}{\pi}\right)(1-s)\frac{\rho\rho'}{\pi^2} + s + (1-s)\frac{\pi-2\delta}{\pi}\right]$$

$$(2c)\quad M_{af} = M_{afo} \frac{4}{1-(1-\frac{2s}{3})(\frac{2\delta}{\pi})^2}$$

$$\times \left[\frac{s\pi}{\delta}\left(\frac{\rho\delta^2}{\pi^3} - \frac{\rho^3}{3\pi^3}\right) + \frac{\rho(\pi-2\delta)}{\pi^2}\right]$$

In these relations L_{ao} and M_{afo} are respectively the maximum values of L_a and M_{af}.

During the second mode of operation, the two sets of brushes have distincts positions. The former one remains at the position $\rho_1 = \delta$ while the position of the latter varies from δ to $\pi - \delta$ (figure 7) in order to insure equal currents into the two sets.

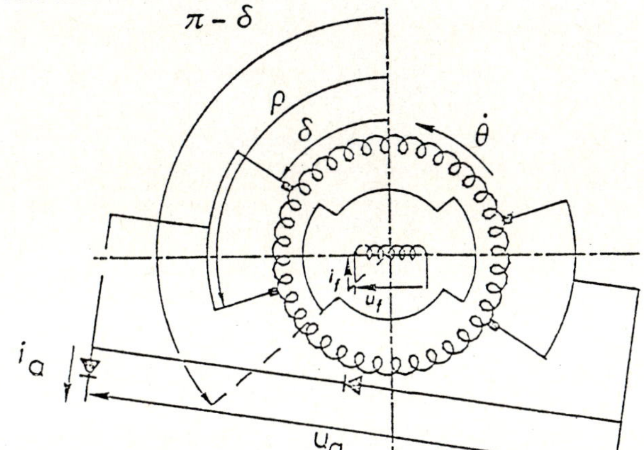

FIGURE 7

The following equations are obtained for the electrical behaviour during this mode :

$$(3a)\quad u_f = R_f\, i_f$$

$$(3b)\quad u_a = \frac{1}{2} \Big| \dot{\theta}\{[\frac{\partial}{\partial\rho_1}(M_{a1f}) + \frac{\partial}{\partial\rho_2}(M_{a2f})]i_f - \frac{1}{4}$$

$$- \frac{1}{4}[\frac{\partial}{\partial\rho_1}(L_{a1}) + \frac{\partial}{\partial\rho_2}(L_{a2})]i_a - \frac{1}{2}[\frac{\partial}{\partial\rho_1}(L_{a1a2})$$

$$+ \frac{\partial}{\partial\rho}(L_{a1a2})]i_a\} - (R_a + R_{a1a2})i_a \Big|_{\rho_1=\delta,\rho_2=\rho}$$

$$(3c)\quad 0 = \Big| \dot{\theta}[\frac{\partial}{\partial\rho_1}(M_{a1f}) - \frac{\partial}{\partial\rho_2}(M_{a2f}) \Big| i_f$$

$$- \frac{1}{4}\, \dot{\theta} \Big| \frac{\partial}{\partial\rho_1}(L_{a1}) - \frac{\partial}{\partial\rho_2}(L_{a2}) \Big| i_a$$

$$- \frac{1}{2}\, \dot{\theta} \Big| \frac{\partial}{\partial\rho_1}(L_{a1a2}) - \frac{\partial}{\partial\rho_2}(L_{a1a2}) \Big| i_a \Big|_{\rho_1=\delta,\rho_2=\rho}$$

with (for $\delta < \rho_1, \rho_2' < \pi-\delta$ and $j\varepsilon|1,2|$:

$$(4a)\quad L_{aj} = L_{ao} \frac{1}{1-(1-s)(\frac{2\delta}{\pi})^3}$$

$$\times\left[\frac{24\delta}{\pi}(1-s)\frac{\rho}{\pi} - \frac{24\delta}{\pi}(1-s)\frac{\rho^2}{\pi^2} + 1 - \frac{2\delta}{\pi}(1-s)(3+(\frac{2\delta}{\pi})^2)\right]$$

$$(4b)\quad L_{a1a2} = L_{ao} \frac{1}{1-(1-s)(\frac{2\delta}{\pi})^3}\quad 4\, \frac{\rho_2-\rho_1}{\pi}^3 - 6\, \frac{\rho_2-\rho_1}{}^2$$

$$- \frac{24\delta}{\pi}(1-s)\frac{\rho_1\rho_2}{\pi} + \frac{12\delta}{\pi}(1-s)\frac{\rho_1+\rho_2}{\pi}$$

$$+ 1 - \frac{2\delta}{\pi}(1-s)(3 + (\frac{2\delta}{\pi})^2)$$

$$(4c)\quad M_{ajf} = M_{afo} \frac{4}{1-(1-\frac{2s}{3})(\frac{2\delta}{\pi})^2}$$

$$\times\quad \frac{\rho_j}{\pi} - (\frac{\rho_j}{\pi})^2 - (1-\frac{2s}{3})\frac{\delta^2}{\pi^2}$$

One can demonstrate that the equivalent DC machine model remains valid when taking saliency into account by showing that, according to the switching logic of the rectifier, the equations obtained uniquely fix the positions of the brushes at points corresponding to the maximum in induced voltage. This is done by determining, for each working point, the voltage accross a fictitious set of brushes with zero current, as a function of its position, and by observing that this voltage is monotonously decreasing from ρ to $\rho+\pi$ (increasing from $\rho-\pi$ to ρ)

Figure 8 gives the evolution of the voltage accross this fictitious set of brushes for three working points corresponding respectively to the first mode of operation, to the limit between the two modes of operation and to the second mode of operation.

2.3. Output characteristics

Figure 9 shows the evolution of the output characteristic for various value of the parameter s which characterises the saliency, s=1 corresponding to a non salient pole generator.

Figures 10 and 11 allow to compare the theoretical and experimental output characteristics of the 18 phases laboratory test machine and of a 44 phases exciter with a power rating of 1.2 MVA in the linear (not saturated) part of the magnetic characteristic.

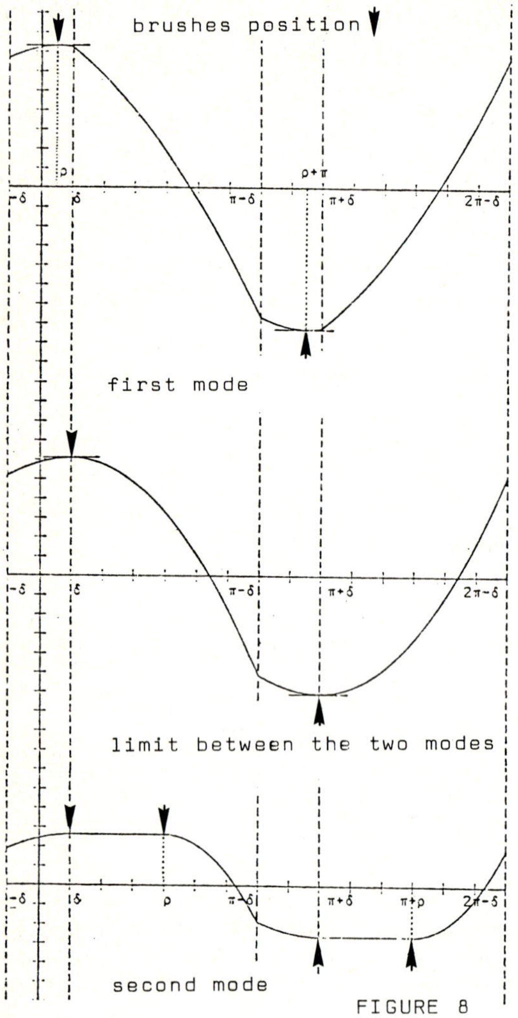

FIGURE 8

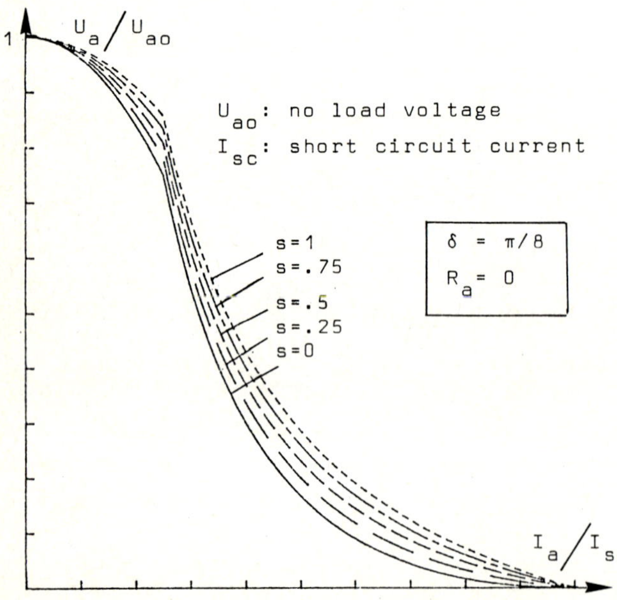

FIGURE 9

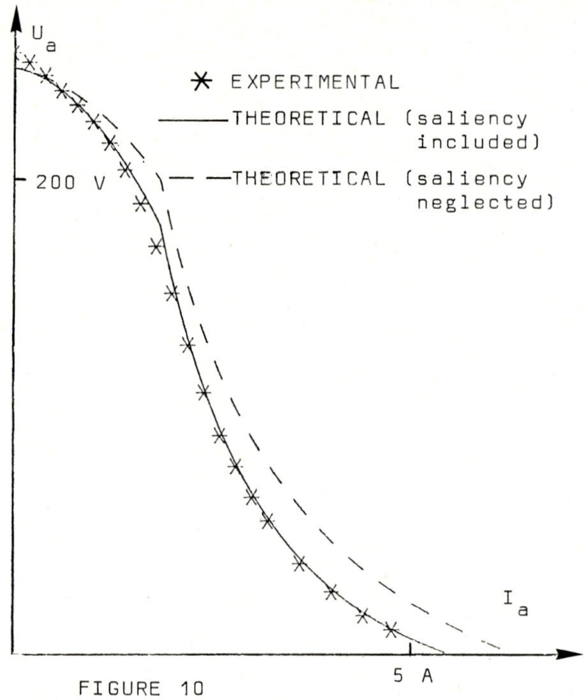

FIGURE 10

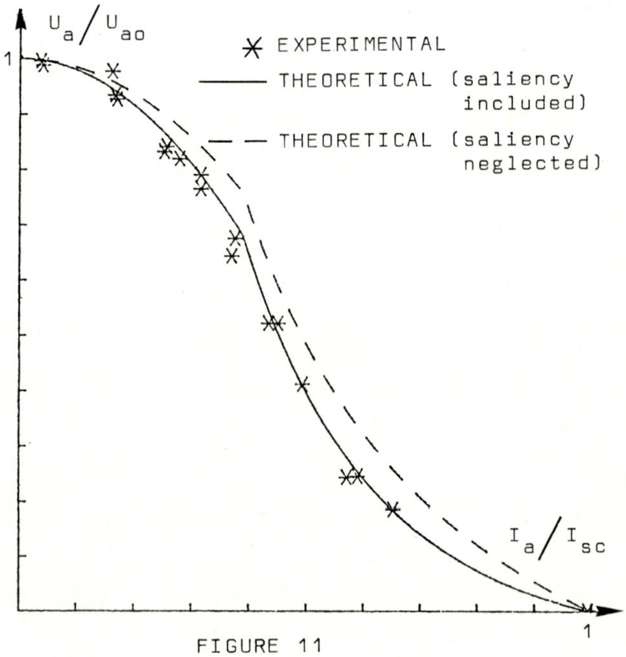

FIGURE 11

The substantial improvement of precision resulting from taking saliency into account is made clearly apparent by the theoretical curves neglecting saliency also drawn on these figures.

3. TRANSIENT BEHAVIOUR

The extension of the model to the transient behaviour relies on the same assumptions as for the basic model |5|. Transients are supposed to be sufficiently slow with respect to commutation phenomena so that :

- the brushes remain in the interval where they are located in steady state
- the splitting of the brushes into two sets occurs for the same position as in steady state; one of the sets remains at this position and the other positions itself in order to ensure that equal currents flow into the two sets.

3.1. Dynamic equations

With the above assumptions the following sets of equations are obtained when the generator is connected to a resistive inductive load (R_1, L_1) :

- for the first mode $(0 < \rho < \delta)$:

$$\begin{vmatrix} L_f & M_{af} & \frac{\partial M_{af}}{\partial\rho} \\ M_{af} & L_a+L_1 & \frac{1}{2}\frac{\partial L_a}{\partial\rho} \\ \frac{\partial M_{af}}{\partial\rho} & \left.\frac{\partial L_{aa'}}{\partial\rho'}\right|_{\rho'=\rho} & \left.\frac{\partial^2 L_{aa'}}{\partial\rho\,\partial\rho'}\right|_{\rho'=\rho} \end{vmatrix} \begin{vmatrix} \frac{di_f}{dt} \\ \frac{di_a}{dt} \\ i_a\frac{d\rho}{dt} \end{vmatrix} = \begin{vmatrix} u_f-R_f\,i_f \\ -u_d-(R_a+R_1)i_a-\dot\theta\left(\frac{\partial M_{af}}{\partial\rho}i_f+\frac{\partial L_a}{\partial\rho}\frac{i_a}{2}\right) \\ -\dot\theta\left.\left[\frac{\partial^2 M_{af}}{\partial\rho^2}i_f+\frac{\partial^2 L_{aa'}}{\partial\rho'^2}\right]\right|_{\rho'=\rho} i_a \end{vmatrix}$$

in these equations L_a and M_{af} are given by equations (2)

- for the second mode $(0 < \rho_1 , \rho_2 < \pi-\delta)$:

$$\begin{vmatrix} L_f & \frac{M_{af}(\delta)+M_{af}(\rho)}{2} & \frac{1}{2}\left.\frac{\partial M_{a2f}}{\partial\rho}\right|_{\rho_2=\rho} \\ \frac{M_{af}(\delta)+M_{af}(\rho)}{2} & L_1+\frac{L_a(\delta)+L_a(\rho)+2L_{a1a2}(\delta,\rho)}{4} & \frac{1}{4}\left[\frac{1}{2}\frac{\partial L_{a2}}{\partial\rho}+\frac{\partial L_{a1a2}}{\partial\rho_2}\right]_{\substack{\rho_1=\delta\\ \rho_2=\rho}} \\ M_{af}(\delta)-M_{af}(\rho) & \frac{L_a(\delta)-L_a(\rho)}{2} & \frac{1}{2}\left[\frac{\partial L_{a1a2}}{\partial\rho_2}-\frac{1}{2}\frac{\partial L_{a2}}{\partial\rho_2}\right]_{\substack{\rho_1=\delta\\ \rho_2=\rho}} \end{vmatrix} \begin{vmatrix} \frac{di_f}{dt} \\ \frac{di_a}{dt} \\ i_a\frac{d\rho}{dt} \end{vmatrix} =$$

$$\begin{vmatrix} u_f - R_f\,i_f \\ \left[-\left(R_1+\frac{R_a+R_{aa'}}{2}\right)i_a-\frac{\dot\theta}{2}\left(\frac{\partial M_{af}}{\partial\rho_1}+\frac{\partial M_{af}}{\partial\rho_2}\right)i_f-\frac{\dot\theta}{4}\left(\frac{1}{2}\frac{\partial L_{a1}}{\partial\rho_1}+\frac{1}{2}\frac{\partial L_{a2}}{\partial\rho_2}+\frac{\partial L_{a1a2}}{\partial\rho_1}+\frac{\partial L_{a1a2}}{\partial\rho_2}\right)i_a\right]_{\substack{\rho_1=\delta\\ \rho_2=\rho}} \\ \left[-\dot\theta\left(\frac{\partial M_{af}}{\partial\rho_1}-\frac{\partial M_{af}}{\partial\rho_2}\right)i_f-\frac{\dot\theta}{2}\left(\frac{1}{2}\frac{L_{a1}}{1}-\frac{1}{2}\frac{\partial L_{a2}}{\partial\rho_2}+\frac{\partial L_{a1a2}}{\partial\rho_1}-\frac{\partial L_{a1a2}}{\partial\rho_2}\right)i_a\right]_{\substack{\rho_1=\delta\\ \rho_2=\rho}} \end{vmatrix}$$

In these equations L_a, L_{a1a2} and M_{af} are given by equations (4).

3.2. Sudden loading on R.L. load

Figures 12 and 13 show the comparison between experimental and computed results for the sudden loading at constant speed of the 18 phase laboratory test machine. The very good agreement between theory and experiment confirms the validity of the assumptions made in extending the improved model to transient analysis.

Results computed without saliency effects are also presented on these figures in order to show the improvement of precision resulting from taking saliency into account.

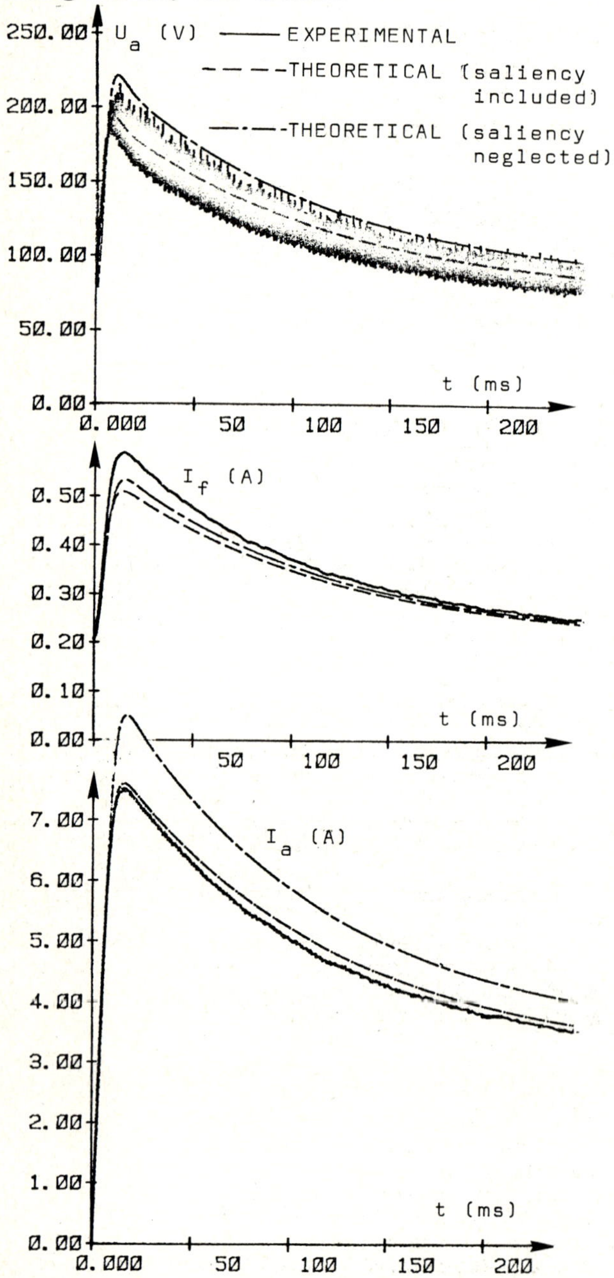

FIGURE 12

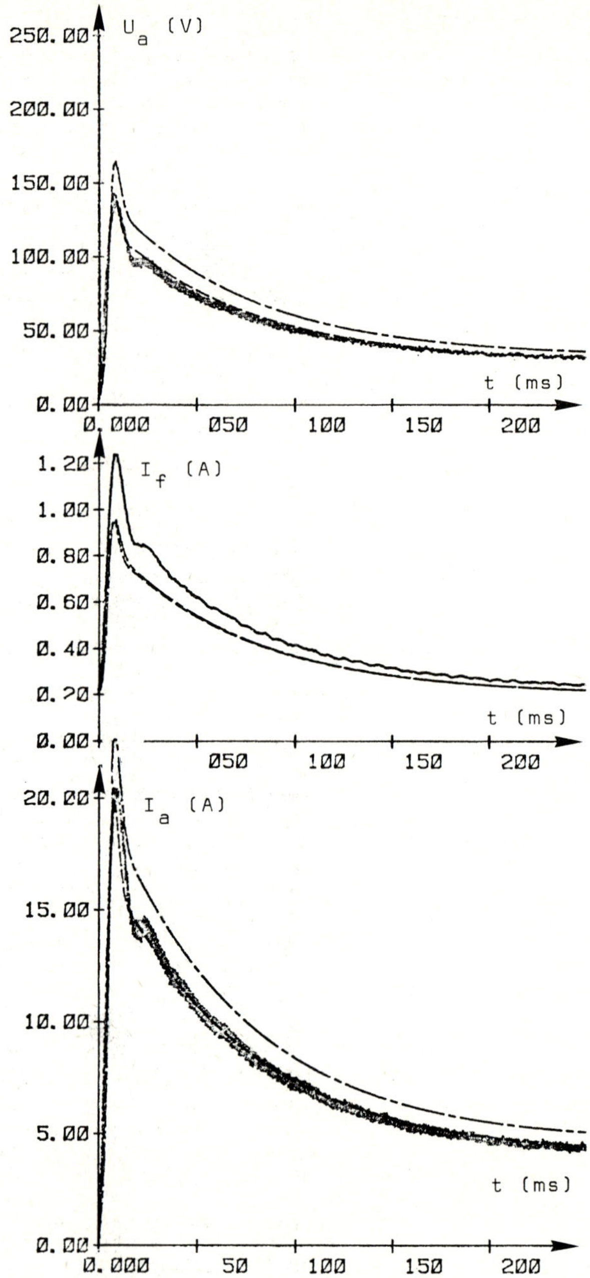

FIGURE 13

4. CONCLUSION

This paper has introduced saliency effects in the DC machine model of rectified output generators with a large number of phases.

It has been shown that taking these effects into account substantially increases the precision of the model.

All theoretical results have been validated by experiments made on two machines with very different power ratings.

5. REFERENCES

|1| L. Pierrat, J. Debernardi, J.D. Thomas, "Excitation des grands turbo-alternateurs. Développements récents. Performances et essais. Perspectives d'évolution" Rapport CIGRE, 11-07, session 1978, 30 août au 3 septembre.

|2| J. Debernardi, R. Gillet, "Système d'excitation compact pour alternateur à 4 pôles" Journées internationales d'étude des centrales électriques modernes, AIM, Liège, octobre 1974, rapport 75.

|3| L. Pierrat, F. Labrique, B. Mauhin, "Influence of the shape of the inductor field on the behaviour of rectified output alternators with a large number of phases" Proceedings of the 16th UPEC Conference, University of Sheffield, Sheffield, 7-9 April 1981.

|4| L. Pierrat, F. Labrique, B. Mauhin, "Transient behaviour of rectified output alternators with a large number of phases" Proceedings of the ICEM conference, chair of electrical machines of the university of Budapest, Budapest, 8-13 september 1982.

|5| F. Labrique, "Modélisation des génératrices à redresseur à grand nombre de phases" Thèse de doctorat, Département d'Electricité, Université Catholique de Louvain, juillet 1983.

ELECTRICAL MACHINES AND CONVERTERS - Modelling and Simulation
H. Buyse and J. Robert (editors)
Elsevier Science Publishers B.V. (North-Holland)/© IMACS, 1984

STATOR WINDINGS FOR D.C. BRUSHLESS MOTORS WITH CONTROLLED POWER ELECTRONICS

Alain MOUILLET, Georges EID

Laboratoire d'Electromécanique
Université de Technologie de COMPIEGNE
B.P. 233 - 60206 COMPIEGNE CEDEX - FRANCE

D.C. brusless motors with transistorized power supplies have been developed during last years[1,2,3,4]. For technical reasons, it is obvious that the power windings are inserted in the stator slots and the inductor (permanent magnet or electromagnet) is on the rotor core.
The present paper concerns the association of the stator coils and the choice of the transistorized supply to get the maximum electromagnetic torque, and thus, the maximum mechanical one.
It is shown, first, that, in order to use fewer transistors for better reliability, the solution consists of the association of the different coils using the stator or polygon connection, and then, because of the influence of the electromotive forces, that the stator winding must have an odd number of coils, and the distribution of the airgap magnetic field must be alternative and rectangular.

1. ASSUMPTIONS

We suppose that the motor is bipolar and the stator winding composed of n diametral independent coils inserted into 2 n slots, continuously fed by the electronic supply. (The extension of the results to the multipolar machine is easy : the mechanical angle of the bipolar motor is replaced by the electrical one).

The positive direction of the magnetic axis of each coil is determined by the Maxwell's rule and the geometric angle between the axis of two consecutive coils is then $\frac{\pi}{n}$ (Figure 1). We call $\vec{R}$ (modulus R) the rotor magnetic field and $\vec{S}$ (modulus S) the stator one.

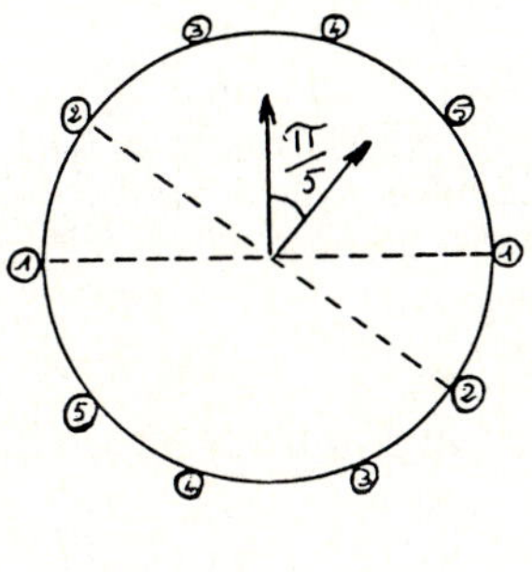

Figure 1

2. MAXIMUM ELECTROMAGNETIC TORQUE

It is well-known that the rotor magnetic field $\vec{R}$ and the stator one $\vec{S}$ must be at right angle to get the maximum electromagnetic torque (for a motor, $\vec{S}$ is leading $\vec{R}$).

The modulus and the direction of $\vec{S}$ depend on the distribution of the currents in the stator coils, which depends itself on $\vec{R}$, because of the induced electromotive forces. The previous condition needs to modify this distribution so as to follow the rotation of $\vec{R}$: it is realized, as in the classical D.C. motors, by commutating the current in every stator coil, one after the other. However, it cannot be realized at every time because of the finite number of coils (it would be possible only if there was an infinite number).

So, the only reasonable solution is to see one's way to get the upper mean value of the electromagnetic torque. For this purpose, the commutation of the current in the concerned coil must take place when the angle between $\vec{S}$ and $\vec{R}$ is $\frac{\pi}{n} - \frac{\pi}{2n}$; that is easily proved. This angle becomes $\frac{\pi}{n} + \frac{\pi}{2n}$ and decreases till the next commutation takes place.

3. GENERATION OF THE STATOR MAGNETIC FIELD $\vec{S}$

In the present chapter, we study first how to generate the rotating fields $\vec{S}$, without taking in account the rotor field $\vec{R}$. Let us consider (always for a bipolar motor) the space distribution of the n coil magnetic axis, and number them from 1 to n (clockwise for instance) ; we measure the angles of the positive direction of theese axis, in the same way, with respect to the diametrical plane which holds the coil number 1.

The axis of this coil and this plane are perpendicular, and the angle between the axis of the next coils and this same plane is, for the coil number p $(1 < p < n)$:

$$\frac{\pi}{2} + (p - 1)\frac{\pi}{n} \qquad (1)$$

The distribution of theese axis is symetrical with regard to a precise direction which is, either the bisectrix of the angle between the axis number $\frac{n}{2}$ and $\frac{n}{2}+1$ if n is even, or the axis number $\frac{n+1}{2}$ if n is odd. (Two examples are given on Figure 2).

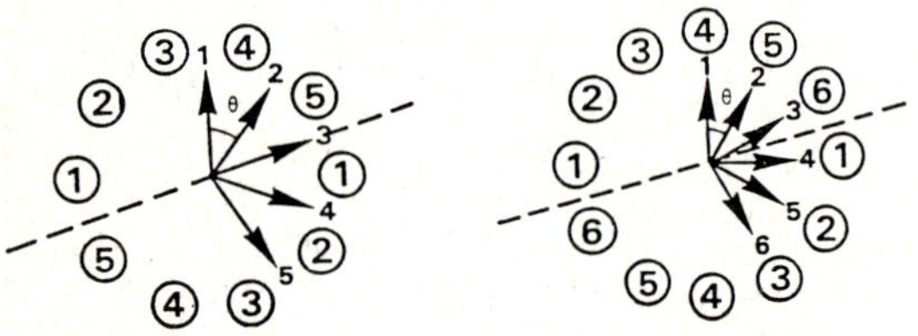

$n = 5\ (\theta = \frac{\pi}{5})$ $n = 6\ (\theta = \frac{\pi}{6})$

Figure 2

Let us feed the n coils with identical current intensities, so that the own magnetic field of a coil falls in with positive direction of its magnetic axis (first case). Then, the resulting magnetic field and the former axis of symmetry have the same direction. It is also true if the current intensities are only equal in two symmetrical coils with regard to this axis of symmetry, and so, for all the coils (second case).

We suppose now that the current is reversed in the first coil and keeps the same intensity in the first case, or this intensity is modified so as to respect the criterium of symmetry in the second case. The magnetic field of this coil has then an opposite direction ; the axis of symmetry and the resulting magnetic field rotates of a $\frac{\pi}{n}$ angle clockwise. Then, doing the same for the second coil, the stator field rotates again of a $\frac{\pi}{n}$ angle, and so on.

How to realize it when using the electronic supply ?

A first solution consists of independent supplies for every coil. To switch on and switch off the current as it is required, it needs transistors used in commutation and supporting the maximum intensity of the coil current. We have an inverter shown on the following figure for one coil (Figure 3 ; the arrow drawn on the coil corresponds to the positive direction of the current ; i.e. its own magnetic field falls in with the positive direction of its magnetic axis).

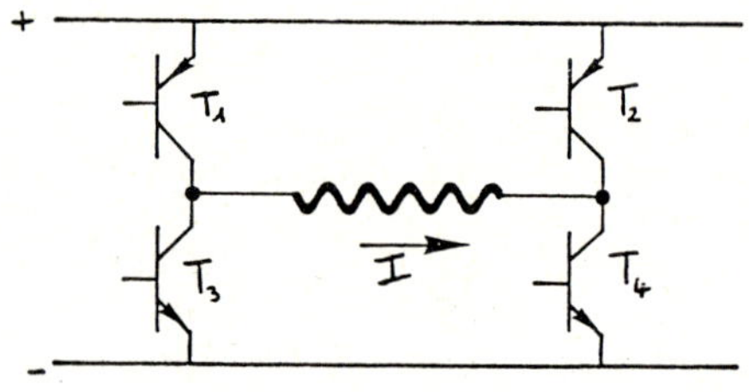

Figure 3

T_1 and T_2 are Darlingtons power PNP transistors and T_3, T_4 NPN ones. When T_1 and T_4 are conducting, T_3 and T_2 are off state. To reverse the current, T_1 and T_4 are then off and simultaneously, T_2 and T_3 become conducting.

Such device needs 4n transistors which commutation must be designed carefully. If T_2 and T_3 switch on when T_1 and T_4 are not yet shut off, the DC supply is short-circuited and the transistors can be destroyed.

So, in order to use fewer transistors for better reliability, the second solution consists of the association of the different coils which offers two possibilities : the star or polygon connection.

Let us represent the different coils like on Figure 4. We use the previous examples (n = 5 and n = 6).

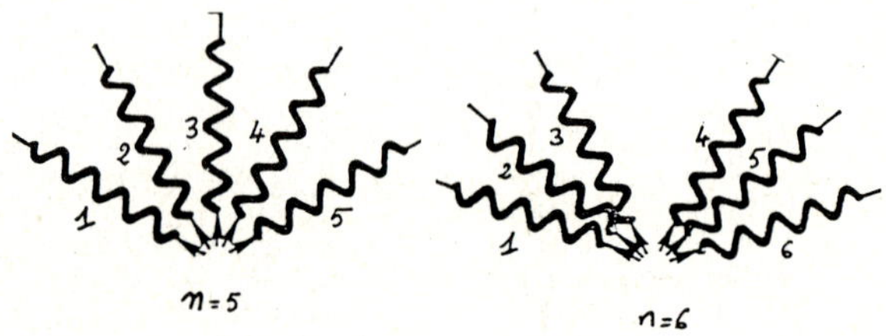

Figure 4

The arrow corresponds to the positive direction of the current in a coil which generates its own magnetic field falling in with the positive direction of its magnetic axis, and each coil is represented by this last direction.

1) Star connection :

Representing the positive pole of the DC supply by (+) and the negative one by (-), the following windings can be realized :

a) <u>odd number of coils (ex : n = 5)</u> :

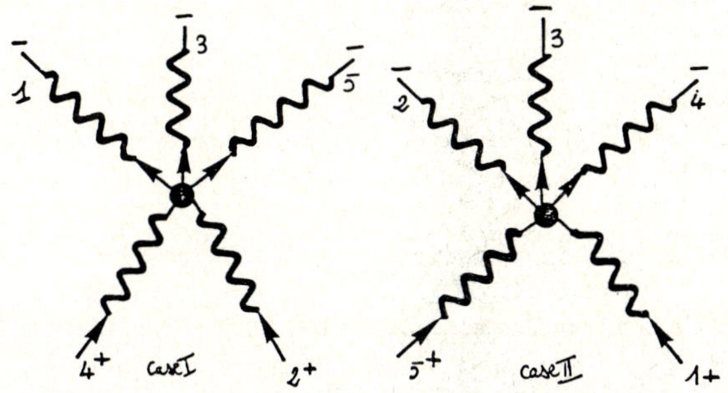

Figure 5

The current intensities are not the same ones in the different coils, but they are equal in two coils which are symmetrical with respect to the coil number $\frac{n+1}{2}$ (here, 3). The resulting magnetic field falls in with the positive direction of the magnetic axis of this coil, as it is required. The commutation consists of the reversing of the current in the first coil.

In case I, the "free" end of the first coil is then connected from (-) to (+), and the resulting magnetic field rotates correctly (angle and amplitude). Figure 6 shows the new distribution of the fields.

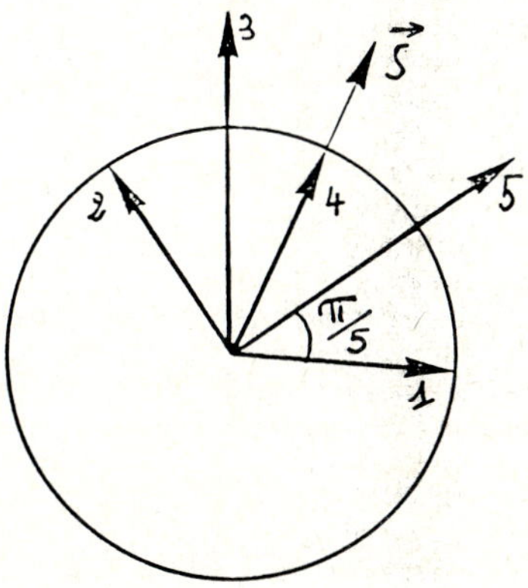

Figure 6

In case II, the free end of the first coil must be connected from (+) to (-), but on the example the current of the coil number 5 is the sum of the other currents and the fields of the different coils never satisfy the condition of symmetry as shown on Figure 7. This case must be excluded.

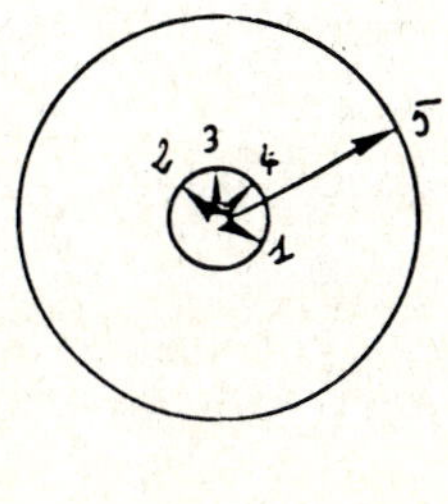

Figure 7

Practically, the electronic supply is connected to the star-connected stator winding as shown on Figure 8 (the direction of the currents corresponds to those defined on Figure 4).

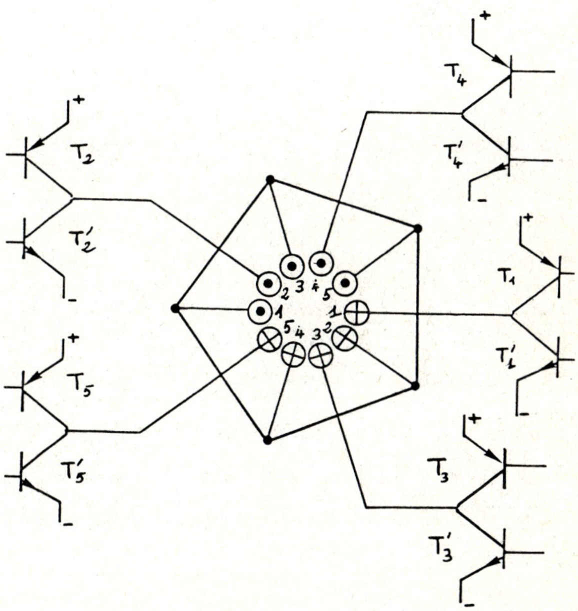

Figure 8

It represents the case I of Figure 5 : T'_1, T'_3, T'_5, T_2 and T_4 are conducting while T_1, T_3, T_5, T'_2 and T'_4 are off state. After the commutation, T'_1 is off and T_1 conducting.

In this device, the total number of transistors and the number of transistors commutating together are divided by two. There are only two transistors for each coil. However, the DC supply can be always short-circuited.

b) <u>even number of coils (ex : n = 6)</u> :

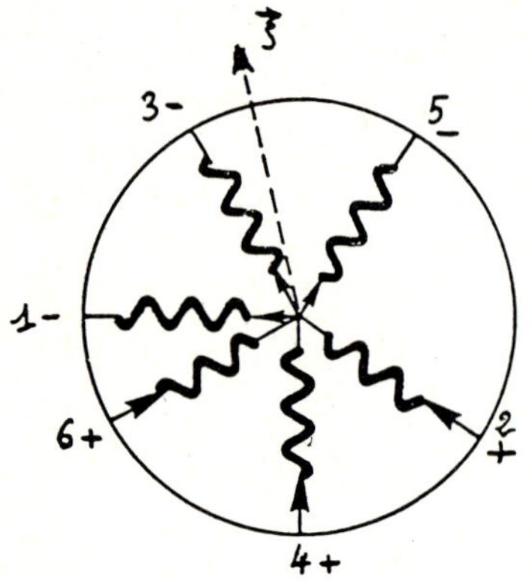

Figure 9

Figure 9 shows that all the current intensities, and then the field intensities, are equal, but, when the free end of the first coil is connected from (-) to (+), the condition of symmetry will never be satisfied (Figure 10). So, the resulting field will not rotate of a $\frac{\pi}{n}$ angle (here, $\frac{\pi}{6}$) and its amplitude will be modified.

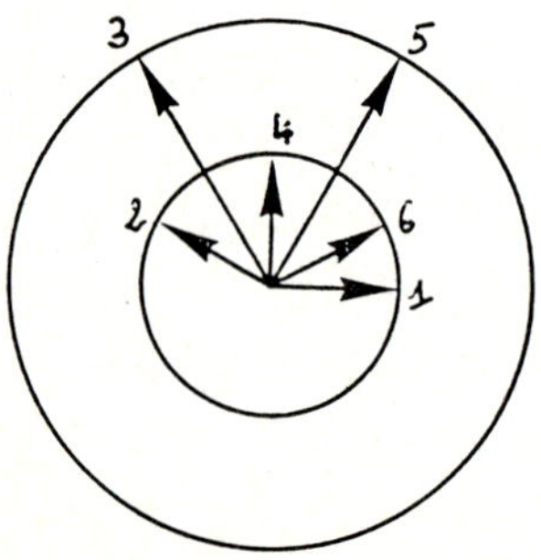

Figure 10

After the next commutation (i.e. the free end of the second coil connected from (+) to (-), the distribution of the fields is right again (Figure 11).

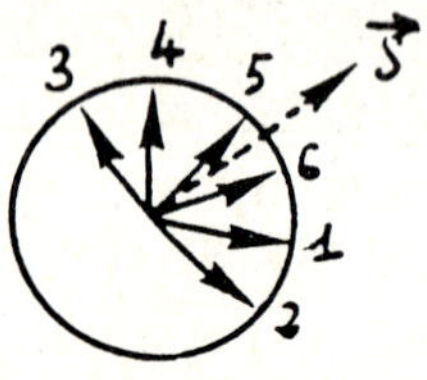

Figure 11

So, with the star connected winding with an even number of coils, the resulting magnetic field rotates of a $\frac{2\pi}{n}$ angle. It needs always two transistors for each coils, but two currents are reversed simultaneously (i.e. two transistors become conducting together while two ones are shut off) and the risk of short-circuiting the DC supply is more important.

c) <u>summary of the results about the star-connection</u> :

- it needs two transistors for each coil, and the half of them are conducting at every time ;
- with an odd number of coils, the magnetic field $\vec{S}$ rotates of a $\frac{\pi}{n}$ angle, and two transistors commutate together on the same coil ;
- with an even number of coils, $\vec{S}$ rotates of a $\frac{2\pi}{n}$ angle and four transistors commutate together on two coils.

2) <u>Polygon connection</u> :

a) <u>odd number of coils (ex : n = 5)</u> :

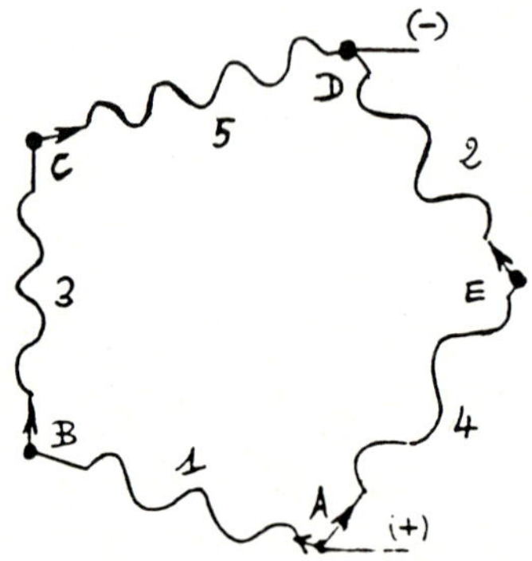

Figure 12

Figure 12 corresponds to the case of Figure 4. In this example, the current intensities are equal, on the one hand, in the coils number 1, 3 and 5, and, on the other hand, in the coils number 2 and 4 : the condition of symmetry is assured.

When the commutation takes place, the end A of the first coil, connected to (+) is shut off, and the end B of this coil is connected to the same pole, as shown on Figure 13.

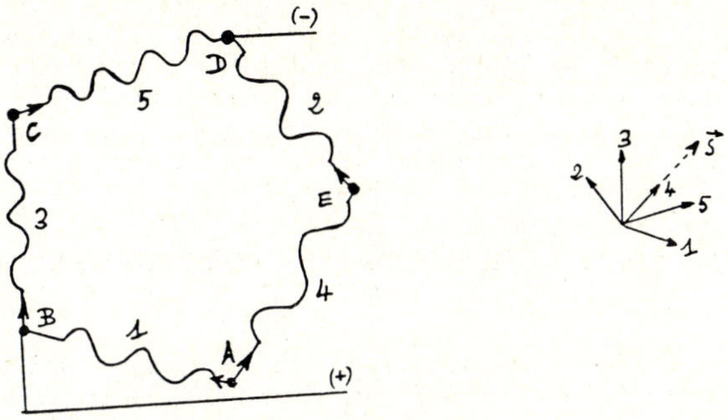

Figure 13

The new distribution of the fields is right, and the resulting magnetic fields keeps the same amplitude and has rotated of a $\frac{\pi}{n}$ angle (here, $\frac{\pi}{5}$). For the next commutation, the pôle (-) is connected from D to E, and so on.

The polygon connection and its associated electronic supply are represented on Figure 14.

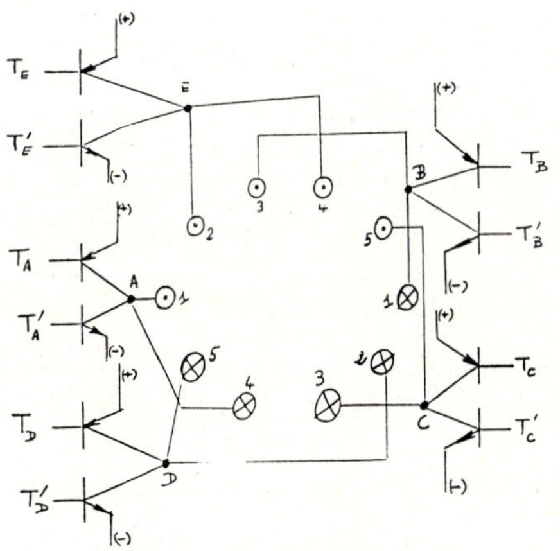

Figure 14

The case of Figure 12 corresponds to T'_D and T_A conducting and the other transistors off. So, each transistor is choosen so as to support the total current of the motor. To inverse the current in the first coil (i.e. the case of Figure 13), T_A is shut off and T_B switched on.

Such device also needs only two transistors for each coil, but when the current is switched over in the required coil, the DC supply is not short-circuited.

b) even number of coils (ex : n = 6) :

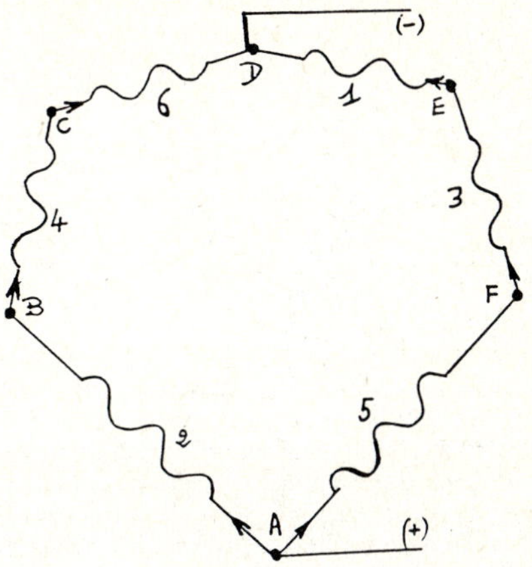

Figure 15

Figure 15 corresponds also to the case of Figure 5. The current intensities are equal in the different coils like in the star connection. To reverse the current in the first coil, the pole (-) is connected from D to E, and then, the condition of symmetry is not satisfied. Then, when the current is switched over in the second coil (i.e. (+) connected from A to B), the distribution of the fields becomes right again (Figure 16).

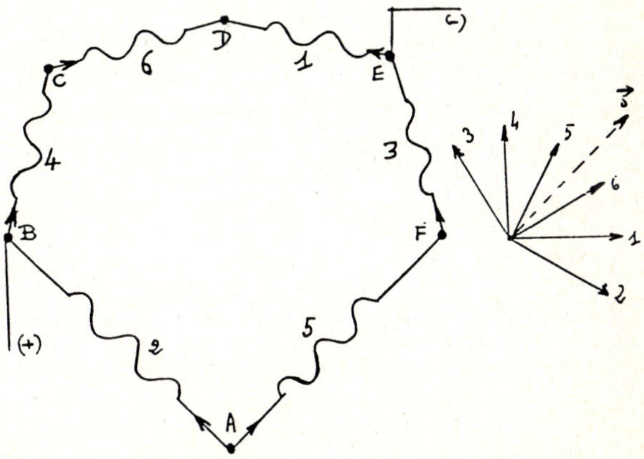

Figure 16

So, with an even number of coils, the resulting field $\vec{S}$ must rotate of a $\frac{2\pi}{n}$ angle. Four transistors commutate together, but the DC supply is never short-circuited.

c) summary of the results about the polygon connection :

The results are the same ones like in the star connection, but two transistors only are conducting at every time and the DC supply is never short-circuited.

4. INTERACTION BETWEEN THE ROTOR MAGNETIC FIELD AND THE STATOR CURRENTS

We study now the influence of the electromotive forces (i.e. of the rotor field $\vec{R}$) on the distribution of the stator currents, and thus, on the stator field $\vec{S}$. It needs to know the direction of theese electromotive forces induced in the stator coils. Figure 17 sums up all the possible solutions with regard to the rotor position (The arrows drawn inside the coil correspond to the positive direction defined in the previous chapters).

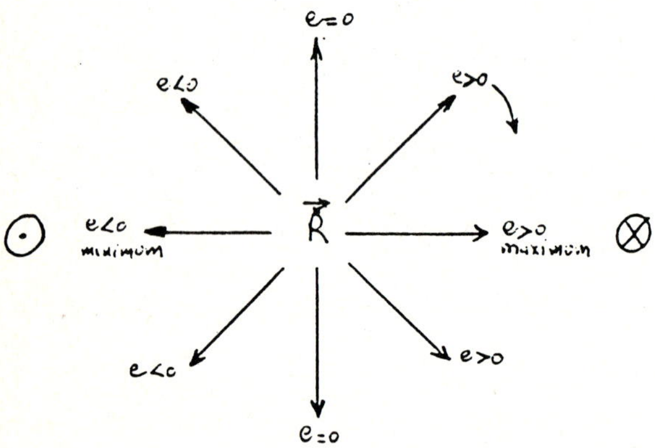

Figure 17

The present chapter concerns the influence of rotor field $\vec{R}$ on the stator currents between two commutations. So, we assume that the connection between the stator winding and the DC supply through the power electronics is given, and we take into account the rotation of $\vec{R}$ only.

1) Star connection :

a) odd number of coils :

Figure 18 corresponds to the first case of Figure 5. The rotation of $\vec{R}$ is represented and its positions when the stator current is reversed in the required coils are noted (1) and (2). The direction of the electromotive forces is noted (→).

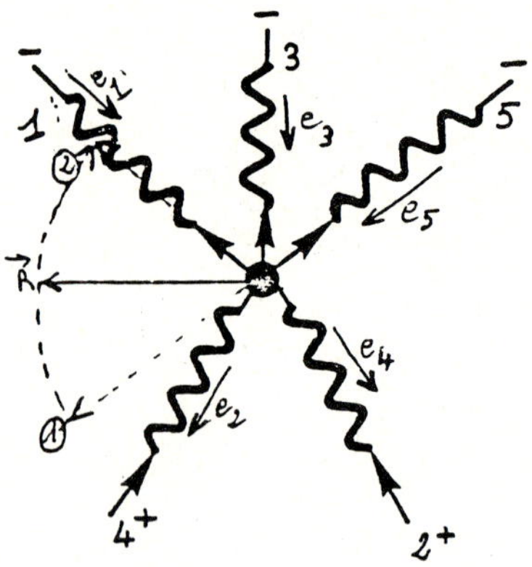

Figure 18

These electromotive forces, in the coils which the magnetic fields are symmetrical with respect to the resulting field $\vec{S}$, are equal only when $\vec{R}$ and $\vec{S}$ are at right angle. This is the only position of $\vec{R}$ which satisfies the condition of symmetry. Otherwise, theese electromotive forces are not equal and thus, they are time-dependant like the current intensities : the amplitude and the position of the stator field $\vec{S}$ are then also time-dependant.

This phenomena would not exist if the induced electromotive forces were constant between two commutations. Figure 18 shows that theese e.m.f. are null in the required coil when the current is switched over. Thus, we can conclude that the electromotive forces must be alternative and rectangular, that means the distribution of the rotor magnetic flux density, in the airgap of the motor at constant velocity, is also alternative and rectangular.

b) even number of coils :

Figure 19 corresponds to the case of Figure 9.

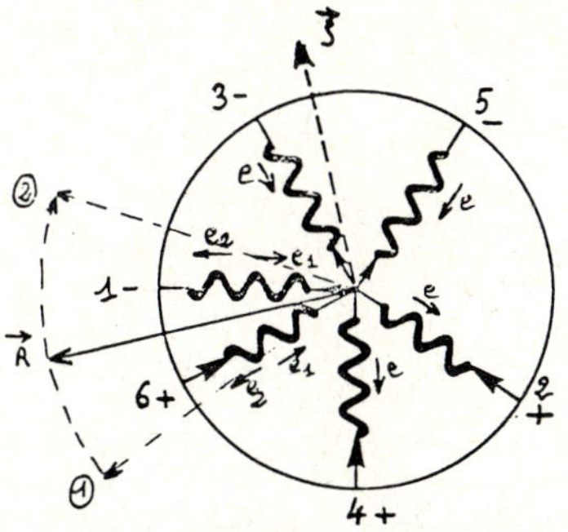

Figure 19

The rotor field $\vec{R}$ rotates of a $\frac{2\pi}{n}$ angle between two commutations. Figure 19 shows that the electromotive force is reversed in two coils during the same time, that means they are brakes during a part of this time and the electromagnetic torque decreases. Moreover, the electromotive forces in the required coil are not null when the current switches over (in the example, the e.m.f. favours the reversal of the current in the coil number 2 and does not favour it in the first coil). At last, the condition of symmetry is never satisfied for the currents.

Here, the phenomena is also the same one with alternative rectangular electromotive forces. So, a stator winding with an even number of coils must be excluded.

2) Polygon connection :

a) odd number of coils :

Figure 20 corresponds to the first case of Figure 4, using the device of Figure 12.

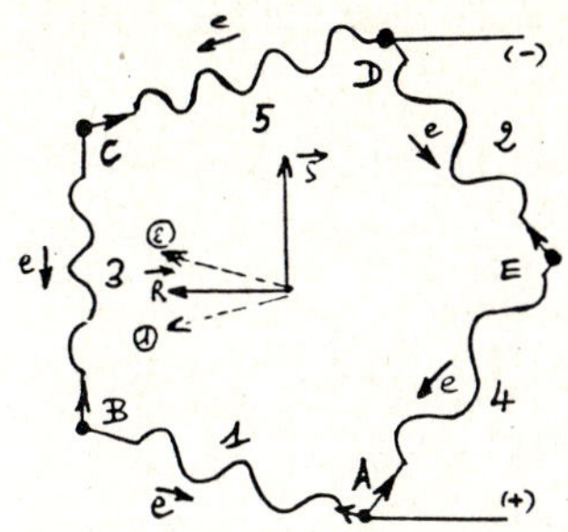

Figure 20

The electromotive forces are time-dependant, and the current intensities are not constant in the different coils, but the condition of symmetry is satisfied. So, only the amplitude of the stator magnetic field $\vec{S}$ is time-dependant. The current is also reversed, in the required coil, when the induced electromotive force is null.

Thus, rectangular alternative e.m.f. are also the solution for the polygon connection.

However, if all e.m.f. have the same amplitude E, one can see, on the example, that it is possible to have : $3\ E > V > 2\ E$ (2)
(V is the DC applied voltage).

Figure 21 corresponding to Figure 20 (r is the resistance of one coil), there is a circulating current inside the motor.

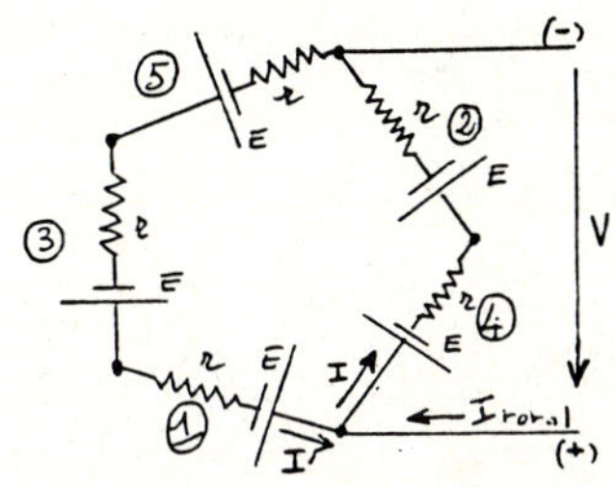

Figure 21

In a general case, for a polygon connection with (2p + 1) coils, condition (2) is written :

$$(p + 1)\ E > V > p\ .\ E \quad (3)$$

The machine is always a motor if :

$$p\ E\ I > (p + 1)\ E\ I' \quad (4)$$

with : $V = (p + 1)\ [E - rI'] = p\ [E = rI]$ (5)

Relation (4) becomes then :

$$E < \frac{2}{2p+1} V \qquad (6)$$

Relations (3) and (6) show finally that :

- if $\frac{V}{p+1} < E < \frac{2}{2p+1} V$, the motor works with a circulating current ;
- if $E < \frac{V}{p+1}$, there is no circulating current.

The theoretical efficiency defined as :

$$\eta = \frac{p\,E\,I - (p+1)\,E\,I'}{V\,(I - I')}$$

can be written with relation (5) :

$$\eta = \frac{E}{V} \cdot \frac{2V - (2p+1)\,E}{\frac{(2p+1)\,V}{p\,(p+1)} - 2\,E} \qquad (7)$$

The minimum value, for $E = \frac{2}{2p+1} V$, is zero, and the maximum one, for $E = \frac{V}{p+1}$, is :

$$\eta_{max} = \frac{p}{p+1} \qquad (8)$$

When the number of coils is little, this efficiency is low, but always better than 50 % ($p \geqslant 1$). On the example, with p = 2, this efficiency is 66 %. It increases with the number of coils, and is practically equal to 1 when this number is very high.

However, for this maximum value, I' = 0. So, the specific power of such machine is very low. For all the coils work together at any time without a circulating current, it needs that the induced electromotive forces have not the same value in the different coils at every time, but they must be always alternative and rectangular. Theese conditions can be resumed by the next relation :

$$(p+1)\,E' = p\,E < V \qquad (9)$$

or :

$$E' = \frac{p}{p+1}\,E < \frac{V}{p+1} \qquad (10)$$

Furthermore, the coils connected in series between the two poles (+) and (-) are alternated-numbered and not consecutive-numbered. The solution is then a distribution of the rotor magnetic field in the airgap of the motor like on Figure 22.

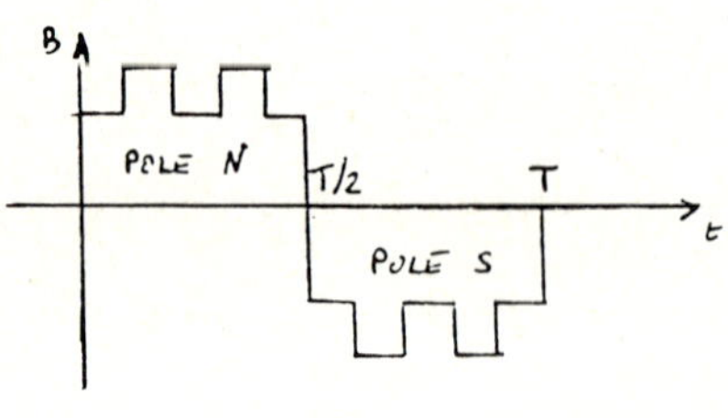

Figure 22 (ex : n = 5)

The theoretical efficiency is :

$$\eta = \frac{p\,E\,I + (p+1)\,E'I'}{V\,(I + I')} \qquad (11)$$

with : $V = (p+1)\,(E' + rI') = p\,(E + rI)$ (12)

Using relations (10) and (12), relation (11) becomes :

$$\eta = p\,\frac{E}{V} \qquad (13)$$

which is the same one like in the classical DC motors.

b) even number of coils :

Figure 23 always corresponds to the case of Figure 9.

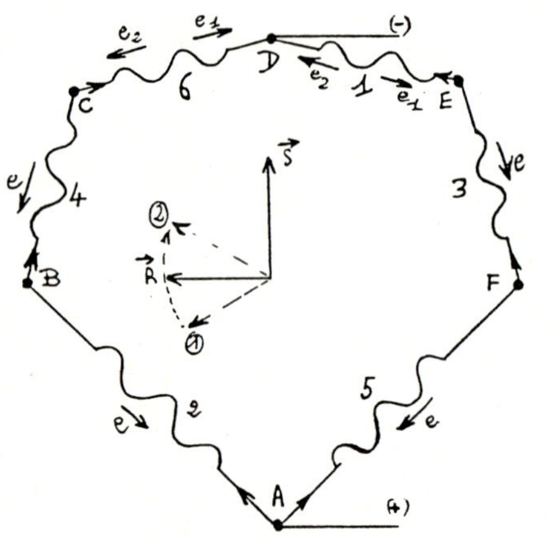

Figure 23

The phenomenas are identical like in the star connection, even with rectangular alternative currents (classical or modified like Figure 22). Here, an even number of coils must be also excluded.

5. CONCLUSION

From the previous studies, we think that the design of DC brushless motors with transistorized power supplies needs a stator winding with an odd number of coils for a bipolar machine. Rectangular alternative currents provide the best results for such motors : the classical rectangular time function is convenient with the star connection, and the rectangular one, modified like on Figure 22, is right for the polygon connection.

REFERENCES :

[1] G. EID, A. MOUILLET : "DC brushless motors with a rare earth permanent magnet rotor" (International Conference on Permanent Magnet Electrical Machines - KATOWICE - 1979 October 9-11)

[2] T. JANIK : "DC brushless motors with a rare earth permanent magnet and electronic commutator" (International Conference on Permanent Magnet Electrical Machines - KATOWICE October 9-11)

[3] G. GANGADHARAN, K.R. VENKATESH : "Novel type of transistorized brushless DC motors" (I.E.C.I. JOURNAL - EL. October 1977)

[4] M. KANT, J.P. VILAIN : "Logadyne an open winding machine with controlled power electronics" (Proceedings of the International Conference on Electrical Machines : Part 2 - ATHENS - 1980 September 15-17).

ELECTRICAL MACHINES AND CONVERTERS - Modelling and Simulation
H. Buyse and J. Robert (editors)
Elsevier Science Publishers B.V. (North-Holland)/© IMACS, 1984

A NEW METHOD FOR SIMULATING THE MACHINE-CONVERTER ASSOCIATION - STUDY OF THE COUPLING BETWEEN MAGNETIC STRUCTURE AND ELECTRONIC CIRCUIT

Jean FAUCHER, Gérard DROUET

Laboratoire d'Electrotechnique et d'Electronique Industrielle
E.R.A. au C.N.R.S.n° 536
E.N.S.E.E.I.H.T.
2 Rue Camichel - 31071 TOULOUSE Cédex - FRANCE

The development of new structures for electrical machines fed by static converters has made it necessary to bring into play appropriate models.

The authors present a method for simulating the coupling between an electronic circuit and an electromagnetic actuator drive. The proposed method is based on an insertion of an equivalent magnetic circuit of the machine in the simulation of the overall system.

The numerical resolution is then considered and the authors suggest an original algorithm. The method is applied to the study of different machines.

1. INTRODUCTION

The recent development of new structures for alternating current electrical machines associated with static converter has made it necessary to bring into play more appropriate simulation tools than the traditional models. The currently used methods are usually based on a Park transformation wich is only applicable for sinusoidal machines, i.e. with a sinusoidal spatial distribution of both m.m.f. and flux density in the airgap. As such useful method can no longer provide satisfactory results for special machines, an alternative approach must then be investigated.

The proposed method consists of machine modelling by means of an equivalent magnetic circuit and inserting it in an overall machine/converter drive simulation.

The coupling between the converter circuit and the magnetic circuit of the actuator remains the essential problem of such a simulation. It follows that the authors insist chiefly on the analysis of the electro-magnetic coupling and propose two methods for a numerical solution.

2. ANALYSIS

The figure 1 presents the schematic and general diagram of an electric circuit coupled with a magnetic circuit. The introduction of an interface block allows to consider separately the electric output of the electric circuit and the electric input of the magnetic circuit.An extra advantage brought by the insertion of this interface is to consider only "two wires" terminations. The equations of the interface then describe the connections existing between the two circuits (nodes and loops kirchoff laws).

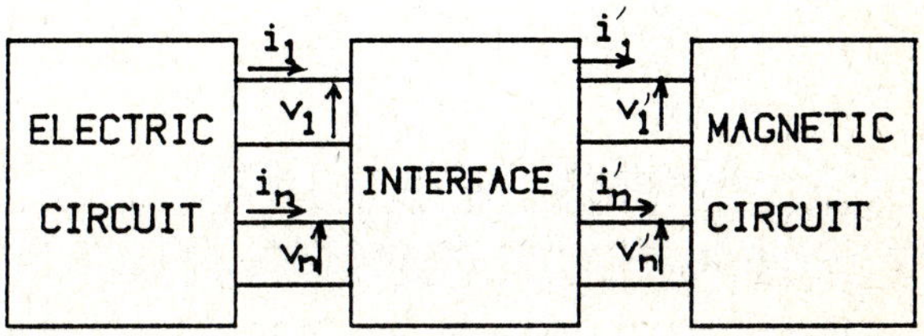

figure 1

Notations

s is the number of internal state variables (y_i ; i = 1 to s) of the electric circuit.

s' is the total number of state variables of the magnetic circuit block. Among these variables, the n'electric currents i'_j (j = 1 to n') will be distinguished from the mechanical positions state variables x'_ℓ (ℓ = 1 to m') with the relation :

$$s' = n' + m' \qquad (1)$$

The electric circuit is described in such a way that the electric terminals are independent. The magnetic nature of the load implies that these terminals are to be considered as connected to n inductive branches.

The electric currents i_k ($k = 1$ to n) are then state variables of the overall system.

The interface algebraic equations are :

$$i' = f'_j(i_1, \dots i_n) \quad \text{n'equations} \quad (2)$$

$$v_k = g_k(v'_1, \dots v'_n) \quad \text{n equations} \quad (3)$$

with :

$$n \leq n'$$

The magnetic flux linkages λ'_j ($j = 1$ to n') are tied to terminal voltages v'_j :

$$\frac{d\lambda'_j}{dt} = v'_j \quad (4)$$

Flux linkages λ_k ($k = 1$ to n) are defined from voltages v_k :

$$\frac{d\lambda_k}{dt} = v_k \quad (5)$$

The flux linkages λ_k are tied to flux linkages λ'_k by relation (3) :

$$\lambda_k = g_k(\lambda'_1, \dots \lambda'_n) \quad (6)$$

The state variables of the overall system are then choosen as :

s electric variables y_i

n electric currents i_k

m' mechanical positions x'_ℓ

It can be noticed that mechanical load takes place in the system only by means of mechanical positions. The corresponding mechanical circuit generally described by differential equations is considered to be out of the system and connected to it by mechanical positions and associated magnetic forces.

By considering no eddy currents occuring in the magnetic parts, a complete solution of the magnetic circuit can be computed from all electric currents i'_j and all mechanical positions x'_ℓ :

$$\lambda'_j = F_j(i'_1, \dots i'_{n'}; x'_1, \dots x'_{m'}) \quad (7)$$

By replacing i'_j by expression (2) in equation (7) and replacing λ'_j by expression (7) in equation (6), the flux linkage λ_k becomes :

$$\lambda_k = G_k[i_1, \dots i_n; x'_1 \dots x'_{m'}] \quad (8)$$

The time-derivative of λ_k can then be written as :

$$\frac{d\lambda_k}{dt} = \sum_{j=1}^{n} \frac{\partial \lambda_k}{\partial i_j} \frac{di_j}{dt} + \sum_{j=1}^{m} \frac{\partial \lambda_k}{\partial x'_j} \frac{dx_j}{dt} \quad (9)$$

The overall system can then be described by the general state differential equation :

$$X^\bullet = AX + BU \quad (10)$$

with :

$$X = \begin{bmatrix} y_1 \\ y_s \\ i_1 \\ i_n \end{bmatrix} \quad (11)$$

The source matrix U includes both mechanical speeds $\frac{dx_\ell}{dt}$ and electric sources.

The coefficient of matrixes A and B are dependent on the actual state of the system by means of partial derivatives of flux linkages $\partial\lambda_k/\partial i_j$ (k and j each other from 1 to n) and $\partial\lambda_k/\partial x'_j$ (k from 1 to n and j from 1 to m'). These coefficients are determined by a resolution of the magnetic circuit (generally non linear).

3. SIMULATION METHODS

The differential euqations system (10) is solved numerically by using a Runge-Kutta method or other appropriate step by step integration method.

The step of time for integration (STI) is quite difficult to be previously choosen. Sufficiently short to take into account fast variations of variables and sufficiently large in view not to increase the time computation in a prohibitive manner.

The essential problem for solving the system under study is the computation of partial derivatives of flux linkages. This computation must be made at each integration step. Owing to the generally non analytical model of magnetic circuit, the partial derivatives coefficients are calculated by making a small and defined increasing in the only concerned state variable and determining the corresponding increasing of magnetic fluxes.

Two methods are presented for the calculation of partial derivatives.

3.1. Independent steps method (IS)

The small increasings of state variables (ISV) are completely independent of the integration steps. They are only choosen to give accurate results.

3.2. Dependent steps method (DS)

For this method, ISV used for calculation of partial derivatives are identical to the ISV at each integration step.

For an easy presentation of the method, an elementary differential equation is considered :

$$\frac{dx}{dt} = g(t, x) \quad (12)$$

x_k is the estimated value of state variable x at the step K ($t = t_k$). The function $g(t, x)$ is supposed to include a partial derivative coefficient $\partial f(x)/\partial x$ The ISV for partial derivative computation is choosen as :

$$\Delta x = x_k - x_{k-1} \qquad (13)$$

then : $$\left(\frac{\partial f}{\partial x}\right)_k = \frac{f(x_k) - f(x_{k-1})}{\Delta x} \qquad (14)$$

The advantage of this method is to imply only a determination of f(x) at each step of integration instead of two for the IS method, in consequence an important decreasing in the computing time is obtained.

Nevertheless, there are two important restrictions for the application of the DS method.

First, the ISV are evoluting with time and can become too small for an accurate determination of partial derivatives. The best solution is then to use the IS method until the DS method can again be used with a good accuracy.

The second restriction is that when more than one state variable of the magnetic system is determined from the differential equations, there is no interest to use the DS method.

In view to explain that, a system is considered which is described by two state variables x and y :

$$\frac{dx}{dt} = f_x(x, y)$$
$$\frac{dy}{dt} = f_y(x, y) \qquad (15)$$

The functions f_x and f_y include partial derivatives coefficients :

$$\frac{\partial g(x,y)}{\partial x} \quad \text{and} \quad \frac{\partial g(x,y)}{\partial y}$$

$g(x,y)$ is a non linear function of x and y

The DS method is applied. At the step K the partial derivatives are :

$$\left(\frac{\partial g}{\partial x}\right)_k = \frac{g(x_k, y_k) - g(x_{k-1}, y_k)}{x_k - x_{k-1}} \qquad (16)$$

$$\left(\frac{\partial g}{\partial y}\right)_k = \frac{g(x_k, y_k) - g(x_k, y_{k-1})}{y_k - y_{k-1}}$$

and at the next step :

$$\left(\frac{\partial g}{\partial x}\right)_{k+1} = \frac{g(x_{k+1}, y_{k+1}) - g(x_k, y_{k+1})}{x_{k+1} - x_k} \qquad (17)$$

$$\left(\frac{\partial g}{\partial y}\right)_{k+1} = \frac{g(x_{k+1}, y_{k+1}) - g(x_{k+1}, y_k)}{y_{k+1} - y_k}$$

It can immediately be verified that neither of the values of $g(x,y)$ needed in relations (17) at step $k+1$ have been previously used for calculations at step K. The DS method then brings no advantage in such a case. However, there is an important type of magnetic systems described by more than one state variable which can have advantage of the DS method : magnetic systems where the time evolutions of state variables used for calculation of partial derivatives except one are analytically known. The previous system (15) is considered where $y(t)$ is an analytically known time function. The partial derivatives at step k can be calculated as :

$$\left(\frac{\partial g}{\partial x}\right)_k = \frac{g(x_k, y_k) - g(x_{k-1}, y_k)}{x_k - x_{k-1}} \qquad (18)$$

$$\left(\frac{\partial g}{\partial y}\right)_k = \frac{g(x_k, y_{k+1}) - g(x_k, y_k)}{y_{k+1} - y_k}$$

and at step $k+1$

$$\left(\frac{\partial g}{\partial x}\right)_{k+1} = \frac{g(x_{k+1}, y_{k+1}) - g(x_k, y_{k+1})}{x_{k+1} - x_k}$$

$$\left(\frac{\partial g}{\partial y}\right)_{k+1} = \frac{g(x_{k+1}, y_{k+2}) - g(x_{k+1}, y_{k+1})}{y_{k+2} - y_{k+1}} \qquad (19)$$

By considering equations 18, it can be noticed that only the quantities $g(x_k, y_k)$ and $g(x_k, y_{k+1})$ are to be calculated at the step k , that implies to know y_{k+1} at step K from the knowledge of the function $y(t)$

4. TWO EXAMPLES OF APPLICATION OF DS AND IS METHODS

4.1. Variable inductance

This elementary system is studied as an easy way to present the resolution methods. The overall system described by figure 2 is constituted by a saturable inductance which is variable with an α' mechanical position variable (the position of an iron plunger for example) and fed by a voltage source through a resistance R.

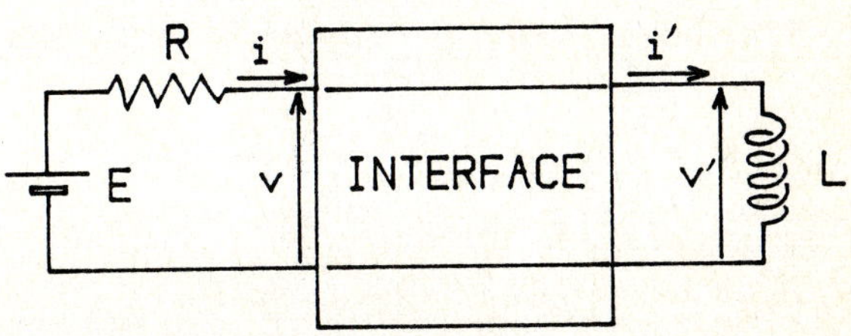

figure 2

The flux linkage variation Law

$$\lambda'(i',\alpha') = \frac{\lambda'_M(i') + \lambda'_m(i')}{2} + \frac{\lambda'_M(i') - \lambda'_m(i')}{2} \cos\alpha' \tag{20}$$

The functions $\lambda'_M(i')$ and $\lambda'_m(i')$ are presented on figure (3) (linear interpolations between values obtained by a computation of the magnetic field).

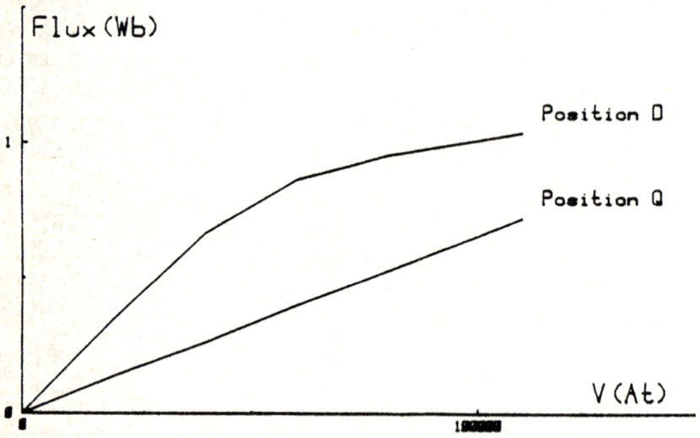

figure 3

The equations of interface are :

$$i' = i \qquad \lambda = \lambda' \tag{21}$$

and the electrical differential equation is :

$$e(t) = Ri + \frac{\partial\lambda}{\partial i}\frac{di}{dt} + \frac{\partial\lambda}{\partial\alpha}\frac{d\alpha}{dt} \tag{22}$$

$e(t)$ and $\alpha(t)$ are given functions :

$$e(t) = E \quad ; \quad \alpha(t) = \Omega t$$

The final differential equation to be solved then becomes :

$$\frac{di}{dt} = -\frac{1}{\frac{\partial\lambda}{\partial i}}\left[Ri + \Omega\frac{\partial\lambda}{\partial\alpha} - E\right] \tag{23}$$

The integration is made by an order 2 Runge-Kutta method. The obtained results are presented on figure 4.

The two methods have been applied. The results are quite identical but the increasing of the computing time from DS method to IS method is about 80 %.

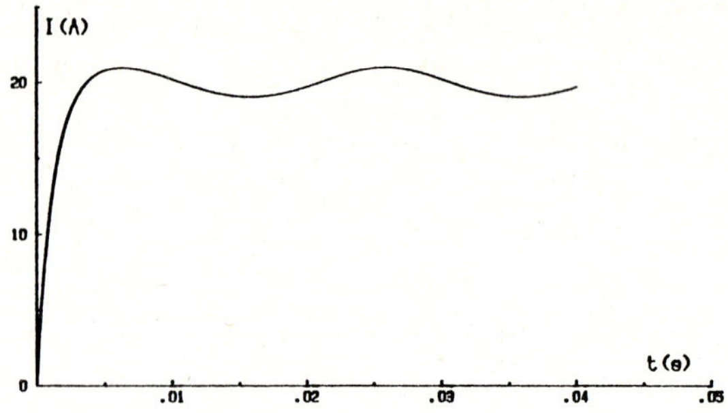

figure 4

4.2. Simulation of the current commutation in current source inverter fed reluctance machines

The first considered machine is an actual prototype called MRV1 and designed in TOULOUSE (FRANCE) at the L.E.E.I. This machine develops a 150 N.m pull out torque. The basic system configuration of the static converter/reluctance motor drive is shown on figure 5. A current controlled trhee-phase rectifier supplies smoothed d.c. current at the three-phase inverter. The system is autopiloted by means of a detector of the rotor position.

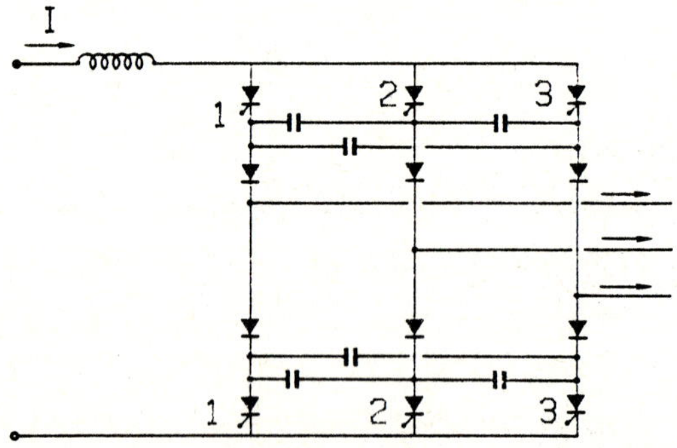

figure 5

The following hypothetical considerations will be made :

- . steady state operation
- . link d.c. current perfectly smoothed and regulated
- . the semiconductor units constitute perfect switches
- . eddy currents are neglected.

During commutations of phase currents, the phase windings of the machine are connected to commutation capacitors and the equivalent system shown on figure 7 can then be considered for a reference commutation from phase 2 to phase 3.

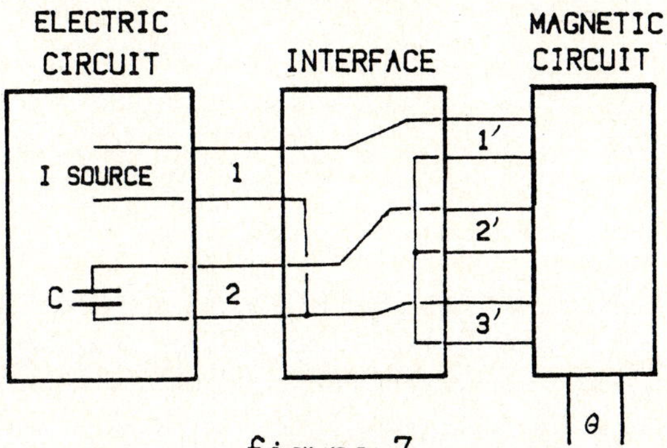

figure 7

The reluctance machine is represented by an equivalent magnetic circuit. The essential elements of the magnetic circuit i.e. teeth, are characterized before the simulation from a local numerical field computation taking into account the actual B, H curve of the used material. The machine model is then able to provide the values of flux linkages $\lambda'_1, \lambda'_2, \lambda'_3$ from given values of input currents i'_1, i'_2, i'_3 and rotor position θ .

The differential equations system to be solved is :

$$i_1 = I \qquad v_1 = \frac{d\lambda_1}{dt} \tag{24}$$

$$\frac{dv_2}{dt} = \frac{i_2}{C} \qquad v_2 = \frac{d\lambda_2}{dt}$$

The state variable are v_2, i_1, i_2 and θ . The interface equations are :

$$\begin{cases} i'_1 = i_1 \\ i'_2 = i_2 \\ i'_3 = -i_1 - i_2 \end{cases} \tag{25}$$

$$\begin{cases} v_1 = v'_1 - v'_3 \\ v_2 = v'_2 - v'_3 \end{cases} \qquad \begin{cases} \lambda_1 = \lambda'_1 - \lambda'_3 \\ \lambda_2 = \lambda'_2 - \lambda'_3 \end{cases} \tag{26}$$

The complete differential equations system then becomes :

$$\frac{di_2}{dt} = \frac{1}{\partial\lambda_2/\partial i_2}\left[v_2 - \frac{\partial\lambda_2}{\partial\theta}\Omega\right] \tag{27}$$

$$\frac{\partial v_2}{\partial t} = \frac{i_2}{C}$$

The state variables to be calcultaed are reduced to i_2 and v_2 because i_1 and θ are analytically known.

The IS method is applied associated with a Runge-Kutta integration method (order 2).

The figure 8 presents the evolution of the maximal capacitor voltage obtained at the end of the commutation. The comparisons between experimental and computed results can be considered sufficiently good to qualify the method.

The second machine is also a vernier reluctance machine but with a stator field winding. This is a projected prototype for which a 400 N.m pull out torque is expected.

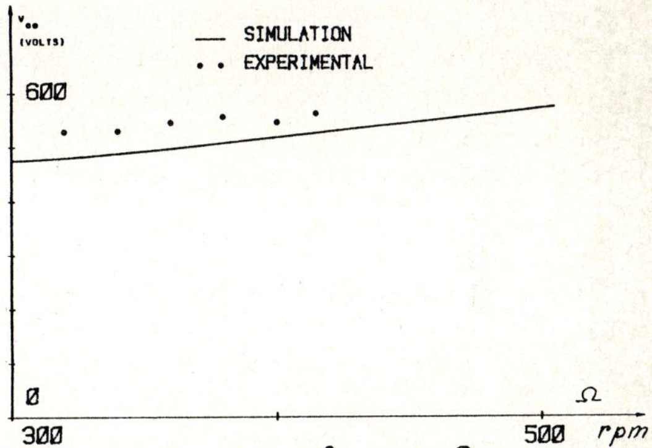

figure 8

The presented results concern a comparison between the DS and IS methods. The system is identical to the precedent except an extra state variable : the field current i_e which is considered as constant. The differential equations are the same as equations (27) and then the IS method can be applied because i_1 and θ are analytically known. The figure 9 presents the computed evolution of capacitor voltage with time. The results obtained by means of both methods are very close, for example the obtained final voltage v_{co} is :

$v_{co} = 630$ v at $t = 6.32\ 10^{-4}$ s for IS method

$v_{co} = 633$ v at $t = 6.29 . 10^{-4}$ s for DS method

However, as it was hoped, the computation times are quite different : the DS computing time is 90 % longer than the IS computing time (respectively 19 sec. and 10 sec. of CPU time for an HP 1000 F computer).

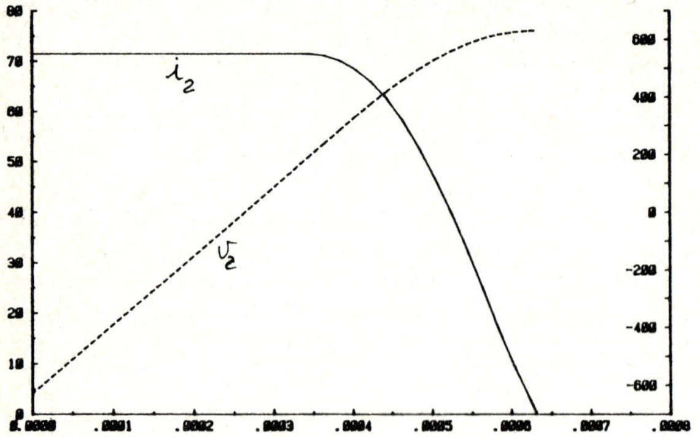

figure 9

5. CONCLUSION

A new method for simulation of an electromagnetic device behaviour has been proposed. It is essentially based on an insertion of a magnetic equivalent circuit of the magnetic device into an overall machine/electric circuit simulation.

The main advantages of this technique is first a magnetic model as close as possible to the real magnetic system allowing no restrictive assumptions such as sinusoïdal repartitions, secondly the capacity to evalue directly the influence of construction parameters on the behaviour.

Two methods have been proposed for numerical integration : the dependent step method (DS) and the independent step method (IS).

The DS method is only applicable when the magnetic device state is defined by only one unknown state variable, the evolution of other variables with time being previously given.

The DS method, when it can be applied, gives the same accuracy as the IS method but with a large decreasing in the computing time.

A comparison between simulation and experimental results for a vernier type reluctance machine has indicated a good accuracy of the model.

ELECTRICAL MACHINES AND CONVERTERS - Modelling and Simulation
H. Buyse and J. Robert (editors)
Elsevier Science Publishers B.V. (North-Holland)/© IMACS, 1984

NUMERICAL SIMULATION OF A RELUCTANCE MACHINE-CONVERTER SET. VARIABLE SPEED DRIVE APPLICATION.

P.DESESQUELLES C.GLAIZE S.ALLANO A.BERRACHE

Laboratoire d'Electrotechnique Bât.214
91405 ORSAY FRANCE

A method of simulation applied to the association of a one-phase reluctance motor and an original static thyristor converter is introduced. To study the functioning ditions, a second order differential equation in flux is established. Solution is obtained through a 4th order Runge-Kutta method.
Thanks to the velocity of the simulation, a feedback program of optimisation is envisaged in order to determine through a large scale of speed-drives the correct inputs to be adopted with regards to specific functioning conditions. One such input is given as an example, at the end of this lecture.

I. PREAMBLE

Advances in the design of reluctance machines come as a result of the development of power electronics and of static converters in particular. In addition to the possibility of variable speed drive, these elements yield a great flexibility in the supply of power and an accurate setting of performances. As a counter part the associated control system may have to be elaborate and require thorough determination.

Concerning our experimental set at the "Laboratoire d'Electrotechnique des Universités Paris VI et Paris XI", the reluctance machine is an axial airgap one 1,2 and should be able to operate up to 10 000 r.p.m. for an input power of 10 kw.

It is a one-phase assembly supplied by an original thyristor converter introduced in [3].

Control of the whole assembly is obtained though the setting of four parameters (i.e. supply voltage E, a commutation voltage Vo, two thyristors gating angles Do, D_c).

This structure has already been extensively studied from the quality point of vue [4]. This lecture however aims at a quantitative description of the functioning based on a numerical simulation.

II. FUNCTIONING OF THE MACHINE-CONVERTER ASSOCIATION :

In the general diagram of the unit (fig. 2.1), the machine is figured by a variable inductance. Drive power is supplied by the electrical source E, via the four thyristors T1, T2, T3 and T4, controlled by a microprocessor in accordance with the mechanical rotor/stator angle This angle is measured by an optical encoder placed on the shaft. Capacitor C is used both as a reactive power reserve and for thyristor commutation.

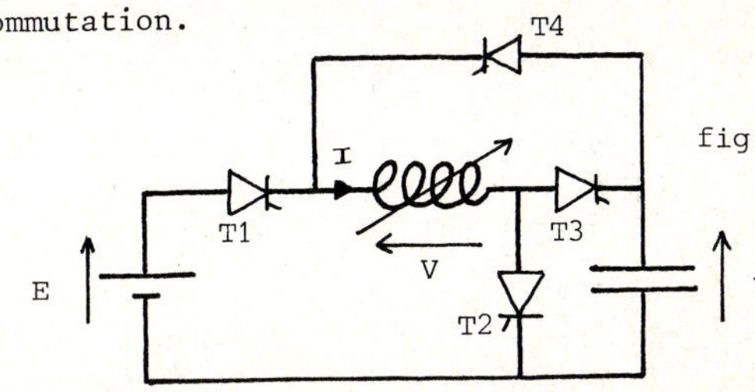

fig. 2.1

For a detailed description of the functioning of the converter and control, of which we only give here the aspects related to the simulation, we refer the readers to lectures treating this specified aspect [3,4].

The full cycle includes three phases, plus one pause when no current is injected in the coil :

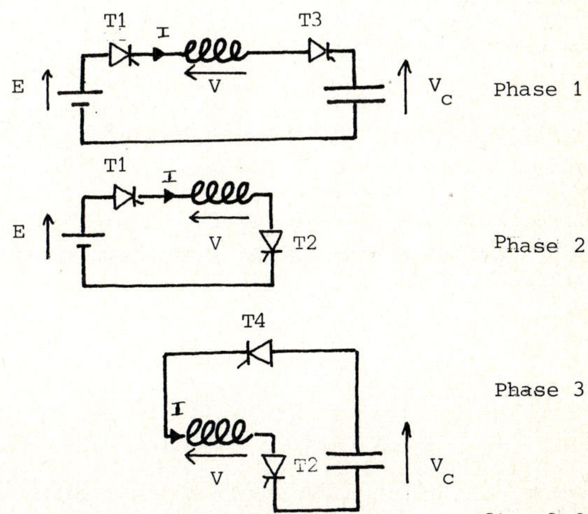

fig. 2.2

Through a proper control of the thyristors, these three phases occur as follows :

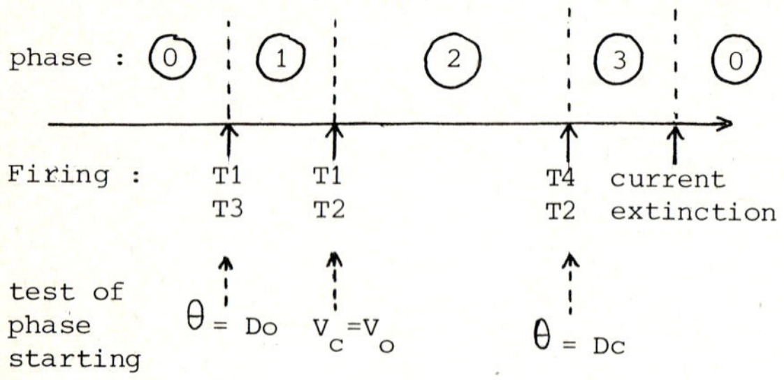

fig. 2.3

Permanent measure of θ and V_c permits the generation of the four trigger pulses under control of a microprocessorised command described in [5,6,7].
Aspects of current I and voltage V_c are given below :

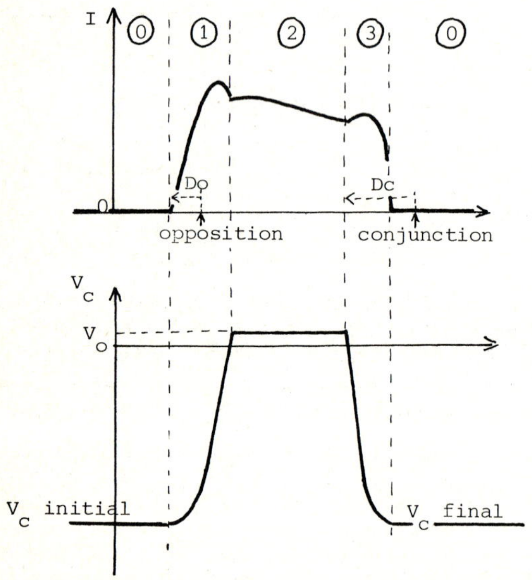

fig. 2.4

Refering to figure 2.2, it can easily be verified that each of the three phases satisfies equation (2.1) :

$$\ddot{\phi} + R\dot{I} + \alpha . \frac{I}{c} = 0 \qquad (2.1)$$

where α = 1 for phases (1) and (3)
α = 0 for phase (2)
ϕ is the global flux through the coil
R is the total resistance (coil, source).

III. SOLUTION MODE

III. (1) Modelling of the machine :

To solve equation (2.1.) we have opted for the elimination of I in order to obtain a second order differential equation in flux. To do so, we based ourselves on the flux-current feature of this type of machine. Iron-losses are neglected, and then depending on the relative rotor/stator angle θ we obtain :

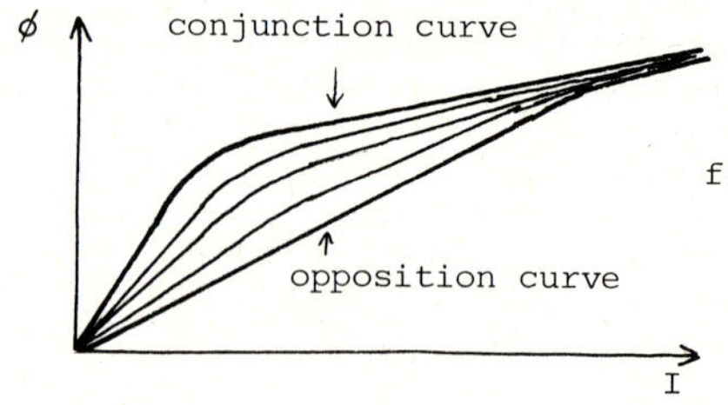

fig. 3.1

For each value of θ, a curve ϕ(I) is obtained between the two extreme curves of opposition and conjunction :

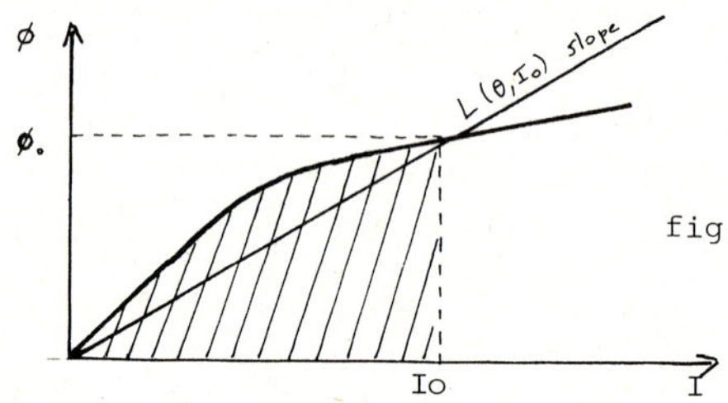

fig. 3.2

Considering a given operating condition (θ, ϕ_o, I_o) it is then easy to get Io from ϕ_o and vice-versa with the help of coefficient $L(\theta,\phi_o)$ defined as the slope of the straight line reaching both the origin and (ϕ_o, I_o).

$$L(\theta,\phi) = \frac{\phi(\theta,I)}{I} \qquad (3.1)$$

$L(\theta,\Phi)$ will henceforth be called inductance. We shall call reluctance $\mathcal{R}(\theta,\phi)$:

$$\mathcal{R}(\theta,\phi) = \frac{1}{L(\theta,\phi)} \qquad (3.2)$$

Eventually, it can be demonstrated [8] that the instant torque given by the machine is :

$$\Gamma(\theta,I) = \frac{\partial}{\partial\theta}\Big/_{I} W'(\theta,I) \qquad (3.3)$$

Where $W'(\theta,I)$ is the state function Coenergy, whose value is equal to the surface stripped on figure 3.2.

III. (2) Analytic expressions of the inductance

In order to withdraw I from equation (2.1) one must find the analytic formulations of $L(\theta,\phi)$ and its derived functions. We have considered a particular case represented on figure 3.3 where the variations of the inductance under non-saturated working conditions can be estimated as sinusoidal with θ (figure 3.3bis).

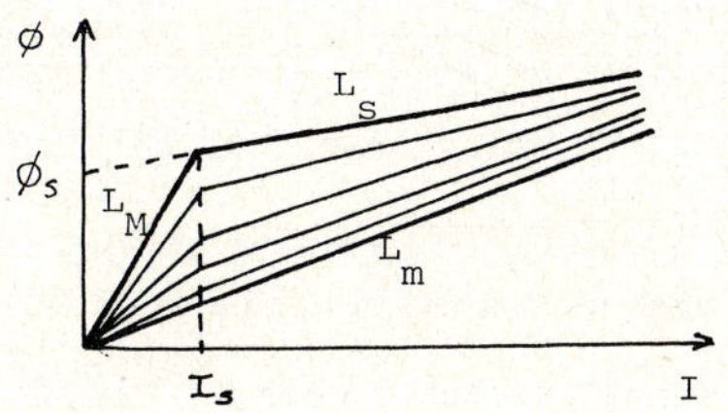

fig. 3.3

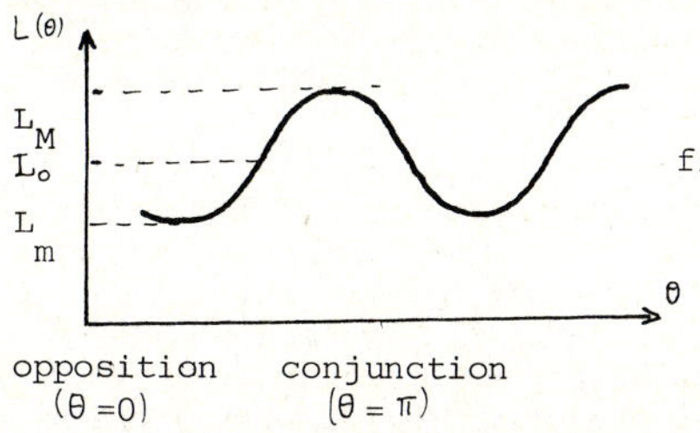

fig. 3.3bis

Others shapes of $\phi(I)$ feature, not presented here, have been developped to study the influence of the geometry of such machines upon the electrical and mechanical functioning.
We only present here this model wich is a first approximation of the real inductance.

Expressions of the inductance $L(\theta,\phi)$ and its derived functions together with the expressions of the instant torque are condensed in the following chart :

Model	Sinusoïdal inductance variations	
mode	non saturated	saturated
L	$L_o - l\cdot\cos\theta$	$(L_{os}+l_s\cos\theta)(1-\frac{\phi_s}{2}\cdot\frac{1-\cos\theta}{\phi})^{-1}$
$\frac{\partial L}{\partial\theta}\,I$	$l.\sin\theta$	$(-l_s+\frac{\phi_s}{2}\cdot\frac{L}{\phi})\sin\theta$
$\frac{\partial L}{\partial I}\,\theta$	0	$-\frac{\phi_s}{2}\cdot(1-\cos\theta)\,\frac{L^2}{\theta^2}$
Γ torque	$\frac{\rho}{2}I^2 l\cdot\sin\theta$	$\frac{\rho}{4}\cdot\sin\theta\left[\phi_s(2I-I_s)-2\cdot l_s\cdot I^2\right]$

with :

$$L_o = \frac{L_M + L_m}{2} \qquad l = \frac{L_M - L_m}{2}$$

$$L_{os} = \frac{L_m + L_s}{2} \qquad l_s = \frac{L_m - L_s}{2}$$

$$\phi_s = (L_M - L_s)\cdot I_s$$

III. (3) 2nd order differential equation in flux

With these expressions of $L(\theta,\phi)$ and its derived functions at hand, equation (2.1) can be rewriten, eliminating variable I.
Therefore the relation giving I from ϕ is used in (3.1) form and its derived form :

$$\dot{I} = \frac{\dot\phi}{L} - \phi\cdot\frac{\dot L}{L^2} \qquad (3.5)$$

Equations (2.1) then writes :

$$\ddot\phi + \frac{R}{L}\cdot\dot\phi + \left(\frac{\alpha}{L\cdot C} - R\frac{\dot L}{L^2}\right)\cdot\phi = 0 \qquad (3.6)$$

We choosed to express $\dot L$ as follows :

$$\dot L = \frac{\partial L}{\partial\theta/I}\cdot\dot\theta + \frac{\partial L}{\partial I/\theta}\cdot\dot I \qquad (3.7)$$

Eliminating $\dot I$ using (3.5), it comes :

$$\dot L = \frac{\partial L}{\partial\theta/I}\,\dot\theta + \frac{\partial L}{\partial I/\theta}\cdot\left(\frac{\dot\phi}{L} - \phi\cdot\frac{\dot L}{L^2}\right) \qquad (3.8)$$

i.e. :

$$\dot L = \frac{\frac{\partial L}{\partial\theta/I}\cdot\dot\theta + \frac{\partial L}{\partial I/\theta}\cdot\frac{\dot\phi}{L}}{1 + \frac{\phi}{L^2}\,\frac{\partial L}{\partial I/\theta}} \qquad (3.9)$$

And (3.6) becomes :

$$\ddot\phi + \frac{R}{L}\cdot\dot\phi + \left[\frac{\alpha}{L\,c} - \frac{R}{L^2}\cdot\left(\frac{\frac{\partial L}{\partial\theta/I}\cdot\dot\theta + \frac{\partial L}{\partial I/\theta}\cdot\frac{\dot\phi}{L}}{1 + \frac{\phi}{L^2}\cdot\frac{\partial L}{\partial I/\theta}}\right)\right]\cdot\phi = 0$$

(3.10)

Reducing to a common denominator and rearranging it comes :

$$\ddot{\phi} + \frac{R}{Lx}\dot{\phi} + \left(\frac{\alpha}{L\,C} - \frac{R}{L^2 x}\cdot\frac{\partial L}{\partial\theta/I}\cdot\dot{\theta}\right)\phi = 0 \quad (3.11)$$

with: $x = 1 + \frac{\phi}{L^2}\cdot\frac{\partial L}{\partial I/\theta}$.

We choosed this formulation to solve the equation numericaly. Others formulations are equally possible (i.e using $\frac{\partial L}{\partial \phi/\theta}$ instead of $\frac{\partial L}{\partial I}/\theta$).

But although they may seem more simple, equation (3.11) requires the most simple analytical expressions and then is more interesting in term of computing time.

III. (4)(a) Input - output :

When starting up the execution, the operator may, according to his wishes, change the characteristic features of the assembly. They are permanently stored on a data disc file and are the following :

- Values of L_M, L_m and L_s
- Saturation current I_s
- Coil loops number
- Capacitance of C
- Resistance R
- Thyristor voltage drop V_{Th}
- Supply voltage
- Firing angles
- Commutation voltage Vo
- Speed

When the simulation is performed, instantaneous results (such as torque, current, flux, capacitor voltage, flux-current cycle) can be drawn on a graphic display. Average and R.M.S. values are given in a chart (average torque, average and R.M.S. current, Joule Losses, inverse polarisation time of thyristor T_3).

III. 4. (b) Simulation progress :

With the parameters as previously defined, the program achieves the resolution of (3.11) on a complete cycle under steady-state condition (i.e. final value of Vc is equal to initial value, see picture 2.4). Resolution of (3.11) is done step by step by Runge-Kutta's formulas. At each step, $L(\theta,\phi)$ and the two derived functions $\frac{\partial L}{\partial\theta/I}$ and $\frac{\partial L}{\partial I/\theta}$ are tabulated according chart I formulas. Runge-Kutta's method has been choosed, due to the small number of intregation steps needed to solve (3.11). We have proved that the solution can be obtained with a relative error less than 10^{-3} with only 5 to 6 steps by phase. The optimal step Ho is then equal to 1/20 of the characteristic period $(2\pi\sqrt{L_0 C})$ for phases (1) and (3) and 1/10 of the cycle period for phase (2).

Initial values of the two state-functions ϕ and $\dot{\phi}$ are easely calculable at the beginning of each phase, except for for the value of at begining of phase (1). This value is linked to the initial capacitor voltage $V_c(0)$, a priori unknown, although unique to ensure steady state conditions. It can only be determined by successive tries following an algorithm based at the beginning upon a dichtomic method, and then to obtain a good precision, the algorithm uses a method based upon NEWTON'S one. The program roughly calculates 10 to 15 cycles before reaching the exact value of Vc (o).

Concerning the ends of phases, when the next test condition (pict. 2.3) is met, the integration step is recomputed following a proportional parts method .

IV. SIMULATION RESULTS :

Comparison between simulation and experimental results has shown a good concordance up to a frequence of 600 Hz (1500 r.p.m.). Correct values of $V_c(0)$ and Imax are found by the program within an error margin of roughly 10 %. Resistive para meters R and V_{th} are very sensitive and must be well predetermined.
On our prototype which has been essentialy designed for measurement and not for industrial use, the yoke is very large. Consequently, for higher frequencies, iron losses can not be neglected. A new modelling of the machine has to be used to take these losses into account.
Particular operating points were already published [6]. We give here global results.

IV. 1 Variable speed drive regulation :

To obtain a variable speed drive, the micro computer has to settle the four parameters controlling the converter : E, Vo, Do and Dc. The performance of our microcomputer [7] is not sufficient to calculate these parameters "on line". So we had to predetermine the variations of these parameters which are functions of the speed and the torque (or current). The simulation is short enough (0.35 s on a 4 Mips machine : UNIVAC 1100) to iterate with an optimisation algorithm. Computed laws are to be entered lately in the microcomputer by means of polynomial form.

IV. 2 Optimisation criteria :

To optimize a forced-commutated thyristor converter it is interesting to maximize the mean torque for a given peak current. On the other hand, to reduce Joule-losses the ratio : mean torque/squared RMS current has to be maximised.

IV. 2. a) Peak current optimisation :

Operating mode is the one described above. For a given torque and within a given speed range, the program calculates the parameters E, Vo, Do and Dc as follows ;

- Do is always equal to 0 (i.e. current-starts growing for exact opposition angle).

- D_c is calculated so that current extinction occurs for the exact conjunction angle.

In that way no negative torque is applied on the shaft.

- Vo is calculated so that the reverse voltage application time on thyristor T_3 is equal to manufacturer specification (typically 30µs).
In that way the overcurrent observed on picture 2.1 is minimised.

- E is calculated to ensure the value of the desired mean torque, following a NEWTON' like method.
The main loop of the optimisation program concerns E. Each time an acceptable value of E is found, D_c and V_o are adjusted. Then E is modified again until final concordance. Typical computing time for a law given over ten points (speed rate) is less than 30 sec. Results are given at the end of the lecture.

IV.2.b) Joule-losses optimisation :

To minimise Joule-losses under non saturated operating conditions, the optimal current shape is a Dirac's pulse centered on the position where inductance variation is maximal [3]. To approach this shape, the converter can operate in a different way than described above. Just using phase (1) and (3) one can supply the machine with current arches [4]. The modelling is quite the same than above.
The only parameters to be settled in that operating mode are E and Do. E. is calculated in the same way as before, and Do is calculated so that the arch is centered on $\theta = \frac{\pi}{2}$ where for our model $\frac{\partial L}{\partial \theta}$ is maximal.
Results are also given at the end.

IV.2.c) General optimisation :

For other kind of criteria and for a rather general optimisation we think it would be preferable to use general optimisation algorithm like for instance the SIMPLEX algorithm.

V. CONCLUSION :

We have discussed here about the setting of a numerical tool wich has prooved to be precious for the predetermination of the operating control of a reluctance machine-converter set. First of all it permitted us to determine the components of the assembly. Then, thanks to its velocity, we have been able to calculate the variation laws of the command parameters under different optimisation criteria.

Since we did not take iron-losses into account, the concordance with experimental results is limited to frequencies where these losses are weak. In fact reluctance machines are actually developed for low frequencies domains : for instance we are testing a 200kw, 500Hz (1500r.p.m.) three-phase machine. The modelling we discussed above is well adapted for this case and has been helpfull to design a converter [3].

Development is now aiming toward the influence of the geometry of such machines. Using others models of inductance $L(\theta,\phi)$, we hope to determine the best shape of $L(\theta,\phi)$ to minimise the torque ondulation and the converter ratings.

REFERENCES :

[1] J.P. BASTOS, R. GOYET et J. LUCIDARME

Performances intrinséques des machines à reluctance variable à disques imbriqués.
Revue physique Appliquée 15 (1980) 45-54

[2] J.P. BASTOS, R. GOYET, J. LUCIDARME, G.QUICHAUD et F. RIOUX-DAMIDAU

"Performance of a multidisc variable reluctance machine "
ICEM 82. Budapest Pp 254-257

[3] C. GLAIZE

"Etude générale et optimisation des modes d'alimentation des machines à reluctance variable. Application aux machines polydiscoïdes" Thèse d'Etat ENSET Déc.83.

[4] S.ALLANO et C.GLAIZE

"Etude systématique de la commande d'une alimentation pulsée pour systèmes magnétiques de forte puissance".
Revue de Physique appliquée 17 (1982) 37-44

[5] C. GLAIZE, S. ALLANO et C.SOL

"Role du microprocesseur dans la commande numérique d'un ensemble convertisseur-machine à reluctance variable"
Journée SEE du 10 Juin 1982 PpXI. 1-XI-6

[6] C.GLAIZE, P.DESESQUELLES and S.ALLANO

"Reluctance machine supply form high power converter and its control supply"
Proc. Motorcorn Sept. 82 Genève Pp25-36

[7] C.GLAIZE, S.ALLANO and C.SOL

"Amicrocomputer for the control of an inverter. Application to a reluctance motor supply".
Proc. of Microelectronics in Power electronics. Darmstadt Oct. 1982 Pp381-384

[8] J.MEISEL

"Principles of electromechanical-energy conversion"
Mac Graw-Hill Co. 1966

Example 1 : Peak current optimisation

For three different values of the mean torque, we present the associated command laws ensuring the best ratio mean torque / maximal current.

Capacitance C is here 60 μf.

The different quantities are first resumed in a chart. Then they may be displayed.

```
PLAGE DE VITESSE        100  1600    (TR/MN).
VALEUR DU COUPLE DESIREE (N.M) ?
>15
VALEUR DU TO ADMISSIBLE (US) ?
>30
```

TR/MN	GAMMA	IEFF	IMAX	IMOY	NUF1	PE	PJ	PTH	PMECA	ZERO.I	TQ (US)	GAMMA / IEFF2	GAMMA / IMAX2	VCN	E	D0	DC
100	14.88	113	179	79	-924	828	449	198	156	998	28	5.92	0.47	151	10.6	0	-27
250	14.95	112	180	78	-926	1089	438	195	391	1003	35	6.09	0.46	192	14.6	0	-63
400	14.94	110	179	76	-922	1343	425	191	626	1001	32	6.28	0.46	182	18.9	0	-104
550	14.91	109	179	75	-920	1596	412	187	858	1001	31	6.45	0.46	183	23.7	0	-143
700	14.95	107	180	73	-922	1856	403	183	1096	1000	30	6.62	0.46	185	29.0	0	-182
850	14.96	106	180	72	-925	2115	394	180	1332	1000	30	6.78	0.46	190	35.0	0	-221
1000	15.13	106	183	72	-939	2408	394	179	1585	1011	36	6.86	0.45	233	41.7	0	-254
1150	14.97	104	182	70	-938	2635	380	174	1803	1005	34	7.04	0.45	233	49.2	0	-299
1300	15.09	104	184	69	-946	2910	375	172	2054	1004	33	7.18	0.45	238	57.8	0	-338
1450	15.04	102	185	68	-951	3154	367	169	2284	1003	32	7.31	0.44	246	67.5	0	-377
1600	14.87	101	186	66	-958	3401	360	166	2491	1001	31	7.37	0.43	255	79.0	0	-417

```
PLAGE DE VITESSE        100  1600    (TR/MN).
VALEUR DU COUPLE DESIREE (N.M) ?
>10
VALEUR DU TO ADMISSIBLE (US) ?
>30
```

TR/MN	GAMMA	IEFF	IMAX	IMOY	NUF1	PE	PJ	PTH	PMECA	ZERO.I	TQ (US)	GAMMA / IEFF2	GAMMA / IMAX2	VCN	E	D0	DC
100	9.98	86	136	60	-705	512	257	150	105	1000	28	6.94	0.54	115	8.7	0	-26
250	9.94	84	136	59	-699	654	248	146	260	1000	30	7.17	0.54	123	11.7	0	-65
400	9.94	83	136	57	-699	801	241	144	417	1000	30	7.37	0.54	128	15.0	0	-104
550	9.97	82	136	56	-700	960	235	140	574	997	28	7.59	0.54	125	18.7	0	-146
700	9.97	81	137	55	-702	1098	229	138	731	996	28	7.77	0.54	132	22.8	0	-185
850	9.97	80	137	54	-706	1248	224	135	887	998	29	7.94	0.53	142	27.4	0	-224
1000	9.95	79	138	53	-710	1396	220	134	1042	1003	31	8.08	0.53	156	32.4	0	-260
1150	9.98	79	138	52	-715	1550	216	131	1202	999	30	8.25	0.52	159	38.4	0	-301
1300	9.98	78	139	52	-721	1701	212	129	1359	1000	29	8.39	0.52	166	45.0	0	-340
1450	10.00	77	141	50	-729	1854	208	126	1519	988	27	8.58	0.51	168	53.2	0	-388
1600	9.94	77	140	50	-732	1998	206	126	1666	1004	31	8.64	0.50	195	61.6	0	-418

```
PLAGE DE VITESSE        100  1600    (TR/MN).
VALEUR DU COUPLE DESIREE (N.M) ?
>5
VALEUR DU TO ADMISSIBLE (US) ?
>30
```

TR/MN	GAMMA	IEFF	IMAX	IMOY	NUF1	PE	PJ	PTH	PMECA	ZERO.I	TQ (US)	GAMMA / IEFF2	GAMMA / IMAX2	VCN	E	D0	DC
100	4.97	60	96	42	-498	286	128	106	52	1000	32	6.94	0.53	92	6.9	0	-26
250	4.97	59	96	41	-492	357	123	103	130	995	26	7.20	0.55	78	9.0	0	-69
400	4.97	59	96	40	-493	430	120	101	208	996	29	7.40	0.54	88	11.4	0	-108
550	4.97	58	96	40	-494	502	117	99	286	995	29	7.59	0.54	91	14.0	0	-147
700	5.04	58	98	40	-502	585	117	99	369	1008	37	7.69	0.53	119	17.0	0	-179
850	4.99	57	97	39	-501	654	113	96	445	1002	34	7.91	0.53	116	20.3	0	-223
1000	4.98	56	97	38	-502	726	110	94	521	1003	33	8.07	0.52	118	23.9	0	-262
1150	4.97	55	98	37	-505	799	108	93	598	1002	32	8.23	0.52	120	28.1	0	-301
1300	4.98	55	98	37	-509	876	106	91	677	1002	31	8.38	0.52	124	32.9	0	-340
1450	4.98	55	99	36	-513	952	104	90	757	1001	31	8.52	0.51	129	38.4	0	-380
1600	4.99	54	100	35	-519	1028	103	89	836	1000	30	8.65	0.50	135	44.9	0	-420

```
DESIREZ VOUS SUPERPOSER UNE COURBE SUPPLEMENTAIRE ?
>N

CHOISIR LA COURBE DESIREE
E       --> 1       IEFF --> 6        PTH     --> 11
VCM     --> 2       IMAX --> 7        PMECA   --> 12
DC      --> 3       IMOY --> 8        NUF1    --> 13
C       --> 4       PTOT --> 9        ZERO.I  --> 14
C/IEFF--> 5         PJ   -->10        TQ(US)  --> 15
0 POUR SORTIR.
```

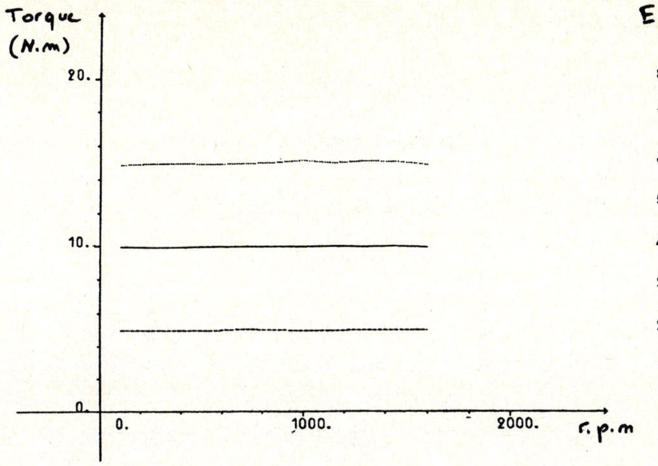
Torque
(N.m)
20.
10.
0.
0.
1000.
2000.
r.p.m

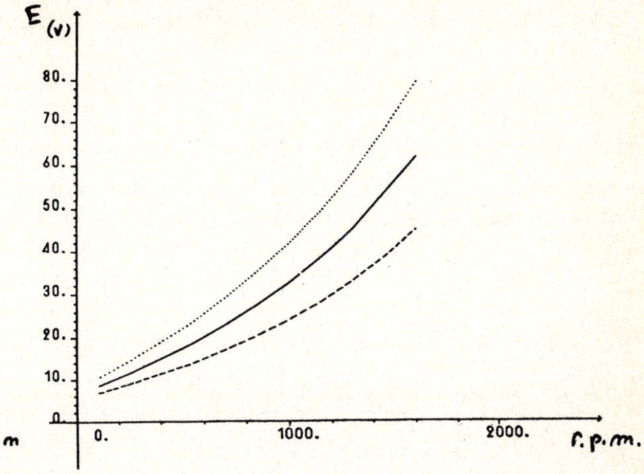
E (v)
80.
70.
60.
50.
40.
30.
20.
10.
0.
0.
1000.
2000.
r.p.m.

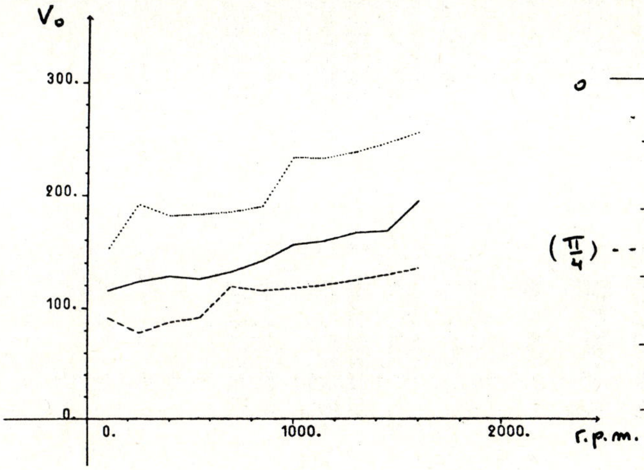
V₀
300.
200.
100.
0.
0.
1000.
2000.
r.p.m.

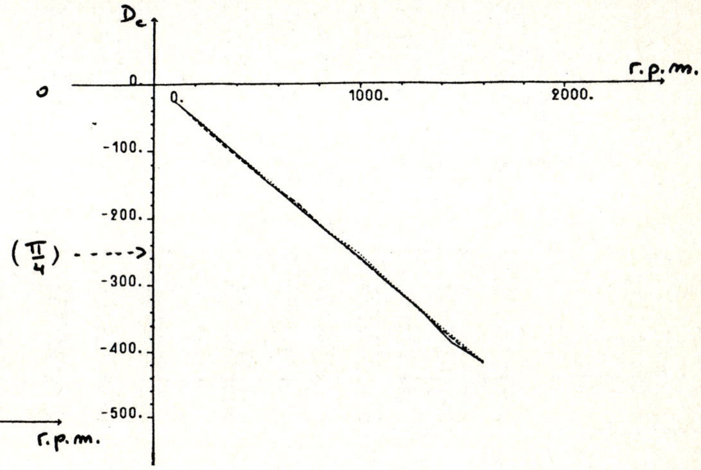
Dc
r.p.m.
0
0.
0.
1000.
2000.
-100.
-200.
(π/4) ----->
-300.
-400.
-500.

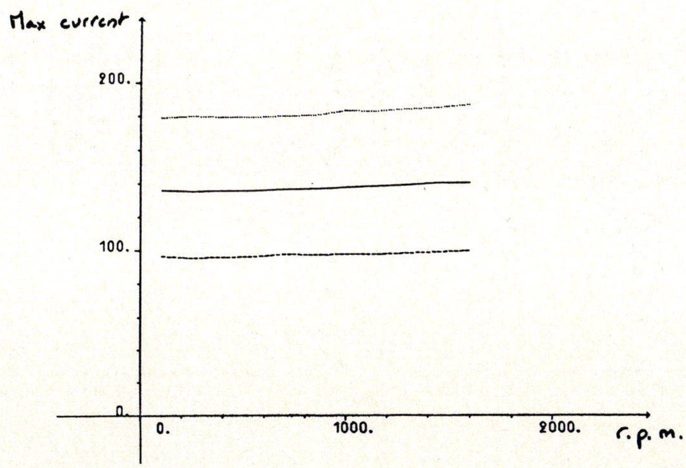
Max current
200.
100.
0.
0.
1000.
2000.
r.p.m.

Exemple 2 :

Two examples of Joule losses optimisation are given for two values of capacitance C.

For a desired torque of 10 N.m. over a 600-1600 r.p.m. range, we present the associated command E and Do, and the shapes of R.M.S. and maximal currents.

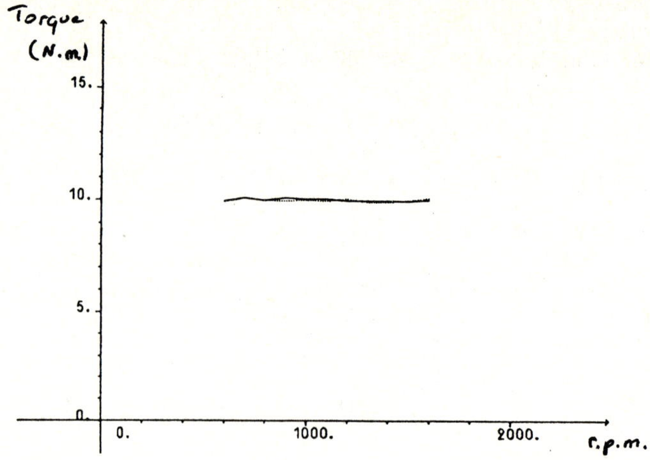

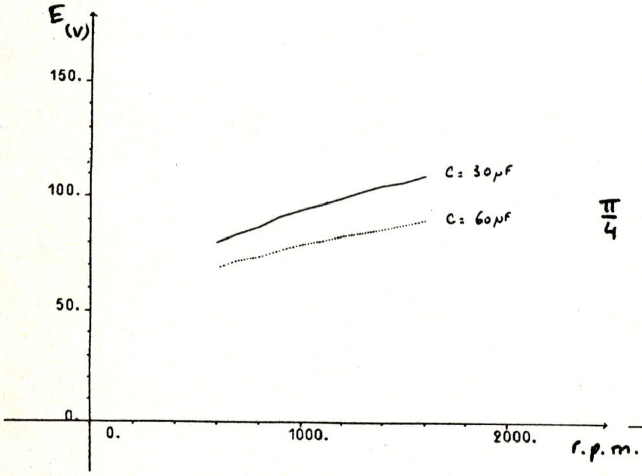

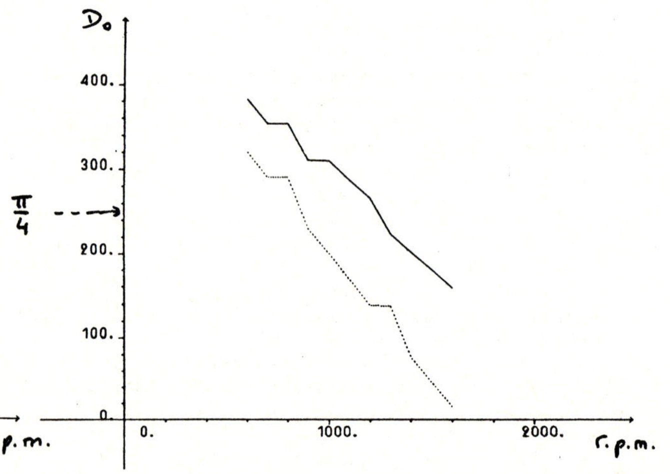

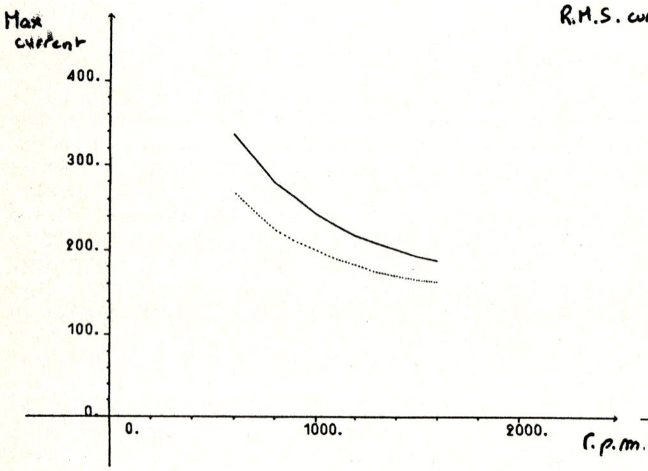

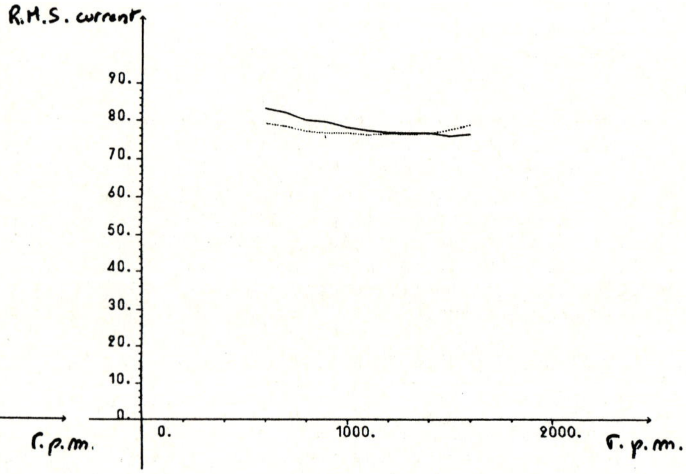

ELECTRICAL MACHINES AND CONVERTERS - Modelling and Simulation
H. Buyse and J. Robert (editors)
Elsevier Science Publishers B.V. (North-Holland)/

A SCALED SIMULATION OF A DARRIEUS TURBINE AIMED AT SOLVING POWER FLUCTUATION PROBLEMS

G. OLIVIER G.-E. APRIL

Ecole Polytechnique de Montréal, P.O. Box 6079, Station A, Montréal, Canada, H3C-3A7

The Darrieus turbine is one of many wind powered machines now under investigation for electrical power generation. It has a number of advantages, mainly mechanical, over other machines, but is also plagued with some problems that must be overcome for the machine to be usefull. This paper proposes a simple control scheme to tackle the most troublesome of said problems, namely the one associated with a wildly fluctuating torque, by making use of rotor inertia as a buffer, so allowing constant power to be drawn despite short term-variations in available torque.

INTRODUCTION

The Darrieus wind turbine, is a mechanically simple machine to extract power from the ever present wind. A 4 MW unit is now under development by the Institute of Research of Hydro-Québec (IREQ) conjointly with the National Research Council of Canada. Besides the fact that this type of machine is well adapted to the rough canadian winters, several other advantages make the Darrieus turbine extremely attractive.

First, it requires no mechanism to align it with the wind, being fully symmetrical in the horizontal plane. Second, the axis being vertical, the power can be extracted at ground level so that it is not necessary to construct a tower capable of supporting heavy electrical machinery. Third, no moving part protrudes above the top of the rotor support, so that a guyed structure is entirely feasible. This last feature alone can make an enormous difference in the cost of the installation. However, for all its virtues, the Darrieus wind turbine is also plagued with deficiencies some of which must be overcome before the machine can be of any practical use.

In the first place, the Darrieus turbine is less efficient than horizontal axis machines so, for a given power output, this type of machine is larger. This, however, is not as important as it may seem since the larger machine will, in all likelyhood, cost less than a horizontal axis machine of the same power capability, and efficiency is of little concern when the "fuel" is free and abundant.

The second problem is more troublesome and must be solved for most applications. It has to do with the rather objectionnable torque-position relationship illustrated in Fig. 1 [1]. As can be seen, the torque varies widely as the machine turns, even becoming negative for two brief periods within each cycle. In instances

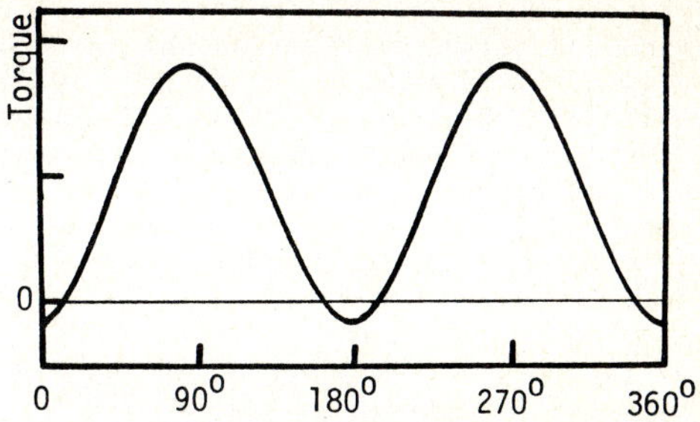

Fig. 1: Typical aerodynamic torque vs blade angular position [1].

where electrical power is desired, this is of real concern since most users would object to their lights fluctuating wildly in brightness at a rate of one or two cycles per second. Furthermore, such power fluctuations, if injected in the network, could have disastrous destabilizing effects.

Thirdly, the machine is not normally self-starting. That can be easily implied from the torque-position curve, Fig. 1, since, should the machine not have enough kinetic energy when it enters the negative torque section of its cycle, it would back off and come to rest at the zero torque position. During start-up, kinetic energy is low or zero, and the machine just sits there, unless some peculiar wind condition just happens, turning or twisting wind for example.

This paper describes a test bench assembled to study the oscillating torque problem and the start-up and braking of the turbine. The proposed test bench consists of a wound-rotor induction generator coupled to an eight thyristor converter. Other schemes are currently being evaluated.

THE TEST BENCH

The test set-up, Fig. 2, is actually a scaled simulation of Hydro-Québec's Magdalen Islands wind powered generator [2]. It consists of:

1) PRIME MOVER:
A 6 kW, 1500 rpm separately excited dc motor programmed through a four quadrant current source ac-dc converter, is used to simulate the torque generated by the aerogenerator, as measured after the speed multiplying gear-box. A gear-box ratio of 59.67:1 has been assumed so that at a speed of 1489 rpm a torque pulsation lasts for exactly one second.

2) INERTIA
A large flywheel simulates (to scale) the inertia of the aerogenerator blades. It consists of a steel plate, ten inches [25 cm] thick and sixteen inches [40 cm] in diameter, dynamically balanced, mounted on a shaft, and supported by heavy duty ball-bearings. The dimensions of the flywheel must be such that the ratio of kinetic energy (at nominal speed) to nominal power is maintained the same in both the real and simulated systems. This ratio measured in seconds, has been established at approximately 12.5 for our installation while it was approximately 16 on the real system. The smaller inertia means that our results are conservative, since a larger ratio naturally tends to smooth out the output power.

3) GENERATOR
A wound-rotor induction generator rated 4 kW, 5 kVA at 1200 rpm has been used. The rotor could be changed to a squirrel cage or synchronous type within the same frame.

4) CONTROL
The rotor control circuit consists of a three-phase, full-wave, diode rectifier and a dc-ac, three-phase, 8 thyristor, naturally commutated inverter of a novel type with microprocessor controlled equidistant firing pulse generator [4].

5) FEEDBACK
A speed pick up, made up of a magnetic transducer placed in close proximity to a sixty-toothed gear on the high speed shaft, produces an interrupt request to the microcomputer for every tooth.

THE PRIME MOVER

A dc motor has been used to simulate the torque encountered at the output of the gear-box. The desired torque is illustrated in Fig. 1. Since the torque produced by a separately excited dc machine is proportional to armature current, it is possible to obtain the required torque by injecting in the armature of the motor a current of the appropriate wave shape. However, to achieve such a control, a fast four quadrant converter was needed. Furthermore, it was mandatory that no dead time occurs during transitions between rectifier and inverter modes. The chosen configuration is an anti-parallel connection of two Graetz bridges, Fig. 2. The circulating currents are limited by four inductors and the motor is connected symmetrically in the middle of the circuit [3].

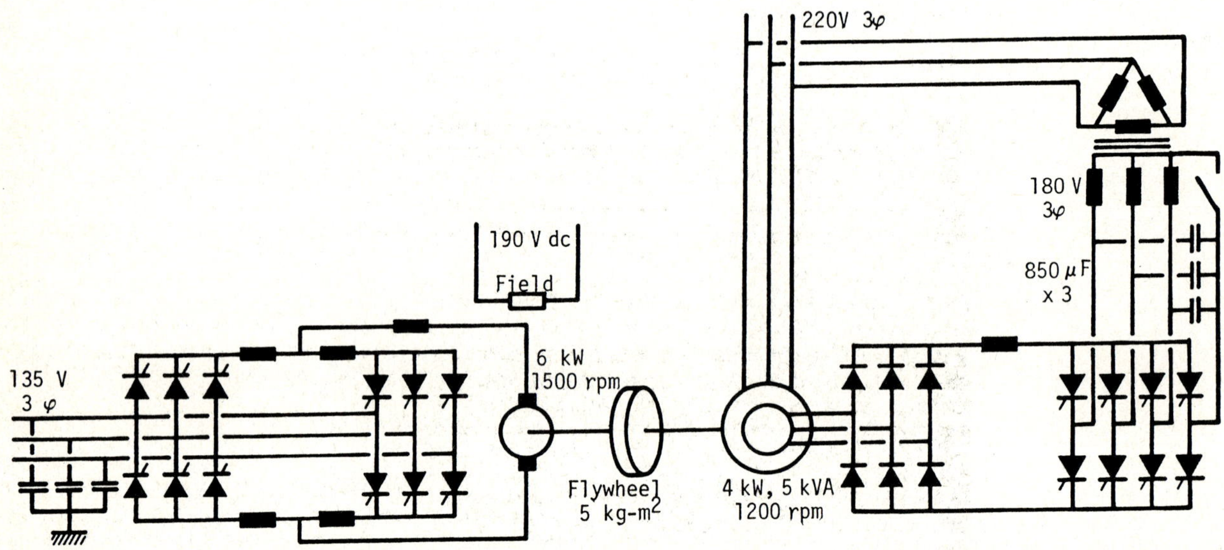

Fig. 2: Schematic of the test bench.

Two microcomputers are used to control the converter. The first one produces the firing pulses for the twelve thyristors. To ensure correct operation, the firing angles of the two bridges must obey the following equation:

$$\alpha_1 + \alpha_2 = 180° \qquad (1)$$

The firing angle values are determined according to a digital reference number, E_R, obtained from a second microcomputer through an input port.

The second microcomputer controls the armature current. The instantaneous reference current, i_D, is computed from an amplitude reference, I_D, and a modulation index, K_M, both read on input ports (operator inputs):

$$i_D = I_D*(0.8 + K_M \sin \theta) \qquad (2)$$

where θ is the instantaneous position of the blades. Then the actual measured current, i_o, is compared to the desired one and the voltage reference, E_R, is computed using the following algorithm:

$$E_R = K_E*n + i_D*R_{EQ} + K_P*(i_D - i_o) \qquad (3)$$

where: n is the speed,
K_E is chosen in such a way that the term K_E*n is equal to the back emf of the motor,
R_{EQ} is the equivalent output impedance of the converter,
K_P is the proportional gain.

THE GENERATOR

The wound-rotor induction generator is, in our opinion, one of the best possible choices for Darrieus type aerogenerators attached to the power grid. This statement is based on several reasons.

-With this scheme, in normal operation, only a small percentage (< 20%) of the total power has to be converted. The rest is directly generated at network frequency. A direct consequence of that, is less harmonic distortion in the line currents.
-Because of the presence of a diode bridge in the rotor circuit, rather than a thyristor converter, this configuration is more reliable than the synchronous generator scheme.
-This system does not require any additional means or special algorithm to start the aerogenerator.
-Dynamic breaking is easily obtained by the standard dc injection method down to zero speed.
-The well-known reactive power consumption problem near synchronous speed for slip-recovery drives can be solved by the use of a modified 8 thyristor bridge rather than the conventional Graetz converter [4].

THE GENERATOR CONTROL CIRCUIT

The schematic of the generator circuit appears in Fig. 2. Except for the use of a modified eight thyristor converter, this is the standard configuration of a slip-recovery drive. The speed of the drive is controlled by varying the input voltage to the inverter through control of the thyrisror firing angles . A three phase transformer is used to match the inverter voltages with the line voltages. This transformer would not be required if the machine had unity voltage ratio. A capacitor bank was connected on the line side of the inverter, first to supply reactive power and second, to provide an artificial neutral point for the inverter.

The equivalent circuit of the induction machine with the inverter appears in Fig. 3. The first part is the conventional equivalent circuit of an induction motor. The voltage sources E_γ and E_3 represent the diode drop and the inverter voltage. The rotor resistance has been increased by 3% to account for the additional losses caused by the fact that rotor currents are not sinusoidal. Typical torque-speed characteristics are shown in Fig. 4.

Simple relations couple the rotor and stator power, P_R and P_S, to the shaft power, P_M.

$$P_R = s * P_M / (1-s) \qquad (4)$$

and

$$P_S = - P_M / (1-s) \qquad (5)$$

where s is the slip.

Therefore, electrical power is produced only if the speed is higher than the synchronous speed and the power produced in the rotor is relatively small if the speed is maintained close to the synchronous speed.

The 8 thyristor converter is very well suited to this application because, as compared to the standard Graetz converter, it generates very little reactive power at low dc voltages. Since, the slip is maintained low, this is the

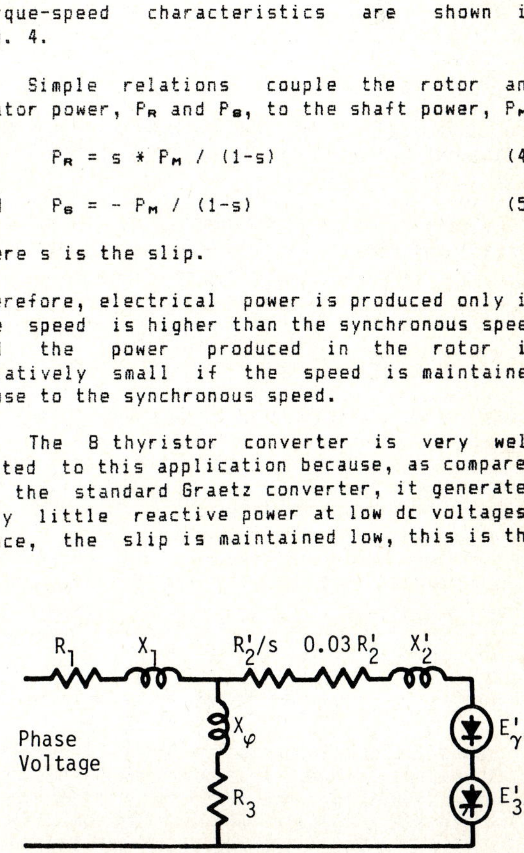

Fig. 3: Equivalent circuit of the induction generator.

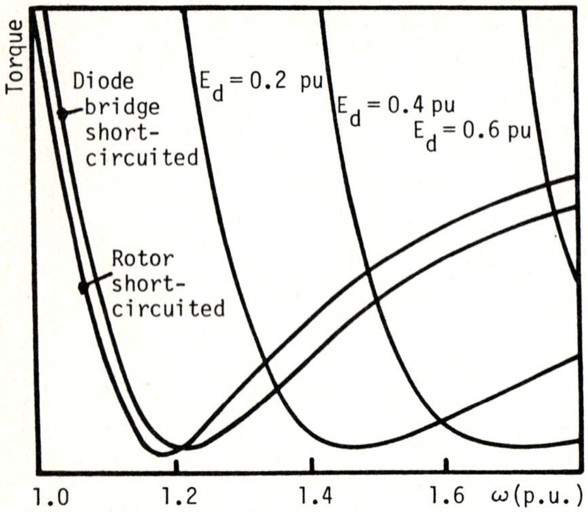

Fig. 4: Typical torque-speed characteristics for a wound rotor induction generator.

usual condition. However, this converter requires a neutral point which is not always available. Tests have shown that the converter operates satisfactorily with an artificial neutral. Moreover, a wye-connected capacitor bank large enough to correct the power factor provides a neutral point with sufficiently low impedance.

As in the motor circuit, two micro-computers are used to control the induction generator. The first one generates the firing pulses to the converter while the second regulates the instantaneous power produced by the generator.

POWER CONTROL SCHEME

Various control schemes have been suggested for the Darrieus wind generator. Most are complicated and based on complex analysis of all the dynamics involved. However, simple reasoning can lead to a very simple scheme giving identical results.

Two obvious statements hold the key to the problem.

1) If constant power is being drawn from a machine while the power supplied to it is time variant, the kinetic energy of the machine (hence the speed) will fluctuate.

2) If constant power is being drawn from a machine into a system with no storage capability, the output of the system is going to remain constant. Simply stated, if the input power to the generator shaft is maintained constant, the electrical output is also going to remain constant.

The first statement means that, since the torque generated is varying, the only way constant power can be obtained is by letting the amount of kinetic energy within the system, vary accordingly. The second statement simply means that, given a source whose speed is time variant, it is sufficient to maintain the product of speed and torque constant on the input shaft, to get constant electrical power out.

Using a wound-rotor machine, it has been shown that the shaft torque is essentially proportional to the current in the dc-link, provided the speed is sufficiently high [4]. The problem reduces to one of maintaining the current in the dc link inversely prorportional to the instantaneous rotational speed, i.e. directly proportional to the rotational period, which is a quantity very easily measured.

The software to implement this algorithm is very similar to the one written for the control of the current in the dc machine. First the reference current, i_R, is computed. If the speed is less than the synchronous speed, 1200 rpm, the dc link current, i_R, is simply maintained constant. At higher speeds, the instantaneous value of i_R is computed in such a way as to maintain constant the product $i_R \times n$.

$$i_R = P_D \qquad \text{for } n < 1200 \text{ rpm} \qquad (6)$$

$$i_R = P_D * (1200 / n) \qquad \text{for } n > 1200 \text{ rpm} \qquad (7)$$

where n is the instantaneous speed and P_D a current reference corresponding to the desired power level read on an input port (operator input). Afterwards, a reference voltage, e_{IN}, is computed for the pulse generating microcomputer.

$$e_{IN} = -K_E * |n - n_B| + K_P * (i_R - i_O) + R_{EQ} * i_R \qquad (8)$$

where i_O, R_{EQ}, K_E and K_P have the same definition as in (3) but refer to the induction machine.

For less demanding applications, a further consideration may be called upon to simplify the problem even more. Given the characteristics of the machine being simulated, one can easily see that, for a rotational speed of 30 rpm at the blades, (one cycle per second power pulse frequency), the kinetic energy variation will be in the order of ± 3% leading to speed variation of ± 1.5% only. Under these conditions, simply maintaining a constant current would lead to ± 1.5% power fluctuations which, in most cases, would be acceptable. By comparison a squirrel cage generator, operating at 10% slip, would, under the same conditions, produce power fluctuations of approximately ± 25%, which is in accordance with the results obtained by Hydro-Québec at the Magdalen Islands site [1].

EXPERIMENTAL RESULTS

Fig. 5 shows the armature current of the dc machine together with the instantaneous power when the rotor of the induction generator is short-circuited. The variations of the generated power are similar to those found in the experimental data from the Magdalen Islands installation [1], proving that the dc drive operates satisfactorily.

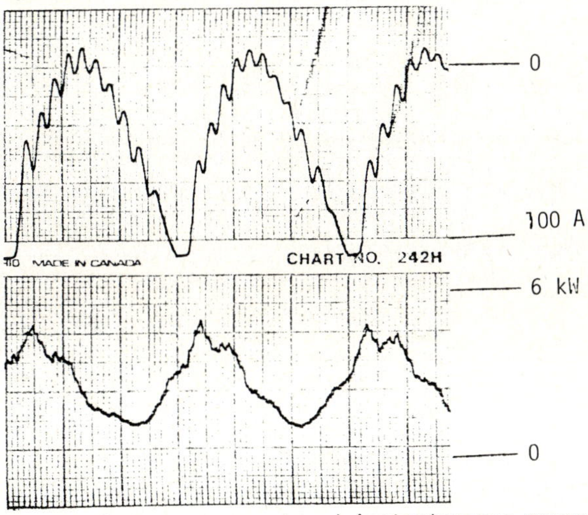

Fig. 5: Armature current and instantaneous power with short-circuited rotor. Chart speed: 25 mm/s.

The following record, Fig. 6, was obtained with the power controller in operation. The effects of the oscillating torque have completely disappeared. In this particular case, the speed was 1405 rpm, the dc-link current was 25 A and the inverter voltage 15.5 V. The total generated power was 2.8 kW while the power coming from the rotor circuit reached 400 W.

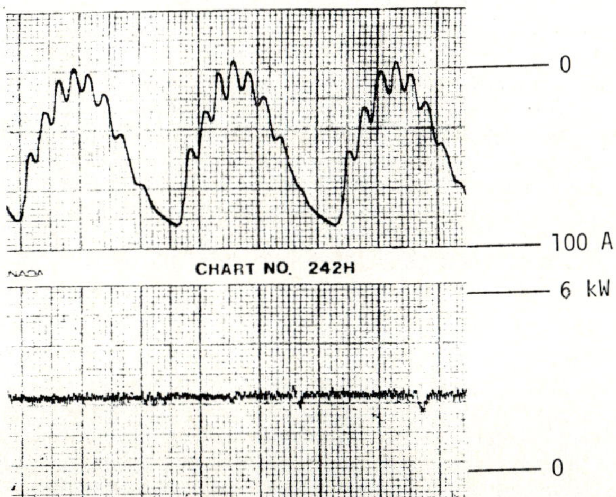

Fig. 6: Armature current and instantaneous power with power controller. Chart speed: 25 mm/s.

The traces of Figures 7 and 8 illustrate respectively a start-up of the test bench and dynamic braking. In the first case, the induction machine is used as a motor to accelerate the flywheel and the dc motor. During the acceleration period the rotor current is maintained at its nominal value. When the speed reaches 700 rpm the dc machine picks-up and the acceleration rate increases, the flywheel being accelerated simultaneously by the two machines. At 1200 rpm, the induction generator begins to produce power and the acceleration rate decreases. The equilibrium point is reached at 1428 rpm. Finally, the last figure shows the effect of a step increase in the power reference, i_R. The power immediately jumps from 2.4 kW to over 6 kW and the machines slow down from 1428 rpm to 1300 rpm in less than 3 seconds. However, since the desired power is larger than the power available from the dc machine, the power cannot be maintained at that level. In other words, if the desired power is larger than the energy available from the wind, the aerogenerator brakes down to a speed little higher than the synchronous speed.

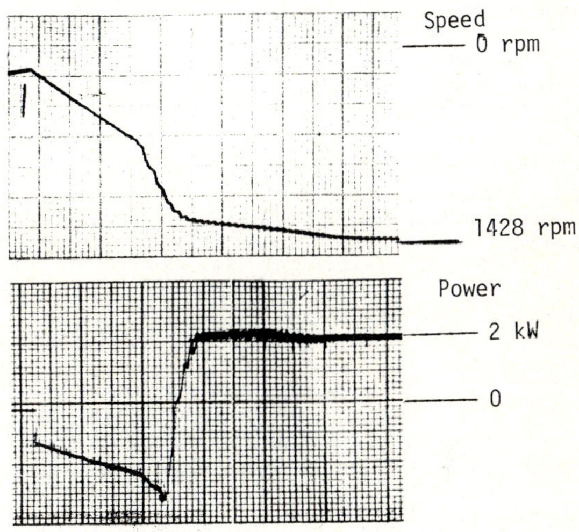

Fig. 7: Speed and power during start-up.

It should be noted that the set-up used was by no means optimized since time constraints forced us to use available machines as much as possible. As a consequence, some components had to be added to the test bench that would not otherwise be needed. The most obvious of these is the control circuit transformer whose presence was required by the fact that the available machine had a winding ratio of about 3 to 1 between stator and rotor, while a ratio of 1:1 is preferred [4]. This transformer is the cause of losses and reactive power consumption that would not otherwise be present in the energy balance. Another cause for reduced

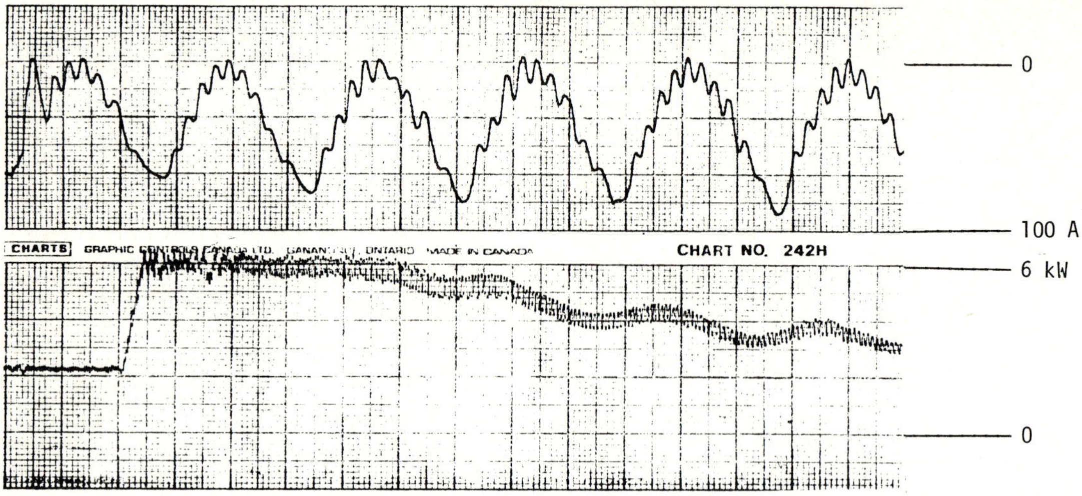

Fig. 8: Dynamic braking.. Chart speed: 25 mm/s.

efficiency is the fact that, again due to the winding ratio, the power electronics components are forced to work at lower voltages and therferore higher current levels than necessary. This leads to extra losses since power semiconductor (diode and thyristor) losses are essentially proportional to current and are little affected by the voltages involved as long as single devices can handle these voltages.

CONCLUSION

A simple control algorithm has been evaluated to regulate power flow from a time variant prime mover, in this case, a Darrieus wind turbine. The scheme has been tested on a simulation of the machine, using an induction generator, but could easily be adapted to other types of generators. The only data required by the control algorithm is the instantaneous speed of the blades which can easily be derived from a toothed wheel on the shaft, magnetically or optically coupled to a microcomputer interrupt input .

ACKNOWLEDGEMENTS

The work described in this paper was supported by the Research Institute of Hydro-Québec, IREQ, under contract CDT P-783.

REFERENCES

[1] Beaulieu G., Massé B. and Dubé B.; "Darrieus Wind Turbine Power Train Dynamics", Wind and Solar Energy Conference, Kansas City, April 1982.

[2] Dessaint L. A., Mukhedkar D., Jeanson G. and Nakra H.; "Modèle hybride d'un aérogénérateur à axe vertical pour la simulation en temps réel", 1982 Canadian Conference on Industrial Computer Systems, May 3-5, 1982, Hamilton, Otario, Canada, pp. 59.1-59.5.

[3] Moltgen G., "Line Commutated Thyristor Converters", Pitman Publishing and Siemens Co., Germany, 1972.

[4] Olivier G., Stefanovic V. and April G.-E., "Evaluation of Phase-Commutated Converters for Slip-Power Control in Induction Drives", IEEE Trans. on Ind. App., Vol. IA-19, No. 1, Jan.-Feb. 1983, pp. 105-112.

AUTHOR INDEX